OXFORD MATHEMATICAL MONOGRAPHS

OXFORD MATHEMATICAL MONOGRAPHS

A. Belleni-Moranti: *Applied semigroups and evoultion equations*
A.M. Arthurs: *Complementary variational principles* 2nd edition
M. Rosenblum and J. Rovnyak: *Hardy classes and operator theory*
J.W.P. Hirschfeld: *Finite projective spaces of three dimensions*
A. Pressley and G. Segal: *Loop groups*
D.E. Edmunds and W.D. Evans: *Spectral theory and differential operators*
Wang Jianhua: *The theory of games*
S. Omatu and J.H. Seinfeld: *Distributed parameter systems: theory and applications*
J. Hilgert, K.H. Hofmann, and J.D. Lawson: *Lie groups, convex cones, and semigroups*
S. Dineen: *The schwarz lemma*
S.K. Donaldson and P.B. Kronheimer: *The geometry of four-manifolds*
D.W. Robinson: *Elliptic operators and Lie groups*
A.G. Werschulz: *The computational complexity of differential and integral equations*
L. Evens: *Cohomology of groups*
G. Effinger and D.R. Hayes: *Additive number theory of polynomials*
J.W.P. Hirschfeld and J.A. Thas: *General Galois geometries*
P.N. Hoffman and J.F. Humpherys: *Projective representations of the symmetric groups*
I. Györi and G. Ladas: *The oscillation theory of delay differential equations*
J. Heinonen, T. Kilpelainen, and O. Martio: *Non-linear potential theory*
B. Amberg, S. Franciosi, and F. de Giovanni: *Products of groups*
M.E. Gurtin: *Thermomechanics of evolving phase boundaries in the plane*
I. Ionescu and M. Sofonea: *Functional and numerical methods in viscoplasticity*
N. Woodhouse: *Geometric quantization* 2nd edition
U. Grenander: *General pattern theory*
J. Faraut and A. Koranyi: *Analysis on symmetric cones*
I.G. Macdonald: *Symmetric functions and Hall polynomials* 2nd edition
B.L.R. Shawyer and B.B. Watson: *Borel's methods of summability*
M. Holschneider: *Wavelets: an analysis tool*
Jacques Thévenaz: *G-algebras and modular representation theory*
Hans-Joachim Baues: *Homotopy type and homology*
P.D.D'Eath: *Black holes: gravitational interactions*
R. Lowen: *Approach spaces: the missing link in the topology–uniformity–metric traid*
Nguyen Dinh Cong: *Topological dynamics of random dynamical systems*
J.W.P Hirschfeld: *Projective geometries over finite fields* 2nd edition
K. Matsuzaki and M. Taniguchi: *Hyperbolic manifolds and Kleinian groups*
David E. Evans and Yasuyuki Kawahigashi: *Quantum symmetries on operator algebras*
Norbert Klingen: *Arithmetical similarities: prime decomposition and finite group theory*
Isabelle Catto, Claude Le Bris, and Pierre-Louis Lions: *The mathematical theory of thermo-dynamic limits: Thomas–Fermi type models*
D. McDuff and D. Salamon: *Introduction to symplectic topology* 2nd edition
William M. Goldman: *Complex hyberbolic geometry*
Charles J. Colbourn and Alexander Rosa: *Triple systems*
V.A. Kozlov, V.G. Maz'ya and A.B. Movchan: *Asymptotic analysis of fields in multi-structures*
Gérard A. Maugin: *Nonlinear waves in elastic crystals*
George Dassios and Ralph Kleinman: *Low frequency scattering*

Low Frequency Scattering

GEORGE DASSIOS

Division of Applied Mathematics
Department of Chemical Engineering
University of Patras

and

RALPH KLEINMAN

Late of the Department of Mathematical Sciences
University of Delaware

CLARENDON PRESS • OXFORD
2000

OXFORD
UNIVERSITY PRESS
Great Clarendon Street, Oxford OX2 6DP

Oxford University Press is a department of the University of Oxford.
It furthers the University's objective of excellence in research, scholarship,
and education by publishing worldwide in

Oxford New York

Athens Auckland Bangkok Bogotá Buenos Aires Calcutta
Cape Town Chennai Dar es Salaam Delhi Florence Hong Kong Istanbul
Karachi Kuala Lumpur Madrid Melbourne Mexico City Mumbai
Nairobi Paris São Paulo Singapore Taipei Tokyo Toronto Warsaw

and associated companies in Berlin Ibadan

Oxford is a registered trade mark of Oxford University Press
in the UK and in certain other countries

Published in the United States
by Oxford University Press, Inc., New York

First published 2000

A catalogue record for this book is available from the British Library

Library of Congress Cataloging in Publication Data

Dassios, G. (George)
Low frequency scattering / George Dassios, Ralph Kleinman.
(Oxford mathematical monographs)
Includes bibliographical references and index.
1. Scattering (Physics) 2. Scattering (Mathematics)
I. Kleinman, Ralph, 1929–1998. II. Title. III. Series.
QC20.7.S3D38 1999 539.7′58–dc21 99-32310

ISBN 0 19 853678 X

Typeset by Newgen Imaging Systems (P) Ltd., Chennai, India
Printed by Thomson Press, India

PREFACE

This book is intended as a monograph on low frequency methods in acoustic, electromagnetic and elastic wave scattering. One of the goals is to provide the active researcher with a comprehensive presentation of the basic techniques as well as results in these three physical areas. At the same time, this material is presented in sufficient detail to provide the basis for an advanced course in low frequency scattering theory. The emphasis is on physically significant problems in three dimensions and it includes scattering of incident waves from bounded objects of various constitutive material. A thorough treatment of penetrable and impenetrable surfaces is presented including the boundary value problems with Dirichlet (soft), Neumann (hard) and Robin (impedance) conditions in acoustics; perfectly conducting and Leontovich (impedance) conditions in electromagnetics, Dirichlet (rigid) and Neumann (cavity) conditions in elasticity as well as transmission problems for homogeneous scatterers in all three cases.

The basic differential equations governing the various fields as well as the radiation, boundary and transmission conditions are presented. Integral representations are provided and the basic scattering and reciprocity theorems which are used in developing low frequency expansions are also included. The intention is to present results in a form which is easily usable in connection with concrete physical problems. Thus, while every effort was made to ensure that these results were mathematically correct in a classical setting, no attempt was made to present them in modern abstract form. A thorough study is given of the static or Rayleigh approximations (leading low frequency terms) emphasizing the physical and geometrical information these terms contain for the scattering problem. Moreover, the full low frequency expansions for all cases are presented including both differential and integral equation formulations for the determination of the coefficients in the expansions. For geometries where separation of variables is available the explicit construction of low frequency expansions, for all the boundary and transmission problems, is possible and this is shown in detail for spheres, ellipsoids and their degenerate forms such as spheroids, discs and needles, all being convex shapes.

The reader should be familiar with the basics of partial differential equations and some acquaintance with at least one of the physical problems is recommended. The common features of the mathematical formulation of the three physical problems treated and the similarities in the low frequency approximation techniques make it easy for the reader with familiarity in only one area to understand all three. The level of the presentation is such that even those readers who are unfamiliar with the physical problems can use this book as an introduction to scattering problems. Although much of this material is available in the research literature, it has never been collected in one place using uniform notation and presentation. The book is intended to provide a systematic treatment of low frequency theory as well as a compilation of known results. This treatment will reveal gaps in current knowledge and suggest topics for further research. Nevertheless, in our effort to be as complete as possible techniques were developed and results were obtained for which no knowledge of previously published work was available. In this sense, a small percentage of the material included appears herein for the first time.

A list of the symbols used in the text is presented which includes the dimensions of all physical quantities on the list. An attempt was made to use consistent notation in the three areas of physical application in order to ease the comparison of results. Symbols used locally in the book are not listed and no special symbols were invented for letters with standard meaning, e.g. the Greek letter μ is used in three different cases as the magnetic permeability in electromagnetics, as one of the Lamé constants in elasticity and as one of the ellipsoidal coordinates in the theory of ellipsoidal harmonics. The expressions 'potential' and 'harmonic function' are used as synonymous. The sections in each chapter are prefixed by either A (Acoustic), EM (ElectroMagnetic) or E (Elastic) so that the reader can more easily identify the particular area of application; for example, Section 3.A.1 is the first section on acoustics in Chapter 3. Furthermore, the reader interested, for example, in electromagnetic scattering, can follow the EM sections throughout the book. Exercises are included at the end of each chapter which either elaborate on the material presented or require the reader to supply the detailed steps which have been omitted in some calculations. These exercises constitute an integral part of the book but are not always straightforward.

Chapter 1 provides the basic theory of acoustic, electromagnetic and elastic waves. This includes the governing equations, the boundary, transmission and radiation conditions, statements of the basic scattering problems and energy considerations. Chapter 2 involves integral representations of the solutions, their far field form, expressions for the scattering amplitude and all kinds of cross sections, as well as the fundamental theorems connecting them. The first two chapters provide a general introduction to scattering theory without any references to low frequencies. The complete low frequency expansions for each one of the four scattering problems in acoustics, the three problems in electromagnetics and the three problems in elasticity are derived in Chapter 3. Chapter 4 contains an analysis of the transmission problems, in all three physical regimes, based on volume rather than on surface integrals. It turns out that this is a more effective way to treat transmission problems at low frequencies. As far as the authors know, the electromagnetic transmission problem has not been treated in this fashion elsewhere. One of the core chapters is Chapter 5, where each of the ten scattering problems is analyzed in detail, and the boundary value problems determining the leading low frequency near field approximations are reduced, in most cases, to standard potential problems. This chapter discusses some ingenious techniques which were developed mainly in the 1960s and 1970s by many eminent researchers. The integral equation formulation of scattering problems is developed in Chapter 6 and it has been kept as compact as possible because of the already sufficient bibliography available covering the integral equations approach to scattering problems (Colton and Kress 1983). Finally, concrete results for the case of a sphere are given in Chapter 7 and corresponding results for a triaxial ellipsoid are provided in Chapter 8. To the authors' knowledge, this is the first attempt to collect analytic results for scattering by an ellipsoid. In order to state the exact solutions in a compatible form and notation, all results given in these last two chapters were rederived for each one of the problems. In the cases where higher order results exist in the literature, references to these sources are provided. With the exception of acoustic scattering problems for the sphere, where point source excitation was also included, all other results in Chapters 7 and 8 refer to plane wave incidence.

Although basic methods and techniques are included, not all aspects of low frequency scattering are discussed in this book. Topics left out include two-dimensional problems (Kleinman and Vainberg 1994; MacCamy 1997), Kelvin inversion methods (Dassios and Kleinman 1989*a*,*b*), inversion of global low frequency data (Charalambopoulos and Dassios 1992), thermoelastic scattering problems (Dassios and Kostopoulos 1990*a*,*b*), inhomogeneous media (Colton and Kress 1998), anisotropic media and, of course, potential scattering.

An effort has been made to build a fairly complete bibliography which would include all the significant contributions to low frequency scattering as well as some important entries to the general theory of scattering. The help of our colleagues J. Boersma, D.S. Jones, R. Kress, R.F. Millar, T.B.A. Senior, B. Sleeman, C.T. Tai, J. Van Bladel and P. Werner in making the bibliography as complete as possible is greatly acknowledged. Undoubtedly, there will be some inadvertent omissions for which the authors apologize. Obviously, exact and accurate references to all bibliographical entries lie beyond our knowledge. The authors express their appreciation to the many colleagues who assisted, through fruitful discussions and suggestions, in improving and completing the presentation in many instances.

This book was written during numerous three- to five- week periods where the authors enjoyed the privileges provided by the academic atmosphere of the University of Delaware, the University of Patras, the Oberwolfach Mathematical Institute, the Institute of Chemical Engineering in Patras, and the Institute for the Mathematics of Waves in Delaware. Financial support from AGARD, from AFOSR, from the Institute for the Mathematics of Waves, from the Oberwolfach Mathcmatical Institute and from Volkswagen is acknowledged. Special thanks go to Mrs Pam Haverland for the excellent job she did with the processing of the manuscript.

Patras G.D.
Newark R.K.

February 1998

Postscript

The project of writing this book started in July 1991 and the manuscript was actually finished on February 13, 1998. Six days after that, on February 19, Professor Ralph Kleinman suddenly collapsed with a massive stroke and died four days later. Collaborating with Ralph was a most pleasant, memorable experience and an intellectual challenge. His loss will be strongly felt by all colleagues and friends who had the privilege to know him personally. Ralph had intended his dedication to be to his son, Jack, a geologist and adventurer, who died in a kayaking accident on February 12, 1994 at age 32. Eleni, Konstantinos, Theodoros, Vicky and Jan were the silent and patient supporters in our long periods of intense efforts. I would like to thank them cordially for their help and understanding. Professors Tom Angell, Gary Roach and Fadil Santosa offered irreplaceable help during the final stage of this project.

George Dassios
January 1999

CONTENTS

NOTATION

General notation

$V^- =$ scatterer

$S = \partial V^- =$ boundary of the scatterer

$|S| =$ surface area of S

$V^+ = (V^- \cup S)^c =$ exterior domain

$|V^-| =$ volume of the scatterer

$\pm =$ indices to indicate the physical parameters and the fields in $V^\pm$

$A \subset V^+ =$ support of the singularities of the incident field

$\hat{\boldsymbol{x}}_1, \hat{\boldsymbol{x}}_2, \hat{\boldsymbol{x}}_3 =$ Cartesian orthonormal basis

$\hat{\boldsymbol{r}}, \hat{\boldsymbol{\theta}}, \hat{\boldsymbol{\varphi}} =$ spherical orthonormal basis

$\hat{\boldsymbol{\rho}}, \hat{\boldsymbol{\mu}}, \hat{\boldsymbol{\nu}} =$ ellipsoidal orthornormal basis

$\boldsymbol{r} = x_1\hat{\boldsymbol{x}}_1 + x_2\hat{\boldsymbol{x}}_2 + x_3\hat{\boldsymbol{x}}_3 =$ Cartesian coordinates

$\boldsymbol{r} = r\hat{\boldsymbol{r}} + \theta\hat{\boldsymbol{\theta}} + \varphi\hat{\boldsymbol{\varphi}} =$ spherical coordinates

$\boldsymbol{r} = \rho\hat{\boldsymbol{\rho}} + \mu\hat{\boldsymbol{\mu}} + \nu\hat{\boldsymbol{\nu}} =$ ellipsoidal coordinates

$a =$ characteristic dimension of the scatterer (radius of sphere circumscribing S)

$a_1, a_2, a_3 =$ ellipsoidal semiaxis

$h_1, h_2, h_3 =$ ellipsoidal semifocal distances

$\boldsymbol{r}' =$ variable of integration

$\hat{\boldsymbol{r}} =$ direction of observation (unit vector)

$\boldsymbol{r}_0 =$ location of point source

$\hat{\boldsymbol{k}} =$ direction of propagation

$t =$ time variable

$f =$ scalar function

$\boldsymbol{f} =$ vector function

$\tilde{\boldsymbol{f}} =$ dyadic function

$\tilde{\tilde{\boldsymbol{f}}} =$ tetradic function

$\tilde{\boldsymbol{I}} = \hat{\boldsymbol{x}}_1\hat{\boldsymbol{x}}_1 + \hat{\boldsymbol{x}}_2\hat{\boldsymbol{x}}_2 + \hat{\boldsymbol{x}}_3\hat{\boldsymbol{x}}_3 = \hat{\boldsymbol{r}}\hat{\boldsymbol{r}} + \hat{\boldsymbol{\theta}}\hat{\boldsymbol{\theta}} + \hat{\boldsymbol{\varphi}}\hat{\boldsymbol{\varphi}} = \hat{\boldsymbol{\rho}}\hat{\boldsymbol{\rho}} + \hat{\boldsymbol{\mu}}\hat{\boldsymbol{\mu}} + \hat{\boldsymbol{\nu}}\hat{\boldsymbol{\nu}} =$ the identity dyadic

$\boldsymbol{ab} : \boldsymbol{cd} = (\boldsymbol{a} \cdot \boldsymbol{d})(\boldsymbol{b} \cdot \boldsymbol{c}) =$ double contraction of dyads

$T =$ time period

$\lambda =$ space period (wavelength)

$\omega =$ angular frequency (temporal wave density)

$k =$ wave number (spatial wave density)

$\hat{\boldsymbol{n}}$ = outward unit vector on S

$B(\boldsymbol{r}, a) = \{\boldsymbol{r}' \in \mathbb{R}^3 \mid |\boldsymbol{r}' - \boldsymbol{r}| < a\}$ = open ball centered at $\boldsymbol{r}$ with radius a

$\partial B(\boldsymbol{r}, a) = \{\boldsymbol{r}' \in \mathbb{R}^3 \mid |\boldsymbol{r}' - \boldsymbol{r}| = a\}$ = sphere centered at $\boldsymbol{r}$ with radius a

$S^2 = \{\hat{\boldsymbol{r}} \in \mathbb{R}^3 \mid |\hat{\boldsymbol{r}}| = 1\}$ = the unit sphere in $\mathbb{R}^3$

$e^{-i\omega t}$ = harmonic time dependence

$\mathcal{F}(\boldsymbol{r}, t) = e^{-i\omega t} f(\boldsymbol{r})$ = script letters denote time dependent fields; nonscript letters denote time independent fields

$\boldsymbol{D} = \hat{\boldsymbol{\theta}}\frac{\partial}{\partial\theta} + \frac{\hat{\boldsymbol{\varphi}}}{\sin\theta}\frac{\partial}{\partial\varphi}$ = angular part of the gradient operator

$\nabla = \hat{\boldsymbol{r}}\frac{\partial}{\partial r} + \frac{1}{r}\boldsymbol{D}$ = spherical gradient operator

$\nabla^2 = \nabla \cdot \nabla$ = Laplace's operator

$\mathbb{B} = \boldsymbol{D} \cdot \boldsymbol{D} = \frac{1}{\sin^2\theta}\left[\left(\sin\theta\frac{\partial}{\partial\theta}\right)\left(\sin\theta\frac{\partial}{\partial\theta}\right) + \frac{\partial^2}{\partial\varphi^2}\right]$ = Beltrami's operator

$h(x) = h_0^{(1)}(x) = \frac{e^{ix}}{ix}$ = radiative spherical Hankel function of zero order

$\delta(\boldsymbol{r})$ = Dirac's delta measure ($[\delta] = [L^{-3}, M^0, T^0]$)

$\alpha(\boldsymbol{r})$ = subtended solid angle at $\boldsymbol{r}$ on S

$\mathcal{W}(\boldsymbol{r}, t)$ = time dependent energy density function ($[\mathcal{W}] = [L^{-1}, M^1, T^{-2}]$)

$W(\boldsymbol{r})$ = frequency dependent energy density function ($[W] = [L^{-1}, M^1, T^{-2}]$)

$\sigma(\hat{\boldsymbol{r}})$ = differential scattering cross section ($[\sigma] = [L^2, M^0, T^0]$)

σ_s = scattering cross section ($[\sigma_s] = [L^2, M^0, T^0]$)

σ_a = absorption cross section ($[\sigma_a] = [L^2, M^0, T^0]$)

$\sigma_e = \sigma_s + \sigma_a$ = extinction cross section ($[\sigma_e] = [L^2, M^0, T^0]$)

$C^n(V)$ = space of functions defined in V with continuous derivatives of order n

$j_n(x)$ = spherical Bessel functions

$h_n^{(1,2)}(x)$ = spherical Hankel functions of the first and second kind

$P_n^m(x)$ = associated Legendre functions

$P_n(x)$ = Legendre polynomials

$Y_n^m(\hat{\boldsymbol{r}})$ = surface spherical harmonics

$I_0^1(\rho)$ = harmonic elliptic integral of degree zero

$I_1^i(\rho)(i = 1, 2, 3)$ = harmonic elliptic integrals of degree one

$\mathbb{E}_1^i(\rho, \mu, \nu)(i = 1, 2, 3)$ = interior ellipsoidal harmonics of degree one

$\mathbb{F}_1^i(\rho, \mu, \nu)(i = 1, 2, 3)$ = exterior ellipsoidal harmonics of degree one

$I_2^i(\rho)(i = 1, 2, 3, 4, 5)$ = harmonic elliptic integrals of degree two

$\mathbb{E}_2^i(\rho, \mu, \nu)(i = 1, 2, 3, 4, 5)$ = interior ellipsoidal harmonics of degree two

$\mathbb{F}_2^i(\rho, \mu, \nu)(i = 1, 2, 3, 4, 5)$ = exterior ellipsoidal harmonics of degree two

C = capacity

$\widetilde{\boldsymbol{Q}}$ = polarization tensor
$Q_i (i = 1, 2, 3)$ = principal polarizations
$\widetilde{\boldsymbol{P}}$ = electric polarizability tensor
$P_i (i = 1, 2, 3)$ = principal electric polarizabilities
$\widetilde{\boldsymbol{W}}$ = virtual mass tensor
$W_i (i = 1, 2, 3)$ = principal virtual masses
$\widetilde{\boldsymbol{M}}$ = magnetic polarizability tensor
$M_i (i = 1, 2, 3)$ = principal magnetic polarizabilities
$\widetilde{\boldsymbol{X}}$ = general polarizability tensor
$X_i (i = 1, 2, 3)$ = principal general polarizabilities

Notation in acoustics

ρ = mass density ($\rho > 0$, $[\rho] = [L^{-3}, M^1, T^0]$)
γ = mean compressibility (relative volume reduction per unit increase of surface pressure)($\gamma \geq 0$, $[\gamma] = [L^1, M^{-1}, T^2]$)
δ = compressional viscosity ($\delta > 0$ for lossy media,
$\delta = 0$ for lossless media, $[\delta] = [L^{-1}, M^1, T^{-1}]$)
Z = acoustic impedance (ratio of the excess pressure over the normal component of the particle velocity on S) ($Z > 0$ for resistive surfaces, $Z = 0$ for non-resistive surfaces, $[Z] = [L^{-2}, M^1, T^{-1}]$)
$\nu = \dfrac{1}{Z^+}\sqrt{\dfrac{\rho^+}{\gamma^+}}$ = Robin's parameter for the impedance problem (dimensionless)
$\beta = \dfrac{\rho^+}{\rho^-}(1 - i\omega\delta^-\gamma^-)$ = relative mass density ($\operatorname{Im}\beta \geq 0$, dimensionless)
$\beta\eta^2 = \dfrac{\gamma^-}{\gamma^+}$ = relative compressibility ($\beta\eta^2 \geq 0$, dimensionless)
$k = k^+ = \omega\sqrt{\gamma^+\rho^+}$ = wave number in V^+($k \geq 0$, $[k] = [L^{-1}, M^0, T^0]$)
$k^- = \dfrac{\omega\sqrt{\gamma^-\rho^-}}{\sqrt{1 - i\omega\delta^-\gamma^-}}$ = wave number in V^- ($\operatorname{Im} k^- \geq 0$, $[k^-] = [L^{-1}, M^0, T^0]$)
$\eta = \dfrac{k^-}{k^+} = \sqrt{\dfrac{\gamma^-\rho^-}{\gamma^+\rho^+}}\dfrac{1}{\sqrt{1 - i\omega\delta^-\gamma^-}}$ = relative index of refraction ($\operatorname{Im}\eta \geq 0$, dimensionless)
$c = \dfrac{1}{\sqrt{\gamma\rho}}\sqrt{\dfrac{2(1 + \omega^2\delta^2\gamma^2)}{1 + \sqrt{1 + \omega^2\delta^2\gamma^2}}}$ = phase velocity ($c > 0$, $[c] = [L^1, M^0, T^{-1}]$)
$\Phi(\boldsymbol{r}, t) = e^{-i\omega t}\phi(\boldsymbol{r})$ = time dependent velocity potential
$\phi(\boldsymbol{r})$ = velocity potential ($[\phi] = [L^2, M^0, T^{-1}]$)
$\mathcal{U}(\boldsymbol{r}, t) = e^{-i\omega t}u(\boldsymbol{r})$ = time dependent (excess) pressure field

$\mathcal{I}(\boldsymbol{r}, t) =$ power flux vector ($[\mathcal{I}] = [L^0, M^1, T^{-3}]$)

$u(\boldsymbol{r}) = \dfrac{i\omega\rho}{1 - i\omega\delta\gamma}\phi(\boldsymbol{r}) =$ (excess) pressure field ($[u] = [L^{-1}, M^1, T^{-2}]$)

$\boldsymbol{V}(\boldsymbol{r}, t) = e^{-i\omega t}\boldsymbol{v}(\boldsymbol{r}) = \nabla\Phi(\boldsymbol{r}, t)$ = time dependent velocity field

$\boldsymbol{v}(\boldsymbol{r}) = \nabla\phi(\boldsymbol{r}) =$ velocity field ($[v] = [L^1, M^0, T^{-1}]$)

$u^i : (\mathbb{R}^3 - A) \to \mathbb{C} =$ incident field

$u : V^+ \to \mathbb{C} =$ scattered field

$u^+ = u^i + u : (V^+ - A) \to \mathbb{C} =$ total field

$u^- : V^- \to \mathbb{C} =$ interior field

$G^{\pm}(\boldsymbol{r}, \boldsymbol{r}') = \dfrac{e^{ik|\boldsymbol{r}-\boldsymbol{r}'|}}{ik|\boldsymbol{r}-\boldsymbol{r}'|} =$ fundamental solution (dimensionless)

$g(\hat{\boldsymbol{r}}) =$ scattering amplitude (normalized to the same dimensions as u)

$g(\hat{\boldsymbol{r}}, \hat{\boldsymbol{k}}) =$ scattering amplitude for plane wave incidence along $\hat{\boldsymbol{k}}$

$F_n(\hat{\boldsymbol{r}}) =$ coefficients of the far field expansion of u ($[F_n] = [L^{n-1}, M^1, T^{-2}]$)

$\boldsymbol{I}(\boldsymbol{r}) = \frac{1}{\rho\omega}\mathrm{Im}[u^*(\boldsymbol{r})\nabla u(\boldsymbol{r})] =$ power flux vector ($[I] = [L^0, M^1, T^{-3}]$)

$u_n^{\pm}(\boldsymbol{r}) = n$th order low frequency approximation of $u^{\pm}$ ($[u_n^{\pm}] = [L^{n-1}, M^1, T^{-2}]$)

$f_n^{\pm}(\boldsymbol{r}) =$ known part of $u_n^{\pm}$ (the part that is dependent on $u_0^{\pm}, u_1^{\pm}, \ldots, u_{n-1}^{\pm}$)

$v_n^{\pm}(\boldsymbol{r}) =$ unknown part of $u_n^{\pm}$

$A_n(\hat{\boldsymbol{r}}) = n$th order low frequency approximation of $g(\hat{\boldsymbol{r}})$ ($[A_n] = [L^{n-1}, M^1, T^{-2}]$)

Notation in electromagnetics

$\epsilon =$ electric permitivity ($[\epsilon] = [L^{-3}, M^{-1}, T^2, Q^2]$)

$\mu =$ magnetic permeability ($[\mu] = [L^1, M^1, T^0, Q^{-2}]$)

$\sigma =$ conductivity ($\sigma \to 0$ for a perfect insulator, $\sigma \to +\infty$ for a perfect conductor) ($[\sigma] = [L^{-3}, M^{-1}, T^1, Q^2]$)

$Z = \sqrt{\dfrac{\mu}{\epsilon}}\dfrac{1}{\sqrt{1 + i\frac{\sigma}{\epsilon\omega}}} =$ characteristic impedance ($[Z] = [L^2, M^1, T^{-1}, Q^{-2}]$)

$Y = \dfrac{1}{Z} =$ characteristic admittance ($[Y] = [L^{-2}, M^{-1}, T^1, Q^2]$)

$Z_s =$ relative surface impedance (dimensionless)

$c = \dfrac{1}{\sqrt{\mu\epsilon}}\sqrt{\dfrac{2}{1 + \sqrt{1 + \frac{\sigma^2}{\epsilon^2\omega^2}}}} =$ phase velocity ($[c] = [L^1, M^0, T^{-1}, Q^0]$)

$k^+ = k = \omega\sqrt{\mu^+\epsilon^+} =$ wave number in V^+ ($[k] = [L^{-1}, M^0, T^0, Q^0]$)

$k^- = \omega\sqrt{\mu^-\epsilon^-}\sqrt{1 + i\dfrac{\sigma^-}{\epsilon^-\omega}} =$ wave number in V^- ($[k^-] = [L^{-1}, M^0, T^0, Q^0]$)

$\eta = \dfrac{k^-}{k^+} = \sqrt{\dfrac{\mu^-\epsilon^-}{\mu^+\epsilon^+}}\sqrt{1 + i\dfrac{\sigma^-}{\epsilon^-\omega}} =$ relative index of refraction (dimensionless)

$\rho =$ volume charge density ($[\rho] = [L^{-3}, M^0, T^0, Q^1]$)

ρ_s = surface charge density ($[\rho_s] = [L^{-2}, M^0, T^0, Q^1]$)
$\hat{\boldsymbol{a}} = \hat{\boldsymbol{b}} \times \hat{\boldsymbol{k}}$ = direction of polarization of the incident electric field
$\hat{\boldsymbol{b}} = \hat{\boldsymbol{k}} \times \hat{\boldsymbol{a}}$ = direction of polarization of the incident magnetic field
$\hat{\boldsymbol{p}}$ = unit electric dipole moment ($[p] = [L^1, M^0, T^0, Q^1]$)
$\hat{\boldsymbol{m}}$ = unit magnetic dipole moment ($[m] = [L^2, M^0, T^{-1}, Q^1]$)
$\mathcal{E}(\boldsymbol{r}, t) = e^{-i\omega t}\boldsymbol{E}(\boldsymbol{r})$ = time dependent electric field
$\boldsymbol{E}(\boldsymbol{r})$ = electric field ($[E] = [L^1, M^1, T^{-2}, Q^{-1}]$)
$\mathcal{H}(\boldsymbol{r}, t) = e^{-i\omega t}\boldsymbol{H}(\boldsymbol{r})$ = time dependent magnetic field
$\boldsymbol{H}(\boldsymbol{r})$ = magnetic field ($[H] = [L^{-1}, M^0, T^{-1}, Q^1]$)
$\mathcal{D}(\boldsymbol{r}, t) = e^{-i\omega t}\boldsymbol{D}(\boldsymbol{r})$ = time dependent displacement field
$\boldsymbol{D}(\boldsymbol{r})$ = displacement field ($[D] = [L^{-2}, M^0, T^0, Q^1]$)
$\mathcal{B}(\boldsymbol{r}, t) = e^{-i\omega t}\boldsymbol{B}(\boldsymbol{r})$ = time dependent magnetic induction field
$\boldsymbol{B}(\boldsymbol{r})$ = magnetic induction field ($[B] = [L^0, M^1, T^{-1}, Q^{-1}]$)
$\boldsymbol{E}_e^i, \boldsymbol{H}_e^i : (\mathbb{R}^3 - \{\boldsymbol{r}_0\}) \to \mathbb{C}^3$ = incident electric dipole fields
$\boldsymbol{E}_m^i, \boldsymbol{H}_m^i : (\mathbb{R}^3 - \{\boldsymbol{r}_0\}) \to \mathbb{C}^3$ = incident magnetic dipole fields
$\mathcal{J}(\boldsymbol{r}, t) = e^{-i\omega t}\boldsymbol{J}(\boldsymbol{r})$ = time dependent current density
$\boldsymbol{J}(\boldsymbol{r})$ = current density ($[J] = [L^{-2}, M^0, T^{-1}, Q^1]$)
$\boldsymbol{J}_s(\boldsymbol{r})$ = surface current density ($[J_s] = [L^{-1}, M^0, T^{-1}, Q^1]$)
$\phi(\boldsymbol{r})$ = potential function ($[\phi] = [L^2, M^1, T^{-2}, Q^{-1}]$)
$\boldsymbol{E}^i, \boldsymbol{H}^i : (\mathbb{R}^3 - A) \to \mathbb{C}^3$ = incident fields
$\boldsymbol{E}, \boldsymbol{H} : V^+ \to \mathbb{C}^3$ = scattered fields
$\left.\begin{aligned} \boldsymbol{E}^+ &= \boldsymbol{E}^i + \boldsymbol{E} \\ \boldsymbol{H}^+ &= \boldsymbol{H}^i + \boldsymbol{H} \end{aligned}\right\}$ = total fields
$\boldsymbol{E}^-, \boldsymbol{H}^- : V^- \to \mathbb{C}^3$ = interior fields
$\widetilde{\boldsymbol{G}}(\boldsymbol{r}, \boldsymbol{r}')$ = fundamental dyadic solution (dimensionless)
$\boldsymbol{g}_e(\hat{\boldsymbol{r}})$ = electric scattering amplitude (normalized to the same dimensions as $\boldsymbol{E}$)
$\boldsymbol{g}_m(\hat{\boldsymbol{r}})$ = magnetic scattering amplitude (normalized to the same dimensions as $\boldsymbol{H}$)
$\boldsymbol{g}_e(\hat{\boldsymbol{r}}, \hat{\boldsymbol{k}})$ = electric scattering amplitude for plane wave incidence along $\hat{\boldsymbol{k}}$
$\boldsymbol{g}_m(\hat{\boldsymbol{r}}, \hat{\boldsymbol{k}})$ = magnetic scattering amplitude for plane wave incidence along $\hat{\boldsymbol{k}}$
$\boldsymbol{F}_n^e(\hat{\boldsymbol{r}})$ = coefficients of the far field expansion of $\boldsymbol{E}$ ($[F_n^e] = [L^{n+1}, M^1, T^{-2}, Q^{-1}]$)
$\boldsymbol{F}_n^m(\hat{\boldsymbol{r}})$ = coefficients of the far field expansion of $\boldsymbol{H}$
($[F_n^m] = [L^{n-1}, M^0, T^{-1}, Q^{-1}]$)
$\boldsymbol{S}(\boldsymbol{r}, t)$ = complex Poynting vector ($[S] = [L^0, M^1, T^{-3}, Q^0]$)
$\boldsymbol{S}(\boldsymbol{r}) = \mathrm{Re}\{\boldsymbol{E}(\boldsymbol{r}) \times \boldsymbol{H}^*(\boldsymbol{r})\}$ = Poynting power flux vector ($[S] = [L^0, M^1, T^{-3}, Q^0]$)
$\boldsymbol{E}_n^{\pm}(\boldsymbol{r})$ = nth order low frequency approximation of $\boldsymbol{E}^{\pm}$ ($[E_n^{\pm}] = [L^{n+1}, M^1, T^{-2}, Q^{-1}]$)
$\boldsymbol{H}_n^{\pm}(\boldsymbol{r})$ = nth order low frequency approximation of $\boldsymbol{H}^{\pm}$ ($[H_n^{\pm}] = [L^{n-1}, M^0, T^{-1}, Q^1]$)
$\boldsymbol{F}_{en}^{\pm}(\boldsymbol{r})$ = known part of $\boldsymbol{E}_n^{\pm}$ (the part that is dependent on $\boldsymbol{E}_0^{\pm}, \boldsymbol{E}_1^{\pm}, \ldots, \boldsymbol{E}_{n-1}^{\pm}$)

$F^{\pm}_{mn}(\boldsymbol{r})$ = known part of $\boldsymbol{H}^{\pm}_{n}$ (the part that is dependent on $\boldsymbol{H}^{\pm}_{0}, \boldsymbol{H}^{\pm}_{1}, \ldots, \boldsymbol{H}^{\pm}_{n-1}$)
$\boldsymbol{h}^{+}_{n}(\boldsymbol{r})$ = auxiliary modifications of $\boldsymbol{F}^{+}_{mn}(\boldsymbol{r})$
$\boldsymbol{U}^{\pm}_{en}(\boldsymbol{r})$ = unknown part of $\boldsymbol{E}^{\pm}_{n}$
$\boldsymbol{U}^{\pm}_{mn}(\boldsymbol{r})$ = unknown part of $\boldsymbol{H}^{\pm}_{n}$
$\phi^{\pm}_{en}(\boldsymbol{r})$ = auxiliary scalar potentials for $\boldsymbol{U}^{\pm}_{en}$ ($[\phi_{en}] = [L^{n+2}, M^1, T^{-2}, Q^{-1}]$)
$\phi^{\pm}_{mn}(\boldsymbol{r})$ = auxiliary scalar potentials for $\boldsymbol{U}^{\pm}_{mn}(\boldsymbol{r})$ ($[\phi_{mn}] = [L^{n}, M^1, T^{-2}, Q^{-1}]$)
$\boldsymbol{V}^{\pm}_{en}(\boldsymbol{r})$ = auxiliary vector potentials for $\boldsymbol{U}^{\pm}_{en}$ ($[V_{en}] = [L^{n+1}, M^1, T^{-2}, Q^{-1}]$)
$\boldsymbol{V}^{\pm}_{mn}(\boldsymbol{r})$ = auxiliary vector potentials for $\boldsymbol{U}^{\pm}_{mn}$ ($[V_{mn}] = [L^{n-1}, M^1, T^{-2}, Q^{-1}]$)
$\boldsymbol{A}_{en}(\hat{\boldsymbol{r}})$ = nth order low frequency approximation of $\boldsymbol{g}_e$ ($[A_{en}] = [L^{n+1}, M^1, T^{-2}, Q^{-1}]$)
$\boldsymbol{A}_{mn}(\hat{\boldsymbol{r}})$ = nth order low frequency approximation of $\boldsymbol{g}_m$ ($[A_{mn}] = [L^{n-1}, M^0, T^{-1}, Q^{1}]$)

Notation in elasticity

ρ = mass density ($[\rho] = [L^{-3}, M^1, T^0]$)
λ, μ = Lamé constants ($\mu > 0$, $2\mu + 3\lambda > 0$, $[\lambda, \mu] = [L^{-1}, M^1, T^{-2}]$)
λ_p = longitudinal wavelength ($[\lambda_p] = [L^1, M^0, T^0]$)
λ_s = transverse wavelength ($[\lambda_s] = [L^1, M^0, T^0]$)
$c_p = \sqrt{\dfrac{\lambda + 2\mu}{\rho}}$ = longitudinal phase velocity ($[c_p] = [L^1, M^0, T^{-1}]$)
$c_s = \sqrt{\dfrac{\mu}{\rho}}$ = transverse phase velocity ($[c_s] = [L^1, M^0, T^{-1}]$)
k_p = longitudinal wave number ($[k_p] = [L^{-1}, M^0, T^0]$)
k_s = transverse wave number ($[k_s] = [L^{-1}, M^0, T^0]$)
$\eta_p = \dfrac{k_p^-}{k_p^+} = \dfrac{c_p^+}{c_p^-}$ = index of refraction for the longitudinal wave (dimensionless)
$\eta_s = \dfrac{k_s^-}{k_s^+} = \dfrac{c_s^+}{c_s^-}$ = index of refraction for the transverse wave (dimensionless)
$\tau = \dfrac{k_p}{k_s} = \dfrac{c_s}{c_p}$ = relative low frequency parameter ($0 < \tau < 1$, dimensionless)
α_p = amplitude of the longitudinal part of the incident plane wave ($\alpha_p \geq 0$, $[\alpha_p] = [L^1, M^0, T^0]$)
α_s = amplitude of the transverse part of the incident plane wave ($\alpha_s \geq 0$, $[\alpha_s] = [L^1, M^0, T^0]$)
$\hat{\boldsymbol{b}}$ = direction of polarization of the incident plane transverse field

$$\left.\begin{aligned}\Delta^e &= c_s^2\nabla^2 + (c_p^2 - c_s^2)\nabla(\nabla\cdot)\\ &= \frac{\mu}{\rho}\nabla^2 + \frac{\lambda+\mu}{\rho}\nabla(\nabla\cdot)\\ &= c_p^2\nabla(\nabla\cdot) - c_s^2\nabla\times(\nabla\times)\end{aligned}\right\} = \text{Lamé operator } ([\Delta^e] = [L^0, M^0, T^{-2}])$$

$\tilde{\boldsymbol{e}}(\boldsymbol{r})$ = strain dyadic (dimensionless)
$\tilde{\boldsymbol{\tau}}(\boldsymbol{r})$ = stress dyadic ($[\tau] = [L^{-1}, M^1, T^{-2}]$)
$\boldsymbol{t}(\boldsymbol{r}, \hat{\boldsymbol{n}})$ = surface traction field ($[t] = [L^{-1}, M^1, T^{-2}]$)

$\boldsymbol{T}(\partial_{\boldsymbol{r}}, \hat{\boldsymbol{n}})$ = surface traction operator ($[T] = L^{-2}, M^1, T^{-2}]$)
$\mathcal{U}(\boldsymbol{r}, t) = e^{-i\omega t}\boldsymbol{u}(\boldsymbol{r})$ = time dependent displacement field
$\mathcal{P}(\boldsymbol{r}, t)$ = elastic power flux vector ($[\mathcal{P}] = [L^0, M^1, T^{-3}]$)
$\boldsymbol{u}(\boldsymbol{r})$ = displacement field ($[u] = [L^1, M^0, T^0]$)
$\boldsymbol{u}^p(\boldsymbol{r})$ = longitudinal (irrotational) part of $\boldsymbol{u}$
$\boldsymbol{u}^s(\boldsymbol{r})$ = transverse (solenoidal) part of $\boldsymbol{u}$
$\boldsymbol{u}^i : (\mathbb{R}^3 - A) \to \mathbb{C}^3$ = incident field
$\boldsymbol{u} : V^+ \to \mathbb{C}^3$ = scattered field
$\boldsymbol{u}^+ = \boldsymbol{u}^i + \boldsymbol{u} : (V^+ - A) \to \mathbb{C}^3$ = total field
$\boldsymbol{u}^- : V^- \to \mathbb{C}^3$ = interior field
$G_p(\boldsymbol{r}, \boldsymbol{r}')$ = scalar fundamental solution for the longitudinal field (dimensionless)
$G_s(\boldsymbol{r}, \boldsymbol{r}')$ = scalar fundamental solution for the transverse field (dimensionless)
$\tilde{\boldsymbol{\Gamma}}(\boldsymbol{r}, \boldsymbol{r}')$ = fundamental dyadic solution ($[\Gamma] = [L^0, M^{-1}, T^2]$)
$\tilde{\boldsymbol{\Gamma}}^p(\boldsymbol{r}, \boldsymbol{r}')$ = longitudinal part of $\tilde{\boldsymbol{\Gamma}}$
$\tilde{\boldsymbol{\Gamma}}^s(\boldsymbol{r}, \boldsymbol{r}')$ = transverse part of $\tilde{\boldsymbol{\Gamma}}$
$g_r(\hat{\boldsymbol{r}})$ = radial scattering amplitude (normalized to the same dimensions as $\boldsymbol{u}$)
$g_\theta(\hat{\boldsymbol{r}}), g_\varphi(\hat{\boldsymbol{r}})$ = tangential scattering amplitudes (normalized to the same dimensions as $\boldsymbol{u}$)
$g_r(\hat{\boldsymbol{r}}, \hat{\boldsymbol{k}})$ = radial scattering amplitude for plane wave incidence along $\hat{\boldsymbol{k}}$
$g_\theta(\hat{\boldsymbol{r}}, \hat{\boldsymbol{k}}), g_\varphi(\hat{\boldsymbol{r}}, \hat{\boldsymbol{k}})$ = tangential scattering amplitudes for plane wave incidence along $\hat{\boldsymbol{k}}$
$\tilde{\boldsymbol{H}}_p^+$ = longitudinal surface displacement dyadic ($[H_p] = [L^3, M^0, T^0]$)
$\tilde{\boldsymbol{H}}_s^+$ = transverse surface displacement dyadic ($[H_s] = [L^3, M^0, T^0]$)
$\boldsymbol{K}_p^+$ = longitudinal surface traction ($[K_p] = [L^1, M^1, T^{-2}]$)
$\boldsymbol{K}_s^+$ = transverse surface traction ($[K_s] = [L^1, M^1, T^{-2}]$)
$\boldsymbol{L}_p^-$ = longitudinal volume displacement ($[L_p] = [L^4, M^0, T^0]$)
$\boldsymbol{L}_s^-$ = transverse volume displacement ($[L_s] = [L^4, M^0, T^0]$)
$\boldsymbol{F}_n^p(\hat{\boldsymbol{r}})$ = coefficients of the far field expansion of $\boldsymbol{u}^p$ ($[F_n^p] = [L^{n+1}, M^0, T^0]$)
$\boldsymbol{F}_n^s(\hat{\boldsymbol{r}})$ = coefficients of the far field expansion of $\boldsymbol{u}^s$ ($[F_n^s] = [L^{n+1}, M^0, T^0]$)
$\boldsymbol{P}(\boldsymbol{r}) = \omega \mathrm{Im}\{\boldsymbol{u}^*(\boldsymbol{r}) \cdot \tilde{\boldsymbol{\tau}}(\boldsymbol{r})\}$ = power flux vector ($[P] = [L^0, M^1, T^{-3}]$)
$\boldsymbol{u}_n^\pm(\boldsymbol{r})$ = nth order low frequency approximation of $\boldsymbol{u}^\pm$ ($[u_n^\pm] = [L^{n+1}, M^0, T^0]$)
$\boldsymbol{f}_n^\pm(\boldsymbol{r})$ = known part of $\boldsymbol{u}_n^\pm$ (the part that is dependent on $\boldsymbol{u}_0^\pm, \boldsymbol{u}_1^\pm, \ldots, \boldsymbol{u}_{n-1}^\pm$)
$\boldsymbol{v}_n^\pm(\boldsymbol{r})$ = unknown part of $\boldsymbol{u}_n^\pm$
$A_{rn}(\hat{\boldsymbol{r}})$ = nth order low frequency approximation of g_r ($[A_{rn}] = [L^{n+1}, M^0, T^0]$)
$A_{\theta n}(\hat{\boldsymbol{r}})$ = nth order low frequency approximation of g_θ ($[A_{\theta n}] = [L^{n+1}, M^0, T^0]$)
$A_{\varphi n}(\hat{\boldsymbol{r}})$ = nth order low frequency approximation of g_φ ($[A_{\varphi n}] = [L^{n+1}, M^0, T^0]$)
$\boldsymbol{A}_n^\pm$ = vector Papkovich potentials for $\boldsymbol{u}_n^\pm$ ($[A] = [L^{n+1}, M^0, T^0]$)
$B_n^\pm$ = scalar Papkovich potentials for $\boldsymbol{u}_n^\pm$ ($[B] = [L^{n+2}, M^0, T^0]$)

1

BASIC THEORY

1.A The basic theory of acoustic waves

1.A.1 *Field equations*

An acoustic wave in an irrotational, homogeneous, isotropic, compressible fluid medium is characterized by a variation in the ambient pressure field. This variation is called the **excess acoustic pressure field**, and is denoted by $\mathcal{U}(\boldsymbol{r}, t)$, a scalar function of position $\boldsymbol{r} = (x_1, x_2, x_3)$ and time. The excess pressure is measured in units of force per unit area. The velocity of propagation of an acoustic wave is denoted by $\boldsymbol{V}(\boldsymbol{r}, t)$, a vector valued function of position and time.

The basic equations of linearized wave propagation of sound waves which relate the velocity and the excess pressure are

$$\dot{\mathcal{U}} = -\frac{1}{\gamma}\nabla \cdot \boldsymbol{V} \tag{1.1}$$

$$\rho\dot{\boldsymbol{V}} = -\nabla\mathcal{U} + \delta\nabla\nabla \cdot \boldsymbol{V} \tag{1.2}$$

where a dot over a letter indicates differentiation with respect to t.

Equation (1.1) combines the **equation of continuity** and the **equation of state**. It embodies the physical principle that the flow of material out of any volume element will reduce the pressure, and for small compression the rate of pressure change is proportional to the divergence of the velocity field. The proportionality constant $1/\gamma$ is called the **compressibility modulus**, while γ is called the **mean compressibility** and expresses relative volume reduction per unit increase in surface pressure. The parameter γ is measured in units of inverse pressure.

Equation (1.2) is the irrotational part of the linearized Navier–Stokes equation for a fluid. The positive real constant ρ is the **mass density** and the non-negative real constant δ is the **compressional viscosity** which describes the rate of change of mass per unit length and represents losses, the conversion of mechanical energy to heat. For **lossless media** $\delta = 0$, while **lossy media** are characterized by $\delta > 0$. Substituting (1.1) into (1.2) leads to the equation

$$\rho\dot{\boldsymbol{V}} = -\nabla\mathcal{U} - \gamma\delta\nabla\dot{\mathcal{U}} \tag{1.3}$$

which, when substituted into the time derivative of (1.1) yields the equation that governs sound wave propagation

$$\ddot{\mathcal{U}} = \frac{1}{\gamma\rho}\nabla^2\mathcal{U} + \frac{\delta}{\rho}\nabla^2\dot{\mathcal{U}} \tag{1.4}$$

For lossless media, (1.4) reduces to the classical wave equation

$$\ddot{\mathcal{U}} = \frac{1}{\gamma\rho}\nabla^2\mathcal{U} \tag{1.5}$$

On the other hand, substituting (1.1) into the time derivative of (1.2) yields

$$\ddot{\boldsymbol{V}} = \frac{1}{\gamma\rho}\nabla(\nabla\cdot\boldsymbol{V}) + \frac{\delta}{\rho}\nabla(\nabla\cdot\dot{\boldsymbol{V}}) \tag{1.6}$$

The fact that the medium is **irrotational media irrotational** means that

$$\nabla\times\boldsymbol{V} = \boldsymbol{0} \tag{1.7}$$

hence

$$\nabla(\nabla\cdot\boldsymbol{V}) = \nabla^2\boldsymbol{V} \tag{1.8}$$

and there exists a scalar time dependent **velocity potential** $\Phi(\boldsymbol{r}, t)$ such that

$$\boldsymbol{V} = \nabla\Phi \tag{1.9}$$

The velocity potential Φ describes the rate of change of the area. Introducing (1.9) into (1.6) yields

$$\ddot{\boldsymbol{V}} = \frac{1}{\gamma\rho}\nabla^2\boldsymbol{V} + \frac{\delta}{\rho}\nabla^2\dot{\boldsymbol{V}} \tag{1.10}$$

which may be further rewritten using (1.9) as

$$\nabla\ddot{\Phi} = \nabla\left(\frac{1}{\gamma\rho}\nabla^2\Phi + \frac{\delta}{\rho}\nabla^2\dot{\Phi}\right) \tag{1.11}$$

The velocity potential is not uniquely defined since we may add any function independent of position $\boldsymbol{r}$. We assume that the appropriate function is added so that not only is (1.11) fulfilled but also that

$$\ddot{\Phi} = \frac{1}{\gamma\rho}\nabla^2\Phi + \frac{\delta}{\rho}\nabla^2\dot{\Phi} \tag{1.12}$$

Thus, we see that the excess pressure, velocity and velocity potential all satisfy the same equation (compare (1.4), (1.10) and (1.12)). Equations (1.1) and (1.9) combine to give the relation between velocity potential and excess pressure

$$\dot{\mathcal{U}} = -\frac{1}{\gamma}\nabla^2\Phi \tag{1.13}$$

Now assume that all field quantities have a harmonic time dependence with **angular** or **circular frequency** ω

$$\mathcal{U}(\boldsymbol{r}, t) = u(\boldsymbol{r})e^{-i\omega t} \tag{1.14}$$

$$\boldsymbol{V}(\boldsymbol{r}, t) = \boldsymbol{v}(\boldsymbol{r})e^{-i\omega t} \tag{1.15}$$

$$\Phi(\boldsymbol{r}, t) = \phi(\boldsymbol{r})e^{-i\omega t} \tag{1.16}$$

where u, $\boldsymbol{v}$, and ϕ are complex valued functions of $\boldsymbol{r}$ and ω but independent of t. These definitions require some interpretation since the physical quantities, pressure $\mathcal{U}$ and velocity $\boldsymbol{V}$, are real whereas in (1.14)–(1.16) they take on complex values. Throughout this book we follow the widely accepted convention of working with the complex pressure, velocity and velocity potential with the understanding that physical quantities will correspond to the real parts, e.g. $\mathrm{Re}\,\mathcal{U} = \mathrm{Re}\{ue^{-i\omega t}\}$.

Introducing (1.14) and (1.16) into (1.4) and (1.12) we find the governing field equations

$$(\nabla^2 + k^2)\begin{Bmatrix} u \\ \phi \end{Bmatrix} = 0 \tag{1.17}$$

where

$$k^2 = \frac{\omega^2 \gamma \rho}{1 - i\omega\gamma\delta} \tag{1.18}$$

This relation is called the **dispersion** or **characteristic relation** for the medium of propagation. The parameter k is called the **wave number** or the **propagation constant** and is not completely defined by (1.18) until we specify the branch of the square root. This choice is dictated from physical considerations. If u is a plane wave propagating in the **direction** $\hat{\boldsymbol{k}}$ (see section 1.A.5), then the complex time dependent field may be written as

$$\begin{aligned}\mathcal{U}(\boldsymbol{r}, t) &= e^{ik\hat{\boldsymbol{k}}\cdot\boldsymbol{r} - i\omega t} \\ &= \exp\left[i\left(\hat{\boldsymbol{k}}\cdot\boldsymbol{r} - \frac{\omega}{\mathrm{Re}\,k}t\right)\mathrm{Re}\,k - \hat{\boldsymbol{k}}\cdot\boldsymbol{r}\,\mathrm{Im}\,k\right]\end{aligned} \tag{1.19}$$

As $\boldsymbol{r}$ increases in the direction of propagation we see from (1.19) that we must choose the branch of the square root so that $\mathrm{Im}\,k \geq 0$ otherwise the wave would increase in intensity as it traverses through the medium whereas physically the intensity should diminish as the wave transverses a lossy medium. The $\mathrm{Im}\,k$ therefore measures the rate at which energy is dissipated. Equation (1.19) is also used to define phase velocity as follows. The **phase fronts** are defined to be the surfaces of constant phase

$$\hat{\boldsymbol{k}}\cdot\boldsymbol{r} - \frac{\omega}{\mathrm{Re}\,k}t = \mathrm{const} \tag{1.20}$$

which are planes (hence **plane waves**). The unit normal to the phase front is $\hat{\boldsymbol{k}}$ and the velocity with which the front moves in the direction of the normal, $\hat{\boldsymbol{k}}\cdot(d/dt)\boldsymbol{r}$, is obtained by differtiating (1.20) with respect to time obtaining

$$\hat{\boldsymbol{k}}\cdot\frac{d\boldsymbol{r}}{dt} = \frac{\omega}{\mathrm{Re}\,k} = c \tag{1.21}$$

where the quantity c is called the **phase velocity**. By virtue of (1.18), the phase velocity may be expressed as

$$c = \frac{1}{\sqrt{\gamma\rho}}\sqrt{\frac{2(1 + \omega^2\delta^2\gamma^2)}{1 + \sqrt{1 + \omega^2\delta^2\gamma^2}}} \tag{1.22}$$

which reduces to

$$c = \frac{1}{\sqrt{\gamma\rho}} \tag{1.23}$$

for lossless media. The angular frequency ω is related to the **temporal period** T by

$$\omega = \frac{2\pi}{T} \tag{1.24}$$

The **spatial period** or **wavelength** λ is the distance the phase front travels in one time period and is given by

$$\lambda = cT = \frac{2\pi c}{\omega} = \frac{2\pi}{\operatorname{Re} k} \tag{1.25}$$

For more general time harmonic waves we may write

$$\mathcal{U}(\boldsymbol{r}, t) = |\mathcal{U}(\boldsymbol{r})| e^{i\Theta(\boldsymbol{r}) - i\omega t} \tag{1.26}$$

The phase fronts are the surfaces

$$\Theta(\boldsymbol{r}) - \omega t = \text{constant} \tag{1.27}$$

with unit normal $\nabla\Theta/|\nabla\Theta|$. The phase velocity c, the velocity in the direction of the normal, is obtained by differentiating with respect to time

$$\frac{d\Theta}{dt} = \nabla\Theta \cdot \frac{d\boldsymbol{r}}{dt} - \omega = 0 \tag{1.28}$$

and therefore

$$c = \frac{\nabla\Theta}{|\nabla\Theta|} \cdot \frac{d\boldsymbol{r}}{dt} = \frac{\omega}{|\nabla\Theta|} \tag{1.29}$$

Suppressing the time factor in (1.13) we obtain

$$i\omega u = \frac{1}{\gamma} \nabla^2 \phi \tag{1.30}$$

which, with (1.17) and (1.18), leads to the following relation between excess pressure and velocity potential

$$u = \frac{i\omega\rho}{1 - i\omega\delta\gamma} \phi = \frac{ik^2}{\omega\gamma} \phi \tag{1.31}$$

With this result we can express the time harmonic velocity field as

$$\boldsymbol{v} = \nabla\phi = \frac{1 - i\omega\delta\gamma}{i\omega\rho} \nabla u \tag{1.32}$$

Having derived the field equations in a homogeneous isotropic medium we now consider the case in which waves propagate in a region consisting of two such media. We assume that the field quantities in both media have the same harmonic time dependence. This is possible since the material parameters of both media are taken to be independent of time.

The class of scattering problems we consider in this book concerns the manner in which a bounded obstacle, denoted by V^- with boundary $S = \partial V^-$, perturbs an acoustic wave originating in V^+, the unbounded exterior of $\bar{V}^- = V^- \cup S$. The obstacle V^- is a nonempty bounded open set, not necessarily simply connected, with boundary S sufficiently smooth so as to allow the applicability of the Gauss–Green theorems and the existence of boundary values of field quantities in the classical sense. When the scattering obstacle is impenetrable, the acoustic field exists only in V^+ and boundary conditions must be imposed on S. When the scatterer is penetrable we consider it to be another homogeneous fluid characterized by different constitutive parameters ρ, δ and γ as well as the derived expressions, c and k. We will affix superscripts $+$ or $-$ to these parameters as well as to the field quantities to distinguish between those in V^+ and those in V^-. One notable exception is the wave number in the exterior V^+ which, because it appears so often, will remain the unsuperscripted k. Moreover, the medium in V^+ is considered to be lossless, i.e.

$$\delta^+ = 0 \tag{1.33}$$

hence in V^+ we have

$$k = k^+ = \omega\sqrt{\gamma^+\rho^+} > 0 \tag{1.34}$$

$$u^+ = i\omega\rho^+\phi^+ \tag{1.35}$$

$$\boldsymbol{v}^+ = \frac{1}{i\omega\rho^+}\nabla u^+ \tag{1.36}$$

and the Helmholtz equation is satisfied by both velocity potential and excess pressure

$$(\nabla^2 + k^2)\begin{Bmatrix} u^+ \\ \phi^+ \end{Bmatrix} = 0 \tag{1.37}$$

The medium in V^- is, in general, not assumed to be lossless. Hence

$$\delta^- \geq 0 \tag{1.38}$$

We introduce the dimensionless **relative index of refraction** η to be the ratio of wave numbers in V^- and V^+

$$\eta = \frac{k^-}{k} = \sqrt{\frac{\gamma^-\rho^-}{\gamma^+\rho^+}}\frac{1}{\sqrt{1 - i\omega\delta^-\gamma^-}} \tag{1.39}$$

where the choice of the branch of the square root is such that $\operatorname{Im}\eta \geq 0$ and $\operatorname{Im}k^- \geq 0$. For relatively small dissipation, (1.39) implies that

$$\eta = \sqrt{\frac{\gamma^-\rho^-}{\gamma^+\rho^+}}\left(1 + \frac{i\omega\delta^-\gamma^-}{2}\right) + O((\delta^-)^2), \quad \delta^- \to 0^+ \tag{1.40}$$

while for lossless scatters ($\delta^- = 0$)

$$\eta = \sqrt{\frac{\gamma^-\rho^-}{\gamma^+\rho^+}} = \frac{c^+}{c^-} \tag{1.41}$$

Now we write the governing equations in V^- as

$$(\nabla^2 + \eta^2 k^2)\begin{Bmatrix} u^- \\ \phi^- \end{Bmatrix} = 0 \tag{1.42}$$

$$u^- = \frac{i\omega\rho^-}{1 - i\omega\delta^-\gamma^-}\phi^- \tag{1.43}$$

and

$$\boldsymbol{v}^- = \frac{1 - i\omega\delta^-\gamma^-}{i\omega\rho^-}\nabla u^- \tag{1.44}$$

Of course, the case of lossless scatterers corresponds to $\delta^- = 0$.

In summary, the governing field equations are the wave equation in the time domain

$$\ddot{\mathcal{U}}(\boldsymbol{r}, t) = \frac{1}{\gamma\rho}\nabla^2\mathcal{U}(\boldsymbol{r}, t) + \frac{\delta}{\rho}\nabla^2\dot{\mathcal{U}}(\boldsymbol{r}, t) \tag{1.45}$$

and the Helmholtz equation in the **frequency** or **spectral domain**

$$\left(\nabla^2 + \frac{\omega^2\gamma\rho}{1 - i\omega\delta\gamma}\right)u(\boldsymbol{r}) = 0 \tag{1.46}$$

and the same equations hold for the velocity potential as well as each Cartesian component of the velocity. The relation between frequency, wave number and constitutive parameters is given by the dispersion relation (1.18) which may also be obtained by substituting the time harmonic plane wave, $e^{ik\hat{\boldsymbol{k}}\cdot\boldsymbol{r} - i\omega t}$ (see section 1.A.5), into equation (1.45).

In the subsequent development we will mainly consider the excess pressure u in the frequency domain although every relation involving u may be translated into a relation involving ϕ through (1.31) or into a relation for $\boldsymbol{v}$ through (1.32).

1.A.2 *Boundary conditions*

The class of scattering problems we are concerned with involves the determination of how V^- perturbs some known incident wave. We denote this incident wave by u^i and its precise nature is discussed below (1.A.5). The total excess pressure field that exists in V^+ is denoted by u^+. When V^- consists of another fluid medium there will also be a field in V^- denoted by u^-. The conditions relating u^+ and u^- on S are called **transmission conditions** and they are discussed in the next section. When no field exists in V^- we say that the scatterer is **impenetrable** and a variety of **boundary conditions** on S are used to model situations in which the effect of the incident field is not felt in V^-. We confine attention to three different boundary conditions:

The Dirichlet or soft or pressure release surface:

$$u^+(\boldsymbol{r}) = 0, \quad \boldsymbol{r} \in S \tag{1.47}$$

The Neumann or hard surface:

$$\frac{\partial}{\partial n} u^+(\boldsymbol{r}) = 0, \quad \boldsymbol{r} \in S \tag{1.48}$$

where $\partial/\partial n$ denotes the derivative in the direction of the normal on S into V^+.

The Robin or impedance surface:

$$\left(\frac{\partial}{\partial n} + i\frac{\omega\rho^+}{Z^+}\right) u^+(\boldsymbol{r}) = 0, \quad \boldsymbol{r} \in S \tag{1.49}$$

where Z^+ is the **acoustic impedance** measured in units of pressure per unit velocity. Equation (1.49) may also be written

$$\left(\frac{\partial}{\partial n} + ik\nu\right) u^+(\boldsymbol{r}) = 0, \quad \boldsymbol{r} \in S \tag{1.50}$$

where the dimensionless parameter ν is given by

$$\nu = \frac{1}{Z^+}\sqrt{\frac{\rho^+}{\gamma^+}} \tag{1.51}$$

These boundary conditions have the following physical interpretation. The function u^+ denotes the excess pressure field of the medium surrounding the scatterer. A soft surface offers no resistance to pressure; hence, the scatterer yields in such a way as to maintain zero pressure on its boundary. On the other hand, a hard surface admits no local displacements and therefore the normal component of the velocity field (being proportional to $\hat{\boldsymbol{n}} \cdot \nabla u^+$) should vanish. Finally, a surface with finite impedance has an intermediate behavior between the soft and the hard surface. In fact, the impedance boundary condition can be written as

$$Z^+ \frac{\partial}{\partial n}\left(\frac{u^+}{i\omega\rho^+}\right) = -u^+ \tag{1.52}$$

or, using (1.31)

$$Z^+ \hat{\boldsymbol{n}} \cdot \nabla\phi^+ = -u^+ \tag{1.53}$$

or

$$Z^+ v_n^+ = -u^+ \tag{1.54}$$

where v_n^+ stands for the normal component of the velocity field on S. Relation (1.54) indicates that the normal velocity is proportional to the decrease of pressure at any point on S, the proportionality constant being the coefficient of acoustic impedance. Hence, an impedance boundary condition describes a balance between the pressure and the normal velocity field. In other words, the surface stress, caused by the normal velocity, compensates for the excess pressure. In fact, the physical meaning of the acoustic impedance is

$$Z = \frac{\text{excess pressure}}{\text{normal velocity}} \tag{1.55}$$

and hence it is measured in units of pressure per unit velocity.

As $Z^+ \to 0$ the normal velocity is incapable of producing any pressure on the surface, which is therefore a 'soft' surface. On the other hand, as $1/Z^+ \to 0$ the pressure produces no normal velocity, i.e. the surface undergoes no local displacements and therefore is a 'hard' surface.

If we denote the total fields satisfying Dirichlet (soft), Neumann (hard) and Robin (impedance) boundary conditions by $u^+(\boldsymbol{r}; 0)$, $u^+(\boldsymbol{r}; \infty)$ and $u^+(\boldsymbol{r}; Z^+)$ respectively, it is interesting to ask how $u^+(\boldsymbol{r}; Z^+)$ may be considered as a perturbation of $u^+(\boldsymbol{r}; 0)$ or $u^+(\boldsymbol{r}; \infty)$ for small or large values of Z^+. In fact the two limiting cases are quite different. When $Z^+ \to \infty$ or $1/Z^+ \to 0$ we have a so-called regular perturbation problem so that

$$u^+(\boldsymbol{r}; Z^+) = u^+(\boldsymbol{r}; \infty) + O\left(\frac{1}{Z^+}\right), \quad Z^+ \to \infty \tag{1.56}$$

However, when $Z^+ \to 0$ we have a singular perturbation problem whose asymptotics are much more complicated, see Kirsch (1985) for details.

1.A.3 *Transmission conditions*

When the **disturbance** or **incident wave** in V^+ is transmitted into V^-, the scatterer is **penetrable** and the excess pressure u^-, the velocity potential ϕ^-, and the velocity $\boldsymbol{v}^-$ in V^- are governed by relations (1.38)–(1.44). However, the two fluids meet at the boundary S and the conditions relating the excess pressure in V^+ and V^- at S, the **transmission conditions**, are

$$u^+(\boldsymbol{r}) = u^-(\boldsymbol{r}), \quad \boldsymbol{r} \in S \tag{1.57}$$

$$\frac{\partial}{\partial n}u^+(\boldsymbol{r}) = \beta\frac{\partial}{\partial n}u^-(\boldsymbol{r}), \quad \boldsymbol{r} \in S \tag{1.58}$$

where β is complex for lossy and real for lossless scatterers. Condition (1.57) states that the excess pressure field is continuous across S. Condition (1.58) results from requiring continuity of the normal component of the velocity field across S. This may be seen from the expressions relating velocity and excess pressure (1.36) and (1.44), as follows

$$\hat{\boldsymbol{n}} \cdot \boldsymbol{v}^+ = \frac{1}{i\omega\rho^+}\frac{\partial}{\partial n}u^+ = \hat{\boldsymbol{n}} \cdot \boldsymbol{v}^- = \frac{1 - i\omega\delta^-\gamma^-}{i\omega\rho^-}\frac{\partial}{\partial n}u^- \tag{1.59}$$

or

$$\frac{\partial}{\partial n}u^+ = \beta\frac{\partial}{\partial n}u^- \tag{1.60}$$

where

$$\beta = \frac{\rho^+}{\rho^-}(1 - i\omega\delta^-\gamma^-) \tag{1.61}$$

is a dimensionless constant.

Note that for lossless scatterers only, β represents the ratio of the mass densities. Nevertheless, for both lossy and lossless scatterers embedded in a lossless medium the product of β and η^2 is always real and it represents the ratio of mean compressibilities, as follows

$$\begin{aligned}\beta\eta^2 &= \frac{\rho^+}{\rho^-}(1 - i\omega\delta^-\gamma^-)\frac{\rho^-\gamma^-}{\rho^+\gamma^+(1 - i\omega\delta^-\gamma^-)}\\ &= \frac{\gamma^-}{\gamma^+}\end{aligned} \tag{1.62}$$

We remark that if the exterior medium were lossy, then this ratio would no longer be real and β would also have to be redefined. However, we always consider V^+ to be lossless.

If the transmission conditions (1.57) and (1.58) are written in terms of the velocity potential ϕ, then they read as follows

$$\beta\phi^+(\boldsymbol{r}) = \phi^-(\boldsymbol{r}), \quad \boldsymbol{r} \in S \tag{1.63}$$

$$\frac{\partial}{\partial n}\phi^+(\boldsymbol{r}) = \frac{\partial}{\partial n}\phi^-(\boldsymbol{r}), \quad \boldsymbol{r} \in S \tag{1.64}$$

Consequently, the excess pressure is continuous across S but its normal derivative has a jump discontinuity, while the velocity potential has a jump discontinuity at S but its normal derivative is continuous.

Remark: The general transmission problem is a two-parameter problem involving β and η. For lossless scatterers, the special cases of equal densities, where $\beta = 1$, or equal wave numbers, where $\eta = 1$, or equal compressibilities, where $\beta\eta^2 = 1$ furnish one-parameter transmission problems. Of course, if both $\beta = \eta = 1$, then the medium exhibits no discontinuity in its physical parameters and therefore no scattering occurs.

Observe that parameter β occurs in the transmission conditions (1.58) and (1.63). We now discuss the limiting cases when $\beta \to 0, \infty$ or $\rho^- \to \infty, 0$. Rewriting equation (1.43) which relates excess pressure and velocity potential in V^- with the help of (1.61) as

$$u^- = \frac{i\omega\rho^+}{\beta}\phi^- \tag{1.65}$$

we observe that if we assume that ϕ^- is bounded in V^- (no sources in V^-), then

$$\lim_{\beta\to\infty} u^- = 0 \quad \text{in } V^- \tag{1.66}$$

On the other hand, under the assumption that u^- is bounded in V^- we infer that

$$\lim_{\beta \to 0} \phi^- = 0 \quad \text{in } V^- \tag{1.67}$$

In the limiting case as $\beta \to \infty$ (or $\rho^- \to 0$), (1.65) implies that $u^- \to 0$ and the boundary conditions (1.57), (1.58), (1.63) and (1.64) become

$$u^+ = 0, \quad \phi^+ = 0 \quad \text{on } S \tag{1.68}$$

$$\frac{\partial}{\partial n} u^- = 0, \quad \frac{\partial}{\partial n} \phi^+ = \frac{\partial}{\partial n} \phi^- \quad \text{on } S \tag{1.69}$$

where we have assumed that ϕ^- and the normal derivative of u^+ remain bounded on S as $\beta \to \infty$.

Therefore, the problem of determining u^+ or ϕ^+ in V^+ becomes one of solving an exterior Dirichlet problem for the total field u^+ or ϕ^+. While $u^- = 0$ in V^-, the velocity potential ϕ^- is nonzero. This interior velocity potential is the solution of the interior Helmholtz equation (1.42) with boundary condition

$$\frac{\partial}{\partial n} \phi^- = \frac{\partial}{\partial n} \phi^+ \quad \text{on } S, \tag{1.70}$$

The right-hand side is known once the exterior Dirichlet problem has been solved. This interior Neumann problem is always solvable but not uniquely so at a countable set values, k_n^N, $n = 1, 2, 3, \ldots$. The numbers $-(k_n^N)^2$, $n = 1, 2, 3, \ldots$ are known as **interior Neumann eigenvalues**.

When $\beta \to 0$ (or $\rho^- \to \infty$), (1.67) implies that $\phi^- = 0$ in V^- and the transmission conditions become, assuming that ϕ^+ and the normal derivative of u^- remain bounded on S as $\beta \to 0$,

$$u^+ = u^-, \quad \phi^- = 0 \quad \text{on } S \tag{1.71}$$

$$\frac{\partial}{\partial n} u^+ = 0, \quad \frac{\partial}{\partial n} \phi^+ = 0 \quad \text{on } S \tag{1.72}$$

Now the problem of finding the exterior fields is reduced to solving an exterior Neumann problem for u^+ or ϕ^+. The interior velocity potential is zero; however, the interior pressure is not. Rather, it is a solution of the interior Dirichlet problem for the Helmholtz equation (1.42) with the boundary condition

$$u^- = u^+ \quad \text{on } S \tag{1.73}$$

where the right-hand side is known once the exterior Neumann problem has been solved. This interior Dirichlet problem is always solvable but not uniquely so at a countable set of values k_n^D, $n = 1, 2, 3, \ldots$. The numbers $-(k_n^D)^2$, $n = 1, 2, 3, \ldots$ are known as **interior Dirichlet eigenvalues**.

We remark that the eigenvalues of the interior Dirichlet and interior Neumann problems may coincide. However, when they do, the corresponding eigenfunctions must differ.

It should be emphasized that in these limiting cases we cannot assume that both u^- and ϕ^- vanish since this would imply that u^+ and ϕ^+ had homogeneous Cauchy data on S and therefore they would also vanish identically. Therefore, when $\beta \to \infty$, the transmission problem reduces to an exterior Dirichlet problem and the interior pressure field vanishes but the interior velocity potential is nonzero. When $\beta \to 0$, the transmission problem reduces to an exterior Neumann problem and the interior velocity potential vanishes but the interior pressure is nonzero. However, to find the scattered field in V^+ it is unnecessary to find the nonvanishing interior fields.

1.A.4 *Radiation conditions*

Scattering problems always involve an unbounded domain which has infinity as part of its boundary. Any condition on that particular part of the boundary has to be given in asymptotic form, as $r \to +\infty$ where r is the magnitude of the position vector, $r = |\boldsymbol{r}|$. For the development of the acoustic scattering problem we use the following condition, due to Sommerfeld (1912),

$$\lim_{r\to\infty} r\left(\frac{\partial}{\partial r}u(\boldsymbol{r}) - iku(\boldsymbol{r})\right) = 0, \quad \hat{\boldsymbol{r}} \in S^2 \tag{1.74}$$

where u stands for the scattered pressure field and the convergence is taken to be uniform over all directions $\hat{\boldsymbol{r}} = \boldsymbol{r}/r$. The set S^2 denotes the surface of the unit ball in $\mathbb{R}^3$. The **Sommerfeld radiation condition** (1.74) specifies the appropriate geometric attenuation of the scattered field and imposes the outgoing character of the scattered wave. It provides necessary condition when formulating the scattering problem as a well-posed exterior boundary value problem. The velocity potential and all Cartesian components of the velocity field satisfy the same radiation condition as the excess pressure. For more on radiation conditions and related theorems see Müller (1955, 1956), Stoker (1956), Wilcox (1959), Dassios (1988a) and the references therein.

1.A.5 *Incident fields and the fundamental solution*

The **incident field** in a scattering problem is the field that would exist in $\mathbb{R}^3$ if there were no scatterer present. We consider incident fields which are plane waves or point sources or superpositions of plane waves and/or point sources. Let $A \subset V^+$ be a bounded domain which contains all point sources. If there are no point sources, then A is empty. All incident fields, u^i, that we consider are solutions of

$$(\nabla^2 + k^2)u^i(\boldsymbol{r}) = 0, \quad \boldsymbol{r} \in \mathbb{R}^3 \backslash A \tag{1.75}$$

In particular, a plane incident field will be a wave of the form

$$u^i(\boldsymbol{r}) = e^{ik\hat{\boldsymbol{k}}\cdot\boldsymbol{r}}, \quad \boldsymbol{r} \in \mathbb{R}^3 \tag{1.76}$$

which propagates in the direction of $\hat{\boldsymbol{k}}$.

Plane waves satisfy the Helmholtz equation (1.37) at all points in $\mathbb{R}^3$ but they do not satisfy the radiation condition (1.74). On the other hand, the field due to a point source at $\boldsymbol{r}_0$ will satisfy the radiation condition but will be a solution of the Helmholtz equation only in $\mathbb{R}^3\backslash\{\boldsymbol{r}_0\}$. Such a field will be a function of two points, $G(\boldsymbol{r}, \boldsymbol{r}_0)$, and is a **fundamental solution** of the Helmholtz equation. Explicitly,

$$G(\boldsymbol{r}, \boldsymbol{r}_0) = \frac{e^{ik|\boldsymbol{r}-\boldsymbol{r}_0|}}{ik|\boldsymbol{r}-\boldsymbol{r}_0|} = h(k|\boldsymbol{r}-\boldsymbol{r}_0|) \tag{1.77}$$

where $h = h_0^{(1)}$ is the spherical Hankel function of the first kind and order zero. This function is a solution of the equation

$$(\nabla_{\boldsymbol{r}}^2 + k^2)G(\boldsymbol{r}, \boldsymbol{r}_0) = -\frac{4\pi}{ik}\delta(\boldsymbol{r}-\boldsymbol{r}_0) \tag{1.78}$$

where δ denotes the Dirac point measure. The reason for considering this particular form of G is not only that it coincides with the spherical Hankel function $h_0^{(1)}$ but, even more importantly, because it satisfies the radiation condition and moreover defines a dimensionless fundamental solution. Note that the dimensions of the Dirac measure are considered to be inverse volumes so that its volume integral is dimensionless.

All the acoustic incident fields considered in this book are of the form (1.76) or (1.77) or a linear combination of such fields for different incident directions $\hat{\boldsymbol{k}}$ or source locations $\boldsymbol{r}_0$. Note that

$$\begin{aligned} G(\boldsymbol{r}, \boldsymbol{r}_0) &= \frac{e^{ikr_0 - ik\hat{\boldsymbol{r}}_0\cdot\boldsymbol{r}}}{ikr_0} + O\left(\frac{1}{r_0^2}\right) \\ &= h(kr_0)e^{-ik\hat{\boldsymbol{r}}_0\cdot\boldsymbol{r}} + O\left(\frac{1}{r_0^2}\right), \; r_0 \to \infty \end{aligned} \tag{1.79}$$

Since

$$e^{-ik\hat{\boldsymbol{r}}_0\cdot\boldsymbol{r}} = \lim_{r_0\to\infty} ikr_0 e^{-ikr_0} G(\boldsymbol{r}, \boldsymbol{r}_0) \tag{1.80}$$

we may consider the plane wave propagating in the direction $-\hat{\boldsymbol{r}}_0$ as a modified point source at $\boldsymbol{r}_0$ as $r_0 \to \infty$.

1.A.6 *Basic scattering problems*

Here we collect the results of the previous sections and define the basic mathematical problems encountered in acoustic scattering. The physical meaning of the parameters involved has been explained in preceding sections and is not repeated here. In all the following problems we specify an incident field u^i with an exterior wave number k and a surface S which bounds the scatterer, V^-. Moreover, we seek functions u^+ and u, the **total** and **scattered fields**, respectively, related by

$$u = u^+ - u^i \text{ in } V^+ \tag{1.81}$$

where the function u satisfies the Helmholtz equation

$$(\nabla^2 + k^2)u = 0 \text{ in } V^+ \tag{1.82}$$

and the radiation condition

$$\lim_{r\to\infty} r\left(\frac{\partial u}{\partial r} - iku\right) = 0, \quad \text{uniformly in } \hat{\boldsymbol{r}} \in S^2 \tag{1.83}$$

In addition, one of the following conditions must be satisfied.

The Dirichlet problem:

$$u^+(\boldsymbol{r}) = 0, \quad \boldsymbol{r} \in S \tag{1.84}$$

The Neumann problem:

$$\frac{\partial}{\partial n} u^+(\boldsymbol{r}) = 0, \quad \boldsymbol{r} \in S \tag{1.85}$$

The Robin problem:

For given ν

$$\frac{\partial}{\partial n} u^+(\boldsymbol{r}) + ik\nu u^+(\boldsymbol{r}) = 0, \quad \boldsymbol{r} \in S \tag{1.86}$$

The transmission problem:

For given η and β, in addition to u^+ we need to find u^- which satisfies the Helmholtz equation

$$(\nabla^2 + k^2\eta^2)u^-(\boldsymbol{r}) = 0, \quad \boldsymbol{r} \in V^- \tag{1.87}$$

and

$$u^+(\boldsymbol{r}) = u^-(\boldsymbol{r}), \quad \frac{\partial}{\partial n} u^+(\boldsymbol{r}) = \beta \frac{\partial}{\partial n} u^-(\boldsymbol{r}), \quad \boldsymbol{r} \in S \tag{1.88}$$

Here, we have formulated the problems in terms of the excess pressure. Equivalent mathematical formulations in terms of the velocity potential may be obtained easily using the expressions relating excess pressure and velocity potential, (1.35) and (1.43).

All four of these problems are well-posed boundary value problems for any C^2-surface S, i.e. each one possesses a unique and stable classical solution (Colton and Kress 1983; Jones 1986).

By classical solutions we mean

(i) $u^+ \in C^2(V^+) \cap C^0(V^+ \cup S)$ for the Dirichlet problem

(ii) $u^+ \in C^2(V^+) \cap C^1(V^+ \cup S)$ for the Neumann and Robin problems, and

(iii) $u^+ \in C^2(V^+) \cap C^1(V^+ \cup S), u^- \in C^2(V^-) \cap C^1(V^- \cup S)$ for the transmission problem.

For surfaces with a finite number of corners and edges an additional condition of local finite energy must be imposed. This is written as an energy condition (see next section, 1.A.7).

$$\int_\Omega (|\nabla u(\boldsymbol{r})|^2 + |ku(\boldsymbol{r})|^2) dv(\boldsymbol{r}) < +\infty \tag{1.89}$$

where Ω is any bounded subdomain of V^+ (Meixner 1949). Condition (1.89) replaces the assumption of differentiability up to the boundary and all the boundary conditions that involve differentiation on the boundary can be applied only when these normal derivatives exist (i.e. almost everywhere). More about this can be found in Van Bladel (1995) and the references therein.

1.A.7 *Energy functionals*

Here, we develop expressions for energy density and power flux from the basic equations (1.1) and (1.2) which we rewrite as

$$\gamma\dot{\mathcal{U}} + \nabla\cdot\boldsymbol{V} = 0 \tag{1.90}$$

$$\nabla\mathcal{U} = \delta\nabla\nabla\cdot\boldsymbol{V} - \rho\dot{\boldsymbol{V}} \tag{1.91}$$

As described earlier, we allow $\mathcal{U}$ and $\boldsymbol{V}$ to assume complex values although the physical quantities are recovered by taking the real parts. Multiplying (1.90) by $\mathcal{U}^*$ we obtain

$$\gamma\mathcal{U}^*\dot{\mathcal{U}} + \nabla\cdot(\mathcal{U}^*\boldsymbol{V}) = \boldsymbol{V}\cdot\nabla\mathcal{U}^* \tag{1.92}$$

Substituting the complex conjugate of (1.91) into (1.92) and rearranging terms we find

$$\rho\boldsymbol{V}\cdot\dot{\boldsymbol{V}}^* + \gamma\mathcal{U}^*\dot{\mathcal{U}} + \nabla\cdot(\mathcal{U}^*\boldsymbol{V}) = \delta\boldsymbol{V}\cdot\nabla(\nabla\cdot\boldsymbol{V}^*) \tag{1.93}$$

To this last equation we add its complex conjugate leading to the relation

$$\frac{\partial}{\partial t}\left(\frac{\rho}{2}|\boldsymbol{V}|^2 + \frac{\gamma}{2}|\mathcal{U}|^2\right) + \nabla\cdot\mathrm{Re}\{\mathcal{U}\boldsymbol{V}^*\} = \delta\mathrm{Re}\{\boldsymbol{V}\cdot\nabla(\nabla\cdot\boldsymbol{V}^*)\} \tag{1.94}$$

Recognizing the terms $(\rho/2)|\boldsymbol{V}|^2$ as **kinetic energy** and $(\gamma/2)|\mathcal{U}|^2$ as **potential energy** we define the **energy density function**

$$\mathcal{W}(\boldsymbol{r},t) = \frac{\rho}{2}|\boldsymbol{V}(\boldsymbol{r},t)|^2 + \frac{\gamma}{2}|\mathcal{U}(\boldsymbol{r},t)|^2 \tag{1.95}$$

which is measured in units of energy per unit volume. Next, we define the **power flux vector**

$$\mathcal{I}(\boldsymbol{r},t) = \mathrm{Re}\{\mathcal{U}(\boldsymbol{r},t)\boldsymbol{V}^*(\boldsymbol{r},t)\} \tag{1.96}$$

measured in units of power per unit area. Then (1.94), now written as

$$\frac{\partial}{\partial t}\mathcal{W}(\boldsymbol{r},t) + \nabla\cdot\mathcal{I}(\boldsymbol{r},t) = \delta\mathrm{Re}\{\boldsymbol{V}(\boldsymbol{r},t)\cdot\nabla(\nabla\cdot\boldsymbol{V}^*(\boldsymbol{r},t))\} \tag{1.97}$$

is the **energy conservation law** which states that the time rate of change of energy density plus the spatial power flux is equal to the energy dissipated in the medium represented by $\delta\mathrm{Re}\{\boldsymbol{V}\cdot\nabla(\nabla\cdot\boldsymbol{V}^*)\}$ and measured in units of power per unit volume.

Introducing the velocity potential (1.9) and the harmonic time dependence $e^{-i\omega t}$ the energy density function becomes

$$W(\boldsymbol{r}) = \frac{\rho}{2}|\nabla\phi(\boldsymbol{r})|^2 + \frac{\gamma}{2}|u(\boldsymbol{r})|^2 \tag{1.98}$$

With (1.18) and (1.31), this may be written in terms of the velocity potential as

$$W = \frac{\rho}{2}\left\{|\nabla\phi|^2 + \frac{|k|^4|\phi|^2}{\gamma\rho\omega^2}\right\} \tag{1.99}$$

or in terms of the excess pressure

$$W = \frac{\gamma}{2}\left\{\frac{\gamma\rho\omega^2|\nabla u|^2}{|k|^4} + |u|^2\right\} \tag{1.100}$$

For a lossless medium these expressions become

$$W = \frac{\rho}{2}\{|\nabla\phi|^2 + |k\phi|^2\} \tag{1.101}$$

and

$$W = \frac{\gamma}{2}\left\{\left|\frac{\nabla u}{k}\right|^2 + |u|^2\right\} \tag{1.102}$$

In the limit as $k \to 0$, the expressions for the energy density in terms of velocity potential reduce to

$$W_0(\boldsymbol{r}) = \frac{\rho}{2}|\nabla\phi(\boldsymbol{r})|^2 \tag{1.103}$$

which is the integrand of the **Dirichlet integral** of Potential Theory (Kellogg 1953). This expression does not follow from the representation of the energy density in terms of excess pressure since as $k \to 0$ the local disturbances tend to equilibrium and the excess pressure will vanish. Introducing the velocity potential and $e^{-i\omega t}$ time dependence into the expression for the power flux vector (1.96) we obtain

$$\boldsymbol{I}(\boldsymbol{r}) = \mathrm{Re}\{u(\boldsymbol{r})\nabla\phi^*(\boldsymbol{r})\} \tag{1.104}$$

which with (1.31) may be written as

$$\boldsymbol{I} = \omega\rho\,\mathrm{Im}\left\{\frac{\phi^*\nabla\phi}{1 - i\omega\delta\gamma}\right\} = \frac{1}{\omega\rho}\,\mathrm{Im}\{(1 - i\omega\delta\gamma)u^*\nabla u\} \tag{1.105}$$

The power flux in a direction $\hat{\boldsymbol{a}}$ is $\hat{\boldsymbol{a}} \cdot \boldsymbol{I}$ and the power flow through a surface $\partial\Omega$ with outward normal $\hat{\boldsymbol{n}}$ is

$$\int_{\partial\Omega} \hat{\boldsymbol{n}} \cdot \boldsymbol{I} ds = \omega\rho\,\mathrm{Im}\int_{\partial\Omega} \frac{\phi^*\nabla\phi}{1 - i\omega\delta\gamma} ds = \frac{1}{\omega\rho}\,\mathrm{Im}\int_{\partial\Omega} (1 - i\omega\delta\gamma)u^*\nabla u ds \tag{1.106}$$

For a lossless medium, set $\delta = 0$ in (1.106).

1.A.8 *Problems in acoustics*

1. Show that if the time dependent field

$$\mathcal{U}(\boldsymbol{r}, t) = \mathrm{Re}\{u(\boldsymbol{r})e^{-i\omega t}\}$$

solves the equation

$$\ddot{\mathcal{U}} = \frac{1}{\gamma\rho}\nabla^2\mathcal{U} + \frac{\delta}{\rho}\nabla^2\dot{\mathcal{U}}$$

then the time independent field $u(\boldsymbol{r})$ has to satisfy the Helmholtz equation

$$(\nabla^2 + k^2)u = 0$$

with

$$k^2 = \frac{\omega^2\gamma\rho}{1 - i\omega\delta\gamma}$$

2. Use the definition of the phase velocity

$$c = \frac{\omega}{\mathrm{Re}\, k}$$

and the dispersion relation

$$\gamma\rho\omega^2 = k^2(1 - i\omega\delta\gamma)$$

to show that

$$c = \frac{1}{\sqrt{\gamma\rho}}\sqrt{\frac{2(1 + \omega^2\delta^2\gamma^2)}{1 + \sqrt{1 + \omega^2\delta^2\gamma^2}}}$$

Find the leading asymptotic correction to the phase velocity of a lossless medium

$$c_0 = \frac{1}{\sqrt{\gamma\rho}}$$

for small δ.

3. Find the interior Dirichlet and the interior Neumann eigenvalues for (i) the ball

$$B(\mathbf{0}, a) = \{\boldsymbol{r} \in \mathbb{R}^3 \mid |\boldsymbol{r}| < a\}$$

and (ii) the rectangle

$$R(a, b) = [0, a] \times [0, b] \subset \mathbb{R}^2$$

4. Use the results of Problem 3 to show that it is possible for a real number to be both interior Dirichlet and interior Neumann eigenvalue, but the corresponding eigenfunctions are different for a ball or a rectangle.
5. Prove in general that it is not possible to have coincidence of an interior eigenvalue and the corresponding eigenfunctions for both the Dirichlet and the Neumann problem.

6. Consider the transmission problems for the excess pressure fields $u^{\pm}$ and for the velocity potentials $\phi^{\pm}$. Assume that ϕ^{-} and the normal derivative of u^{+} remain bounded on S as $\beta \to \infty$, and that ϕ^{+} and the normal derivative of u^{-} remain bounded on S as $\beta \to 0$. Prove that in both limiting cases of $\beta \to \infty$ and $\beta \to 0$, if u^{-} and ϕ^{-} vanish identically in the interior V^{-} of the scatterer, then u^{+} and ϕ^{+} also vanish identically in the exterior medium V^{+}.
7. Verify that the fundamental solution $G(\boldsymbol{r}, \boldsymbol{r}_0)$, as given by (1.77), satisfies equation (1.78). Derive the asymptotic form (1.79).
8. Given the energy density function

$$\mathcal{W} = \frac{\rho}{2}|\nabla\Phi|^2 + \frac{\gamma}{2}|\mathcal{U}|^2$$

where Φ is the time dependent velocity potential and $\mathcal{U}$ is the time dependent excess pressure field, prove by direct differentiation that

$$\dot{\mathcal{W}} = \delta \mathrm{Re}\{(\nabla\Phi)\cdot\nabla(\nabla^2\Phi^*)\} - \nabla\cdot\mathrm{Re}\{\mathcal{U}\nabla\Phi^*\}$$

1.EM The basic theory of electromagnetic waves

1.EM.1 *Field equations*

The fundamental equations in electromagnetics in the time domain consist of the **Maxwell equations**, which are the differential form of **Faraday's law**

$$\nabla \times \mathcal{E}(\boldsymbol{r}, t) = -\dot{\mathcal{B}}(\boldsymbol{r}, t) \tag{1.107}$$

and **Ampere's law**

$$\nabla \times \mathcal{H}(\boldsymbol{r}, t) = \dot{\mathcal{D}}(\boldsymbol{r}, t) + \mathcal{J}(\boldsymbol{r}, t) \tag{1.108}$$

where $\mathcal{E}$ is the **electric field**, $\mathcal{B}$ is the **magnetic induction**, $\mathcal{H}$ is the **magnetic field**, $\mathcal{D}$ is the **electric displacement** and $\mathcal{J}$ is the **conduction current density**. The quantity $\mathcal{E}$ is measured in units of force per unit charge, $\mathcal{B}$ is measured in units of mass per unit charge per unit time, $\mathcal{H}$ in units of charge per unit length per unit time, $\mathcal{D}$ in units of charge per unit area, and $\mathcal{J}$ in units of charge per unit area per unit time. The time harmonic form of these equations is obtained by assuming a harmonic time dependence

$$\mathcal{E}(\boldsymbol{r}, t) = \boldsymbol{E}(\boldsymbol{r})e^{-i\omega t} \tag{1.109}$$
$$\mathcal{B}(\boldsymbol{r}, t) = \boldsymbol{B}(\boldsymbol{r})e^{-i\omega t} \tag{1.110}$$
$$\mathcal{H}(\boldsymbol{r}, t) = \boldsymbol{H}(\boldsymbol{r})e^{-i\omega t} \tag{1.111}$$
$$\mathcal{D}(\boldsymbol{r}, t) = \boldsymbol{D}(\boldsymbol{r})e^{-i\omega t} \tag{1.112}$$
$$\mathcal{J}(\boldsymbol{r}, t) = \boldsymbol{J}(\boldsymbol{r})e^{-i\omega t} \tag{1.113}$$

The quantities $\boldsymbol{E}, \boldsymbol{B}, \boldsymbol{H}, \boldsymbol{D}$ and $\boldsymbol{J}$ are complex vector valued functions of position and are universally employed in electromagnetics. Physical quantities are obtained by taking the real parts of the quantities in (1.109)–(1.113). Substituting (1.109)–(1.113) into the time dependent Maxwell equations leads to the equations

$$\nabla \times \boldsymbol{E}(\boldsymbol{r}) = i\omega \boldsymbol{B}(\boldsymbol{r}) \tag{1.114}$$

$$\nabla \times \boldsymbol{H}(\boldsymbol{r}) = -i\omega \boldsymbol{D}(\boldsymbol{r}) + \boldsymbol{J}(\boldsymbol{r}) \tag{1.115}$$

For a linear, homogeneous and isotropic medium we have the following constitutive relations

$$\mathcal{D} = \epsilon \mathcal{E}, \quad \boldsymbol{D} = \epsilon \boldsymbol{E} \tag{1.116}$$

$$\mathcal{B} = \mu \mathcal{H}, \quad \boldsymbol{B} = \mu \boldsymbol{H} \tag{1.117}$$

$$\mathcal{J} = \sigma \mathcal{E}, \quad \boldsymbol{J} = \sigma \boldsymbol{E} \tag{1.118}$$

where ϵ is the **electric permittivity** or **dielectric constant**, measured in capacity per unit length, μ is the **magnetic permeability**, measured in inductance per unit length, and σ is the **conductivity**, measured in capacity per unit length per unit time, of the media.

With the constitutive relations, equations (1.114) and (1.115) may be written as

$$\nabla \times \boldsymbol{E} = i\omega\mu \boldsymbol{H} \tag{1.119}$$

$$\nabla \times \boldsymbol{H} = (-i\omega\epsilon + \sigma)\boldsymbol{E} \tag{1.120}$$

from which we easily deduce that

$$\nabla \cdot \boldsymbol{E} = \nabla \cdot \boldsymbol{H} = 0 \tag{1.121}$$

Equations (1.119)–(1.121) provide the standard form of the time independent Maxwell equations in terms of the medium parameters ϵ, μ, σ and the angular frequency ω.

The perfectly conducting medium ($\sigma \to \infty$)
Under the assumption that $\nabla \times \boldsymbol{H}$ remains bounded as $\sigma \to \infty$ (i.e. the medium becomes perfectly conducting) equation (1.120) implies that $\boldsymbol{E}$ vanishes and equation (1.119) then implies that $\boldsymbol{H}$ also vanishes. Consequently, no fields exist within a perfectly conducting medium.

The nonconducting medium ($\sigma = 0$)
A nonconducting medium does not allow any motion of charge and therefore $\boldsymbol{J} = \boldsymbol{0}$. In this case, the medium is usually called a **perfect dielectric**.

For a general **conducting medium**, $\sigma \neq 0$, substitution of the plane wave, $\boldsymbol{E} = \boldsymbol{a}e^{ik\hat{\boldsymbol{k}}\cdot\boldsymbol{r}}$ (see section 1.EM.5) into the equation

$$\nabla \times (\nabla \times \boldsymbol{E}) = (\epsilon\mu\omega^2 + i\mu\sigma\omega)\boldsymbol{E} \tag{1.122}$$

obtained by combining (1.119) and (1.120) leads to the dispersion relation

$$\mu\epsilon\omega^2 + i\mu\sigma\omega = k^2 \tag{1.123}$$

This provides the definition of the wave number, k, as

$$k = \sqrt{\epsilon\mu\omega^2 + i\mu\sigma\omega} \tag{1.124}$$

and we choose that branch of the square root such that $\operatorname{Im} k \geq 0$. A plane wave propagating in a conducting medium travels with a phase velocity given by

$$c = \frac{\omega}{\operatorname{Re} k} \tag{1.125}$$

and it is dissipated at a rate equal to $\operatorname{Im} k$ (see section 1.A.1 for a more complete discussion of phase velocity). Obviously, if $\sigma = 0$, then the wave experiences no dissipation.

From (1.123) we may deduce that in a conductive medium

$$\operatorname{Re} k = \omega\sqrt{\epsilon\mu}\left(\frac{1 + \sqrt{1 + \sigma^2/\epsilon^2\omega^2}}{2}\right)^{1/2} \tag{1.126}$$

and

$$c = \frac{1}{\sqrt{\epsilon\mu}}\left(\frac{2}{1 + \sqrt{1 + \sigma^2/\epsilon^2\omega^2}}\right)^{1/2} \tag{1.127}$$

In a lossless medium, $\sigma = 0$, and we see that

$$k = \omega\sqrt{\epsilon\mu} \tag{1.128}$$

and

$$c = \frac{1}{\sqrt{\epsilon\mu}} \tag{1.129}$$

We define the **characteristic impedance** of the medium, measured in inductance per unit length, by

$$Z = \frac{\mu\omega}{\sqrt{\epsilon\mu\omega^2 + i\mu\sigma\omega}} \tag{1.130}$$

and the **characteristic admittance** by

$$Y = \frac{1}{Z} = \frac{\sqrt{\epsilon\mu\omega^2 + i\mu\sigma\omega}}{\mu\omega} \tag{1.131}$$

and note that the square root convention we have adopted implies that $\operatorname{Im} Y \geq 0$ and $\operatorname{Im} Z \leq 0$. In terms of these quantities, the Maxwell equations (1.119) and (1.120) take the form

$$\nabla \times \boldsymbol{E} = ikZ\boldsymbol{H}, \quad \nabla \times \boldsymbol{H} = -ikY\boldsymbol{E} \tag{1.132}$$

Rewriting (1.122) and the corresponding equation for $\boldsymbol{H}$ in terms of k, taking into account (1.121) and (1.123) we obtain

$$(\nabla^2 + k^2)\boldsymbol{E} = \boldsymbol{0}, \quad (\nabla^2 + k^2)\boldsymbol{H} = \boldsymbol{0} \tag{1.133}$$

Hence, all the Cartesian components of $\boldsymbol{E}$ and $\boldsymbol{H}$ satisfy the Helmholtz equation with the same wave number.

We distinguish the constitutive parameters and the field quantities in the different media in V^+ and V^- by adding superscripts $+$ or $-$ as appropriate. Moreover, the medium in the unbounded domain V^+ will be assumed lossless ($\sigma^+ = 0$). The only exception to the superscript convention will be the wave number in V^+

$$k = \omega\sqrt{\epsilon^+\mu^+} \tag{1.134}$$

which will remain unsuperscripted. Employing the relative index of refraction $\eta = k^-/k$ we may write the governing Maxwell equations in V^+ and V^- as

$$\nabla \times \boldsymbol{E}^+ = ikZ^+\boldsymbol{H}^+, \quad \nabla \times \boldsymbol{H}^+ = -ikY^+\boldsymbol{E}^+ \quad \text{in } V^+ \tag{1.135}$$

$$\nabla \times \boldsymbol{E}^- = ik\eta Z^-\boldsymbol{H}^-, \quad \nabla \times \boldsymbol{H}^- = -ik\eta Y^-\boldsymbol{E}^- \quad \text{in } V^- \tag{1.136}$$

where the impedances and admittances are defined in (1.130) and (1.131) with the constitutive parameters appropriately superscripted.

1.EM.2 *Boundary conditions*

The class of scattering problems we are concerned with involves the determination of how V^- perturbs some known incident wave. We denote this incident wave by $(\boldsymbol{E}^i, \boldsymbol{H}^i)$ and its precise nature is discussed in section 1.EM.5. When the incident wave penetrates into V^- conditions relating the total fields $(\boldsymbol{E}^+, \boldsymbol{H}^+)$ and the interior fields $(\boldsymbol{E}^-, \boldsymbol{H}^-)$ on S must be imposed and they are discussed in the next section. When no fields exist in V^- we say that the scatterer is **impenetrable** and we consider two such cases where boundary conditions must be imposed on the fields on S. These are the following.

The perfectly conducting surface:

$$\hat{\boldsymbol{n}} \times \boldsymbol{E}^+(\boldsymbol{r}) = \boldsymbol{0}, \quad \hat{\boldsymbol{n}} \cdot \boldsymbol{H}^+(\boldsymbol{r}) = 0, \quad \boldsymbol{r} \in S. \tag{1.137}$$

We remark that the quantities not determined by the boundary conditions may be interpreted as surface charge, ρ_s and surface current, $\boldsymbol{J}_s$; that is,

$$\rho_s(\boldsymbol{r}) = \hat{\boldsymbol{n}} \cdot \boldsymbol{D}^+(\boldsymbol{r}), \quad \boldsymbol{r} \in S \tag{1.138}$$

$$\boldsymbol{J}_s(\boldsymbol{r}) = \hat{\boldsymbol{n}} \times \boldsymbol{H}^+(\boldsymbol{r}), \quad \boldsymbol{r} \in S \tag{1.139}$$

The surface charge density ρ_s is measured in charge per unit area, while the surface current density is a tangential vector function whose magnitude measures the charge moving past the normal to a curve on S per unit length per unit time.

The impedance surface:

$$\hat{\boldsymbol{n}} \times (\hat{\boldsymbol{n}} \times \boldsymbol{E}^+(\boldsymbol{r})) = -Z_s Z^+(\hat{\boldsymbol{n}} \times \boldsymbol{H}^+(\boldsymbol{r})), \quad \boldsymbol{r} \in S \tag{1.140}$$

or equivalently,

$$\hat{\boldsymbol{n}} \times \boldsymbol{E}^+(\boldsymbol{r}) = Z_s Z^+ \hat{\boldsymbol{n}} \times (\hat{\boldsymbol{n}} \times \boldsymbol{H}^+(\boldsymbol{r})), \quad \boldsymbol{r} \in S \tag{1.141}$$

The dimensionless parameter Z_s denotes the **surface impedance** relative to the characteristic impedance Z^+ of the medium and may vary on S.

The impedance condition was introduced by Leontovich (1948) in connection with scattering by a conductive half space and it describes a relationship between the tangential component of the electric and the tangential component of the magnetic field on the plane interface of the two half spaces. Its use as a boundary condition has been extended to scattering by bounded obstacles in situations where the obstacle is not perfectly conducting but the exterior field will not penetrate deeply into the scatterer (Senior 1960*b*) and the radii of curvature of S are large relative to the wavelength λ. The impedance condition accounts for highly absorbing coating layers and it provides a method to simulate the material properties of a scattering surface (Senior and Volakis 1995). Since it involves only the exterior fields the more difficult transmission problem involving interior fields is replaced by a simpler exterior boundary value problem.

The impedance problem will reduce to the perfect conductor problem as the surface impedance approaches zero.

1.EM.3 *Transmission conditions*

If the conductivity σ^- in V^- is finite (possibly 0), then the incident wave in the lossless medium V^+ is transmitted into V^-. The fields in V^- and V^+ are governed by Maxwell's equation with appropriate constitutive parameters. The two different media meet at the boundary S and the conditions relating the fields in V^+ and V^- on S are

$$\hat{\boldsymbol{n}} \times \boldsymbol{E}^+(\boldsymbol{r}) = \hat{\boldsymbol{n}} \times \boldsymbol{E}^-(\boldsymbol{r}) \tag{1.142}$$

$$\epsilon^+\hat{\boldsymbol{n}} \cdot \boldsymbol{E}^+(\boldsymbol{r}) = \epsilon^-\left(1 + \frac{i\sigma^-}{\epsilon^-\omega}\right)\hat{\boldsymbol{n}} \cdot \boldsymbol{E}^-(\boldsymbol{r}) \tag{1.143}$$

or

$$Y^+\hat{\boldsymbol{n}} \cdot \boldsymbol{E}^+(\boldsymbol{r}) = \eta Y^-\hat{\boldsymbol{n}} \cdot \boldsymbol{E}^-(\boldsymbol{r}) \tag{1.144}$$

$$\hat{\boldsymbol{n}} \times \boldsymbol{H}^+(\boldsymbol{r}) = \hat{\boldsymbol{n}} \times \boldsymbol{H}^-(\boldsymbol{r}) \tag{1.145}$$

$$\mu^+\hat{\boldsymbol{n}} \cdot \boldsymbol{H}^+(\boldsymbol{r}) = \mu^-\hat{\boldsymbol{n}} \cdot \boldsymbol{H}^-(\boldsymbol{r}) \tag{1.146}$$

or

$$Z^+\hat{\boldsymbol{n}} \cdot \boldsymbol{H}^+(\boldsymbol{r}) = \eta Z^-\hat{\boldsymbol{n}} \cdot \boldsymbol{H}^-(\boldsymbol{r}) \tag{1.147}$$

where $\boldsymbol{r} \in S$ in all six equations.

Using the definitions of impedance, admittance and index of refraction, conditions (1.142) and (1.145) state that each component of the tangential electric and the tangential magnetic field is continuous across S. Conditions (1.143) and (1.146) state that the normal components of both the electric displacement as well as the magnetic induction fields are also continuous as we cross S (Jackson 1962; Jones 1979*c*; Müller 1969; Stratton 1941).

1.EM.4 *Radiation conditions*

In addition to the field equations and boundary or transmission conditions, the scattered field in V^+ must represent energy being radiated outward. In electromagnetics this condition is ensured by requiring compliance with the **Silver–Müller radiation conditions** (Müller 1948; Silver 1949)

$$\lim_{r\to\infty} [\boldsymbol{r} \times (\nabla \times \boldsymbol{E}) + ikr\boldsymbol{E}] = \boldsymbol{0} \tag{1.148}$$

$$\lim_{r\to\infty} [\boldsymbol{r} \times (\nabla \times \boldsymbol{H}) + ikr\boldsymbol{H}] = \boldsymbol{0} \tag{1.149}$$

uniformly over all directions $\hat{\boldsymbol{r}}$. They can also be written as

$$Z^+\hat{\boldsymbol{r}} \times \boldsymbol{H} + \boldsymbol{E} = o\left(\frac{1}{r}\right) \tag{1.150}$$

$$Y^+\hat{\boldsymbol{r}} \times \boldsymbol{E} - \boldsymbol{H} = o\left(\frac{1}{r}\right) \tag{1.151}$$

which should hold uniformly in $\hat{\boldsymbol{r}}$ as $r \to +\infty$. Note that in the far field both fields $\boldsymbol{E}$ and $\boldsymbol{H}$ are asymptotically orthogonal to $\hat{\boldsymbol{r}}$ that is, the fields have no radial components. The radiation conditions can be replaced by the Sommerfeld condition (1.74) for each Cartesian component of $\boldsymbol{E}$ and $\boldsymbol{H}$. Furthermore, it can be shown that each one of the conditions (1.148) and (1.149) implies the other. Hence, they are not independent conditions but are connected via the Maxwell equations and it is only necessary to impose one of them.

1.EM.5 *Incident fields and the fundamental solution*

We consider as an incident field any pair $(\boldsymbol{E}^i, \boldsymbol{H}^i)$ of either entire solutions of the homogeneous exterior Maxwell equation (1.135), or fields that solve the exterior Maxwell equation everywhere in V^- and also in $V^+\backslash A$, where A is the support of the sources of the incident waves. We assume all sources originate in V^+. In particular, for plane wave incidence in the direction of $\hat{\boldsymbol{k}}$ we define the normalized fields

$$\boldsymbol{E}^i(\boldsymbol{r}) = \hat{\boldsymbol{a}} e^{ik\hat{\boldsymbol{k}}\cdot\boldsymbol{r}} \tag{1.152}$$

$$\boldsymbol{H}^i(\boldsymbol{r}) = \hat{\boldsymbol{b}} Y^+ e^{ik\hat{\boldsymbol{k}}\cdot\boldsymbol{r}} \tag{1.153}$$

where the **polarization vectors** $\hat{\boldsymbol{a}}$ and $\hat{\boldsymbol{b}}$ and the direction of propagation $\hat{\boldsymbol{k}}$ form a right-handed orthonormal system (i.e. $\hat{\boldsymbol{a}} \times \hat{\boldsymbol{b}} = \hat{\boldsymbol{k}}$). Note that plane waves do not satisfy the Silver–Müller radiation conditions as can easily be seen by direct substitution into (1.148) or (1.149). Physically the radiation condition describes the geometric attenuation of the field far from its source and since plane waves can be thought of as (normalized) point sources, as the source point recedes to infinity, the field point is 'close' to the source of the plane wave where the radiation condition is not applicable.

Other types of incident fields that occur in practical applications are the electromagnetic fields radiated from an **electric dipole** of unit moment $\hat{\boldsymbol{p}}$, or a **magnetic dipole** of unit moment $\hat{\boldsymbol{m}}$.

In particular, the fields radiated from an electric dipole, located at the point $\boldsymbol{r}_0$, with unit electric dipole moment $\hat{\boldsymbol{p}}$ are given by

$$\boldsymbol{E}_e^i(\boldsymbol{r}, \boldsymbol{r}_0) = \frac{ik}{4\pi\epsilon^+}(k^2\widetilde{\boldsymbol{I}} + \nabla_{\boldsymbol{r}}\nabla_{\boldsymbol{r}}) \cdot [G^+(\boldsymbol{r}, \boldsymbol{r}_0)\hat{\boldsymbol{p}}] \tag{1.154}$$

$$\boldsymbol{H}_e^i(\boldsymbol{r}, \boldsymbol{r}_0) = \frac{k\omega}{4\pi}\nabla_{\boldsymbol{r}} \times [G^+(\boldsymbol{r}, \boldsymbol{r}_0)\hat{\boldsymbol{p}}] \tag{1.155}$$

where $G^+(\boldsymbol{r}, \boldsymbol{r}_0)$ is the fundamental solution of the Helmholtz equation, (1.77) in V^+ and $\widetilde{\boldsymbol{I}}$ denotes the unit dyadic. These fields satisfy the equations

$$\nabla_{\boldsymbol{r}} \times \boldsymbol{H}_e^i(\boldsymbol{r}) + i\omega\epsilon^+\boldsymbol{E}_e^i(\boldsymbol{r}) = -i\omega\delta(\boldsymbol{r} - \boldsymbol{r}_0)\hat{\boldsymbol{p}} \tag{1.156}$$

$$\nabla_{\boldsymbol{r}} \times \boldsymbol{E}_e^i(\boldsymbol{r}) - i\omega\mu^+\boldsymbol{H}_e^i(\boldsymbol{r}) = \boldsymbol{0} \tag{1.157}$$

Similarly, the fields due to a magnetic dipole of unit dipole moment $\hat{\boldsymbol{m}}$, at the point $\boldsymbol{r}_0$, will radiate the following fields

$$\boldsymbol{E}_m^i(\boldsymbol{r}, \boldsymbol{r}_0) = -\frac{k\omega\mu^+}{4\pi}\nabla_{\boldsymbol{r}} \times [G^+(\boldsymbol{r}, \boldsymbol{r}_0)\hat{\boldsymbol{m}}] \tag{1.158}$$

$$\boldsymbol{H}_m^i(\boldsymbol{r}, \boldsymbol{r}_0) = \frac{ik}{4\pi}(k^2\widetilde{\boldsymbol{I}} + \nabla_{\boldsymbol{r}}\nabla_{\boldsymbol{r}}) \cdot [G^+(\boldsymbol{r}, \boldsymbol{r}_0)\hat{\boldsymbol{m}}] \tag{1.159}$$

which satisfy the equations

$$\nabla_{\boldsymbol{r}} \times \boldsymbol{H}_m^i(\boldsymbol{r}) + i\omega\epsilon^+\boldsymbol{E}_m^i(\boldsymbol{r}) = \boldsymbol{0} \tag{1.160}$$

$$\nabla_{\boldsymbol{r}} \times \boldsymbol{E}_m^i(\boldsymbol{r}) - i\omega\mu^+\boldsymbol{H}_m^i(\boldsymbol{r}) = i\omega\mu^+\delta(\boldsymbol{r} - \boldsymbol{r}_0)\hat{\boldsymbol{m}} \tag{1.161}$$

For $\boldsymbol{E}_e^i, \boldsymbol{H}_e^i, \boldsymbol{E}_m^i, \boldsymbol{H}_m^i$ the support of the singularities, A, is $\{\boldsymbol{r}_0\}$. All the incident electromagnetic fields considered in this book will be plane waves of the form (1.152) and (1.153), or dipoles of the form (1.154), (1.155), (1.158) and (1.159), or a linear combination of such fields for differing incident directions $\hat{\boldsymbol{k}}$, polarizations $\hat{\boldsymbol{a}}, \hat{\boldsymbol{b}}$, source locations $\boldsymbol{r}_0$, or dipole moments $\hat{\boldsymbol{p}}, \hat{\boldsymbol{m}}$. Dipole incident waves will satisfy the radiation conditions. Note that

$$\boldsymbol{E}_e^i(\boldsymbol{r}) = \frac{-k^2}{4\pi\epsilon^+}\hat{\boldsymbol{r}}_0 \times (\hat{\boldsymbol{r}}_0 \times \hat{\boldsymbol{p}})\frac{e^{ikr_0}}{r_0}e^{-ik\hat{\boldsymbol{r}}_0\cdot\boldsymbol{r}} + O\left(\frac{1}{r_0^2}\right), \quad r_0 \to \infty \tag{1.162}$$

so that

$$\hat{\boldsymbol{r}}_0 \times (\hat{\boldsymbol{r}}_0 \times \hat{\boldsymbol{p}})e^{-ik\hat{\boldsymbol{r}}_0\cdot\boldsymbol{r}} = \lim_{r_0\to\infty}\left(-\frac{4\pi\epsilon^+}{k^2}\right)r_0e^{-ikr_0}\boldsymbol{E}_e^i(\boldsymbol{r}) \tag{1.163}$$

Thus, we may consider the electric plane wave propagating in the direction $-\hat{\boldsymbol{r}}_0$, polarized in the direction $\hat{\boldsymbol{r}}_0 \times (\hat{\boldsymbol{r}}_0 \times \hat{\boldsymbol{p}})$, as a modified electric dipole at $\boldsymbol{r}_0$ as $\boldsymbol{r}_0 \to \infty$.

The dipole fields defined in this section may also be expressed in terms of the **fundamental dyadic solution** of the vector Helmholtz equation

$$\nabla_{\boldsymbol{r}} \times (\nabla_{\boldsymbol{r}} \times \widetilde{\boldsymbol{G}}(\boldsymbol{r}, \boldsymbol{r}_0)) - k^2\widetilde{\boldsymbol{G}}(\boldsymbol{r}, \boldsymbol{r}_0) = \frac{4\pi}{ik}\delta(\boldsymbol{r} - \boldsymbol{r}_0)\widetilde{\boldsymbol{I}} \tag{1.164}$$

where

$$\widetilde{\boldsymbol{I}} = \hat{\boldsymbol{x}}_1\hat{\boldsymbol{x}}_1 + \hat{\boldsymbol{x}}_2\hat{\boldsymbol{x}}_2 + \hat{\boldsymbol{x}}_3\hat{\boldsymbol{x}}_3 \tag{1.165}$$

denotes the identity dyadic in $\mathbb{R}^3$.

Explicitly,

$$\begin{aligned}\widetilde{\boldsymbol{G}}(\boldsymbol{r}, \boldsymbol{r}_0) &= \frac{1}{k^2}(\nabla_{\boldsymbol{r}}\nabla_{\boldsymbol{r}} + k^2\widetilde{\boldsymbol{I}})G(\boldsymbol{r}, \boldsymbol{r}_0) \\ &= \frac{1}{(ik|\boldsymbol{r} - \boldsymbol{r}_0|)^2}\Bigg[k^2(\boldsymbol{r} - \boldsymbol{r}_0)(\boldsymbol{r} - \boldsymbol{r}_0) \\ &\quad + (1 - ik|\boldsymbol{r} - \boldsymbol{r}_0|)\left(\widetilde{\boldsymbol{I}} - 3\frac{(\boldsymbol{r} - \boldsymbol{r}_0)(\boldsymbol{r} - \boldsymbol{r}_0)}{|\boldsymbol{r} - \boldsymbol{r}_0|^2}\right)\Bigg]G(\boldsymbol{r}, \boldsymbol{r}_0) \\ &\quad + G(\boldsymbol{r}, \boldsymbol{r}_0)\widetilde{\boldsymbol{I}}\end{aligned} \tag{1.166}$$

where $G(\boldsymbol{r}, \boldsymbol{r}_0)$ is the fundamental solution for the Helmholtz equation (1.77). In terms of this fundamental dyadic the electric and magnetic dipoles are

$$\boldsymbol{E}_e^i(\boldsymbol{r}, \boldsymbol{r}_0) = \frac{ik^3}{4\pi\epsilon^+}\widetilde{\boldsymbol{G}}(\boldsymbol{r}, \boldsymbol{r}_0) \cdot \hat{\boldsymbol{p}} \tag{1.167}$$

and

$$\boldsymbol{H}_m^i(\boldsymbol{r}, \boldsymbol{r}_0) = \frac{ik^3}{4\pi}\widetilde{\boldsymbol{G}}(\boldsymbol{r}, \boldsymbol{r}_0) \cdot \hat{\boldsymbol{m}} \tag{1.168}$$

The fields (1.154), (1.155), (1.158) and (1.159) and the equations (1.156), (1.157), (1.160) and (1.161) are connected via the correspondences

$$\begin{aligned}\hat{\boldsymbol{p}} &\leftrightarrow \hat{\boldsymbol{m}} \\ \boldsymbol{D}_e^i &\leftrightarrow \boldsymbol{H}_m^i \\ \boldsymbol{B}_e^i &\leftrightarrow -\boldsymbol{E}_m^i\end{aligned}$$

which transform one set of equations into the other.

1.EM.6 *Basic scattering problems*

Here, we collect the results of the previous sections and write down the basic mathematical problems in electromagnetic scattering. In all problems, we are given $\omega, \epsilon^+, \mu^+$ (hence Z^+, Y^+ and k), $\boldsymbol{E}^i, \boldsymbol{H}^i$ and S and we must determine $\boldsymbol{E}^+ = \boldsymbol{E}^i + \boldsymbol{E}$ and $\boldsymbol{H}^+ = \boldsymbol{H}^i + \boldsymbol{H}$ so that $\boldsymbol{E}$ and $\boldsymbol{H}$ satisfy Maxwell's equation (1.135) in V^+ and the

Silver–Müller radiation conditions (1.148) and (1.149). In addition, the following conditions must be met.

The perfect conductor problem:

$$\hat{\boldsymbol{n}} \times \boldsymbol{E}^{+}(\boldsymbol{r}) = \boldsymbol{0}, \qquad \hat{\boldsymbol{n}} \cdot \boldsymbol{H}^{+}(\boldsymbol{r}) = 0, \quad \boldsymbol{r} \in S \tag{1.169}$$

The impedance problem:

$$\hat{\boldsymbol{n}} \times (\hat{\boldsymbol{n}} \times \boldsymbol{E}^{+}(\boldsymbol{r})) = -Z_s Z^{+} \hat{\boldsymbol{n}} \times \boldsymbol{H}^{+}(\boldsymbol{r}), \quad \boldsymbol{r} \in S \tag{1.170}$$

The transmission problem:

Here, ϵ^{-}, μ^{-} and σ^{-} (hence Z^{-}, Y^{-}, k^{-} and η) are given, and we seek, in addition to $\boldsymbol{E}^{+}$ and $\boldsymbol{H}^{+}$, the fields $\boldsymbol{E}^{-}$ and $\boldsymbol{H}^{-}$ as solutions to Maxwell's equation (1.136) in V^{-}, which satisfy the following conditions on S

$$\left.\begin{aligned} \hat{\boldsymbol{n}} \times \boldsymbol{E}^{+}(\boldsymbol{r}) &= \hat{\boldsymbol{n}} \times \boldsymbol{E}^{-}(\boldsymbol{r}) \\ Y^{+}\hat{\boldsymbol{n}} \cdot \boldsymbol{E}^{+}(\boldsymbol{r}) &= \eta Y^{-} \hat{\boldsymbol{n}} \cdot \boldsymbol{E}^{-}(\boldsymbol{r}) \\ \hat{\boldsymbol{n}} \times \boldsymbol{H}^{+}(\boldsymbol{r}) &= \hat{\boldsymbol{n}} \times \boldsymbol{H}^{-}(\boldsymbol{r}) \\ Z^{+}\hat{\boldsymbol{n}} \cdot \boldsymbol{H}^{+}(\boldsymbol{r}) &= \eta Z^{-} \hat{\boldsymbol{n}} \cdot \boldsymbol{H}^{-}(\boldsymbol{r}) \end{aligned}\right\} \tag{1.171}$$

All three problems are well-posed problems for any C^2-smooth surface S, i.e. each one has a unique and stable classical solution (Colton and Kress 1983; Müller 1969). By a classical solution we mean that $u^{+} \in C^1(V^{+}) \cap C^0(\bar{V}^{+})$ and $u^{-} \in C^1(V^{-}) \cap C^0(\bar{V}^{-})$, where $u^{\pm}$ stands for any Cartesian component of $\boldsymbol{E}^{\pm}$, $\boldsymbol{H}^{\pm}$ in $V^{\pm}$. If S is piecewise C^2, uniqueness will require that the additional finite energy condition

$$\frac{1}{2} \int_{\Omega} (\epsilon |\boldsymbol{E}|^2 + \mu |\boldsymbol{H}|^2) \, dv < +\infty \tag{1.172}$$

be imposed, where Ω stands for every bounded subdomain of V^{+}. This condition is equivalent to the edge condition of Meixner (1954). For more on this condition see Van Bladel (1995) and the references therein.

1.EM.7 *Energy functionals*

Here, we develop expressions for energy density and power flux from the time dependent Maxwell equations. Introducing the constitutive relation (1.116)–(1.118) into (1.107) and (1.108) we obtain

$$\mu \dot{\mathcal{H}} + \nabla \times \mathcal{E} = \boldsymbol{0} \tag{1.173}$$

$$\epsilon \dot{\mathcal{E}} - \nabla \times \mathcal{H} = -\sigma \mathcal{E} \tag{1.174}$$

We allow $\mathcal{E}$ and $\mathcal{H}$ to assume complex values on the understanding that the physically meaningful quantities are recovered by taking the real parts. Multiplying (1.173) by $\mathcal{H}^*$ and (1.174) by $\mathcal{E}^*$ and adding leads to

$$\mu \mathcal{H}^* \cdot \dot{\mathcal{H}} + \epsilon \mathcal{E}^* \cdot \dot{\mathcal{E}} + \mathcal{H}^* \cdot \nabla \times \mathcal{E} - \mathcal{E}^* \cdot \nabla \times \mathcal{H} = -\sigma |\mathcal{E}|^2 \tag{1.175}$$

To this equation we add its complex conjugate and use a basic vector identity to obtain

$$\frac{\partial}{\partial t}\left(\frac{\epsilon}{2}|\mathcal{E}|^2 + \frac{\mu}{2}|\mathcal{H}|^2\right) + \nabla \cdot \mathrm{Re}\{\mathcal{E} \times \mathcal{H}^*\} = -\sigma|\mathcal{E}|^2 \tag{1.176}$$

Identifying $(\epsilon/2)|\mathcal{E}|^2$ and $(\mu/2)|\mathcal{H}|^2$ as **electric** and **magnetic energy**, respectively, we define the **energy density function**

$$\mathcal{W}(\boldsymbol{r}, t) = \frac{\epsilon}{2}|\mathcal{E}(\boldsymbol{r}, t)|^2 + \frac{\mu}{2}|\mathcal{H}(\boldsymbol{r}, t)|^2 \tag{1.177}$$

which is measured in units of energy per unit volume. Next, we define the **power flux vector**, known in electromagnetics as the **complex Poynting vector**

$$\mathcal{S}(\boldsymbol{r}, t) = \mathrm{Re}\{\mathcal{E}(\boldsymbol{r}, t) \times \mathcal{H}^*(\boldsymbol{r}, t)\} \tag{1.178}$$

measured in units of power per unit area. Then (1.176), now written as

$$\frac{\partial}{\partial t}\mathcal{W}(\boldsymbol{r}, t) + \nabla \cdot \mathcal{S}(\boldsymbol{r}, t) = -\sigma|\mathcal{E}(\boldsymbol{r}, t)|^2 \tag{1.179}$$

describes the energy conservation law of electromagnetics. It states that the time rate of change of energy density plus the spatial power flux is equal to the energy dissipated in the medium represented by $-\sigma|\mathcal{E}|^2$ and measured in units of power per unit volume.

For time harmonic fields with time dependence $e^{-i\omega t}$ the energy density function becomes

$$W(\boldsymbol{r}) = \frac{\epsilon}{2}|\boldsymbol{E}(\boldsymbol{r})|^2 + \frac{\mu}{2}|\boldsymbol{H}(\boldsymbol{r})|^2 \tag{1.180}$$

and the power flux is

$$\boldsymbol{S}(\boldsymbol{r}) = \mathrm{Re}\{\boldsymbol{E}(\boldsymbol{r}) \times \boldsymbol{H}^*(\boldsymbol{r})\} \tag{1.181}$$

The energy flux in the direction $\hat{\boldsymbol{a}}$ is $\hat{\boldsymbol{a}} \cdot \boldsymbol{S}$ and the power flow through a surface, $\partial\Omega$, in the direction of the outward normal is

$$\int_{\partial\Omega} \hat{\boldsymbol{n}} \cdot \boldsymbol{S} ds = \mathrm{Re} \int_{\partial\Omega} \hat{\boldsymbol{n}} \cdot (\boldsymbol{E} \times \boldsymbol{H}^*) ds \tag{1.182}$$

1.EM.8 *Problems in electromagnetics*

1. Use appropriate plane wave solutions and the constitutive relations (1.116)–(1.118) with the Maxwell equations (1.114), (1.115) and (1.121) to obtain the dispersion relation

$$k^2 = \epsilon\mu\omega^2 + i\sigma\mu\omega$$

2. Use the dispersion relation

$$\epsilon\mu\omega^2 = k^2 - i\sigma\mu\omega$$

and the definition of the phase velocity as $c = \omega(\mathrm{Re}\, k)^{-1}$ to show that

$$c = \frac{1}{\sqrt{\mu\epsilon}}\sqrt{\frac{2}{1 + \sqrt{1 + \sigma^2/\epsilon^2\omega^2}}}$$

3. Demonstrate the equivalence between the transmission conditions (1.142) and (1.143), and between (1.145) and (1.146)
4. For the Silver–Müller radiation conditions

$$\lim_{r\to\infty}[\boldsymbol{r}\times(\nabla\times\boldsymbol{E})+ikr\boldsymbol{E}]=\boldsymbol{0}$$

$$\lim_{r\to\infty}[\boldsymbol{r}\times(\nabla\times\boldsymbol{H})+ikr\boldsymbol{H}]=\boldsymbol{0}$$

which hold uniformly over directions, prove that

   (i) each one of them implies the other
   (ii) both $\boldsymbol{E}$ and $\boldsymbol{H}$ are asymptotically orthogonal to $\hat{\boldsymbol{r}}$
   (iii) they can be replaced by the Sommerfeld condition

$$\lim_{r\to\infty} r\left[\frac{\partial u}{\partial r}-iku\right]=0$$

uniformly over directions, for each Cartesian component of $\boldsymbol{E}$ and of $\boldsymbol{H}$.

5. If $\boldsymbol{E}_e^i$ is given by equation (1.154) then derive equation (1.155). Hence, verify that the electric dipole fields $\boldsymbol{E}_e^i$ and $\boldsymbol{H}_e^i$ given by (1.154) and (1.155) satisfy the differential equations (1.156) and (1.157).
6. If $\boldsymbol{E}_m^i$ is given by equation (1.158) then derive equation (1.159). Hence, verify that the magnetic dipole fields $\boldsymbol{E}_m^i$ and $\boldsymbol{H}_m^i$ given by (1.158) and (1.159) satisfy the differential equations (1.160) and (1.161).
7. Show that the fields $\boldsymbol{E}_e^i$ and $\boldsymbol{H}_e^i$ given by (1.154) and (1.155), respectively, assume the following expanded form

$$\begin{aligned}
\boldsymbol{E}_e^i(\boldsymbol{r},\boldsymbol{r}_0) &= -\frac{ik}{4\pi\epsilon^+}G^+(\boldsymbol{r},\boldsymbol{r}_0)\\
&\quad\times\left\{\frac{1}{|\boldsymbol{r}-\boldsymbol{r}_0|^2}\left[\widetilde{\boldsymbol{I}}-3\frac{(\boldsymbol{r}-\boldsymbol{r}_0)(\boldsymbol{r}-\boldsymbol{r}_0)}{|\boldsymbol{r}-\boldsymbol{r}_0|^2}\right]\cdot\hat{\boldsymbol{p}}\right.\\
&\quad-\frac{ik}{|\boldsymbol{r}-\boldsymbol{r}_0|}\left[\widetilde{\boldsymbol{I}}-3\frac{(\boldsymbol{r}-\boldsymbol{r}_0)(\boldsymbol{r}-\boldsymbol{r}_0)}{|\boldsymbol{r}-\boldsymbol{r}_0|^2}\right]\cdot\hat{\boldsymbol{p}}\\
&\quad\left.+k^2\frac{\boldsymbol{r}-\boldsymbol{r}_0}{|\boldsymbol{r}-\boldsymbol{r}_0|}\times\left[\frac{\boldsymbol{r}-\boldsymbol{r}_0}{|\boldsymbol{r}-\boldsymbol{r}_0|}\times\hat{\boldsymbol{p}}\right]\right\}\\
\boldsymbol{H}_e^i(\boldsymbol{r},\boldsymbol{r}_0) &= -\frac{i\omega}{4\pi}G^+(\boldsymbol{r},\boldsymbol{r}_0)\left[ik-\frac{1}{|\boldsymbol{r}-\boldsymbol{r}_0|}\right]\frac{\boldsymbol{r}-\boldsymbol{r}_0}{|\boldsymbol{r}-\boldsymbol{r}_0|}\times\hat{\boldsymbol{p}}
\end{aligned}$$

Also find the corresponding expansions for the fields $\boldsymbol{E}_m^i$ and $\boldsymbol{H}_m^i$ given in (1.158) and (1.159).

8. Derive the following asymptotic form for the electric dipole field

$$\boldsymbol{E}_e(\boldsymbol{r}, \boldsymbol{r}_0) = \frac{1}{4\pi\epsilon}(k^2\widetilde{\boldsymbol{I}} + \nabla_{\boldsymbol{r}}\nabla_{\boldsymbol{r}}) \cdot \left(\frac{e^{ik|\boldsymbol{r}-\boldsymbol{r}_0|}}{|\boldsymbol{r}-\boldsymbol{r}_0|}\hat{\boldsymbol{p}}\right)$$

$$= \frac{k^2}{4\pi\epsilon}\hat{\boldsymbol{p}} \cdot (\widetilde{\boldsymbol{I}} - \hat{\boldsymbol{r}}_0\hat{\boldsymbol{r}}_0)\frac{e^{ikr_0}}{r_0}e^{-ik\hat{\boldsymbol{r}}_0\cdot\boldsymbol{r}}$$

$$+ O\left(\frac{1}{r_0^2}\right), \quad r_0 \to \infty$$

Derive the asymptotic form of the corresponding magnetic field

$$\boldsymbol{H}_e(\boldsymbol{r}, \boldsymbol{r}_0) = \frac{\omega}{4\pi i}\nabla_{\boldsymbol{r}} \times \left(\frac{e^{ik|\boldsymbol{r}-\boldsymbol{r}_0|}}{|\boldsymbol{r}-\boldsymbol{r}_0|}\hat{\boldsymbol{p}}\right)$$

9. Given the electromagnetic energy density function

$$\mathcal{W} = \frac{\epsilon}{2}|\mathcal{E}|^2 + \frac{\mu}{2}|\mathcal{H}|^2$$

and the Poynting power flux vector

$$\mathcal{S} = \mathrm{Re}[\mathcal{E} \times \mathcal{H}^*]$$

prove by direct differentiation that

$$\dot{\mathcal{W}} = -\nabla \cdot \mathcal{S} - \sigma|\mathcal{E}|^2$$

10. Show that any pair of functions that satisfy the vector Helmholtz equations (1.133) will not necessarily satisfy Maxwell's equation (1.132) as well.

1. E The basic theory of elastic waves

1. E.1 *Field equations*

Continuum mechanics of solids is concerned with describing how a material deforms under prescribed stresses. Within the vast domain of continuum mechanics, **linear elasticity** occupies that part of the theory that deals with small deformations, which disappear when the imposed stresses are removed leaving the material in its equilibrium configuration. In linear elasticity the Lagrangian, or material, and the Eulerian, or spatial, description of the continuum coincide. If we denote a reference (undeformed) body at time $t = 0$ by $B_0 \subset \mathbb{R}^3$ and its deformed configuration at a later time $t > 0$ by B_t, then a point $\boldsymbol{r}$ of the undeformed body B_0 will be mapped to the point $\boldsymbol{f}(\boldsymbol{r}, t)$ in the deformed configuration B_t, where $\boldsymbol{f}$ is the **deformation map**, defined by $\boldsymbol{f}: B_0 \times [0, \infty) \to B_t \in \mathbb{R}^3$. This map should be smooth enough to preserve the topological structure of the body. In fact, it should be a diffeomorphism (a one-to-one, onto, differentiable mapping with differentiable inverse) which is **orientation preserving** in the sense that the sign of det ∇f is preserved at all points in the body for all time. One of the consequences of this assumption is that no two material points can occupy the same spatial point at the same time. Therefore, it must be one-to-one (so that no two material points occupy the same point in $\mathbb{R}^3$) and orientation preserving (det $\nabla f > 0,\ \forall\ \boldsymbol{r} \in B_0$). For more, as well as results used later in the book, on continuum mechanics and elasticity we refer to Gurtin (1972), Kupradze (1979), Marsden and Hughes (1983) and Ciarlet (1988).

The **displacement field** is defined by

$$\mathcal{U}(\boldsymbol{r}, t) = \boldsymbol{f}(\boldsymbol{r}, t) - \boldsymbol{r} \tag{1.183}$$

and describes the relative change of position of the point $\boldsymbol{r} \in B_0$ at time t.

The **deformation gradient** ∇f generates the **displacement gradient**

$$\nabla \mathcal{U} = \nabla f - \widetilde{\boldsymbol{I}} \tag{1.184}$$

which describes the spatial rate of change of position at any time. These gradients are dyadic fields and may be written in component form as, e.g. $\nabla \mathcal{U} = [(\partial/\partial x_i)\mathcal{U}_j],\ i, j = 1, 2, 3$ where $\mathcal{U}_j$ denotes the Cartesian components of $\mathcal{U}$.

Green's strain tensor is defined as

$$\widetilde{\boldsymbol{e}} = \frac{1}{2}[\nabla \mathcal{U} + (\nabla \mathcal{U})^{\top} + (\nabla \mathcal{U})^{\top} \cdot \nabla \mathcal{U}] \tag{1.185}$$

where the superscript $\top$ denotes transposition, i.e. $(\nabla \mathcal{U})^{\top} = [(\partial/\partial x_j)\mathcal{U}_i]$. The dimensionless dyadic $\widetilde{\boldsymbol{e}}$ measures the deviation between a given deformation and a **rigid deformation** described by the identity dyadic.

For 'small' displacement gradients the linearized theory of elasticity considers the strain tensor to be the linear and symmetric part of the displacement gradient, that is

$$\widetilde{\boldsymbol{e}} = \frac{1}{2}[\nabla \mathcal{U} + (\nabla \mathcal{U})^{\top}] \tag{1.186}$$

and $\widetilde{\boldsymbol{e}}$ now denotes the linearized strain tensor.

Under the action of external forces a deformable body remains in a state of stress. Cauchy described these stresses as follows. Let an infinitesimal oriented surface with a unit normal $\hat{\boldsymbol{n}}$ be located at a point $\boldsymbol{r}$ inside the body. Then, the part of the body that is located in the vicinity of one side of the surface will exert a force on the other side. This force is called **traction** and it is denoted by

$$\boldsymbol{t} = \boldsymbol{t}(\boldsymbol{r}, \hat{\boldsymbol{n}}) \tag{1.187}$$

If we assume (as Cauchy did) that the dependence of $\boldsymbol{t}$ on $\hat{\boldsymbol{n}}$ is linear, then there exists a dyadic $\widetilde{\boldsymbol{\tau}}$ which embodies this dependence via

$$\boldsymbol{t}(\boldsymbol{r}, \hat{\boldsymbol{n}}) = \widetilde{\boldsymbol{\phi}}(\boldsymbol{r}) \cdot \hat{\boldsymbol{n}} \tag{1.188}$$

This is the **stress dyadic** or **stress tensor** which, as a consequence of conservation of angular momentum, is symmetric, i.e. $\widetilde{\boldsymbol{\tau}} = \widetilde{\boldsymbol{\tau}}^{\mathsf{T}}$. In contrast to the strain dyadic which is dimensionless, the stress dyadic has dimensions of force per unit volume. Newton's second law will then take the form

$$\nabla \cdot \widetilde{\boldsymbol{\tau}} = \rho \ddot{\boldsymbol{\mathcal{U}}} \tag{1.189}$$

where ρ denotes **mass density**. This is the **Cauchy equation of motion** expressed in units of force per unit volume.

Up to this point everything relies either upon physical laws, or upon mathematical arguments. A **constitutive relation** is now introduced which is based on experimental evidence and is known as **Hooke's law**. It provides the connection between the strained state of the body and the stresses that caused it. Hooke's law for an isotropic medium is written in terms of the strain tensor as

$$\widetilde{\boldsymbol{\tau}} = \lambda(\operatorname{tr}\widetilde{\boldsymbol{e}})\widetilde{\boldsymbol{I}} + 2\mu\widetilde{\boldsymbol{e}} \tag{1.190}$$

where $\operatorname{tr}\widetilde{\boldsymbol{e}}$ is the trace of the strain tensor, or in terms of the displacement field as

$$\widetilde{\boldsymbol{\tau}} = \lambda(\nabla \cdot \boldsymbol{\mathcal{U}})\widetilde{\boldsymbol{I}} + \mu[\nabla\boldsymbol{\mathcal{U}} + (\nabla\boldsymbol{\mathcal{U}})^{\mathsf{T}}] \tag{1.191}$$

The parameters λ and μ in Hooke's law, which are constants for a homogeneous medium, are called **Lamé constants**. They have dimensions of pressure or energy density and they specify the elastic characteristics of the medium. The physical restrictions upon them are the following:

(i) A given stress results in a strain in the same direction. This is ensured by

$$\mu > 0 \tag{1.192}$$

(ii) A hydrostatic pressure should reduce the volume of a body. This requires that

$$2\mu + 3\lambda > 0 \tag{1.193}$$

(iii) A finite stress always produces a nonzero strain. For this to hold, we need

$$\mu < \infty, \quad 2\mu + 3\lambda < \infty \tag{1.194}$$

Combining these restrictions we arrive at

$$0 < \mu < \infty, \quad 0 < 2\mu + 3\lambda < \infty \tag{1.195}$$

We remark that complex values of the Lamé constants λ and μ account for viscosity, and although the study of viscoelastic media is an important one, here we only consider purely elastic media so that λ and μ are always real, and subject to the restrictions (1.195).

Substituting (1.191) into (1.189) we obtain the **equation of motion** or **Navier equation** of linearized elasticity in terms of displacement

$$\mu \nabla^2 \mathcal{U}(\boldsymbol{r}, t) + (\lambda + \mu) \nabla \nabla \cdot \mathcal{U}(\boldsymbol{r}, t) = \rho \, \ddot{\mathcal{U}}(\boldsymbol{r}, t) \tag{1.196}$$

We may rewrite this equation as

$$\Delta^e \mathcal{U}(\boldsymbol{r}, t) = \ddot{\mathcal{U}}(\boldsymbol{r}, t) \tag{1.197}$$

where

$$\Delta^e = c_p^2 \nabla(\nabla \cdot) - c_s^2 \nabla \times (\nabla \times) = (c_p^2 - c_s^2) \nabla(\nabla \cdot) + c_s^2 \nabla^2 \tag{1.198}$$

is the **Lamé operator** of linear elasticity and the quantities c_p and c_s are defined as

$$c_p^2 = \frac{\lambda + 2\mu}{\rho}, \quad c_s^2 = \frac{\mu}{\rho} \tag{1.199}$$

If we now assume the harmonic time dependence

$$\mathcal{U}(\boldsymbol{r}, t) = \boldsymbol{u}(\boldsymbol{r}) e^{-i\omega t} \tag{1.200}$$

at the fixed frequency ω, and substitute into (1.197) we obtain the following equation for the complex displacement

$$\Delta^e \boldsymbol{u} + \omega^2 \boldsymbol{u} = \boldsymbol{0} \tag{1.201}$$

which is known as the **spectral** or **stationary** or **time independent Navier equation**. Rewriting this equation as

$$\boldsymbol{u} = \frac{c_s^2}{\omega^2} \nabla \times (\nabla \times \boldsymbol{u}) - \frac{c_p^2}{\omega^2} \nabla(\nabla \cdot \boldsymbol{u}) \tag{1.202}$$

it is clear that this defines a Helmholtz decomposition of $\boldsymbol{u}$, i.e.

$$\boldsymbol{u} = \boldsymbol{u}^p + \boldsymbol{u}^s \tag{1.203}$$

where

$$\boldsymbol{u}^p = -\frac{c_p^2}{\omega^2} \nabla(\nabla \cdot \boldsymbol{u}), \quad \text{hence } \nabla \times \boldsymbol{u}^p = \boldsymbol{0} \tag{1.204}$$

and

$$\boldsymbol{u}^s = \frac{c_s^2}{\omega^2} \nabla \times (\nabla \times \boldsymbol{u}), \quad \text{hence } \nabla \cdot \boldsymbol{u}^s = 0 \tag{1.205}$$

Therefore,

$$\begin{aligned} \boldsymbol{u}^p &= -\frac{c_p^2}{\omega^2} \nabla(\nabla \cdot \boldsymbol{u}^p) = -\frac{c_p^2}{\omega^2} \nabla \times (\nabla \times \boldsymbol{u}^p) - \frac{c_p^2}{\omega^2} \nabla^2 \boldsymbol{u}^p \\ &= -\frac{c_p^2}{\omega^2} \nabla^2 \boldsymbol{u}^p \end{aligned} \tag{1.206}$$

or

$$\nabla^2 \boldsymbol{u}^p + \frac{\omega^2}{c_p^2} \boldsymbol{u}^p = \boldsymbol{0} \tag{1.207}$$

and

$$\boldsymbol{u}^s = \frac{c_s^2}{\omega^2} \nabla \times (\nabla \times \boldsymbol{u}^s) = -\frac{c_s^2}{\omega^2} \nabla^2 \boldsymbol{u}^s$$

or

$$\nabla^2 \boldsymbol{u}^s + \frac{\omega^2}{c_s^2} \boldsymbol{u}^s = \boldsymbol{0} \tag{1.208}$$

These equations are also written as

$$\nabla^2 \boldsymbol{u}^s + k_s^2 \boldsymbol{u}^s = \boldsymbol{0} \tag{1.209}$$

$$\nabla^2 \boldsymbol{u}^p + k_p^2 \boldsymbol{u}^p = \boldsymbol{0} \tag{1.210}$$

where

$$k_s = \frac{\omega}{c_s}, \quad k_p = \frac{\omega}{c_p} \tag{1.211}$$

The displacement is thus seen to consist of two separate waves, a **shear** or **transverse** or **S wave**, $\boldsymbol{u}^s$, which describes all the equivoluminal twisting motion and a **compressional** or **longitudinal** or **P wave**, $\boldsymbol{u}^p$, which describes all the dilatational motion. The quantities c_s and c_p are the **phase velocities** of the two waves and k_s and k_p are the corresponding **wave numbers**.

The main complication in elastic scattering, compared with acoustic or electromagnetic scattering, lies in the fact that the two vector wave fields which comprise any displacement field not only travel with different phase velocities but are also independent of each other. That is, if one of the wave fields is given, then there is no way to recover the other, in contrast to the case of electromagnetics. The longitudinal and transverse waves in an elastic medium belong to orthogonal complements of the solution space of the spectral Navier equation (1.201). They only interact or couple when they meet a spatial discontinuity such as the surface of a scatterer. This interaction is called **mode conversion** in elastic wave theory.

Another important difficulty stems from the fact that the Lamé operator and the surface traction operator are both dependent upon the material parameters.

At any point of a surface, with unit normal $\hat{\boldsymbol{n}}$, a pressure is exerted from one side of the surface to the other. This force is called the **surface traction** at that point with respect to the particular orientation $\hat{\boldsymbol{n}}$, and is given in terms of $\boldsymbol{u}$, using (1.191), as

$$\boldsymbol{t}(\boldsymbol{r}) = \widetilde{\boldsymbol{\tau}}(\boldsymbol{r}) \cdot \hat{\boldsymbol{n}} = \boldsymbol{T}(\partial_{\boldsymbol{r}}, \hat{\boldsymbol{n}}) \boldsymbol{u}(\boldsymbol{r}) \tag{1.212}$$

where the **surface traction operator** $\boldsymbol{T}(\partial_{\boldsymbol{r}}, \hat{\boldsymbol{n}})$ assumes the form

$$\boldsymbol{T}(\partial_{\boldsymbol{r}}, \hat{\boldsymbol{n}}) = 2\mu \hat{\boldsymbol{n}} \cdot \nabla + \lambda \hat{\boldsymbol{n}}(\nabla \cdot) + \mu \hat{\boldsymbol{n}} \times (\nabla \times) \tag{1.213}$$

The operator $\boldsymbol{T}$ maps local displacements to local tractions on a surface, and it is material dependent.

1.E.2 *Boundary conditions*

Just as in the cases of acoustic and electromagnetic scattering we are concerned with how an **incident elastic wave** $\boldsymbol{u}^i$ (see section 1.E.4) is perturbed by a bounded inclusion, V^-. The **total displacement** in V^+ will be denoted by $\boldsymbol{u}^+$ where $\boldsymbol{u}^+ = \boldsymbol{u}^i + \boldsymbol{u}$ and $\boldsymbol{u}$ is the **scattered displacement field**. The total field $\boldsymbol{u}^+$ will satisfy the spectral Navier equation (1.201) at all points in V^+ with the exception of the singularities (if any) of the incident field $\boldsymbol{u}^i$. In the Navier equation the Lamé operator Δ^{e+} will be expressed in terms of the phase velocities c_p^+ and c_s^+ in V^+. The physical characteristics of the inclusion are embodied in the boundary conditions at its surface. There are two kinds of material with very different physical characteristics which do not allow waves to penetrate into the inclusion.

The rigid surface:

$$\boldsymbol{u}^+(\boldsymbol{r}) = \boldsymbol{0}, \quad \boldsymbol{r} \in S \tag{1.214}$$

The cavity or stress-free surface:

$$\boldsymbol{T}^+(\partial_{\boldsymbol{r}}, \hat{\boldsymbol{n}})\boldsymbol{u}^+(\boldsymbol{r}) = \boldsymbol{0}, \quad \boldsymbol{r} \in S \tag{1.215}$$

At one extreme, consider a body whose surface cannot be deformed by the stresses generated by the incident displacement field. Such a body is characterized as a **rigid scatterer** and the corresponding boundary condition is (1.214). A rigid scatterer cannot be deformed but it effects the deformation of the surrounding elastic medium giving rise to a scattered wave.

At the other extreme, the physical nature of the inclusion is such that its surface is incapable of sustaining any tractions. This is the case of a **cavity** or **stress-free surface** and the corresponding boundary condition is (1.215). The surface of a cavity undergoes small variations in such a way as to coincide with the local surface of vanishing traction. The scattered wave is generated by these boundary variations.

Thus, the Dirichlet boundary condition in elastic scattering describes a rigid surface, while the Neumann-type boundary condition describes a cavity. This characterization of the scattering surface is opposite to that in acoustics where the Dirichlet condition corresponds to a soft surface and the Neumann condition to a hard surface.

1.E.3 *Transmission conditions*

An intermediate state between the rigid surface, with zero surface displacement, and the cavity, with zero surface traction, corresponds to an inclusion with a surface on which both a nonvanishing displacement and a nonvanishing traction field can exist. Such a body allows for the elastic disturbances to penetrate into its interior generating the **interior displacement field** $\boldsymbol{u}^-$. The interior field will satisfy the spectral Navier equation (1.201) with the Lamé operator Δ^{e-} expressed in terms of the interior phase velocities c_p^- and c_s^-. The inclusion is assumed to be such that both displacement and stresses will be continuous at the interface. Therefore, we impose the **transmission** or **elastic boundary conditions**

$$\boldsymbol{u}^+(\boldsymbol{r}) = \boldsymbol{u}^-(\boldsymbol{r}), \quad \boldsymbol{r} \in S \tag{1.216}$$

and

$$\boldsymbol{T}^{+}(\partial_{\boldsymbol{r}}, \hat{\boldsymbol{n}})\boldsymbol{u}^{+}(\boldsymbol{r}) = \boldsymbol{T}^{-}(\partial_{\boldsymbol{r}}, \hat{\boldsymbol{n}})\boldsymbol{u}^{-}(\boldsymbol{r}), \quad \boldsymbol{r} \in S \tag{1.217}$$

which ensure the continuity of the displacement and of the surface traction field so that the two media are always in perfect contact. Comparing condition (1.217) with the corresponding condition (1.58) in the acoustic transmission problem we observe that in addition to the complexity of the operator $\boldsymbol{T}$, consisting of a linear combination of normal derivative, divergence and curl, the coefficients in this linear combination differ in V^+ and V^-. This is one of the major complications in dealing with transmission problems in elastic scattering theory. The differences in the two media which, along with the geometry of the surface S will determine the nature of the displacement fields, can be usefully expressed in terms of the dimensionless **relative indices of refraction**

$$\eta_p = \frac{c_p^+}{c_p^-} = \frac{k_p^-}{k_p^+} \tag{1.218}$$

for the longitudinal waves and

$$\eta_s = \frac{c_s^+}{c_s^-} = \frac{k_s^-}{k_s^+} \tag{1.219}$$

for the transverse waves.

1. E.4 *Radiation conditions*

As in all scattering problems, the scattered field in V^+ must have the property that the wave propagates away from the scatterer and undergoes geometric attenuation as it does so. This property not only makes physical sense, but is needed in order to model the scattering process as well-posed mathematical problems. In elastic wave propagation we have seen that there are two propagating waves, a longitudinal and a transverse wave. The **radiation conditions** on the scattered waves that guarantee the well-posedness of the scattering problems are

$$\lim_{r\to\infty} \boldsymbol{u}^p(\boldsymbol{r}) = \boldsymbol{0}, \quad \lim_{r\to\infty} r\left(\frac{\partial}{\partial r} - ik_p\right)\boldsymbol{u}^p(\boldsymbol{r}) = \boldsymbol{0} \tag{1.220}$$

$$\lim_{r\to\infty} \boldsymbol{u}^s(\boldsymbol{r}) = \boldsymbol{0}, \quad \lim_{r\to\infty} r\left(\frac{\partial}{\partial r} - ik_s\right)\boldsymbol{u}^s(\boldsymbol{r}) = \boldsymbol{0} \tag{1.221}$$

where all limits are to hold uniformly over all directions. These conditions are due to Kupradze and are essentially the Sommerfeld radiation condition for each Cartesian component of the longitudinal and transverse waves. We denote the wave numbers in V^+ as k_p and k_s, omitting the superscript $+$; however, a superscript $-$ will be used to denote the wave numbers in V^-

1. E.5 *Incident fields and the fundamental solution*

Incident elastic waves are displacement fields that originate in V^+. Just as in acoustics and electromagnetics we consider only plane waves or point sources or linear combinations of plane waves and point sources. Elastic plane waves are of the form

$$\boldsymbol{u}^i(\boldsymbol{r}) = \alpha_p \hat{\boldsymbol{k}} e^{ik_p \hat{\boldsymbol{k}}\cdot\boldsymbol{r}} + \alpha_s \hat{\boldsymbol{b}} e^{ik_s \hat{\boldsymbol{k}}\cdot\boldsymbol{r}} \tag{1.222}$$

where $\hat{\boldsymbol{k}} \cdot \hat{\boldsymbol{b}} = 0$, while α_p and α_s denote the **amplitudes** of the incident longitudinal and transverse fields, respectively. By taking $\alpha_p \neq 0, \alpha_s = 0$, or $\alpha_p = 0, \alpha_s \neq 0$ the incident field is restricted either to a P wave or an S wave.

A point source at the point $\boldsymbol{r}_0 \in V^+$, denoted by $\widetilde{\boldsymbol{\Gamma}}(\boldsymbol{r}, \boldsymbol{r}_0)$, is a **fundamental solution** of the spectral Navier equation

$$\rho(\Delta^e + \omega^2)\widetilde{\boldsymbol{\Gamma}}(\boldsymbol{r}, \boldsymbol{r}_0) = -4\pi\delta(\boldsymbol{r} - \boldsymbol{r}_0)\widetilde{\boldsymbol{I}} \tag{1.223}$$

and is measured in units of length per unit force. This is a symmetric dyadic mapping a unit force at the point $\boldsymbol{r}_0$ to the displacement field at $\boldsymbol{r}$ generated by this force. Explicitly, it is given by

$$\widetilde{\boldsymbol{\Gamma}}(\boldsymbol{r}, \boldsymbol{r}_0) = \widetilde{\boldsymbol{\Gamma}}^p(\boldsymbol{r}, \boldsymbol{r}_0) + \widetilde{\boldsymbol{\Gamma}}^s(\boldsymbol{r}, \boldsymbol{r}_0) \tag{1.224}$$

where

$$\widetilde{\boldsymbol{\Gamma}}^p(\boldsymbol{r}, \boldsymbol{r}_0) = -\frac{ik_p}{\rho\omega^2}\nabla_{\boldsymbol{r}}\nabla_{\boldsymbol{r}} G_p(\boldsymbol{r}, \boldsymbol{r}_0) \tag{1.225}$$

$$\widetilde{\boldsymbol{\Gamma}}^s(\boldsymbol{r}, \boldsymbol{r}_0) = \frac{ik_s}{\rho\omega^2}(\nabla_{\boldsymbol{r}}\nabla_{\boldsymbol{r}} + k_s^2\widetilde{\boldsymbol{I}}) G_s(\boldsymbol{r}, \boldsymbol{r}_0) \tag{1.226}$$

and

$$G_p(\boldsymbol{r}, \boldsymbol{r}_0) = \frac{e^{ik_p|\boldsymbol{r}-\boldsymbol{r}_0|}}{ik_p|\boldsymbol{r} - \boldsymbol{r}_0|}, \quad G_s(\boldsymbol{r}, \boldsymbol{r}_0) = \frac{e^{ik_s|\boldsymbol{r}-\boldsymbol{r}_0|}}{ik_s|\boldsymbol{r} - \boldsymbol{r}_0|} \tag{1.227}$$

Differentiating (1.225) and (1.226) we obtain the following expressions in terms of the vector $\boldsymbol{R} = \boldsymbol{r} - \boldsymbol{r}_0$

$$\widetilde{\boldsymbol{\Gamma}}^p(\boldsymbol{r}, \boldsymbol{r}_0) = \frac{i}{(\lambda + 2\mu)k_p R^2}[k_p^2 \boldsymbol{R}\boldsymbol{R} + (1 - ik_p R)(\widetilde{\boldsymbol{I}} - 3\hat{\boldsymbol{R}}\hat{\boldsymbol{R}})]h(k_p R) \tag{1.228}$$

and

$$\begin{aligned}\widetilde{\boldsymbol{\Gamma}}^s(\boldsymbol{r}, \boldsymbol{r}_0) = &-\frac{i}{\mu k_s R^2}[k_s^2 \boldsymbol{R}\boldsymbol{R} + (1 - ik_s R)(\widetilde{\boldsymbol{I}} - 3\hat{\boldsymbol{R}}\hat{\boldsymbol{R}})]h(k_s R) \\ &+\frac{ik_s}{\mu}h(k_s R)\widetilde{\boldsymbol{I}}\end{aligned} \tag{1.229}$$

The fundamental dyadic $\widetilde{\boldsymbol{\Gamma}}$ generates the following surface stress dyadic on any surface with unit normal $\hat{\boldsymbol{n}}_0$

$$\begin{aligned}\boldsymbol{T}_{r_0}\widetilde{\boldsymbol{\Gamma}}(\boldsymbol{r}, \boldsymbol{r}_0) = &\frac{ik_p(1 - ik_p R)}{R}h(k_p R)\hat{\boldsymbol{n}}_0 \cdot \frac{\lambda\widetilde{\boldsymbol{I}} + 2\mu\hat{\boldsymbol{R}}\hat{\boldsymbol{R}}}{\lambda + 2\mu}\hat{\boldsymbol{R}} \\ &+\frac{ik_s(1 - ik_s R)}{R}h(k_s R)[\hat{\boldsymbol{R}}(\widetilde{\boldsymbol{I}} - \hat{\boldsymbol{R}}\hat{\boldsymbol{R}}) + (\widetilde{\boldsymbol{I}} - \hat{\boldsymbol{R}}\hat{\boldsymbol{R}})\hat{\boldsymbol{R}}] \cdot \hat{\boldsymbol{n}}_0\end{aligned}$$

$$
\begin{aligned}
&-\frac{2\mu}{\rho\omega^2 R^3}[ik_p(k_p^2R^2+3ik_pR-3)h(k_pR)\\
&-ik_s(k_s^2R^2+3ik_sR-3)h(k_sR)]\\
&\times\{[\hat{\boldsymbol{R}}(\widetilde{\boldsymbol{I}}-\hat{\boldsymbol{R}}\hat{\boldsymbol{R}})+(\widetilde{\boldsymbol{I}}-\hat{\boldsymbol{R}}\hat{\boldsymbol{R}})\hat{\boldsymbol{R}}]\cdot\hat{\boldsymbol{n}}_0+\hat{\boldsymbol{n}}_0\cdot(\widetilde{\boldsymbol{I}}-3\hat{\boldsymbol{R}}\hat{\boldsymbol{R}})\hat{\boldsymbol{R}}\} \qquad (1.230)
\end{aligned}
$$

where h is given in (1.77) and $\boldsymbol{r}_0$ lies on S.

An elastic dipole radiation field produces a very complicated incident field which can be found in Ben-Menahem and Singh (1981, p. 162).

We remark that the transverse part, $\widetilde{\boldsymbol{\Gamma}}^s$, of the fundamental solution is almost identical to the fundamental solution $\widetilde{\boldsymbol{G}}$ (1.166) of electromagnetics. Indeed, if we take the electromagnetic wave number k to be the transverse wave number k_s, then we have

$$
\widetilde{\boldsymbol{G}}(\boldsymbol{r},\boldsymbol{r}_0)=\frac{\rho\omega^2}{ik_s^3}\widetilde{\boldsymbol{\Gamma}}^s(\boldsymbol{r},\boldsymbol{r}_0) \qquad (1.231)
$$

which establishes the similarity between electromagnetic and transverse elastic waves.

1. E.6 *Basic scattering problems*

Here, we collect the results of the previous sections and write down the basic mathematical scattering problems in elasticity. In all problems we are given the angular frequency w, the density ρ^+, and the Lamé constants λ^+ and μ^+ (hence, k_p, k_s, c_p^+ and c_s^+) in V^+, the incident field $\boldsymbol{u}^i$, and the surface S and we must determine the total field $\boldsymbol{u}^+=\boldsymbol{u}^i+\boldsymbol{u}$ in V^+ so that $\boldsymbol{u}=\boldsymbol{u}^s+\boldsymbol{u}^p$ satisfies the spectral Navier equation (1.201) in V^+, and $\boldsymbol{u}^s$ and $\boldsymbol{u}^p$ satisfy the radiation conditions (1.220) and (1.221). In addition, the following conditions must be met.

The rigid problem:

$$
\boldsymbol{u}^+(\boldsymbol{r})=\boldsymbol{0}, \quad \boldsymbol{r}\in S \qquad (1.232)
$$

The cavity problem:

$$
\boldsymbol{T}^+(\partial_{\boldsymbol{r}},\hat{\boldsymbol{n}})\boldsymbol{u}^+(\boldsymbol{r})=\boldsymbol{0}, \quad \boldsymbol{r}\in S \qquad (1.233)
$$

The transmission problem:

In this problem, the parameters ρ^-, λ^- and μ^- (hence, k_p^-, k_s^-, c_p^- and c_s^-) are given in V^- and we seek, in addition to $\boldsymbol{u}^+$, an interior field $\boldsymbol{u}^-$ in V^- which satisfies the spectral Navier equation in V^- and the following transmission conditions on S

$$
\left.\begin{aligned}
\boldsymbol{u}^+(\boldsymbol{r})&=\boldsymbol{u}^-(\boldsymbol{r})\\
\boldsymbol{T}^+(\partial_{\boldsymbol{r}},\hat{\boldsymbol{n}})\boldsymbol{u}^+(\boldsymbol{r})&=\boldsymbol{T}^-(\partial_{\boldsymbol{r}},\hat{\boldsymbol{n}})\boldsymbol{u}^-(\boldsymbol{r})
\end{aligned}\right\}, \quad \boldsymbol{r}\in S \qquad (1.234)
$$

where $\boldsymbol{T}^+$ and $\boldsymbol{T}^-$ are the traction operators (1.213) with appropriate Lamé constants for V^+ and V^-. The spectral Navier equation in V^+ and V^- must also contain the appropriate density and Lamé constants and the quantities c_p, c_s, k_p and k_s derived from them.

For the well-posedness of these three problems we refer to Knops and Payne (1971), Gurtin (1972), Kupradze (1979), and Marsden and Hughes (1983).

1. E.7 *Energy functionals*

Here, we derive the energy conservation law in linear elasticity and develop expressions for energy density and power flux. Starting with Newton's law, (1.189), we multiply by the conjugate velocity field $\dot{\mathcal{U}}^*$ to obtain

$$\rho\,\dot{\mathcal{U}}^* \cdot \ddot{\mathcal{U}} = (\nabla \cdot \tilde{\tau}) \cdot \dot{\mathcal{U}}^* \tag{1.235}$$

where we permit $\mathcal{U}$ and $\tilde{\tau}$ to assume complex values and recover physical quantities by taking real parts. Making use of the symmetry of $\tilde{\tau}$ and the identity

$$\nabla \cdot (\tilde{\tau} \cdot \dot{\mathcal{U}}^*) = (\nabla \cdot \tilde{\tau}) \cdot \dot{\mathcal{U}}^* + \tilde{\tau} : \nabla \dot{\mathcal{U}}^* \tag{1.236}$$

where

$$\tilde{\tau} : \nabla \dot{\mathcal{U}}^* = \sum_{i,j=1}^{3} \tau_{ij} \frac{\partial}{\partial x_j} \dot{\mathcal{U}}_i^* \tag{1.237}$$

we rewrite (1.235) as

$$\rho\,\dot{\mathcal{U}}^* \cdot \ddot{\mathcal{U}} + \tilde{\tau} : \nabla \dot{\mathcal{U}}^* = \nabla \cdot (\tilde{\tau} \cdot \dot{\mathcal{U}}^*) \tag{1.238}$$

The symmetry of $\tilde{\tau}$ also implies that

$$\tilde{\tau} : \nabla \dot{\mathcal{U}}^* = \tilde{\tau} : (\nabla \dot{\mathcal{U}}^*)^{\top} \tag{1.239}$$

hence in terms of the strain tensor, (1.186),

$$\tilde{\tau} : \nabla \dot{\mathcal{U}}^* = \tilde{\tau} : \dot{\tilde{e}}^* \tag{1.240}$$

Incorporating this relation into (1.238) and adding the complex conjugate of the resulting equation yields

$$\rho[\dot{\mathcal{U}}^* \cdot \ddot{\mathcal{U}} + \dot{\mathcal{U}} \cdot \ddot{\mathcal{U}}^*] + \tilde{\tau} : \dot{\tilde{e}}^* + \tilde{\tau}^* : \dot{\tilde{e}} = \nabla \cdot (\tilde{\tau} \cdot \dot{\mathcal{U}}^* + \tilde{\tau}^* \cdot \dot{\mathcal{U}}) \tag{1.241}$$

It may be shown that $\tilde{\tau} : \dot{\tilde{e}}^*$ is real, that is

$$\tilde{\tau} : \dot{\tilde{e}}^* = \tilde{\tau}^* : \dot{\tilde{e}} \tag{1.242}$$

which allows us to rewrite (1.241) as

$$\frac{\partial}{\partial t}\left[\frac{\rho}{2}|\dot{\mathcal{U}}|^2 + \frac{1}{2}\tilde{\tau} : \tilde{e}^*\right] - \nabla \cdot \mathrm{Re}\{\tilde{\tau} \cdot \dot{\mathcal{U}}^*\} = 0 \tag{1.243}$$

Now we define the **elastic energy density function** as

$$\mathcal{W}(\boldsymbol{r}, t) = \frac{\rho}{2}|\dot{\mathcal{U}}(\boldsymbol{r}, t)|^2 + \frac{1}{2}\tilde{\tau}(\boldsymbol{r}, t) : \tilde{e}^*(\boldsymbol{r}, t) \tag{1.244}$$

measured in units of energy per unit volume, where the first term on the right-hand side is the **kinetic energy** and the second term is the **strain** or **potential energy**. We also define the **elastic power flux vector** as

$$\mathcal{P}(\boldsymbol{r}, t) = -\mathrm{Re}\{\widetilde{\boldsymbol{\tau}}(\boldsymbol{r}, t) \cdot \dot{\mathcal{U}}^*(\boldsymbol{r}, t)\} \tag{1.245}$$

measured in units of power per unit area. In terms of these quantities (1.243) may be written as

$$\frac{\partial}{\partial t}\mathcal{W}(\boldsymbol{r}, t) + \nabla \cdot \mathcal{P}(\boldsymbol{r}, t) = 0 \tag{1.246}$$

which describes the energy conservation law in linear elasticity and states that the rate of change of the elastic energy density is balanced by the spatial power flux. For time harmonic fields with time dependence $e^{-i\omega t}$ the energy density becomes

$$W(\boldsymbol{r}) = \frac{\rho\omega^2}{2}|\boldsymbol{u}(\boldsymbol{r})|^2 + \frac{1}{2}\widetilde{\boldsymbol{\tau}}(\boldsymbol{r}) : \widetilde{\boldsymbol{e}}^*(\boldsymbol{r}) \tag{1.247}$$

which may also be written as

$$\begin{aligned} W &= \frac{\rho\omega^2}{2}|\boldsymbol{u}|^2 + \frac{\lambda}{2}|\nabla \cdot \boldsymbol{u}|^2 + \mu||\widetilde{\boldsymbol{e}}||^2 \\ &= \frac{\rho\omega^2}{2}|\boldsymbol{u}|^2 + \frac{\lambda}{2}|\nabla \cdot \boldsymbol{u}|^2 + \frac{\mu}{4}||\nabla \boldsymbol{u} + (\nabla \boldsymbol{u})^\top||^2 \end{aligned} \tag{1.248}$$

where we define the norm of the dyadic as

$$||\widetilde{\boldsymbol{e}}||^2 = \sum_{i,j=1}^{3} |e_{ij}|^2 \tag{1.249}$$

In accordance with common usage, we use the same symbols, $\widetilde{\boldsymbol{\tau}}$ and $\widetilde{\boldsymbol{e}}$, to denote both the time dependent and the time independent stress and strain tensors. In the low frequency limit, as $\omega \to 0$, $W(\boldsymbol{r})$ will recover the **elastostatic energy density function**

$$\begin{aligned} W_0(\boldsymbol{r}) &= \frac{\lambda}{2}|\nabla \cdot \boldsymbol{u}(\boldsymbol{r})|^2 + \mu||\widetilde{\boldsymbol{e}}(\boldsymbol{r})||^2 \\ &= \frac{\lambda}{2}|\nabla \cdot \boldsymbol{u}(\boldsymbol{r})|^2 + \frac{\mu}{4}||\nabla \boldsymbol{u}(\boldsymbol{r}) + (\nabla \boldsymbol{u}(\boldsymbol{r}))^\top||^2 \end{aligned} \tag{1.250}$$

For time harmonic fields the power flux vector becomes

$$\boldsymbol{P}(\boldsymbol{r}) = -\mathrm{Re}\{i\omega\widetilde{\boldsymbol{\tau}}(\boldsymbol{r}) \cdot \boldsymbol{u}^*(\boldsymbol{r})\} = \omega\,\mathrm{Im}\{\widetilde{\boldsymbol{\tau}}(\boldsymbol{r}) \cdot \boldsymbol{u}^*(\boldsymbol{r})\} \tag{1.251}$$

which is also written as

$$\boldsymbol{P} = \omega\,\mathrm{Im}\,\{2\mu\boldsymbol{u}^* \cdot \nabla\boldsymbol{u} + \lambda\boldsymbol{u}^*\nabla \cdot \boldsymbol{u} + \mu\boldsymbol{u}^* \times (\nabla \times \boldsymbol{u})\} \tag{1.252}$$

The power flux in the direction $\hat{\boldsymbol{a}}$ is simply $\hat{\boldsymbol{a}} \cdot \boldsymbol{P}(\boldsymbol{r})$ while the power flow through a surface $\partial\Omega$ in the direction of the normal is

$$\int_{\partial\Omega} \hat{\boldsymbol{n}} \cdot \boldsymbol{P} ds = \omega\,\mathrm{Im} \int_{\partial\Omega} \hat{\boldsymbol{n}} \cdot \widetilde{\boldsymbol{\tau}} \cdot \boldsymbol{u}^*\, ds \tag{1.253}$$

1. E.8 *Problems in elasticity*

1. Substitute Hooke's law

$$\widetilde{\boldsymbol{\tau}} = \lambda(\nabla \cdot \boldsymbol{u})\widetilde{\boldsymbol{I}} + \mu[\nabla \boldsymbol{u} + (\nabla \boldsymbol{u})^{\top}]$$

into Newton's law

$$\nabla \cdot \widetilde{\boldsymbol{\tau}} = \rho \ddot{\mathcal{U}}$$

to obtain Navier's equation of linear elasticity

$$\mu \nabla^2 \mathcal{U} + (\lambda + \mu)\nabla\nabla \cdot \mathcal{U} = \rho \ddot{\mathcal{U}}$$

2. Show that

$$\hat{\boldsymbol{n}} \times (\nabla \times \boldsymbol{u}) = (\nabla \boldsymbol{u}) \cdot \hat{\boldsymbol{n}} - \hat{\boldsymbol{n}} \cdot (\nabla \boldsymbol{u})$$

and use Hooke's law

$$\widetilde{\boldsymbol{\tau}} = \lambda(\nabla \cdot \boldsymbol{u})\widetilde{\boldsymbol{I}} + \mu[\nabla \boldsymbol{u} + (\nabla \boldsymbol{u})^{\top}]$$

as well as the definition of traction as

$$\boldsymbol{t}(\boldsymbol{r}) = \widetilde{\boldsymbol{\tau}}(\boldsymbol{r}) \cdot \hat{\boldsymbol{n}}(\boldsymbol{r})$$

to derive the form of the surface traction operator $\boldsymbol{T}(\partial_{\boldsymbol{r}}, \hat{\boldsymbol{n}})$ for which

$$\boldsymbol{T}(\partial_{\boldsymbol{r}}, \hat{\boldsymbol{n}})\boldsymbol{u}(\boldsymbol{r}) = \boldsymbol{t}(\boldsymbol{r})$$

3. Verify that the time dependent displacement field

$$\mathcal{U}(\boldsymbol{r}, t) = \boldsymbol{u}(\boldsymbol{r})e^{-i\omega t}$$

where $\boldsymbol{u}(\boldsymbol{r})$ is the plane wave (1.222) solves the Navier equation (1.197) where

$$\rho\omega^2 = (\lambda + 2\mu)k_p^2 = \mu k_s^2$$

4. Given the function

$$G(\boldsymbol{r}, \boldsymbol{r}_0) = \frac{e^{ik|\boldsymbol{r}-\boldsymbol{r}_0|}}{ik|\boldsymbol{r} - \boldsymbol{r}_0|}$$

find the dyadic $\nabla_{\boldsymbol{r}}\nabla_{\boldsymbol{r}}G(\boldsymbol{r}, \boldsymbol{r}_0)$.

5. Given that $h(x) = e^{ix}/ix$ and $\boldsymbol{R} = R\hat{\boldsymbol{R}} = \boldsymbol{r} - \boldsymbol{r}_0$ calculate

   (i) $\nabla h(kR)$
   (ii) $\nabla\nabla h(kR)$
   (iii) $\nabla\nabla\nabla h(kR)$

   when $\nabla = \nabla_{\boldsymbol{r}}$ and when $\nabla = \nabla_{\boldsymbol{r}_0}$.

6. Use Problem 5 to derive expressions (1.228) and (1.229) for the fundamental dyadic $\widetilde{\boldsymbol{\Gamma}}$.
7. Use Problems 5 and 6 to derive expression (1.230) for the surface traction field $T_{r_0}\widetilde{\boldsymbol{\Gamma}}$, generated by the fundamental dyadic.
8. Verify that the fundamental dyadic $\widetilde{\boldsymbol{\Gamma}}$ given by (1.224)–(1.227) satisfies equation (1.223).
9. (i) Use the strain tensor

$$\widetilde{\boldsymbol{e}} = \frac{1}{2}[\nabla \boldsymbol{\mathcal{U}} + (\nabla \boldsymbol{\mathcal{U}})^{\top}]$$

and Hooke's law

$$\widetilde{\boldsymbol{\tau}} = \lambda(\mathrm{tr}\widetilde{\boldsymbol{e}})\widetilde{\boldsymbol{I}} + 2\mu\widetilde{\boldsymbol{e}}$$

to show that $\widetilde{\boldsymbol{\tau}}^* : \widetilde{\boldsymbol{e}} = \widetilde{\boldsymbol{\tau}} : \widetilde{\boldsymbol{e}}^*$

(ii) Show that

$$\nabla \cdot [\widetilde{\boldsymbol{\tau}} \cdot \dot{\boldsymbol{\mathcal{U}}}] = [\nabla \cdot \widetilde{\boldsymbol{\tau}}] \cdot \dot{\boldsymbol{\mathcal{U}}} + \widetilde{\boldsymbol{\tau}} : \nabla \dot{\boldsymbol{\mathcal{U}}}$$

and that

$$\widetilde{\boldsymbol{\tau}} : \nabla \dot{\boldsymbol{\mathcal{U}}} = \widetilde{\boldsymbol{\tau}} : (\nabla \dot{\boldsymbol{\mathcal{U}}})^{\top}$$

10. Show that the strain energy function for $\boldsymbol{u} = (u_1, u_2, u_3) \in \mathbb{R}^3$ is given by

$$\frac{1}{2}\widetilde{\boldsymbol{\tau}} : \widetilde{\boldsymbol{e}}^* = \frac{\lambda}{2}|\nabla \cdot \boldsymbol{u}|^2 + \frac{\mu}{4}\|\nabla \boldsymbol{u} + (\nabla \boldsymbol{u})^{\top}\|^2$$

where $\widetilde{\boldsymbol{\tau}}$ and $\widetilde{\boldsymbol{e}}$ are given in Problem 9,

$$\begin{aligned}\|\nabla \boldsymbol{u} + \nabla \boldsymbol{u}^{\top}\|^2 = 4&\left[\left(\frac{\partial u_1}{\partial x_1}\right)^2 + \left(\frac{\partial u_2}{\partial x_2}\right)^2 + \left(\frac{\partial u_3}{\partial x_3}\right)^2\right] \\ &+2\left[\left(\frac{\partial u_1}{\partial x_2}\right)^2 + \left(\frac{\partial u_2}{\partial x_1}\right)^2\right] \\ &+2\left[\left(\frac{\partial u_2}{\partial x_3}\right)^2 + \left(\frac{\partial u_3}{\partial x_2}\right)^2\right] \\ &+2\left[\left(\frac{\partial u_3}{\partial x_1}\right)^2 + \left(\frac{\partial u_1}{\partial x_3}\right)^2\right]\end{aligned}$$

and

$$\boldsymbol{\mathcal{U}}(\boldsymbol{r}, t) = \boldsymbol{u}(\boldsymbol{r})e^{-i\omega t}$$

11. Show that the power flux vector for the time harmonic displacement field

$$\boldsymbol{\mathcal{U}}(\boldsymbol{r}, t) = \boldsymbol{u}(\boldsymbol{r})e^{-i\omega t}$$

can be expressed in terms of the time independent displacement field as

$$\boldsymbol{P} = \omega\,\mathrm{Im}\{2\mu\boldsymbol{u}^* \cdot \nabla \boldsymbol{u} + \lambda \boldsymbol{u}^* \nabla \cdot \boldsymbol{u} + +\mu \boldsymbol{u}^* \times (\nabla \times \boldsymbol{u})\}$$

2

INTEGRAL REPRESENTATIONS AND SCATTERING THEOREMS

2.A Acoustics

2.A.1 *Integral representations in acoustics*

If u represents the scattered pressure field which satisfies the radiation condition, then a standard procedure (Colton and Kress 1983), wherein Green's second identity is applied to the functions $u(\boldsymbol{r}')$ and $G^+(\boldsymbol{r}, \boldsymbol{r}')$ in the domain $V^+ \backslash B(\boldsymbol{r}, \epsilon)$ and subsequently letting $\epsilon \to 0$ leads to the **integral representation**

$$\alpha(\boldsymbol{r})u(\boldsymbol{r}) = \frac{ik}{4\pi} \int_S \left[u(\boldsymbol{r}') \frac{\partial}{\partial n'} G^+(\boldsymbol{r}, \boldsymbol{r}') - G^+(\boldsymbol{r}, \boldsymbol{r}') \frac{\partial}{\partial n'} u(\boldsymbol{r}') \right] ds(\boldsymbol{r}') \tag{2.1}$$

for the scattered field u which holds for all $\boldsymbol{r}$ and the left-hand side vanishes in V^-.

In (2.1), G^+ represents the fundamental solution of the Helmholtz equation in V^+, (1.77), where the superscript indicates that the wave number k is that of the exterior medium V^+ so that

$$G^+(\boldsymbol{r}, \boldsymbol{r}') = \frac{e^{ik|\boldsymbol{r}-\boldsymbol{r}'|}}{ik|\boldsymbol{r}-\boldsymbol{r}'|} = h(k|\boldsymbol{r}-\boldsymbol{r}'|) \tag{2.2}$$

The function α is defined in all of $\mathbb{R}^3$ to be

$$\alpha(\boldsymbol{r}) = -\lim_{\epsilon \to 0+} \frac{1}{4\pi} \int_{\partial B(\boldsymbol{r},\epsilon) \cap V^+} \frac{\partial}{\partial n'} \frac{1}{|\boldsymbol{r}-\boldsymbol{r}'|} ds(\boldsymbol{r}') \tag{2.3}$$

When $\boldsymbol{r} \in S$, α provides a measure of the solid angle subtended at $\boldsymbol{r}$ by V^+ (Mikhlin 1970). Provided $\boldsymbol{r}$ does not coincide with a point on S for which there is no unique normal, e.g. a corner (in which case the definition (2.3) still holds) α is given explicitly as

$$\alpha(\boldsymbol{r}) = \begin{cases} 0, & \boldsymbol{r} \in V^- \\ \frac{1}{2}, & \boldsymbol{r} \in S \\ 1, & \boldsymbol{r} \in V^+ \end{cases} \tag{2.4}$$

Similarly, an integral representation for the interior field u^- is furnished by

$$(\alpha(\boldsymbol{r}) - 1)u^-(\boldsymbol{r}) = \frac{ik^-}{4\pi} \int_S \left[u^-(\boldsymbol{r}') \frac{\partial}{\partial n'} G^-(\boldsymbol{r}, \boldsymbol{r}') - G^-(\boldsymbol{r}, \boldsymbol{r}') \frac{\partial}{\partial n'} u^-(\boldsymbol{r}') \right] ds(\boldsymbol{r}'), \quad \boldsymbol{r} \in \mathbb{R}^3 \tag{2.5}$$

which in view of (2.4) is now extended from V^- to the whole of $\mathbb{R}^3$. The function G^- is the fundamental solution in V^-, given by (2.2) with wave number k^-. A similar integral representation for the incident field u^i can be obtained from the region of space occupied by V^- in the absence of the scatterer. This will lead to

$$(\alpha(\boldsymbol{r}) - 1)u^i(\boldsymbol{r}) = \frac{ik}{4\pi} \int_S \left[u^i(\boldsymbol{r}') \frac{\partial}{\partial n'} G^+(\boldsymbol{r}, \boldsymbol{r}') - G^+(\boldsymbol{r}, \boldsymbol{r}') \frac{\partial}{\partial n'} u^i(\boldsymbol{r}') \right] ds(\boldsymbol{r}'), \quad \boldsymbol{r} \in \mathbb{R}^3 \tag{2.6}$$

Combining (2.1) and (2.6) we arrive at the integral representation of the total field u^+

$$\alpha(\boldsymbol{r})u^+(\boldsymbol{r}) = u^i(\boldsymbol{r}) + \frac{ik}{4\pi} \int_S \left[u^+(\boldsymbol{r}') \frac{\partial}{\partial n'} G^+(\boldsymbol{r}, \boldsymbol{r}') - G^+(\boldsymbol{r}, \boldsymbol{r}') \frac{\partial}{\partial n'} u^+(\boldsymbol{r}') \right] ds(\boldsymbol{r}') \tag{2.7}$$

Formulae (2.5) and (2.7) furnish the general integral representations in acoustics for the interior and the total fields, respectively. In (2.7) when $\boldsymbol{r} \in V^+$, the total field u^+ can be replaced by the scattered field u, in the integrand.

When the boundary conditions are introduced into (2.7), we obtain the following integral representations for the solutions of the four basic problems which hold for all $\boldsymbol{r}$ in $\mathbb{R}^3$.

The Dirichlet problem:

$$\alpha(\boldsymbol{r})u^+(\boldsymbol{r}) = u^i(\boldsymbol{r}) - \frac{ik}{4\pi} \int_S G^+(\boldsymbol{r}, \boldsymbol{r}') \frac{\partial}{\partial n'} u^+(\boldsymbol{r}')\, ds(\boldsymbol{r}') \tag{2.8}$$

The Neumann problem:

$$\alpha(\boldsymbol{r})u^+(\boldsymbol{r}) = u^i(\boldsymbol{r}) + \frac{ik}{4\pi} \int_S u^+(\boldsymbol{r}') \frac{\partial}{\partial n'} G^+(\boldsymbol{r}, \boldsymbol{r}')\, ds(\boldsymbol{r}') \tag{2.9}$$

The Robin problem:

$$\begin{aligned} \alpha(\boldsymbol{r})u^+(\boldsymbol{r}) &= u^i(\boldsymbol{r}) + \frac{ik}{4\pi} \int_S u^+(\boldsymbol{r}') \left[\frac{\partial}{\partial n'} + \frac{i w \rho^+}{Z^+} \right] G^+(\boldsymbol{r}, \boldsymbol{r}')\, ds(\boldsymbol{r}') \\ &= u^i(\boldsymbol{r}) + \frac{ik}{4\pi} \int_S u^+(\boldsymbol{r}') \left[\frac{\partial}{\partial n'} + ik\nu \right] G^+(\boldsymbol{r}, \boldsymbol{r}')\, ds(\boldsymbol{r}') \end{aligned} \tag{2.10}$$

The transmission problem:

For the transmission problem we need both (2.5) and (2.7) for the integral representations of u^+ and u^- which involve the boundary values of u^+ and u^- and their normal derivatives, although by virtue of the transmission conditions (1.57) and (1.58) there are only two unknown surface quantities, i.e. (2.7) may be rewritten as

$$\alpha(\boldsymbol{r})u^+(\boldsymbol{r}) = u^i(\boldsymbol{r}) + \frac{ik}{4\pi} \int_S \left[u^-(\boldsymbol{r}') \frac{\partial}{\partial n'} G^+(\boldsymbol{r}, \boldsymbol{r}') - \beta G^+(\boldsymbol{r}, \boldsymbol{r}') \frac{\partial}{\partial n'} u^-(\boldsymbol{r}') \right] ds(\boldsymbol{r}') \tag{2.11}$$

2.A.2 *The far field*

Using the asymptotic relations as $r \to \infty$

$$|\boldsymbol{r} - \boldsymbol{r}'| = r - \hat{\boldsymbol{r}} \cdot \boldsymbol{r}' + O\left(\frac{1}{r}\right) \tag{2.12}$$

$$\frac{\boldsymbol{r} - \boldsymbol{r}'}{|\boldsymbol{r} - \boldsymbol{r}'|} = \hat{\boldsymbol{r}} + O\left(\frac{1}{r}\right) \tag{2.13}$$

we find that

$$G^{+}(\boldsymbol{r}, \boldsymbol{r}') = e^{-ik\hat{\boldsymbol{r}}\cdot\boldsymbol{r}'} h(kr) + O\left(\frac{1}{r^2}\right) \tag{2.14}$$

$$\nabla_{\boldsymbol{r}'} G^{+}(\boldsymbol{r}, \boldsymbol{r}') = -ik\hat{\boldsymbol{r}} e^{-ik\hat{\boldsymbol{r}}\cdot\boldsymbol{r}'} h(kr) + O\left(\frac{1}{r^2}\right) \tag{2.15}$$

$$\nabla_{\boldsymbol{r}} G^{+}(\boldsymbol{r}, \boldsymbol{r}') = ik\hat{\boldsymbol{r}} e^{-ik\hat{\boldsymbol{r}}\cdot\boldsymbol{r}'} h(kr) + O\left(\frac{1}{r^2}\right) \tag{2.16}$$

As $r \to \infty$, the integral representation (2.7), together with these asymptotic relations, yields the following representation of the scattered field

$$u(\boldsymbol{r}) = g(\hat{\boldsymbol{r}}) h(kr) + O\left(\frac{1}{r^2}\right), \qquad r \to \infty \tag{2.17}$$

where

$$g(\hat{\boldsymbol{r}}) = -\frac{ik}{4\pi} \int_S \left[\frac{\partial}{\partial n'} u^{+}(\boldsymbol{r}') + ik(\hat{\boldsymbol{r}} \cdot \hat{\boldsymbol{n}}') u^{+}(\boldsymbol{r}')\right] e^{-ik\hat{\boldsymbol{r}}\cdot\boldsymbol{r}'}\, ds(\boldsymbol{r}') \tag{2.18}$$

For plane wave incidence we write $g(\hat{\boldsymbol{r}}, \hat{\boldsymbol{k}})$ instead of $g(\hat{\boldsymbol{r}})$ where

$$g : S^2 \times S^2 \to \mathbb{C} \tag{2.19}$$

The function g has the dimensions of the scattered field and depends only on the **direction of incidence** $\hat{\boldsymbol{k}}$ and the **direction of observation** $\hat{\boldsymbol{r}}$. It is perhaps the most important function in scattering theory. It actually describes the response of the scatterer in the direction of $\hat{\boldsymbol{r}}$ due to a plane wave excitation of incident direction $\hat{\boldsymbol{k}}$. If the incident wave is not planar, then g provides the response of the scatterer in the direction of observation $\hat{\boldsymbol{r}}$ due to the excitation described by the particular incident wave. The function g is known variously as **scattering amplitude**, **scattering coefficient**, **radiation pattern**, **far field pattern** or **radiation function**. Of course, the boundary conditions will reduce the representation (2.18) to the appropriate form for each particular problem.

The importance of the scattering amplitude lies in the fact that it contains all the angular characteristics of the scattered far field and does not change with r. In other words, if r is large enough, then all the characteristics of the interaction of the obstacle with the incident field are contained in g, while the radially dependent function $h(kr)$ is the same for all scatterers. A change of the scatterer, either in its geometry (i.e. S), or in its physics (i.e. boundary conditions) will be felt in g alone. Conversely, it is this particular form of the scattered field, as the product of the obstacle dependent angular function g and the fixed radial function h, that determines whether the observation takes place in the far field or not.

We observe that the general integral representation of the scattered field (2.7) is a combination of the weighted distribution of **monopoles** and **dipoles** on the boundary, whereas the integral representation of the scattering amplitude (2.18) is a weighted distribution of plane waves on the boundary, all propagating in the direction $-\hat{\boldsymbol{r}}$.

Relation (2.17) provides an equivalent form of the radiation condition (1.67). It can be shown that g is an analytic function of its arguments.

2.A.3 *The far field expansion theorem*

The asymptotic formula (2.17) represents the first term of a convergent series representation of the scattered field in the exterior of a sphere completely surrounding the scatterer. In fact, Atkinson (1949) proved that any solution of the Helmholtz equation, which satisfies the radiation condition, has an absolutely and uniformly convergent series representation in inverse powers of the radial distance r, in all space exterior to the circumscribing sphere. In other words,

$$u(\boldsymbol{r}) = h(kr) \sum_{n=0}^{\infty} F_n(\hat{\boldsymbol{r}}) r^{-n}, \qquad r > a \tag{2.20}$$

where a represents the radius of the smallest sphere that includes S in its interior and the convergence of the series is absolute and uniform. The radius a is known as the **characteristic dimension** of the scatterer. Wilcox (1956*a*) proved that all the coefficients of the expansion (2.20) can be evaluated iteratively from the recursion formula

$$2ikn F_n(\hat{\boldsymbol{r}}) = [\mathbb{B} + n(n-1)]\, F_{n-1}(\hat{\boldsymbol{r}}), \qquad n = 1, 2, 3, \ldots \tag{2.21}$$

once the first coefficient F_0 is given. In (2.21), the operator

$$\mathbb{B} = \frac{1}{\sin\theta} \frac{\partial}{\partial\theta}\left(\sin\theta \frac{\partial}{\partial\theta}\right) + \frac{1}{\sin^2\theta} \frac{\partial^2}{\partial\varphi^2} \tag{2.22}$$

defines the **Beltrami operator** which is the angular part of Laplace's operator, while the leading coefficient is the scattering amplitude

$$F_0(\hat{\boldsymbol{r}}) = g(\hat{\boldsymbol{r}}) \tag{2.23}$$

Hence, knowledge of the scattering amplitude g, which provides the scattered field in a neighborhood of infinity, enables the reconstruction of the scattered field up to the circumscribing sphere. This is another illustration of the importance of the scattering amplitude.

Expansion (2.20) replaces the radiation condition with an exact boundary condition on any sphere surrounding the scatterer, and in this respect it can be useful for numerical evaluation of the scattered field.

This representation provides an explicit realization of the **Dirichlet to Neumann map** which relates the Neumann data (normal derivative of u) to the Dirichlet data (value of u) on a surface. If the surface is the circumscribing sphere, then, as explained above, all the coefficients, F_n in (2.20) are known in terms of g and the Dirichlet data for u is obtained by setting $r = a$ in (2.20). On the other hand, the Neumann data is easily found in terms of the same coefficients, F_n, by differentiation, i.e.

$$\frac{\partial}{\partial n}u = \frac{e^{ika}}{ik}\sum_{n=0}^{\infty} F_n\left(\frac{ik}{a^{n+1}} - \frac{n+1}{a^{n+2}}\right) \tag{2.24}$$

Asymptotically, the Dirichlet to Neumann map on a sphere is contained in the radiation condition

$$\frac{\partial}{\partial n}u = iku + O\left(\frac{1}{r^2}\right), \quad r \to \infty \tag{2.25}$$

where the term $O\left(1/r^2\right)$ can be found exactly using (2.24).

2.A.4 *Cross sections*

We now consider the **power flux** or **acoustic intensity** of the scattered field due to a plane incident wave in the radiation zone or far field where the medium is lossless. Hence, the acoustic power flux in the lossless medium, V^+, (see (1.105) for $\delta = 0$) is

$$\boldsymbol{I}(\boldsymbol{r}) = \frac{1}{\rho^+\omega}\,\mathrm{Im}\,\{u^*\nabla u\} \tag{2.26}$$

For an incident plane wave

$$u^i(\boldsymbol{r}) = e^{i\boldsymbol{k}\cdot\boldsymbol{r}} = e^{ik\hat{\boldsymbol{k}}\cdot\boldsymbol{r}} \tag{2.27}$$

the acoustic intensity in the direction of propagation $\hat{\boldsymbol{k}}$ is equal to

$$\hat{\boldsymbol{k}}\cdot\boldsymbol{I}^i(\boldsymbol{r}) = \frac{1}{\rho^+\omega}\,\mathrm{Im}\,\{\hat{\boldsymbol{k}}\cdot i\boldsymbol{k}\} = \frac{k}{\rho^+\omega} \tag{2.28}$$

The scattered wave in the radiation zone, where it propagates in the radial direction, will have the following radial intensity

$$\hat{\boldsymbol{r}}\cdot\boldsymbol{I}(\boldsymbol{r}) = \frac{1}{\rho^+\omega}\,\mathrm{Im}\left\{u^*(\boldsymbol{r})\frac{\partial}{\partial r}u(\boldsymbol{r})\right\}, \quad r \to \infty \tag{2.29}$$

The radiation condition (2.25) together with the fact that

$$u(\boldsymbol{r}) = O\left(\frac{1}{r}\right), \quad r \to \infty \tag{2.30}$$

as we can see from (2.17), implies that

$$\begin{aligned}\hat{\boldsymbol{r}} \cdot \boldsymbol{I}(\boldsymbol{r}) &= \frac{k}{\rho^+ \omega} |u(\boldsymbol{r})|^2 + O\left(\frac{1}{r^3}\right) \\ &= \frac{1}{k\rho^+ \omega r^2} |g(\hat{\boldsymbol{r}})|^2 + O\left(\frac{1}{r^3}\right), \quad r \to \infty \end{aligned} \tag{2.31}$$

Suppose now that the power flux on a large sphere of radius r was the same in every direction and equal to the intensity at $\boldsymbol{r}$ given by (2.31), then the total power flux crossing this sphere would be

$$4\pi r^2 \hat{\boldsymbol{r}} \cdot \boldsymbol{I}(\boldsymbol{r}) = \frac{4\pi}{k\rho^+ \omega} |g(\hat{\boldsymbol{r}})|^2 + O\left(\frac{1}{r}\right), \quad r \to \infty \tag{2.32}$$

This power flux, measured in units of the corresponding power flux of the incident plane wave, will define in the limit as $r \to \infty$ the **differential scattering cross section** as

$$\sigma(\hat{\boldsymbol{r}}) = \lim_{r \to \infty} \frac{4\pi r^2 \hat{\boldsymbol{r}} \cdot \boldsymbol{I}(\boldsymbol{r})}{\hat{\boldsymbol{k}} \cdot \boldsymbol{I}^i(\boldsymbol{r})} = \frac{4\pi}{k^2} |g(\hat{\boldsymbol{r}})|^2 \tag{2.33}$$

which is a function of the observation direction $\hat{\boldsymbol{r}}$ and the incident direction $\hat{\boldsymbol{k}}$.

The function $\sigma(\hat{\boldsymbol{r}})$ is measured in units of area. Its value specifies the amount of power scattered in the direction $\hat{\boldsymbol{r}}$ relative to the incident power flux in the direction of propagation.

In fact, we can define $\sigma(\hat{\boldsymbol{r}})$ as

$$\sigma(\hat{\boldsymbol{r}}) = \lim_{r \to \infty} 4\pi r^2 \frac{|u(\boldsymbol{r})|^2}{|u^i(\boldsymbol{r})|^2} = \frac{4\pi}{k^2} |g(\hat{\boldsymbol{r}})|^2 \tag{2.34}$$

but its physical interpretation comes from (2.33).

The **scattering cross section** or **total cross section** is defined as being the average of $\sigma(\hat{\boldsymbol{r}})$ over all directions as

$$\begin{aligned}\sigma_s &= \frac{1}{4\pi} \int_{S^2} \sigma(\hat{\boldsymbol{r}}) \, ds(\hat{\boldsymbol{r}}) = \frac{1}{k^2} \int_{S^2} |g(\hat{\boldsymbol{r}})|^2 \, ds(\hat{\boldsymbol{r}}) \\ &= \int_{\partial B(\boldsymbol{0}, \infty)} |u(\boldsymbol{r})|^2 \, ds(\boldsymbol{r}) = \frac{1}{k} \operatorname{Im} \int_{\partial B(\boldsymbol{0}, \infty)} u^*(\boldsymbol{r}) \frac{\partial}{\partial r} u(\boldsymbol{r}) \, ds(\boldsymbol{r}) \end{aligned} \tag{2.35}$$

where S^2 is the surface of the unit sphere and $\partial B(\boldsymbol{0}, \infty)$ is the surface of the sphere of asymptotically infinite radius centered at the origin. The last integral representation of σ_s can be used to bring the integration back on S, as follows. Green's identity implies that

$$\int_{V^+ \cap B(\mathbf{0},\infty)} (u^* \nabla^2 u - u \nabla^2 u^*)\, dv$$

$$= \int_{\partial B(\mathbf{0},\infty)} \left(u^* \frac{\partial}{\partial r} u - u \frac{\partial}{\partial r} u^* \right) ds - \int_S \left(u^* \frac{\partial}{\partial n} u - u \frac{\partial}{\partial n} u^* \right) ds \tag{2.36}$$

Since u solves the Helmholtz equation for real k, (2.36) implies that

$$\operatorname{Im} \int_{\partial B(\mathbf{0},\infty)} u^* \frac{\partial}{\partial r} u\, ds = \operatorname{Im} \int_S u^* \frac{\partial}{\partial n} u\, ds \tag{2.37}$$

Hence, from (2.35) we obtain

$$\sigma_s = \frac{1}{k} \operatorname{Im} \int_S u^* \frac{\partial}{\partial n} u\, ds \tag{2.38}$$

This last expression for the scattering cross section is defined in terms of an integral over the surface S of the scattered field and its normal derivative. The corresponding expression involving the total field is used to define the **absorption cross section** as

$$\sigma_a = -\frac{1}{k} \operatorname{Im} \int_S u^{+^*} \frac{\partial}{\partial n} u^+\, ds = \frac{1}{k} \operatorname{Im} \int_S u^+ \frac{\partial}{\partial n} u^{+^*}\, ds \tag{2.39}$$

where the $-$ sign indicates that the power flux is inward. The quantity σ_a defines the total energy absorbed by the scatterer whenever it is penetrable and lossy. Using the transmission conditions (1.57) and (1.58) and Green's second identity we may write

$$\begin{aligned}
\sigma_a &= \frac{1}{k} \operatorname{Im} \int_S u^+ \frac{\partial}{\partial n} u^{+^*}\, ds \\
&= \frac{1}{2ik} \int_S \left[u^- \left(\beta \frac{\partial}{\partial n} u^- \right)^* - u^{-^*} \left(\beta \frac{\partial}{\partial n} u^- \right) \right] ds \\
&= \frac{1}{2ik} \int_{V^-} \nabla \cdot [u^- (\beta \nabla u^-)^* - (u^-)^* (\beta \nabla u^-)]\, dv \\
&= \frac{1}{2ik} \int_{V^-} \{ (\beta^* - \beta) |\nabla u^-|^2 + k^2 [\beta \eta^2 - (\beta \eta^2)^*]\, |u^-|^2 \}\, dv \\
&= \delta^- \frac{\gamma^-}{\rho^-} \sqrt{\frac{\rho^+}{\gamma^+}} \int_{V^-} |\nabla u^-|^2\, dv
\end{aligned} \tag{2.40}$$

where we have used (1.61) and (1.62) to evaluate $\beta^* - \beta$ and $\beta\eta^2 - (\beta\eta^2)^*$. Obviously, for lossless scatterers $\delta^- = 0$ and therefore $\sigma_a = 0$. Similarly, for the Dirichlet and the Neumann problems $\sigma_a = 0$, since

$$u^+ \left(\frac{\partial u^+}{\partial n} \right)^* = 0 \tag{2.41}$$

For the Robin problem, on the other hand

$$\sigma_a = -\frac{1}{k} \operatorname{Im} \int_S u^{+*}(-ik\nu u^+) \, ds = \nu \int_S |u^+|^2 \, ds \tag{2.42}$$

which is not zero. The surface of the scatterer absorbs energy in the case of the Robin boundary condition. This discussion shows that it is always true that σ_a is nonnegative.

Finally, we define the **extinction cross section** to be

$$\sigma_e = \sigma_s + \sigma_a \tag{2.43}$$

which describes the total energy that the body extracts from the incident wave either by radiation, σ_s, or by absorption, σ_a.

Note that in all formulae for the cross sections we have normalized with respect to the incident field, so all cross sections are expressed in units of area.

2.A.5 *Basic scattering theorems*

In this section we derive some important properties of scattered fields which we designate as **basic scattering theorems**. The first results concern **reciprocity**.

The reciprocity theorems:

1. **Point sources:** The scattered field at $\boldsymbol{r}_1$ due to a source at $\boldsymbol{r}_2$ in V^+ is equal to the field at $\boldsymbol{r}_2$ due to a source at $\boldsymbol{r}_1$, i.e.

$$u(\boldsymbol{r}_1, \boldsymbol{r}_2) = u(\boldsymbol{r}_2, \boldsymbol{r}_1) \tag{2.44}$$

   Moreover, the same relation holds for total fields

$$u^+(\boldsymbol{r}_1, \boldsymbol{r}_2) = u^+(\boldsymbol{r}_2, \boldsymbol{r}_1) \tag{2.45}$$

2. **Plane waves:** The scattering amplitude in the direction $\hat{\boldsymbol{r}}$ due to a plane wave in the direction $\hat{\boldsymbol{k}}$ is equal to the scattering amplitude in the direction $-\hat{\boldsymbol{k}}$ due to a plane wave in the direction $-\hat{\boldsymbol{r}}$, i.e.

$$g(\hat{\boldsymbol{r}}, \hat{\boldsymbol{k}}) = g(-\hat{\boldsymbol{k}}, -\hat{\boldsymbol{r}}) \tag{2.46}$$

The proof of these theorems rests on the following identity. If $u_1^+(\boldsymbol{r}')$ and $u_2^+(\boldsymbol{r}')$ are the total fields due to different incident fields, $u_1^i(\boldsymbol{r}')$ and $u_2^i(\boldsymbol{r}')$, then

$$\int_S \left[u_1^+(\boldsymbol{r}') \frac{\partial}{\partial n'} u_2^+(\boldsymbol{r}') - u_2^+(\boldsymbol{r}') \frac{\partial}{\partial n'} u_1^+(\boldsymbol{r}') \right] ds(\boldsymbol{r}') = 0 \tag{2.47}$$

The validity of this identity follows by applying any of the boundary or transmission conditions, together with Green's second identity applied in V^-. Moreover, the same Green's identity may be used to show that

$$\int_S \left[u_1^i(\boldsymbol{r}') \frac{\partial}{\partial n'} u_2^i(\boldsymbol{r}') - u_2^i(\boldsymbol{r}') \frac{\partial}{\partial n'} u_1^i(\boldsymbol{r}') \right] ds(\boldsymbol{r}') = 0 \tag{2.48}$$

and

$$\int_S \left[u_1(\boldsymbol{r}') \frac{\partial}{\partial n'} u_2(\boldsymbol{r}') - u_2(\boldsymbol{r}') \frac{\partial}{\partial n'} u_1(\boldsymbol{r}') \right] ds(\boldsymbol{r}') = 0 \tag{2.49}$$

The radiation condition is also needed to establish the relation (2.49) involving scattered fields. Using these two identities together with (2.47) leads to the identity

$$\int_S \left[u_1^i(\boldsymbol{r}') \frac{\partial}{\partial n'} u_2(\boldsymbol{r}') - u_2(\boldsymbol{r}') \frac{\partial}{\partial n'} u_1^i(\boldsymbol{r}') \right] ds(\boldsymbol{r}')$$
$$= \int_S \left[u_2^i(\boldsymbol{r}') \frac{\partial}{\partial n'} u_1(\boldsymbol{r}') - u_1(\boldsymbol{r}') \frac{\partial}{\partial n'} u_2^i(\boldsymbol{r}') \right] ds(\boldsymbol{r}') \tag{2.50}$$

Choosing $u_j^i(\boldsymbol{r}') = G^+(\boldsymbol{r}_j, \boldsymbol{r}')$, $j = 1, 2$, equation (2.50) together with the representation (2.1) implies that

$$u_2(\boldsymbol{r}_1) = u_1(\boldsymbol{r}_2) \quad \text{or} \quad u(\boldsymbol{r}_1, \boldsymbol{r}_2) = u(\boldsymbol{r}_2, \boldsymbol{r}_1) \tag{2.51}$$

This establishes the reciprocity theorem for point sources.

To establish reciprocity for the scattering amplitudes, note first that (2.18) may be rewritten as

$$g(-\hat{\boldsymbol{r}}, \hat{\boldsymbol{k}}) = -\frac{ik}{4\pi} \int_S \left[e^{ik\hat{\boldsymbol{r}}\cdot\boldsymbol{r}'} \frac{\partial}{\partial n'} u(\boldsymbol{r}', \hat{\boldsymbol{k}}) - u(\boldsymbol{r}', \hat{\boldsymbol{k}}) \frac{\partial}{\partial n'} e^{ik\hat{\boldsymbol{r}}\cdot\boldsymbol{r}'} \right] ds(\boldsymbol{r}') \tag{2.52}$$

Identifying $\hat{\boldsymbol{r}}$ and $\hat{\boldsymbol{k}}$, first with $\hat{\boldsymbol{k}}_1$ and $\hat{\boldsymbol{k}}_2$ and then with $\hat{\boldsymbol{k}}_2$ and $\hat{\boldsymbol{k}}_1$, in this equation and using the identity (2.50) with $u_j^i(\boldsymbol{r}') = e^{ik\hat{\boldsymbol{k}}_j\cdot\boldsymbol{r}'}$, $j = 1, 2$ we find that

$$g(-\hat{\boldsymbol{k}}_1, \hat{\boldsymbol{k}}_2) = g(-\hat{\boldsymbol{k}}_2, \hat{\boldsymbol{k}}_1) \tag{2.53}$$

or, by choosing $-\hat{\boldsymbol{k}}_1 = \hat{\boldsymbol{r}}$ and $\hat{\boldsymbol{k}}_2 = \hat{\boldsymbol{k}}$,

$$g(\hat{\boldsymbol{r}}, \hat{\boldsymbol{k}}) = g(-\hat{\boldsymbol{k}}, -\hat{\boldsymbol{r}}) \tag{2.54}$$

hence (2.46) is established.

Next, we establish a theorem which will be useful in determining low frequency expansions of the scattering amplitude.

The general scattering theorem: Let $u^+(\boldsymbol{r}; \hat{\boldsymbol{k}}_i)$, $i = 1, 2$ define the total field corresponding to a plane wave excitation in the direction $\hat{\boldsymbol{k}}_i$, $i = 1, 2$. Then

$$\begin{aligned}
&\int_S \left[u^{+*}(\boldsymbol{r}; \hat{\boldsymbol{k}}_1) \frac{\partial}{\partial n} u^+(\boldsymbol{r}; \hat{\boldsymbol{k}}_2) - u^+(\boldsymbol{r}; \hat{\boldsymbol{k}}_2) \frac{\partial}{\partial n} u^{+*}(\boldsymbol{r}; \hat{\boldsymbol{k}}_1) \right] ds(\boldsymbol{r}) \\
&\quad = \int_S \left[e^{-ik\boldsymbol{r}\cdot\hat{\boldsymbol{k}}_1} \frac{\partial}{\partial n} e^{ik\boldsymbol{r}\cdot\hat{\boldsymbol{k}}_2} - e^{ik\boldsymbol{r}\cdot\hat{\boldsymbol{k}}_2} \frac{\partial}{\partial n} e^{-ik\boldsymbol{r}\cdot\hat{\boldsymbol{k}}_1} \right] ds(\boldsymbol{r}) \\
&\quad + \int_S \left[e^{-ik\boldsymbol{r}\cdot\hat{\boldsymbol{k}}_1} \frac{\partial}{\partial n} u(\boldsymbol{r}; \hat{\boldsymbol{k}}_2) - u(\boldsymbol{r}; \hat{\boldsymbol{k}}_2) \frac{\partial}{\partial n} e^{-ik\boldsymbol{r}\cdot\hat{\boldsymbol{k}}_1} \right] ds(\boldsymbol{r}) \\
&\quad + \int_S \left[u^*(\boldsymbol{r}; \hat{\boldsymbol{k}}_1) \frac{\partial}{\partial n} e^{ik\boldsymbol{r}\cdot\hat{\boldsymbol{k}}_2} - e^{ik\boldsymbol{r}\cdot\hat{\boldsymbol{k}}_2} \frac{\partial}{\partial n} u^*(\boldsymbol{r}; \hat{\boldsymbol{k}}_1) \right] ds(\boldsymbol{r}) \\
&\quad + \int_S \left[u^*(\boldsymbol{r}; \hat{\boldsymbol{k}}_1) \frac{\partial}{\partial n} u(\boldsymbol{r}; \hat{\boldsymbol{k}}_2) - u(\boldsymbol{r}; \hat{\boldsymbol{k}}_2) \frac{\partial}{\partial n} u^*(\boldsymbol{r}; \hat{\boldsymbol{k}}_1) \right] ds(\boldsymbol{r}) \qquad (2.55)
\end{aligned}$$

The first surface integral on the right-hand side of (2.55) vanishes since plane waves are regular solutions of the Helmholtz equation in V^-. The second and third terms on the right-hand side may be rewritten in terms of scattering amplitudes using (2.18), while the last term on the right may be transformed to an integral over $\partial B(\mathbf{0}, \infty)$ using Green's second identity and then the asymptotic form (2.17) may be employed. With these observations (2.55) may be rewritten as

$$\begin{aligned}
&\int_S \left[u^{+*}(\boldsymbol{r}; \hat{\boldsymbol{k}}_1) \frac{\partial}{\partial n} u^+(\boldsymbol{r}; \hat{\boldsymbol{k}}_2) - u^+(\boldsymbol{r}; \hat{\boldsymbol{k}}_2) \frac{\partial}{\partial n} u^{+*}(\boldsymbol{r}; \hat{\boldsymbol{k}}_1) \right] ds(\boldsymbol{r}) \\
&\quad = -\frac{4\pi}{ik} g(\hat{\boldsymbol{k}}_1, \hat{\boldsymbol{k}}_2) - \frac{4\pi}{ik} g^*(\hat{\boldsymbol{k}}_2, \hat{\boldsymbol{k}}_1) + \frac{2i}{k} \int_{S^2} g^*(\hat{\boldsymbol{r}}, \hat{\boldsymbol{k}}_1) g(\hat{\boldsymbol{r}}, \hat{\boldsymbol{k}}_2)\, ds(\hat{\boldsymbol{r}}) \qquad (2.56)
\end{aligned}$$

which gives the following general scattering theorem

$$\begin{aligned}
&-g(\hat{\boldsymbol{k}}_1, \hat{\boldsymbol{k}}_2) - g^*(\hat{\boldsymbol{k}}_2, \hat{\boldsymbol{k}}_1) \\
&\quad = \frac{1}{2\pi} \int_{S^2} g^*(\hat{\boldsymbol{r}}, \hat{\boldsymbol{k}}_1) g(\hat{\boldsymbol{r}}, \hat{\boldsymbol{k}}_2)\, ds(\hat{\boldsymbol{r}}) \\
&\quad + \frac{ik}{4\pi} \int_S \left[u^{+*}(\boldsymbol{r}; \hat{\boldsymbol{k}}_1) \frac{\partial}{\partial n} u^+(\boldsymbol{r}; \hat{\boldsymbol{k}}_2) - u^+(\boldsymbol{r}; \hat{\boldsymbol{k}}_2) \frac{\partial}{\partial n} u^{+*}(\boldsymbol{r}; \hat{\boldsymbol{k}}_1) \right] ds(\boldsymbol{r}) \qquad (2.57)
\end{aligned}$$

The last integral in (2.57) vanishes for Dirichlet, Neumann and lossless penetrable scatterers. This derivation is due to Twersky (1954).

The optical theorem: If we let $\hat{\boldsymbol{k}}_1 = \hat{\boldsymbol{k}}_2 = \hat{\boldsymbol{k}}$ in (2.57) we obtain

$$-2\,\mathrm{Re}\, g(\hat{\boldsymbol{k}}, \hat{\boldsymbol{k}}) = \frac{1}{2\pi} \int_{S^2} |g(\hat{\boldsymbol{r}}, \hat{\boldsymbol{k}})|^2 \, ds(\hat{\boldsymbol{r}}) + \frac{k}{2\pi}\,\mathrm{Im} \int_S u^+(\boldsymbol{r}; \hat{\boldsymbol{k}}) \frac{\partial}{\partial n} u^{+^*}(\boldsymbol{r}; \hat{\boldsymbol{k}}) \, ds(\boldsymbol{r}) \tag{2.58}$$

or, by virtue of (2.35) and (2.39),

$$\sigma_e = \sigma_s + \sigma_a = -\frac{4\pi}{k^2}\,\mathrm{Re}\, g(\hat{\boldsymbol{k}}, \hat{\boldsymbol{k}}) \tag{2.59}$$

where again $\sigma_a = 0$ for the Dirichlet, Neumann and lossless penetrable scatterers.

Formula (2.59) describes the famous **optical theorem** which states that the total energy that the scatterer removes from the incident field is proportional to the value of the scattering amplitude in the forward direction $\hat{\boldsymbol{k}}$. This relation between the forward scattering amplitude and the total energy scattered may be interpreted as an interference pattern between the incident and the scattered wave which establishes the mechanism of energy transfer.

As a consequence of the optical theorem, the power scattered in the forward direction can never be zero since if it were, then by (2.59) and the fact that $\sigma_a \geq 0$ it follows that $\sigma_s = 0$, or, in view of (2.58)

$$g(\hat{\boldsymbol{r}}, \hat{\boldsymbol{k}}) = 0, \quad \hat{\boldsymbol{r}} \in S^2 \tag{2.60}$$

which in turn implies, by Atkinson's far field theorem, that u is identically zero and therefore no scattering occurs.

The optical theorem embodied in (2.59) is also known as the **forward scattering theorem**, or the **fundamental extinction formula** (Van de Hulst 1981).

2.A.6 *Problems in acoustics*

1. Derive the exterior integral representation (2.1) and the interior integral representation (2.5)
2. For the transmission problem derive the following expression for the absorption cross section

$$\sigma_a = k \frac{\gamma^-}{\gamma^+} \frac{\omega\delta^-\gamma^-}{1 + \omega^2\delta^{-2}\gamma^{-2}} \int_{V^-} |u^-(\boldsymbol{r})|^2 \, dv(\boldsymbol{r}) + \frac{1}{k}\frac{\rho^+}{\rho^-}\omega\delta^-\gamma^-\,\mathrm{Re} \int_S u^-(\boldsymbol{r}) \frac{\partial}{\partial n} u^{-*}(\boldsymbol{r}) \, ds(\boldsymbol{r})$$

3. Show that the coefficients $F_n, n \geq 1$ in the Atkinson expansion

$$u(\boldsymbol{r}) = h(kr) \sum_{n=0}^{\infty} F_n(\hat{\boldsymbol{r}}) r^{-n}, \qquad r > a$$

where a is the radius of the sphere circumscribing S, are expressed in terms of the scattering amplitude $g = f_0$, through the formula

$$2iknF_n(\hat{\boldsymbol{r}}) = [\mathbb{B} + n(n-1)]F_{n-1}(\hat{\boldsymbol{r}}), \qquad n \geq 1$$

where $\mathbb{B}$ denotes the Beltrami operator given by

$$\sin^2\theta\, \mathbb{B} = \left(\sin\theta \frac{\partial}{\partial\theta}\right)\left(\sin\theta \frac{\partial}{\partial\theta}\right) + \frac{\partial^2}{\partial\varphi^2}$$

4. Obtain the asymptotic relations

$$|\boldsymbol{r} - \boldsymbol{r}'| = r - \hat{\boldsymbol{r}} \cdot \boldsymbol{r}' + O\left(\frac{1}{r}\right), \qquad r \to \infty$$

$$\frac{\boldsymbol{r} - \boldsymbol{r}'}{|\boldsymbol{r} - \boldsymbol{r}'|} = \hat{\boldsymbol{r}} + O\left(\frac{1}{r}\right), \qquad r \to \infty$$

and

$$\frac{e^{ik|\boldsymbol{r}-\boldsymbol{r}'|}}{|\boldsymbol{r} - \boldsymbol{r}'|} = \frac{e^{ikr}}{r} e^{-ik\hat{\boldsymbol{r}}\cdot\boldsymbol{r}'} + O\left(\frac{1}{r}\right), \qquad r \to \infty$$

5. Consider the Dirichlet problem for the soft surface S. Prove that if it is true that $\boldsymbol{r} \in S$ implies $-\boldsymbol{r} \in S$, then

$$g(\hat{\boldsymbol{r}}, \hat{\boldsymbol{k}}) = g(\hat{\boldsymbol{k}}, \hat{\boldsymbol{r}})$$

where g stands for the scattering amplitude for plane wave excitation in the direction of $\hat{\boldsymbol{k}}$.

2.EM Electromagnetics

2.EM.1 *Integral representations in electromagnetics*

In this section we derive the fundamental integral representations in electromagnetics. They are obtained as follows.

Gauss' theorem applied to the vector field

$$\boldsymbol{F} = \boldsymbol{f}_1 \times (\nabla \times \boldsymbol{f}_2) - \boldsymbol{f}_2 \times (\nabla \times \boldsymbol{f}_1) \tag{2.61}$$

for $\boldsymbol{f}_1, \boldsymbol{f}_2 \in C^2(V) \cap C^1(\bar{V})$ leads to the integral identity

$$\int_V [\boldsymbol{f}_1 \cdot \nabla \times (\nabla \times \boldsymbol{f}_2) - \boldsymbol{f}_2 \cdot \nabla \times (\nabla \times \boldsymbol{f}_1)]\, dv$$

$$= \int_S [\boldsymbol{f}_2 \times (\nabla \times \boldsymbol{f}_1) - \boldsymbol{f}_1 \times (\nabla \times \boldsymbol{f}_2)] \cdot \hat{\boldsymbol{n}}\, ds \tag{2.62}$$

for every smooth bounded domain V. Identity (2.62) can then be used to derive the **Stratton–Chu representation formula** (1939) by identifying V with $V^+ \backslash B(\boldsymbol{r}, \epsilon), \boldsymbol{f}_1$

with $G^+(\boldsymbol{r}, \hat{\boldsymbol{r}}')\boldsymbol{a}$ where $\boldsymbol{a}$ is an arbitrary constant vector, $\boldsymbol{f}_2$ successively with the scattered fields $\boldsymbol{E}$ and $\boldsymbol{H}$ and using the Maxwell equations. Explicitly

$$\begin{aligned}\alpha(\boldsymbol{r})\boldsymbol{E}(\boldsymbol{r}) = \frac{ik}{4\pi}\int_S &[ikZ^+G^+(\boldsymbol{r},\boldsymbol{r}')(\hat{\boldsymbol{n}}' \times \boldsymbol{H}(\boldsymbol{r}')) \\ &+(\nabla_{\boldsymbol{r}'}G^+(\boldsymbol{r},\boldsymbol{r}'))(\hat{\boldsymbol{n}}' \cdot \boldsymbol{E}(\boldsymbol{r}')) \\ &-(\nabla_{\boldsymbol{r}'}G^+(\boldsymbol{r},\boldsymbol{r}')) \times (\hat{\boldsymbol{n}}' \times \boldsymbol{E}(\boldsymbol{r}'))]\, ds(\boldsymbol{r}')\end{aligned} \tag{2.63}$$

and

$$\begin{aligned}\alpha(\boldsymbol{r})\boldsymbol{H}(\boldsymbol{r}) = \frac{ik}{4\pi}\int_S &[-ikY^+G^+(\boldsymbol{r},\boldsymbol{r}')(\hat{\boldsymbol{n}}' \times \boldsymbol{E}(\boldsymbol{r}')) \\ &+(\nabla_{\boldsymbol{r}'}G^+(\boldsymbol{r},\boldsymbol{r}'))(\hat{\boldsymbol{n}}' \cdot \boldsymbol{H}(\boldsymbol{r}')) \\ &-(\nabla_{\boldsymbol{r}'}G^+(\boldsymbol{r},\boldsymbol{r}')) \times (\hat{\boldsymbol{n}}' \times \boldsymbol{H}(\boldsymbol{r}'))]\, ds(\boldsymbol{r}')\end{aligned} \tag{2.64}$$

which hold for every $\boldsymbol{r} \in \mathbb{R}^3$ where $\alpha(\boldsymbol{r})$ is given by (2.3), G^+ is the fundamental solution in V^+ and $\boldsymbol{E}$ and $\boldsymbol{H}$ denote the scattered electric and magnetic fields, respectively.

We can also derive the interior integral representations, identifying V with $V^-\backslash B(\boldsymbol{r},\epsilon)$, $\boldsymbol{f}_1$ with $G^-(\boldsymbol{r},\boldsymbol{r}')\boldsymbol{a}$, $\boldsymbol{f}_2$ successively with $\boldsymbol{E}^-$ and $\boldsymbol{H}^-$, and using the Maxwell equations. Explicitly

$$\begin{aligned}(\alpha(\boldsymbol{r})-1)\boldsymbol{E}^-(\boldsymbol{r}) = \frac{ik^-}{4\pi}\int_S &[ik^-Z^-G^-(\boldsymbol{r},\boldsymbol{r}')(\hat{\boldsymbol{n}}' \times \boldsymbol{H}^-(\boldsymbol{r}')) \\ &+(\nabla_{\boldsymbol{r}'}G^-(\boldsymbol{r},\boldsymbol{r}'))(\hat{\boldsymbol{n}}' \cdot \boldsymbol{E}^-(\boldsymbol{r}')) \\ &-(\nabla_{\boldsymbol{r}'}G^-(\boldsymbol{r},\boldsymbol{r}')) \times (\hat{\boldsymbol{n}}' \times \boldsymbol{E}^-(\boldsymbol{r}'))]\, ds(\boldsymbol{r}')\end{aligned} \tag{2.65}$$

$$\begin{aligned}(\alpha(\boldsymbol{r})-1)\boldsymbol{H}^-(\boldsymbol{r}) = \frac{ik^-}{4\pi}\int_S &[-ik^-Y^-G^-(\boldsymbol{r},\boldsymbol{r}')(\hat{\boldsymbol{n}}' \times \boldsymbol{E}^-(\boldsymbol{r}')) \\ &+(\nabla_{\boldsymbol{r}'}G^-(\boldsymbol{r},\boldsymbol{r}'))(\hat{\boldsymbol{n}}' \cdot \boldsymbol{H}^-(\boldsymbol{r}')) \\ &-(\nabla_{\boldsymbol{r}'}G^-(\boldsymbol{r},\boldsymbol{r}')) \times (\hat{\boldsymbol{n}}' \times \boldsymbol{H}^-(\boldsymbol{r}'))]\, ds(\boldsymbol{r}')\end{aligned} \tag{2.66}$$

which also hold for $\boldsymbol{r} \in \mathbb{R}^3$ and where the superscript $-$ indicates that all fields and medium parameters are related to V^-.

The incident pair $\boldsymbol{E}^i$ and $\boldsymbol{H}^i$ also satisfy the representations

$$\begin{aligned}(\alpha(\boldsymbol{r})-1)\boldsymbol{E}^i(\boldsymbol{r}) = \frac{ik}{4\pi}\int_S &[ikZ^+G^+(\boldsymbol{r},\boldsymbol{r}')(\hat{\boldsymbol{n}}' \times \boldsymbol{H}^i(\boldsymbol{r}')) \\ &+(\nabla_{\boldsymbol{r}'}G^+(\boldsymbol{r},\boldsymbol{r}'))(\hat{\boldsymbol{n}}' \cdot \boldsymbol{E}^i(\boldsymbol{r}')) \\ &-(\nabla_{\boldsymbol{r}'}G^+(\boldsymbol{r},\boldsymbol{r}')) \times (\hat{\boldsymbol{n}}' \times \boldsymbol{E}^i(\boldsymbol{r}'))]\, ds(\boldsymbol{r}')\end{aligned} \tag{2.67}$$

and

$$(\alpha(\boldsymbol{r}) - 1)\boldsymbol{H}^i(\boldsymbol{r}) = \frac{ik}{4\pi}\int_S [-ikY^+G^+(\boldsymbol{r},\boldsymbol{r}')(\hat{\boldsymbol{n}}' \times \boldsymbol{E}^i(\boldsymbol{r}'))$$
$$+(\nabla_{r'}G^+(\boldsymbol{r},\boldsymbol{r}'))(\hat{\boldsymbol{n}}' \cdot \boldsymbol{H}^i(\boldsymbol{r}'))$$
$$-(\nabla_{r'}G^+(\boldsymbol{r},\boldsymbol{r}')) \times (\hat{\boldsymbol{n}}' \times \boldsymbol{H}^i(\boldsymbol{r}'))]\, ds(\boldsymbol{r}') \qquad (2.68)$$

for every $\boldsymbol{r} \in \mathbb{R}^3$, since incident fields satisfy the Maxwell equations in the region occupied by V^- in the absence of the obstacle.

Adding (2.63) and (2.67), and (2.64) and (2.68) we arrive at the following integral representations of the total field

$$\alpha(\boldsymbol{r})\boldsymbol{E}^+(\boldsymbol{r}) = \boldsymbol{E}^i(\boldsymbol{r}) + \frac{ik}{4\pi}\int_S [ikZ^+G^+(\boldsymbol{r},\boldsymbol{r}')(\hat{\boldsymbol{n}}' \times \boldsymbol{H}^+(\boldsymbol{r}'))$$
$$+(\nabla_{r'}G^+(\boldsymbol{r},\boldsymbol{r}'))(\hat{\boldsymbol{n}}' \cdot \boldsymbol{E}^+(\boldsymbol{r}'))$$
$$-(\nabla_{r'}G^+(\boldsymbol{r},\boldsymbol{r}')) \times (\hat{\boldsymbol{n}}' \times \boldsymbol{E}^+(\boldsymbol{r}'))]\, ds(\boldsymbol{r}') \qquad (2.69)$$

and

$$\alpha(\boldsymbol{r})\boldsymbol{H}^+(\boldsymbol{r}) = \boldsymbol{H}^i(\boldsymbol{r}) + \frac{ik}{4\pi}\int_S [-ikY^+G^+(\boldsymbol{r},\boldsymbol{r}')(\hat{\boldsymbol{n}}' \times \boldsymbol{E}^+(\boldsymbol{r}'))$$
$$+(\nabla_{r'}G^+(\boldsymbol{r},\boldsymbol{r}'))(\hat{\boldsymbol{n}}' \cdot \boldsymbol{H}^+(\boldsymbol{r}'))$$
$$-(\nabla_{r'}G^+(\boldsymbol{r},\boldsymbol{r}')) \times (\hat{\boldsymbol{n}}' \times \boldsymbol{H}^+(\boldsymbol{r}'))]\, ds(\boldsymbol{r}') \qquad (2.70)$$

The Stratton–Chu representation pairs (2.65) and (2.66), (2.69) and (2.70) provide the general integral representations in electromagnetic scattering for the interior and total fields, respectively.

When $\boldsymbol{r} \in V^+$, the field quantities in the integrands in (2.69) and (2.70) can be either total or scattered fields provided the choice is made consistently.

Using the dyadic form of Green's identity

$$\int_V [(\nabla \times (\nabla \times \boldsymbol{f})) \cdot \widetilde{\boldsymbol{F}} - \boldsymbol{f} \cdot (\nabla \times (\nabla \times \widetilde{\boldsymbol{F}}))]\, dv$$
$$= \int_S \hat{\boldsymbol{n}} \cdot [(\nabla \times \boldsymbol{f}) \times \widetilde{\boldsymbol{F}} + \boldsymbol{f} \times (\nabla \times \widetilde{\boldsymbol{F}})]\, ds$$
$$= -\int_S [(\nabla \times \boldsymbol{f}) \cdot (\hat{\boldsymbol{n}} \times \widetilde{\boldsymbol{F}}) - (\hat{\boldsymbol{n}} \times \boldsymbol{f}) \cdot (\nabla \times \widetilde{\boldsymbol{F}})]\, ds \qquad (2.71)$$

we can obtain the following integral representations in terms of the fundamental dyadic $\widetilde{\boldsymbol{G}}$, (1.166), which hold for all $\boldsymbol{r}$ in $\mathbb{R}^3 \backslash S$

$$\alpha(\boldsymbol{r})\boldsymbol{E}^+(\boldsymbol{r}) = \boldsymbol{E}^i(\boldsymbol{r}) - \frac{ik}{4\pi}\int_S [(\nabla_{\boldsymbol{r}'} \times \boldsymbol{E}^+(\boldsymbol{r}')) \cdot (\hat{\boldsymbol{n}}' \times \widetilde{\boldsymbol{G}}^+(\boldsymbol{r},\boldsymbol{r}')) - (\hat{\boldsymbol{n}}' \times \boldsymbol{E}^+(\boldsymbol{r}')) \cdot (\nabla_{\boldsymbol{r}'} \times \widetilde{\boldsymbol{G}}^+(\boldsymbol{r},\boldsymbol{r}'))]\, ds(\boldsymbol{r}') \tag{2.72}$$

$$\alpha(\boldsymbol{r})\boldsymbol{H}^+(\boldsymbol{r}) = \boldsymbol{H}^i(\boldsymbol{r}) - \frac{ik}{4\pi}\int_S [(\nabla_{\boldsymbol{r}'} \times \boldsymbol{H}^+(\boldsymbol{r}')) \cdot (\hat{\boldsymbol{n}}' \times \widetilde{\boldsymbol{G}}^+(\boldsymbol{r},\boldsymbol{r}')) - (\hat{\boldsymbol{n}}' \times \boldsymbol{H}^+(\boldsymbol{r}')) \cdot (\nabla_{\boldsymbol{r}'} \times \widetilde{\boldsymbol{G}}^+(\boldsymbol{r},\boldsymbol{r}'))]\, ds(\boldsymbol{r}') \tag{2.73}$$

and

$$(\alpha(\boldsymbol{r}) - 1)\boldsymbol{E}^-(\boldsymbol{r}) = -\frac{ik^-}{4\pi}\int_S [(\nabla_{\boldsymbol{r}'} \times \boldsymbol{E}^-(\boldsymbol{r}')) \cdot (\hat{\boldsymbol{n}}' \times \widetilde{\boldsymbol{G}}^-(\boldsymbol{r},\boldsymbol{r}')) - (\hat{\boldsymbol{n}}' \times \boldsymbol{E}^-(\boldsymbol{r}')) \cdot (\nabla_{\boldsymbol{r}'} \times \widetilde{\boldsymbol{G}}^-(\boldsymbol{r},\boldsymbol{r}'))]\, ds(\boldsymbol{r}') \tag{2.74}$$

$$(\alpha(\boldsymbol{r}) - 1)\boldsymbol{H}^-(\boldsymbol{r}) = -\frac{ik^-}{4\pi}\int_S [(\nabla_{\boldsymbol{r}'} \times \boldsymbol{H}^-(\boldsymbol{r}')) \cdot (\hat{\boldsymbol{n}}' \times \widetilde{\boldsymbol{G}}^-(\boldsymbol{r},\boldsymbol{r}')) - (\hat{\boldsymbol{n}}' \times \boldsymbol{H}^-(\boldsymbol{r}')) \cdot (\nabla_{\boldsymbol{r}'} \times \widetilde{\boldsymbol{G}}^-(\boldsymbol{r},\boldsymbol{r}'))]\, ds(\boldsymbol{r}') \tag{2.75}$$

An alternative form for the interior representations (2.65) and (2.66) as well as the exterior representations (2.69) and (2.70) may be obtained by applying the curl operator to these representations and then using the Maxwell equations, resulting in, when $\boldsymbol{r} \notin S$

$$(\alpha(\boldsymbol{r}) - 1)\boldsymbol{E}^-(\boldsymbol{r}) = \frac{ik^-}{4\pi}\nabla \times \int_S G^-(\boldsymbol{r},\boldsymbol{r}')(\hat{\boldsymbol{n}}' \times \boldsymbol{E}^-(\boldsymbol{r}'))ds(\boldsymbol{r}') - \frac{Z^-}{4\pi}\nabla \times \left[\nabla \times \int_S G^-(\boldsymbol{r},\boldsymbol{r}')(\hat{\boldsymbol{n}}' \times \boldsymbol{H}^-(\boldsymbol{r}'))\, ds(\boldsymbol{r}')\right] \tag{2.76}$$

$$(\alpha(\boldsymbol{r}) - 1)\boldsymbol{H}^-(\boldsymbol{r}) = \frac{ik^-}{4\pi}\nabla \times \int_S G^-(\boldsymbol{r},\boldsymbol{r}')(\hat{\boldsymbol{n}}' \times \boldsymbol{H}^-(\boldsymbol{r}'))ds(\boldsymbol{r}') + \frac{Y^-}{4\pi}\nabla \times \left[\nabla \times \int_S G^-(\boldsymbol{r},\boldsymbol{r}')(\hat{\boldsymbol{n}}' \times \boldsymbol{E}^-(\boldsymbol{r}'))\ (\boldsymbol{r}')ds(\boldsymbol{r}')\right] \tag{2.77}$$

for the interior fields and

$$\alpha(\boldsymbol{r})\boldsymbol{E}^+(\boldsymbol{r}) = \boldsymbol{E}^i(\boldsymbol{r}) + \frac{ik}{4\pi}\nabla \times \int_S G^+(\boldsymbol{r},\boldsymbol{r}')(\hat{\boldsymbol{n}}' \times \boldsymbol{E}^+(\boldsymbol{r}'))\, ds(\boldsymbol{r}')$$

$$-\frac{Z^+}{4\pi}\nabla \times \left[\nabla \times \int_S G^+(\boldsymbol{r},\boldsymbol{r}')(\hat{\boldsymbol{n}}' \times \boldsymbol{H}^+(\boldsymbol{r}'))\, ds(\boldsymbol{r}')\right] \tag{2.78}$$

$$\alpha(\boldsymbol{r})\boldsymbol{H}^+(\boldsymbol{r}) = \boldsymbol{H}^i(\boldsymbol{r}) + \frac{ik}{4\pi}\nabla \times \int_S G^+(\boldsymbol{r},\boldsymbol{r}')(\hat{\boldsymbol{n}}' \times \boldsymbol{H}^+(\boldsymbol{r}'))\, ds(\boldsymbol{r}')$$

$$+\frac{Y^+}{4\pi}\nabla \times \left[\nabla \times \int_S G^+(\boldsymbol{r},\boldsymbol{r}')(\hat{\boldsymbol{n}}' \times \boldsymbol{E}^+(\boldsymbol{r}'))\, ds(\boldsymbol{r}')\right] \tag{2.79}$$

for the exterior fields. In the integrands in (2.78) and (2.79), $\boldsymbol{E}^+$ and $\boldsymbol{H}^+$ may be replaced with the scattered fields $\boldsymbol{E}$ and $\boldsymbol{H}$ in a consistent way when $\boldsymbol{r} \in V^+$.

When boundary conditions are introduced we obtain the following integral representations of the solutions of the basic problems in electromagnetics.

The perfect conductor problem:

$$\alpha(\boldsymbol{r})\boldsymbol{E}^+\boldsymbol{r}) = \boldsymbol{E}^i(\boldsymbol{r}) + \frac{ik}{4\pi}\int_S [ikZ^+G^+(\boldsymbol{r},\boldsymbol{r}')(\hat{\boldsymbol{n}}' \times \boldsymbol{H}^+(\boldsymbol{r}'))$$

$$+(\nabla_{r'}G^+(\boldsymbol{r},\boldsymbol{r}'))(\hat{\boldsymbol{n}}' \cdot \boldsymbol{E}^+(\boldsymbol{r}'))]\, ds(\boldsymbol{r}')$$

$$= \boldsymbol{E}^i(\boldsymbol{r}) + \frac{Z^+}{4\pi}\int_S [i\omega\rho_s(\boldsymbol{r}')(\nabla_{r'}G^+(\boldsymbol{r},\boldsymbol{r}'))$$

$$-k^2\boldsymbol{J}_s(\boldsymbol{r}')G^+(\boldsymbol{r},\boldsymbol{r}')]\, ds(\boldsymbol{r}') \tag{2.80}$$

$$\alpha(\boldsymbol{r})\boldsymbol{H}^+(\boldsymbol{r}) = \boldsymbol{H}^i(\boldsymbol{r}) - \frac{ik}{4\pi}\int_S (\nabla_{r'}G^+(\boldsymbol{r},\boldsymbol{r}')) \times (\hat{\boldsymbol{n}}' \times \boldsymbol{H}^+(\boldsymbol{r}'))\, ds(\boldsymbol{r}')$$

$$= \boldsymbol{H}^i(\boldsymbol{r}) - \frac{ik}{4\pi}\int_S (\nabla_{r'}G^+(\boldsymbol{r},\boldsymbol{r}')) \times \boldsymbol{J}_s(\boldsymbol{r}')\, ds(\boldsymbol{r}') \tag{2.81}$$

where ρ_s and $\boldsymbol{J}_s$ are the surface charge and current densities given by (1.138) and (1.139), respectively.

The impedance problem:

$$\alpha(\boldsymbol{r})\boldsymbol{E}^{+}(\boldsymbol{r}) = \boldsymbol{E}^{i}(\boldsymbol{r}) + \frac{ik}{4\pi}\int_{S}\Big[-\frac{ik}{Z_s}G^{+}(\boldsymbol{r},\boldsymbol{r}')(\hat{\boldsymbol{n}}'\times(\hat{\boldsymbol{n}}'\times\boldsymbol{E}^{+}(\boldsymbol{r}')))$$
$$+(\nabla_{\boldsymbol{r}'}G^{+}(\boldsymbol{r},\boldsymbol{r}'))(\hat{\boldsymbol{n}}'\cdot\boldsymbol{E}^{+}(\boldsymbol{r}'))$$
$$-(\nabla_{\boldsymbol{r}'}G^{+}(\boldsymbol{r},\boldsymbol{r}'))\times(\hat{\boldsymbol{n}}'\times\boldsymbol{E}^{+}(\boldsymbol{r}'))\Big]\,ds(\boldsymbol{r}') \tag{2.82}$$

$$\alpha(\boldsymbol{r})\boldsymbol{H}^{+}(\boldsymbol{r}) = \boldsymbol{H}^{i}(\boldsymbol{r}) + \frac{ik}{4\pi}\int_{S}[-ikZ_sG^{+}(\boldsymbol{r},\boldsymbol{r}')(\hat{\boldsymbol{n}}'\times(\hat{\boldsymbol{n}}'\times\boldsymbol{H}^{+}(\boldsymbol{r}')))$$
$$+(\nabla_{\boldsymbol{r}'}G^{+}(\boldsymbol{r},\boldsymbol{r}'))(\hat{\boldsymbol{n}}'\cdot\boldsymbol{H}^{+}(\boldsymbol{r}'))$$
$$-(\nabla_{\boldsymbol{r}'}G^{+}(\boldsymbol{r},\boldsymbol{r}'))\times(\hat{\boldsymbol{n}}'\times\boldsymbol{H}^{+}(\boldsymbol{r}'))\,ds(\boldsymbol{r}') \tag{2.83}$$

where Z_s denotes the surface impedance.

The transmission problem:

For the transmission problem we need all the equations (2.65) and (2.66), (2.69) and (2.70) for the interior and exterior fields, although the transmission conditions (1.142)–(1.147) reduce the number of unknown quantities on S, i.e. (2.69) and (2.70) may be rewritten as

$$\alpha(\boldsymbol{r})\boldsymbol{E}^{+}(\boldsymbol{r}) = \boldsymbol{E}^{i}(\boldsymbol{r}) + \frac{ik}{4\pi}\int_{S}[ikZ^{+}G^{+}(\boldsymbol{r},\boldsymbol{r}')(\hat{\boldsymbol{n}}'\times\boldsymbol{H}^{-}(\boldsymbol{r}'))$$
$$+\eta Y^{-}Z^{+}(\nabla_{\boldsymbol{r}'}G^{+}(\boldsymbol{r},\boldsymbol{r}'))(\hat{\boldsymbol{n}}'\cdot\boldsymbol{E}^{-}(\boldsymbol{r}'))$$
$$-(\nabla_{\boldsymbol{r}'}G^{+}(\boldsymbol{r},\boldsymbol{r}'))\times(\hat{\boldsymbol{n}}'\times\boldsymbol{E}^{-}(\boldsymbol{r}'))]\,ds(\boldsymbol{r}') \tag{2.84}$$

$$\alpha(\boldsymbol{r})\boldsymbol{H}^{+}(\boldsymbol{r}) = \boldsymbol{H}^{i}(\boldsymbol{r}) + \frac{ik}{4\pi}\int_{S}[-ikY^{+}G^{+}(\boldsymbol{r},\boldsymbol{r}')(\hat{\boldsymbol{n}}'\times\boldsymbol{E}^{-}(\boldsymbol{r}'))$$
$$+\eta Z^{-}Y^{+}(\nabla_{\boldsymbol{r}'}G^{+}(\boldsymbol{r},\boldsymbol{r}'))(\hat{\boldsymbol{n}}'\cdot\boldsymbol{H}^{-}(\boldsymbol{r}'))$$
$$-(\nabla_{\boldsymbol{r}'}G^{+}(\boldsymbol{r},\boldsymbol{r}'))\times(\hat{\boldsymbol{n}}'\times\boldsymbol{H}^{-}(\boldsymbol{r}'))]\,ds(\boldsymbol{r}') \tag{2.85}$$

2.EM.2 *The far field*

Introducing the asymptotic forms of the fundamental solution $G^{+}(\boldsymbol{r},\boldsymbol{r}')$, (2.14)–(2.16), into the representations (2.69) and (2.70) we obtain

$$\boldsymbol{E}(\boldsymbol{r}) = \boldsymbol{g}_e(\hat{\boldsymbol{r}})h(kr) + O\left(\frac{1}{r^2}\right), \quad r\to\infty, \quad \hat{\boldsymbol{r}}\in S^2 \tag{2.86}$$

$$\boldsymbol{H}(\boldsymbol{r}) = \boldsymbol{g}_m(\hat{\boldsymbol{r}})h(kr) + O\left(\frac{1}{r^2}\right), \quad r\to\infty, \quad \hat{\boldsymbol{r}}\in S^2 \tag{2.87}$$

where the **electric scattering amplitude** is given by

$$g_e(\hat{\boldsymbol{r}}) = \frac{k^2}{4\pi}\int_S [-Z^+(\hat{\boldsymbol{n}}' \times \boldsymbol{H}^+(\boldsymbol{r}')) + (\hat{\boldsymbol{n}}' \cdot \boldsymbol{E}^+(\boldsymbol{r}'))\hat{\boldsymbol{r}}$$
$$-\hat{\boldsymbol{r}} \times (\hat{\boldsymbol{n}}' \times \boldsymbol{E}^+(\boldsymbol{r}'))]e^{-ik\hat{\boldsymbol{r}}\cdot\boldsymbol{r}'}\, ds(\boldsymbol{r}') \tag{2.88}$$

and the **magnetic scattering amplitude** by

$$g_m(\hat{\boldsymbol{r}}) = \frac{k^2}{4\pi}\int_S [Y^+(\hat{\boldsymbol{n}}' \times \boldsymbol{E}^+(\boldsymbol{r}')) + (\hat{\boldsymbol{n}}' \cdot \boldsymbol{H}^+(\boldsymbol{r}'))\hat{\boldsymbol{r}}$$
$$-\hat{\boldsymbol{r}} \times (\hat{\boldsymbol{n}}' \times \boldsymbol{H}^+(\boldsymbol{r}'))]e^{-ik\hat{\boldsymbol{r}}\cdot\boldsymbol{r}'}\, ds(\boldsymbol{r}') \tag{2.89}$$

The structure of the electromagnetic far fields is similar to that in acoustics with a radial dependence common to all scatterers and all the obstacle information contained in the scattering amplitudes.

If we consider the asymptotic form of (2.78) and (2.79), then the scattering amplitudes assume the forms

$$g_e(\hat{\boldsymbol{r}}) = -\frac{k^2}{4\pi}\hat{\boldsymbol{r}} \times \int_S (\hat{\boldsymbol{n}}' \times \boldsymbol{E}^+(\boldsymbol{r}'))e^{-ik\hat{\boldsymbol{r}}\cdot\boldsymbol{r}'}\, ds(\boldsymbol{r}')$$
$$+\frac{k^2Z^+}{4\pi}\hat{\boldsymbol{r}} \times \left[\hat{\boldsymbol{r}} \times \int_S (\hat{\boldsymbol{n}}' \times \boldsymbol{H}^+(\boldsymbol{r}'))e^{-ik\hat{\boldsymbol{r}}\cdot\boldsymbol{r}'}\, ds(\boldsymbol{r}')\right] \tag{2.90}$$

$$g_m(\hat{\boldsymbol{r}}) = -\frac{k^2}{4\pi}\hat{\boldsymbol{r}} \times \int_S (\hat{\boldsymbol{n}}' \times \boldsymbol{H}^+(\boldsymbol{r}'))e^{-ik\hat{\boldsymbol{r}}\cdot\boldsymbol{r}'}\, ds(\boldsymbol{r}')$$
$$-\frac{k^2Y^+}{4\pi}\hat{\boldsymbol{r}} \times \left[\hat{\boldsymbol{r}} \times \int_S (\hat{\boldsymbol{n}}' \times \boldsymbol{E}^+(\boldsymbol{r}'))e^{-ik\hat{\boldsymbol{r}}\cdot\boldsymbol{r}'}\, ds(\boldsymbol{r}')\right] \tag{2.91}$$

In this form it is easily seen that

$$\hat{\boldsymbol{r}} \cdot g_e(\hat{\boldsymbol{r}}) = \hat{\boldsymbol{r}} \cdot g_m(\hat{\boldsymbol{r}}) = 0 \tag{2.92}$$

that is, the scattering amplitudes have no radial components. Moreover, it is a straightforward exercise in vector algebra to show that

$$g_e(\hat{\boldsymbol{r}}) = -Z^+\hat{\boldsymbol{r}} \times g_m(\hat{\boldsymbol{r}}), \quad g_m(\hat{\boldsymbol{r}}) = Y^+\hat{\boldsymbol{r}} \times g_e(\hat{\boldsymbol{r}}) \tag{2.93}$$

which again demonstrates the tangential nature of the scattering amplitudes.

If we explicitly exhibit the fact that the scattering amplitudes depend not only on $\hat{\boldsymbol{r}}$ but also on the polarizations of the incident waves $\hat{\boldsymbol{a}}$ and $\hat{\boldsymbol{b}}$ and the direction of incidence $\hat{\boldsymbol{k}}$, where $\hat{\boldsymbol{a}} \times \hat{\boldsymbol{b}} = \hat{\boldsymbol{k}}$, $\hat{\boldsymbol{a}} \cdot \hat{\boldsymbol{b}} = 0$, then (2.93) can be written as

$$\boldsymbol{g}_e(\hat{\boldsymbol{r}}; \hat{\boldsymbol{k}}, \hat{\boldsymbol{a}}) = -Z^+ \hat{\boldsymbol{r}} \times \boldsymbol{g}_m(\hat{\boldsymbol{r}}; \hat{\boldsymbol{k}}, \hat{\boldsymbol{b}}) \tag{2.94a}$$

$$\boldsymbol{g}_m(\hat{\boldsymbol{r}}; \hat{\boldsymbol{k}}, \hat{\boldsymbol{b}}) = Y^+ \hat{\boldsymbol{r}} \times \boldsymbol{g}_e(\hat{\boldsymbol{r}}; \hat{\boldsymbol{k}}, \hat{\boldsymbol{a}}) \tag{2.94b}$$

A more convenient form of the scattering amplitudes for low frequency purposes is the following (Kleinman and Senior 1986)

$$\begin{aligned}\boldsymbol{g}_e(\hat{\boldsymbol{r}}) = &-\frac{ik^3}{4\pi}\hat{\boldsymbol{r}} \times \left[\hat{\boldsymbol{r}} \times \int_S [-Z^+ \hat{\boldsymbol{r}} \cdot (\hat{\boldsymbol{n}}' \times \boldsymbol{H}^+(\boldsymbol{r}'))\right.\\ &\left. + \hat{\boldsymbol{n}}' \cdot \boldsymbol{E}^+(\boldsymbol{r}')]\boldsymbol{r}' e^{-ik\hat{\boldsymbol{r}}\cdot\boldsymbol{r}'}\, ds(\boldsymbol{r}')\right]\\ &-\frac{ik^3}{4\pi}\hat{\boldsymbol{r}} \times \int_S [\hat{\boldsymbol{r}} \cdot (\hat{\boldsymbol{n}}' \times (\boldsymbol{E}^+(\boldsymbol{r}'))\\ &+ Z^+ \hat{\boldsymbol{n}}' \cdot \boldsymbol{H}^+(\boldsymbol{r}')]\boldsymbol{r}' e^{-ik\hat{\boldsymbol{r}}\cdot\boldsymbol{r}'}\, ds(\boldsymbol{r}')\end{aligned} \tag{2.95}$$

and

$$\begin{aligned}\boldsymbol{g}_m(\hat{\boldsymbol{r}}) = &-\frac{ik^3}{4\pi}\hat{\boldsymbol{r}} \times \left[\hat{\boldsymbol{r}} \times \int_S [Y^+ \hat{\boldsymbol{r}} \cdot (\hat{\boldsymbol{n}}' \times \boldsymbol{E}^+(\boldsymbol{r}'))\right.\\ &\left. + \hat{\boldsymbol{n}}' \cdot \boldsymbol{H}^+(\boldsymbol{r}')]\boldsymbol{r}' e^{-ik\hat{\boldsymbol{r}}\cdot\boldsymbol{r}'}\, ds(\boldsymbol{r}')\right]\\ &-\frac{ik^3}{4\pi}\hat{\boldsymbol{r}} \times \int_S [\hat{\boldsymbol{r}} \cdot (\hat{\boldsymbol{n}}' \times \boldsymbol{H}^+(\boldsymbol{r}'))\\ &- Y^+ \hat{\boldsymbol{n}}' \cdot \boldsymbol{E}^+(\boldsymbol{r}')\big]\boldsymbol{r}' e^{-ik\hat{\boldsymbol{r}}\cdot\boldsymbol{r}'}\, ds(\boldsymbol{r}')\end{aligned} \tag{2.96}$$

The dyadic representations (2.72) and (2.73) provide still another representation for the scattering amplitudes. By virtue of the asymptotic forms

$$\widetilde{\boldsymbol{G}}^+(\boldsymbol{r}, \boldsymbol{r}') = (\widetilde{\boldsymbol{I}} - \hat{\boldsymbol{r}}\hat{\boldsymbol{r}})e^{-ik\hat{\boldsymbol{r}}\cdot\boldsymbol{r}'} h(kr) + O\left(\frac{1}{r^2}\right) \tag{2.97}$$

$$\nabla_{\boldsymbol{r}'} \times \widetilde{\boldsymbol{G}}^+(\boldsymbol{r}, \boldsymbol{r}') = -ik\hat{\boldsymbol{r}} \times (\widetilde{\boldsymbol{I}} - \hat{\boldsymbol{r}}\hat{\boldsymbol{r}})e^{-ik\hat{\boldsymbol{r}}\cdot\boldsymbol{r}'} h(kr) + O\left(\frac{1}{r^2}\right) \tag{2.98}$$

as $r \to \infty$, we obtain

$$\begin{aligned}\boldsymbol{g}_e(\hat{\boldsymbol{r}}) = &\frac{ik}{4\pi}(\widetilde{\boldsymbol{I}} - \hat{\boldsymbol{r}}\hat{\boldsymbol{r}}) \cdot \int_S [\hat{\boldsymbol{n}}' \times (\nabla \times \boldsymbol{E}^+(\boldsymbol{r}'))\\ &+ ik\hat{\boldsymbol{r}} \times (\hat{\boldsymbol{n}}' \times \boldsymbol{E}^+(\boldsymbol{r}'))]e^{-ik\hat{\boldsymbol{r}}\cdot\boldsymbol{r}'}\, ds(\boldsymbol{r}')\end{aligned} \tag{2.99}$$

and

$$g_m(\hat{r}) = \frac{ik}{4\pi}(\widetilde{I} - \hat{r}\hat{r}) \cdot \int_S [\hat{n}' \times (\nabla \times H^+(r')) + ik\hat{r} \times (\hat{r}' \times H^+(r'))] e^{-ik\hat{r}\cdot r'} \, ds(r') \tag{2.100}$$

Of course, all these expressions for g_e and g_m will be simplified if boundary conditions are taken into account.

2.EM.3 *The far field expansion theorem*

The asymptotic expressions (2.86) and (2.87) represent the first terms in the convergent series

$$E^+(r) = h(kr) \sum_{n=0}^{\infty} F_n^e(\hat{r}) r^{-n} \tag{2.101}$$

$$H^+(r) = h(kr) \sum_{n=0}^{\infty} F_n^m(\hat{r}) r^{-n} \tag{2.102}$$

where the convergence is absolute and uniform outside the smallest sphere, $B(\mathbf{0}, a)$, that contains S in its interior (Wilcox 1956*b*).

As in the acoustic case, all the coefficients F_n^e, $n = 1, 2, \ldots$ can be evaluated once F_0^e is known. This is done by using the formulae,

$$\left.\begin{aligned} ik\hat{r} \cdot F_1^e &= -\left[\hat{\theta}\frac{\partial}{\partial\theta} + \frac{\hat{\varphi}}{\sin\theta}\frac{\partial}{\partial\varphi}\right] \cdot F_0^e \\ 2ikn\hat{r} \cdot F_{n+1}^e &= [\mathbb{B} + n(n-1)]\hat{r} \cdot F_n^e \end{aligned}\right\} \tag{2.103}$$

$$2ikn\hat{\theta} \cdot F_n^e = [\mathbb{B} + n(n-1)]\hat{\theta} \cdot F_{n-1}^e + D_\theta \cdot F_{n-1}^e \tag{2.104}$$

$$2ikn\hat{\varphi} \cdot F_n^e = [\mathbb{B} + n(n-1)]\hat{\varphi} \cdot F_{n-1}^e + D_\varphi \cdot F_{n-1}^e \tag{2.105}$$

for $n = 1, 2, 3, \ldots$, where $\mathbb{B}$ is the Beltrami operator (2.22) and the operators D_θ and D_φ are defined by

$$D_\theta \cdot f = 2\frac{\partial}{\partial\theta}(\hat{r} \cdot f) - \frac{1}{\sin^2\theta}(\hat{\theta} \cdot f) - 2\frac{\cos\theta}{\sin^2\theta}\frac{\partial}{\partial\varphi}(\hat{\varphi} \cdot f) \tag{2.106}$$

$$D_\varphi \cdot f = \frac{2}{\sin\theta}\frac{\partial}{\partial\varphi}(\hat{r} \cdot f) + 2\frac{\cos\theta}{\sin^2\theta}\frac{\partial}{\partial\varphi}(\hat{\theta} \cdot f) - \frac{1}{\sin^2\theta}(\hat{\varphi} \cdot f) \tag{2.107}$$

The expansion for H^+ can be obtained from (2.101) via the Maxwell equations.

In (2.103) it is necessary to consider different forms for the recurrence formula involving $\hat{r} \cdot F_n^e$ for $n = 1$ and $n > 1$ because $\hat{r} \cdot F_0^e = 0$. Note also that

$$F_0^e = g_e, \quad F_0^m = g_m \tag{2.108}$$

so that once the electric scattering amplitude is known, both the electric and magnetic fields can be evaluated at all points r with $r > a$.

2.EM.4 *Cross sections*

As in acoustics we now consider power flux in the **radiation zone** or **far field** where the medium is lossless. Recall that this quantity was defined for electromagnetic waves in (1.181) as

$$\boldsymbol{S}(\boldsymbol{r}) = \operatorname{Re}\{\boldsymbol{E}(\boldsymbol{r}) \times \boldsymbol{H}^*(\boldsymbol{r})\} \tag{2.109}$$

For the incident field

$$\boldsymbol{E}^i(\boldsymbol{r}) = \hat{\boldsymbol{a}} e^{ik\hat{\boldsymbol{k}}\cdot\boldsymbol{r}}, \quad \boldsymbol{H}^i(\boldsymbol{r}) = \hat{\boldsymbol{b}} Y^+ e^{ik\hat{\boldsymbol{k}}\cdot\boldsymbol{r}}, \quad \boldsymbol{r} \in \mathbb{R}^3 \tag{2.110}$$

the flux in the direction of propagation is given by

$$\hat{\boldsymbol{k}} \cdot \boldsymbol{S}^i = \hat{\boldsymbol{k}} \cdot (Y^+ \hat{\boldsymbol{k}}) = Y^+ \tag{2.111}$$

In the far field, the asymptotic expressions (2.86) and (2.87) for the scattered wave enable us to write the flux as

$$\begin{aligned}
\hat{\boldsymbol{r}} \cdot \boldsymbol{S}(\boldsymbol{r}) &= \hat{\boldsymbol{r}} \cdot \operatorname{Re}\{\boldsymbol{E} \times \boldsymbol{H}^*\} \\
&= \hat{\boldsymbol{r}} \cdot \operatorname{Re}\{(\boldsymbol{g}_e h) \times (\boldsymbol{g}_m^* h^*)\} + O\left(\frac{1}{r^3}\right) \\
&= \hat{\boldsymbol{r}} \cdot \operatorname{Re}\left\{\frac{1}{(kr)^2} \boldsymbol{g}_e \times (Y^+ \hat{\boldsymbol{r}} \times \boldsymbol{g}_e^*)\right\} + O\left(\frac{1}{r^3}\right) \\
&= \frac{Y^+}{k^2 r^2} \hat{\boldsymbol{r}} \cdot \{|\boldsymbol{g}_e|^2 \hat{\boldsymbol{r}} - (\boldsymbol{g}_e \cdot \hat{\boldsymbol{r}}) \boldsymbol{g}_e^*\} + O\left(\frac{1}{r^3}\right) \\
&= \frac{Y^+}{k^2 r^2} |\boldsymbol{g}_e|^2 + O\left(\frac{1}{r^3}\right), \quad r \to \infty
\end{aligned} \tag{2.112}$$

where we have used the fact that $\boldsymbol{g}_e$ is tangential. Now, as in acoustics, we define the **differential scattering cross section** or **radar cross section** to be the power scattered in the direction $\hat{\boldsymbol{r}}$ relative to the incident power flux in the direction of incidence and given explicitly by

$$\sigma(\hat{\boldsymbol{r}}) = \lim_{r\to\infty} \frac{4\pi r^2 \hat{\boldsymbol{r}} \cdot \boldsymbol{S}(\boldsymbol{r})}{\hat{\boldsymbol{k}} \cdot \boldsymbol{S}^i(\boldsymbol{r})} = \frac{4\pi}{k^2} |\boldsymbol{g}_e(\hat{\boldsymbol{r}})|^2 \tag{2.113}$$

which can also be obtained from the formula

$$\sigma(\hat{\boldsymbol{r}}) = \lim_{r\to\infty} 4\pi r^2 \frac{|\boldsymbol{E}(\boldsymbol{r})|^2}{|\boldsymbol{E}^i(\boldsymbol{r})|^2} = \frac{4\pi}{k^2} |\boldsymbol{g}_e(\hat{\boldsymbol{r}})|^2 \tag{2.114}$$

The **scattering cross section** or **total cross section** in electromagnetics is defined, as in acoustics, as the value of $\sigma(\hat{\boldsymbol{r}})$ averaged over all directions, i.e.

$$\begin{aligned}
\sigma_s &= \frac{1}{4\pi} \int_{S^2} \sigma(\hat{\boldsymbol{r}})\, ds(\hat{\boldsymbol{r}}) = \frac{1}{k^2} \int_{S^2} |\boldsymbol{g}_e|^2\, ds \\
&= \int_{\partial B(\boldsymbol{0},\infty)} |\boldsymbol{E}|^2\, ds = Z^+ \operatorname{Re} \int_{\partial B(\boldsymbol{0},\infty)} \hat{\boldsymbol{n}} \cdot \boldsymbol{E} \times \boldsymbol{H}^*\, ds
\end{aligned} \tag{2.115}$$

The last expression can be used to transfer the integral over $\partial B(\mathbf{0}, \infty)$ to a surface integral over the scatterer S. This is easily done with the help of the identity (2.62). Therefore,

$$\sigma_s = Z^+ \operatorname{Re} \int_S \hat{\boldsymbol{n}} \cdot \boldsymbol{E} \times \boldsymbol{H}^* \, ds \tag{2.116}$$

A similar expression involving total rather than scattered fields is used to define the **absorption cross section** as

$$\begin{aligned}\sigma_a &= -Z^+ \operatorname{Re} \int_S \hat{\boldsymbol{n}} \cdot (\boldsymbol{E}^+ \times \boldsymbol{H}^{+^*}) \, ds \\ &= Z^+ \operatorname{Re} \int_S \hat{\boldsymbol{n}} \cdot (\boldsymbol{H}^+ \times \boldsymbol{E}^{+^*}) \, ds \end{aligned} \tag{2.117}$$

For a perfect conductor where $\hat{\boldsymbol{n}} \times \boldsymbol{E}^+ = \mathbf{0}$ it follows that $\sigma_a = 0$. Therefore, as expected, a perfect conductor does not absorb energy.

For an impedance surface, where

$$\hat{\boldsymbol{n}} \times \boldsymbol{H}^+ = -\frac{1}{Z_s Z^+} \hat{\boldsymbol{n}} \times (\hat{\boldsymbol{n}} \times \boldsymbol{E}^+) \tag{2.118}$$

we have

$$\begin{aligned}\sigma_a &= Z^+ \operatorname{Re} \int_S (\hat{\boldsymbol{n}} \times \boldsymbol{H}^+) \cdot \boldsymbol{E}^{+*} \, ds \\ &= -\operatorname{Re} \int_S \frac{1}{Z_s} [\hat{\boldsymbol{n}} \times (\hat{\boldsymbol{n}} \times \boldsymbol{E}^+)] \cdot \boldsymbol{E}^{+*} \, ds \\ &= \operatorname{Re} \int_S \frac{1}{Z_s} (|\boldsymbol{E}^+|^2 - |\hat{\boldsymbol{n}} \cdot \boldsymbol{E}^+|^2) \, ds \\ &= \int_S \frac{\operatorname{Re} Z_s}{|Z_s|^2} |\boldsymbol{E}_t^+|^2 \, ds \end{aligned} \tag{2.119}$$

where

$$\boldsymbol{E}_t^+ = (\widetilde{\boldsymbol{I}} - \hat{\boldsymbol{n}}\hat{\boldsymbol{n}}) \cdot \boldsymbol{E}^+ \tag{2.120}$$

For constant real impedance

$$\sigma_a = \frac{1}{Z_s} \int_S |\boldsymbol{E}_t^+|^2 \, ds \tag{2.121}$$

whereas for purely imaginary impedance, $\sigma_a = 0$.

In the case of penetrable scatterers we use the transmission conditions and the Maxwell equations to write

$$\sigma_a = Z^+ \operatorname{Re} \int_S \hat{\boldsymbol{n}} \cdot (\boldsymbol{H}^- \times \boldsymbol{E}^{-*})\, ds$$
$$= Z^+ \operatorname{Re} \int_S \frac{\hat{\boldsymbol{n}}}{ik\eta Z^-} \cdot [(\nabla \times \boldsymbol{E}^-) \times \boldsymbol{E}^{-*}]\, ds \qquad (2.122)$$

But k is real as is ηZ^-, hence

$$\sigma_a = \frac{Z^+}{k\eta Z^-} \operatorname{Im} \int_S \hat{\boldsymbol{n}} \cdot [(\nabla \times \boldsymbol{E}^-) \times \boldsymbol{E}^{-*}]\, ds \qquad (2.123)$$

Using the identity (2.62) we transform this into an integral over V^- and then use the Maxwell equations and the constitutive relations to obtain

$$\sigma_a = \frac{Z^+}{k\eta Z^-} \operatorname{Im} \int_{V^-} \boldsymbol{E}^- \cdot \nabla \times (\nabla \times \boldsymbol{E}^{-*})\, dv$$
$$= \frac{kZ^+}{\eta Z^-} (\operatorname{Im} \eta^2) \int_{V^-} |\boldsymbol{E}^-|^2\, dv$$
$$= \sigma^- Z^+ \int_{V^-} |\boldsymbol{E}^-|^2\, dv \qquad (2.124)$$

Hence, $\sigma_a = 0$ for lossless penetrable scatterers.

Exactly as in the acoustic case the **extinction cross section** is defined by

$$\sigma_e = \sigma_s + \sigma_a \qquad (2.125)$$

and measures, in units of area, the total energy that the scatterer removes from the incident wave either by scattering in all directions, (σ_s), or by absorption, (σ_a).

2.EM.5 *The basic scattering theorems*

We develop here the basic scattering theorems in electromagnetics which are counterparts of the similar theorems in acoustics. The first results concern reciprocity.

The reciprocity theorems:

1. **Point sources:** Denote by $\boldsymbol{E}_j(\boldsymbol{r})$ and $\boldsymbol{H}_j(\boldsymbol{r})$ the scattered fields at $\boldsymbol{r}$ due to an electric dipole (see Section 1.EM.5) at $\boldsymbol{r}_j$ with the moment $\hat{\boldsymbol{p}}_j$, $j = 1, 2$. Then the reciprocity theorem states that the component of the fields at $\boldsymbol{r}_1$ in the direction $\hat{\boldsymbol{p}}_1$ due to an electric dipole at $\boldsymbol{r}_2$ with moment $\hat{\boldsymbol{p}}_2$ is the same as the component of the fields at $\boldsymbol{r}_2$ in the direction $\hat{\boldsymbol{p}}_2$ due to an electric dipole at $\boldsymbol{r}_1$ with moment $\hat{\boldsymbol{p}}_1$, i.e.

$$\hat{\boldsymbol{p}}_1 \cdot \boldsymbol{E}_2(\boldsymbol{r}_1) = \hat{\boldsymbol{p}}_2 \cdot \boldsymbol{E}_1(\boldsymbol{r}_2), \quad \hat{\boldsymbol{p}}_1 \cdot \boldsymbol{H}_2(\boldsymbol{r}_1) = \hat{\boldsymbol{p}}_2 \cdot \boldsymbol{H}_1(\boldsymbol{r}_2) \qquad (2.126)$$

A similar result holds for magnetic dipole sources.

2. **Plane waves:** The reciprocity theorem takes a slightly different form. If we denote the electric scattering amplitude due to a plane wave, $\boldsymbol{E}^i = \hat{\boldsymbol{a}}_j e^{ik\hat{\boldsymbol{k}}_j\cdot\boldsymbol{r}}$, $j = 1, 2$, by $\boldsymbol{g}_e(\hat{\boldsymbol{r}}; \hat{\boldsymbol{k}}_j, \hat{\boldsymbol{a}}_j)$ then

$$\hat{\boldsymbol{a}}_1 \cdot \boldsymbol{g}_e(-\hat{\boldsymbol{k}}_1; \hat{\boldsymbol{k}}_2, \hat{\boldsymbol{a}}_2) = \hat{\boldsymbol{a}}_2 \cdot \boldsymbol{g}_e(-\hat{\boldsymbol{k}}_2; \hat{\boldsymbol{k}}_1, \hat{\boldsymbol{a}}_1) \tag{2.127}$$

Moreover, if we set $-\hat{\boldsymbol{k}}_1 = \hat{\boldsymbol{r}}$ and $\hat{\boldsymbol{k}}_2 = \hat{\boldsymbol{k}}$, we find that

$$\hat{\boldsymbol{a}}_1 \cdot \boldsymbol{g}_e(\hat{\boldsymbol{r}}; \hat{\boldsymbol{k}}, \hat{\boldsymbol{a}}_2) = \hat{\boldsymbol{a}}_2 \cdot \boldsymbol{g}_e(-\hat{\boldsymbol{k}}; -\hat{\boldsymbol{r}}, \hat{\boldsymbol{a}}_1) \tag{2.128}$$

The derivation of these reciprocity relations is analogous to the corresponding result in acoustics. It relies on the following identity. If $\boldsymbol{E}_j^+(\boldsymbol{r}')$ and $\boldsymbol{H}_j^+(\boldsymbol{r}')$ denote the total fields at $\boldsymbol{r}'$ due to an incident field $\boldsymbol{E}_j^i(\boldsymbol{r}')$, $j = 1, 2$, where $\boldsymbol{E}_j^i$ may be either plane waves or point sources, then

$$\int_S [\boldsymbol{E}_1^+ \times (\nabla \times \boldsymbol{E}_2^+) - \boldsymbol{E}_2^+ \times (\nabla \times \boldsymbol{E}_1^+)] \cdot \hat{\boldsymbol{n}}\, ds = 0 \tag{2.129}$$

The validity of this identity follows by applying either the boundary conditions or transmission conditions together with the vector identity (2.62). Moreover, this same identity may be used to show that

$$\int_S [\boldsymbol{E}_1^i \times (\nabla \times \boldsymbol{E}_2^i) - \boldsymbol{E}_2^i \times (\nabla \times \boldsymbol{E}_1^i)] \cdot \hat{\boldsymbol{n}}\, ds = 0 \tag{2.130}$$

and

$$\int_S [\boldsymbol{E}_1 \times (\nabla \times \boldsymbol{E}_2) - \boldsymbol{E}_2 \times (\nabla \times \boldsymbol{E}_1)] \cdot \hat{\boldsymbol{n}}\, ds = 0 \tag{2.131}$$

The Silver–Müller radiation condition is also needed to establish the relation (2.131) involving the scattered fields. Using these two relations together with (2.129) leads to the identity

$$\int_S [\boldsymbol{E}_1^i \times (\nabla \times \boldsymbol{E}_2) - \boldsymbol{E}_2 \times (\nabla \times \boldsymbol{E}_1^i)] \cdot \hat{\boldsymbol{n}}\, ds$$
$$= \int_S [\boldsymbol{E}_2^i \times (\nabla \times \boldsymbol{E}_1) - \boldsymbol{E}_1 \times (\nabla \times \boldsymbol{E}_2^i)] \cdot \hat{\boldsymbol{n}}\, ds \tag{2.132}$$

Choosing $\boldsymbol{E}_j^i$ to be the fields due to electric dipoles (1.154) at points $\boldsymbol{r}_j$ with dipole moments $\hat{\boldsymbol{p}}_j$, $j = 1, 2$, equation (2.132), together with the representation (2.72), the Maxwell equations and a little vector manipulation, establishes the reciprocity theorem for point sources.

To establish the reciprocity theorem (2.127) for the scattering amplitudes, first observe that the definition (2.90) of the scattering amplitudes for incident plane waves, $\boldsymbol{E}^i_j = \boldsymbol{a}_j e^{ik\hat{\boldsymbol{k}}_j\cdot\boldsymbol{r}}$ in the direction of $-\hat{\boldsymbol{r}}$ may be written as

$$g_e(-\hat{\boldsymbol{r}};\hat{\boldsymbol{k}}_j,\hat{\boldsymbol{a}}_j) = \frac{k^2}{4\pi}\int_S [\hat{\boldsymbol{r}}\times(\hat{\boldsymbol{n}}'\times\boldsymbol{E}_j(\boldsymbol{r}')) \\ +Z^+\hat{\boldsymbol{r}}\times(\hat{\boldsymbol{r}}\times(\hat{\boldsymbol{n}}'\times\boldsymbol{H}_j(\boldsymbol{r}')))]e^{ik\hat{\boldsymbol{r}}\cdot\boldsymbol{r}'}ds(\boldsymbol{r}') \tag{2.133}$$

This relation, first with $\boldsymbol{r} = \hat{\boldsymbol{k}}_2$ and $j = 1$, then $\hat{\boldsymbol{r}} = \hat{\boldsymbol{k}}_1$ and $j = 2$, together with (2.132), the Maxwell equations and a little vector algebra leads to the reciprocity relation (2.127).

Note that with equation (2.93) relating electric and magnetic scattering amplitudes we may infer that

$$\hat{\boldsymbol{a}}_1\cdot[\hat{\boldsymbol{k}}_1\times g_m(-\hat{\boldsymbol{k}}_1;\hat{\boldsymbol{k}}_2,\hat{\boldsymbol{b}}_2)] = \hat{\boldsymbol{a}}_2\cdot[\hat{\boldsymbol{k}}_2\times g_m(-\hat{\boldsymbol{k}}_2;\hat{\boldsymbol{k}}_1,\hat{\boldsymbol{b}}_1)] \tag{2.134}$$

or, since $\hat{\boldsymbol{k}}_j\times\hat{\boldsymbol{a}}_j = \hat{\boldsymbol{b}}_j$

$$\hat{\boldsymbol{b}}_1\cdot g_m(-\hat{\boldsymbol{k}}_1;\hat{\boldsymbol{k}}_2,\hat{\boldsymbol{b}}_2) = \hat{\boldsymbol{b}}_2\cdot g_m(-\hat{\boldsymbol{k}}_2;\hat{\boldsymbol{k}}_1,\hat{\boldsymbol{b}}_1) \tag{2.135}$$

and this completes the proof of the theorem.

The general scattering theorem: Again, assume plane wave incidence in the directions $\hat{\boldsymbol{k}}_j$, which give rise to the corresponding fields $\boldsymbol{E}^+(\boldsymbol{r};\hat{\boldsymbol{k}}_j,\hat{\boldsymbol{a}}_j), \boldsymbol{H}^+(\boldsymbol{r};\hat{\boldsymbol{k}}_j,\hat{\boldsymbol{b}}_j),\ j = 1, 2$, where $\hat{\boldsymbol{a}}_j, \hat{\boldsymbol{b}}_j$ denote the polarization vectors of $\boldsymbol{E}^i_j$ and $\boldsymbol{H}^i_j$, respectively. Then by rewriting $\boldsymbol{E}^+$ as $\boldsymbol{E}^i + \boldsymbol{E}$ we obtain the following identity

$$\begin{aligned}
&\int_S [\boldsymbol{E}^{+^*}(\boldsymbol{r};\hat{\boldsymbol{k}}_1,\hat{\boldsymbol{a}}_1)\times(\nabla\times\boldsymbol{E}^+(\boldsymbol{r};\hat{\boldsymbol{k}}_2,\hat{\boldsymbol{a}}_2)) \\
&\quad -\boldsymbol{E}^+(\boldsymbol{r};\hat{\boldsymbol{k}}_2,\hat{\boldsymbol{a}}_2)\times(\nabla\times\boldsymbol{E}^{+^*}(\boldsymbol{r};\hat{\boldsymbol{k}}_1,\hat{\boldsymbol{a}}_1))]\cdot\hat{\boldsymbol{n}}\,ds(\boldsymbol{r}) \\
&\quad = \int_S [\boldsymbol{E}^{i^*}(\boldsymbol{r};\hat{\boldsymbol{k}}_1,\hat{\boldsymbol{a}}_1)\times(\nabla\times\boldsymbol{E}^i(\boldsymbol{r};\hat{\boldsymbol{k}}_2,\hat{\boldsymbol{a}}_2)) \\
&\quad\quad -\boldsymbol{E}^i(\boldsymbol{r};\hat{\boldsymbol{k}}_2,\hat{\boldsymbol{a}}_2)\times(\nabla\times\boldsymbol{E}^{i^*}(\boldsymbol{r};\hat{\boldsymbol{k}}_1,\hat{\boldsymbol{a}}_1))]\cdot\hat{\boldsymbol{n}}\,ds(\boldsymbol{r}) \\
&\quad\quad +\int_S [\boldsymbol{E}^{i^*}(\boldsymbol{r};\hat{\boldsymbol{k}}_1,\hat{\boldsymbol{a}}_1)\times(\nabla\times\boldsymbol{E}(\boldsymbol{r};\hat{\boldsymbol{k}}_2,\hat{\boldsymbol{a}}_2)) \\
&\quad\quad -\boldsymbol{E}(\boldsymbol{r};\hat{\boldsymbol{k}}_2,\hat{\boldsymbol{a}}_2)\times(\nabla\times\boldsymbol{E}^{i^*}(\boldsymbol{r};\hat{\boldsymbol{k}}_1,\hat{\boldsymbol{a}}_1))]\cdot\hat{\boldsymbol{n}}\,ds(\boldsymbol{r}) \\
&\quad\quad +\int_S [\boldsymbol{E}^*(\boldsymbol{r};\hat{\boldsymbol{k}}_1,\hat{\boldsymbol{a}}_1)\times(\nabla\times\boldsymbol{E}^i(\boldsymbol{r};\hat{\boldsymbol{k}}_2,\hat{\boldsymbol{a}}_2)) \\
&\quad\quad -\boldsymbol{E}^i(\boldsymbol{r};\hat{\boldsymbol{k}}_2,\hat{\boldsymbol{a}}_2)\times(\nabla\times\boldsymbol{E}^*(\boldsymbol{r};\hat{\boldsymbol{k}}_1,\hat{\boldsymbol{a}}_1))]\cdot\hat{\boldsymbol{n}}\,ds(\boldsymbol{r}) \\
&\quad\quad +\int_S [\boldsymbol{E}^*(\boldsymbol{r};\hat{\boldsymbol{k}}_1,\hat{\boldsymbol{a}}_1)\times(\nabla\times\boldsymbol{E}(\boldsymbol{r};\hat{\boldsymbol{k}}_2,\hat{\boldsymbol{a}}_2)) \\
&\quad\quad -\boldsymbol{E}(\boldsymbol{r};\hat{\boldsymbol{k}}_2,\hat{\boldsymbol{a}}_2)\times(\nabla\times\boldsymbol{E}^*(\boldsymbol{r};\hat{\boldsymbol{k}}_1,\hat{\boldsymbol{a}}_1))]\cdot\hat{\boldsymbol{n}}\,ds(\boldsymbol{r})
\end{aligned} \tag{2.136}$$

The first integral on the right vanishes by virtue of the identity (2.62) applied in V^- and the vector Helmholtz equation for $\boldsymbol{E}^i$. The same procedure in V^+ allows us to transform the last integral on the right to an integral over $\partial B(\mathbf{0}, \infty)$ where the asymptotic forms of $\boldsymbol{E}$ and $\boldsymbol{H}$ may be employed. Using these facts and the Maxwell equations we may rewrite (2.136) as follows

$$\begin{aligned}
&\int_S [\boldsymbol{E}^{+^*}(\boldsymbol{r}; \hat{\boldsymbol{k}}_1, \hat{\boldsymbol{a}}_1) \times Z^+\boldsymbol{H}^+(\boldsymbol{r}; \hat{\boldsymbol{k}}_2, \hat{\boldsymbol{b}}_2) \\
&\quad + \boldsymbol{E}^+(\boldsymbol{r}; \hat{\boldsymbol{k}}_2, \hat{\boldsymbol{a}}_2) \times Z^+\boldsymbol{H}^{+^*}(\boldsymbol{r}; \hat{\boldsymbol{k}}_1, \hat{\boldsymbol{b}}_1)] \cdot \hat{\boldsymbol{n}}\, ds(\boldsymbol{r}) \\
&\quad = \int_S [Z^+\hat{\boldsymbol{a}}_1 \times \boldsymbol{H}(\boldsymbol{r}; \hat{\boldsymbol{k}}_2, \hat{\boldsymbol{b}}_2) \\
&\qquad - \hat{\boldsymbol{b}}_1 \times \boldsymbol{E}(\boldsymbol{r}; \hat{\boldsymbol{k}}_2, \hat{\boldsymbol{a}}_2)] \cdot \hat{\boldsymbol{n}} e^{-ik\hat{\boldsymbol{k}}_1\cdot\boldsymbol{r}}\, ds(\boldsymbol{r}) \\
&\qquad + \int_S [Z^+\hat{\boldsymbol{a}}_2 \times \boldsymbol{H}^*(\boldsymbol{r}; \hat{\boldsymbol{k}}_1, \hat{\boldsymbol{b}}_1) \\
&\qquad - \hat{\boldsymbol{b}}_2 \times \boldsymbol{E}^*(\boldsymbol{r}; \hat{\boldsymbol{k}}_1, \hat{\boldsymbol{a}}_1)] \cdot \hat{\boldsymbol{n}} e^{ik\hat{\boldsymbol{k}}_2\cdot\boldsymbol{r}}\, ds(\boldsymbol{r}) \\
&\qquad + \frac{1}{k^2}\int_{S^2} [Z^+\boldsymbol{g}_e^*(\boldsymbol{r}; \hat{\boldsymbol{k}}_1, \hat{\boldsymbol{a}}_1) \times \boldsymbol{g}_m(\boldsymbol{r}; \hat{\boldsymbol{k}}_2, \hat{\boldsymbol{b}}_2) \\
&\qquad + Z^+\boldsymbol{g}_e(\boldsymbol{r}; \hat{\boldsymbol{k}}_2, \hat{\boldsymbol{a}}_2) \times \boldsymbol{g}_m^*(\boldsymbol{r}; \hat{\boldsymbol{k}}_1, \hat{\boldsymbol{b}}_1)] \cdot \hat{\boldsymbol{r}}\, ds(\hat{\boldsymbol{r}})
\end{aligned} \tag{2.137}$$

or, using the definition of the scattering amplitude (2.90)

$$\begin{aligned}
&Z^+\int_S [\boldsymbol{E}^{+^*}(\boldsymbol{r}; \hat{\boldsymbol{k}}_1, \hat{\boldsymbol{a}}_1) \times \boldsymbol{H}^+(\boldsymbol{r}; \hat{\boldsymbol{k}}_2, \hat{\boldsymbol{b}}_2) \\
&\quad + \boldsymbol{E}^+(\boldsymbol{r}; \hat{\boldsymbol{k}}_2, \hat{\boldsymbol{a}}_2) \times \boldsymbol{H}^{+^*}(\boldsymbol{r}; \hat{\boldsymbol{k}}_1, \hat{\boldsymbol{b}}_1)] \cdot \hat{\boldsymbol{n}}\, ds(\boldsymbol{r}) \\
&\quad = \frac{4\pi}{k^2}[\hat{\boldsymbol{a}}_1 \cdot \boldsymbol{g}_e(\hat{\boldsymbol{k}}_1; \hat{\boldsymbol{k}}_2, \hat{\boldsymbol{a}}_2) + \hat{\boldsymbol{a}}_2 \cdot \boldsymbol{g}_e^*(\hat{\boldsymbol{k}}_2; \hat{\boldsymbol{k}}_1, \hat{\boldsymbol{a}}_1)] \\
&\qquad + \frac{2}{k^2}\int_{S^2} \boldsymbol{g}_e^*(\hat{\boldsymbol{r}}; \hat{\boldsymbol{k}}_1, \hat{\boldsymbol{a}}_1) \cdot \boldsymbol{g}_e(\hat{\boldsymbol{r}}; \hat{\boldsymbol{k}}_2, \hat{\boldsymbol{a}}_2)\, ds(\hat{\boldsymbol{r}})
\end{aligned} \tag{2.138}$$

Equation (2.138) provides the following general scattering theorem of electromagnetism

$$
\begin{aligned}
&-\hat{\boldsymbol{a}}_1 \cdot \boldsymbol{g}_e(\hat{\boldsymbol{k}}_1; \hat{\boldsymbol{k}}_2, \hat{\boldsymbol{a}}_2) - \hat{\boldsymbol{a}}_2 \cdot \boldsymbol{g}_e^*(\hat{\boldsymbol{k}}_2; \hat{\boldsymbol{k}}_1, \hat{\boldsymbol{a}}_1) \\
&\quad = \frac{1}{2\pi} \int_{S^2} \boldsymbol{g}_e^*(\hat{\boldsymbol{r}}; \hat{\boldsymbol{k}}_1, \hat{\boldsymbol{a}}_1) \cdot \boldsymbol{g}_e(\hat{\boldsymbol{r}}; \hat{\boldsymbol{k}}_2, \hat{\boldsymbol{a}}_2)\, ds(\hat{\boldsymbol{r}}) \\
&\quad - \frac{k^2 Z^+}{4\pi} \int_S [\boldsymbol{E}^{+^*}(\boldsymbol{r}; \hat{\boldsymbol{k}}_1, \hat{\boldsymbol{a}}_1) \times \boldsymbol{H}^+(\boldsymbol{r}; \hat{\boldsymbol{k}}_2, \hat{\boldsymbol{b}}_2) \\
&\quad + \boldsymbol{E}^+(\boldsymbol{r}; \hat{\boldsymbol{k}}_2, \hat{\boldsymbol{a}}_2) \times \boldsymbol{H}^{+^*}(\boldsymbol{r}; \hat{\boldsymbol{k}}_1, \hat{\boldsymbol{b}}_1)] \cdot \hat{\boldsymbol{n}}\, ds(\boldsymbol{r})
\end{aligned}
\tag{2.139}
$$

The last integral in (2.139) vanishes for lossless scatterers.

The optical theorem: For $\hat{\boldsymbol{k}}_1 = \hat{\boldsymbol{k}}_2 = \hat{\boldsymbol{k}}, \hat{\boldsymbol{a}}_1 = \hat{\boldsymbol{a}}_2 = \hat{\boldsymbol{a}}$ (2.139) assumes the form

$$
\begin{aligned}
&-2\,\mathrm{Re}\{\hat{\boldsymbol{a}} \cdot \boldsymbol{g}_e(\hat{\boldsymbol{k}}; \hat{\boldsymbol{k}}, \hat{\boldsymbol{a}})\} \\
&= \frac{1}{2\pi} \int_{S^2} |\boldsymbol{g}_e(\hat{\boldsymbol{r}}; \hat{\boldsymbol{k}}, \hat{\boldsymbol{a}})|^2\, ds(\hat{\boldsymbol{r}}) + \frac{k^2}{2\pi} Z^+\, \mathrm{Re} \int_S \hat{\boldsymbol{n}} \cdot \boldsymbol{H}^+(\boldsymbol{r}; \hat{\boldsymbol{k}}, \hat{\boldsymbol{b}}) \times \boldsymbol{E}^{+^*}(\boldsymbol{r}; \hat{\boldsymbol{k}}, \hat{\boldsymbol{a}})\, ds(\boldsymbol{r})
\end{aligned}
\tag{2.140}
$$

or, in view of the definitions of the cross sections (2.115), (2.117) and (2.125) we obtain the following form of the **optical** or **forward scattering theorem**

$$
\sigma_e = \sigma_s + \sigma_a = -\frac{4\pi}{k^2}\,\mathrm{Re}\,\{\hat{\boldsymbol{a}} \cdot \boldsymbol{g}_e(\hat{\boldsymbol{k}}; \hat{\boldsymbol{k}}, \hat{\boldsymbol{a}})\} \tag{2.141}
$$

The comments that follow the acoustical version of the optical theorem hold true for the electromagnetic case as well.

2.EM.6 *Problems in electromagnetics*

1. Derive the exterior Stratton–Chu representation formulae (2.63) and (2.64) and the corresponding interior formulae (2.65) and (2.66) for the electric and magnetic fields.
2. If $\boldsymbol{f}$ denotes a smooth vector field, $\widetilde{\boldsymbol{F}}$ denotes a smooth dyadic field and $\hat{\boldsymbol{n}}$ is the outward unit normal on the smooth surface S bounding the domain V, then prove that

$$
\begin{aligned}
&\int_V [(\nabla \times (\nabla \times \boldsymbol{f})) \cdot \widetilde{\boldsymbol{F}} - \boldsymbol{f} \cdot (\nabla \times (\nabla \times \widetilde{\boldsymbol{F}}))]\, dv \\
&\quad = -\int_S [(\nabla \times \boldsymbol{f}) \cdot (\hat{\boldsymbol{n}} \times \widetilde{\boldsymbol{F}}) - (\hat{\boldsymbol{n}} \times \boldsymbol{f}) \cdot (\nabla \times \widetilde{\boldsymbol{F}})]\, ds
\end{aligned}
$$

3. Use the identity in Problem 2 to derive the representations (2.72) and (2.73) for the exterior fields $\boldsymbol{E}^+$ and $\boldsymbol{H}^+$ and the representations (2.74) and (2.75) for the interior fields $\boldsymbol{E}^-$ and $\boldsymbol{H}^-$

4. Prove that the fundamental dyadic $\widetilde{\boldsymbol{G}}$ of electromagnetics satisfies the following asymptotic forms

$$\widetilde{\boldsymbol{G}}(\boldsymbol{r},\boldsymbol{r}') = (\widetilde{\boldsymbol{I}} - \hat{\boldsymbol{r}}\hat{\boldsymbol{r}})\frac{e^{ikr}}{ikr}e^{-ik\hat{\boldsymbol{r}}\cdot\boldsymbol{r}'} + O\left(\frac{1}{r^2}\right), \quad r \to \infty$$

and

$$\nabla_{r'} \times \widetilde{\boldsymbol{G}}(\boldsymbol{r},\boldsymbol{r}') = -\hat{\boldsymbol{r}} \times \widetilde{\boldsymbol{I}}\frac{e^{ikr}}{r}e^{-ik\hat{\boldsymbol{r}}\cdot\boldsymbol{r}'} + O\left(\frac{1}{r^2}\right)$$

$$= (\hat{\boldsymbol{\theta}}\hat{\boldsymbol{\varphi}} - \hat{\boldsymbol{\varphi}}\hat{\boldsymbol{\theta}})\frac{e^{ikr}}{r}e^{-ik\hat{\boldsymbol{r}}\cdot\boldsymbol{r}'} + O\left(\frac{1}{r^2}\right), \quad r \to \infty$$

5. Derive the expressions (2.88) and (2.89), (2.90) and (2.91) for the scattering amplitudes.
6. Let $\boldsymbol{f}$ be a differentiable vector function defined in a neighborhood of a closed surface S. Prove that

$$\int_S \hat{\boldsymbol{n}} \times \boldsymbol{f}\, ds = \int_S \boldsymbol{r}\hat{\boldsymbol{n}} \cdot (\nabla \times \boldsymbol{f})\, ds$$

where $\hat{\boldsymbol{n}}$ is the outward unit normal on S.
7. Use Problem 6 to derive representations (2.95) and (2.96) for the scattering amplitudes $\boldsymbol{g}_e$ and $\boldsymbol{g}_m$, respectively.
8. Derive the recurrence relations (2.103)–(2.107) that provide the coefficients $\boldsymbol{F}_n^e, n \geq 1$ of Atkinson's expansion (2.101) for the electric field in terms of the leading coefficient $\boldsymbol{F}_0^e = \boldsymbol{g}_e$.
9. Use the recurrence formulae (2.103)–(2.107) and the Maxwell equations to obtain the corresponding recurrence formulae that provide the coefficients $\boldsymbol{F}_n^m, n \geq 1$ of the expansion (2.102) in terms of $\boldsymbol{F}_0^m = \boldsymbol{g}_m$.
10. Show that the scattering cross section can be expressed as an integral over the surface of the scatterer by the formula

$$\sigma_s = Z^+ Re \int_S \hat{\boldsymbol{n}} \cdot \boldsymbol{E} \times \boldsymbol{H}^*\, ds$$

11. Provide all the details that lead to the formula

$$\sigma_a = \frac{1}{k\eta}\frac{Z^+}{Z^-}\mathrm{Im} \int_S \hat{\boldsymbol{n}} \cdot [(\nabla \times \boldsymbol{E}^-)) \times \boldsymbol{E}^{-*}]ds$$

for the absorption cross section in the case of the transmission problem. Then show that σ_a can also be written as a volume integral in the form

$$\sigma_a = Z^+\sigma^- \int_{V^-} |\boldsymbol{E}^-|^2\, dv$$

12. Prove that if the real part of the projection of the electric scattering amplitude on the direction of polarization of an incident plane wave vanishes, then no scattering occurs, i.e. the scattering amplitudes vanish in every direction.

2.E Elasticity

2.E.1 *Integral representations in elasticity*

Integral representations of the displacement field in elasticity are obtained from **Betti's third identity**

$$\rho \int_V [\boldsymbol{u}(\boldsymbol{r}) \cdot \Delta^e \boldsymbol{v}(\boldsymbol{r}) - \boldsymbol{v}(\boldsymbol{r}) \cdot \Delta^e \boldsymbol{u}(\boldsymbol{r})]\, dv(\boldsymbol{r})$$

$$= \int_S [\boldsymbol{u}(\boldsymbol{r}) \cdot \boldsymbol{T}(\partial_{\boldsymbol{r}}, \hat{\boldsymbol{n}})\boldsymbol{v}(\boldsymbol{r}) - \boldsymbol{v}(\boldsymbol{r}) \cdot \boldsymbol{T}(\partial_{\boldsymbol{r}}, \hat{\boldsymbol{n}})\boldsymbol{u}(\boldsymbol{r})]\, ds(\boldsymbol{r}) \qquad (2.142)$$

where S is the boundary of V. Using the standard approach of identifying V with V^+, $\boldsymbol{u}$ with the scattered displacement field, and $\boldsymbol{v}$ with, successively, each column of the fundamental dyadic, we obtain, after combining the three expressions, the following representation of the scattered field

$$\alpha(\boldsymbol{r})\boldsymbol{u}(\boldsymbol{r}) = \frac{1}{4\pi} \int_S [\boldsymbol{u}(\boldsymbol{r}') \cdot \boldsymbol{T}^+(\partial_{\boldsymbol{r}'}, \hat{\boldsymbol{n}}')\widetilde{\boldsymbol{\Gamma}}^+(\boldsymbol{r}, \boldsymbol{r}')$$

$$- \widetilde{\boldsymbol{\Gamma}}^+(\boldsymbol{r}, \boldsymbol{r}') \cdot \boldsymbol{T}^+(\partial_{\boldsymbol{r}'}, \hat{\boldsymbol{n}}')\boldsymbol{u}(\boldsymbol{r}')]\, ds(\boldsymbol{r}') \qquad (2.143)$$

for every $\boldsymbol{r} \in \mathbb{R}^3$, where α is given by (2.3), $\widetilde{\boldsymbol{\Gamma}}^+$ and $\boldsymbol{T}^+\widetilde{\boldsymbol{\Gamma}}^+$ by (1.224)–(1.230) and $\boldsymbol{T}^+$ by (1.213). In deriving this expression it is also necessary to use the fact that both $\boldsymbol{u}$ and $\widetilde{\boldsymbol{\Gamma}}^+$ satisfy the radiation conditions of elasticity (1.220) and (1.221).

The integral representation for the interior displacement field is obtained similarly and it assumes the form

$$(\alpha(\boldsymbol{r}) - 1)\boldsymbol{u}^-(\boldsymbol{r}) = \frac{1}{4\pi} \int_S [\boldsymbol{u}^-(\boldsymbol{r}') \cdot \boldsymbol{T}^-(\partial_{\boldsymbol{r}'}, \hat{\boldsymbol{n}}')\widetilde{\boldsymbol{\Gamma}}^-(\boldsymbol{r}, \boldsymbol{r}')$$

$$- \widetilde{\boldsymbol{\Gamma}}^-(\boldsymbol{r}, \boldsymbol{r}') \cdot \boldsymbol{T}^-(\partial_{\boldsymbol{r}'}, \hat{\boldsymbol{n}}')\boldsymbol{u}^-(\boldsymbol{r}')]\, ds(\boldsymbol{r}') \qquad (2.144)$$

for $\boldsymbol{r} \in \mathbb{R}^3$.

The incident field which satisfies the exterior Navier equation in V^- has the representation

$$(\alpha(\boldsymbol{r}) - 1)\boldsymbol{u}^i(\boldsymbol{r}) = \frac{1}{4\pi} \int_S [\boldsymbol{u}^i(\boldsymbol{r}') \cdot \boldsymbol{T}^+(\partial_{\boldsymbol{r}'}, \hat{\boldsymbol{n}}')\widetilde{\boldsymbol{\Gamma}}^+(\boldsymbol{r}, \boldsymbol{r}')$$

$$- \widetilde{\boldsymbol{\Gamma}}^+(\boldsymbol{r}, \boldsymbol{r}') \cdot \boldsymbol{T}^+(\partial_{\boldsymbol{r}'}, \hat{\boldsymbol{n}}')\boldsymbol{u}^i(\boldsymbol{r}')]\, ds(\boldsymbol{r}') \qquad (2.145)$$

for $\boldsymbol{r} \in \mathbb{R}^3$.

Combining (2.143) and (2.145) results in the following representation of the total field

$$\alpha(\boldsymbol{r})\boldsymbol{u}^+(\boldsymbol{r}) = \boldsymbol{u}^i(\boldsymbol{r}) + \frac{1}{4\pi} \int_S [\boldsymbol{u}^+(\boldsymbol{r}') \cdot \boldsymbol{T}^+(\partial_{\boldsymbol{r}'}, \hat{\boldsymbol{n}}')\widetilde{\boldsymbol{\Gamma}}^+(\boldsymbol{r}, \boldsymbol{r}')$$

$$- \widetilde{\boldsymbol{\Gamma}}^+(\boldsymbol{r} \cdot \boldsymbol{r}') \cdot \boldsymbol{T}^+(\partial_{\boldsymbol{r}'}, \hat{\boldsymbol{n}}')\boldsymbol{u}^+(\boldsymbol{r}')]\, ds(\boldsymbol{r}') \qquad (2.146)$$

for every $\boldsymbol{r} \in \mathbb{R}^3$, where, instead of the total field, $\boldsymbol{u}^+$, we can also use the scattered field, $\boldsymbol{u}$, in the integrand, when $\boldsymbol{r} \in V^+$.

Formulae (2.144) and (2.146) provide the general integral representations for the interior and the total displacement fields, respectively.

The integral representations for the solutions of the two boundary value problems of elasticity assume the following form when the boundary conditions are introduced.

The rigid problem:

$$\alpha(\boldsymbol{r})\boldsymbol{u}(\boldsymbol{r}) = \boldsymbol{u}^i(\boldsymbol{r}) - \frac{1}{4\pi}\int_S \widetilde{\boldsymbol{\Gamma}}^+(\boldsymbol{r},\boldsymbol{r}') \cdot \boldsymbol{T}^+(\partial_{\boldsymbol{r}'},\hat{\boldsymbol{n}}')\boldsymbol{u}^+(\boldsymbol{r}')\,ds(\boldsymbol{r}') \tag{2.147}$$

The cavity problem:

$$\alpha(\boldsymbol{r})\boldsymbol{u}(\boldsymbol{r}) = \boldsymbol{u}^i(\boldsymbol{r}) + \frac{1}{4\pi}\int_S \boldsymbol{u}^+(\boldsymbol{r}') \cdot \boldsymbol{T}^+(\partial_{\boldsymbol{r}'},\hat{\boldsymbol{n}}')\widetilde{\boldsymbol{\Gamma}}^+(\boldsymbol{r},\boldsymbol{r}')\,ds(\boldsymbol{r}') \tag{2.148}$$

The transmission problem:

For the transmission problem we need to use (2.146) for $\boldsymbol{r} \in \bar{V}^+$ and (2.144) for $\boldsymbol{r} \in \bar{V}^-$. However, the four boundary values which occur in the integrands may be reduced to two using the transmission conditions (1.216) and (1.217).

2.E.2 *The far field*

In order to obtain the asymptotic form of the displacement field in the **radiation zone** or **far field** we need the following asymptotic formulae as $r \to \infty$,

$$\widetilde{\boldsymbol{\Gamma}}^p(\boldsymbol{r},\boldsymbol{r}') = \frac{ik_p}{\lambda^+ + 2\mu^+} e^{-ik_p\hat{\boldsymbol{r}}\cdot\boldsymbol{r}'} h(k_p r)\hat{\boldsymbol{r}}\hat{\boldsymbol{r}} + O\left(\frac{1}{r^2}\right) \tag{2.149}$$

$$\widetilde{\boldsymbol{\Gamma}}^s(\boldsymbol{r},\boldsymbol{r}') = \frac{ik_s}{\mu^+} e^{-ik_s\hat{\boldsymbol{r}}\cdot\boldsymbol{r}'} h(k_s r)(\widetilde{\boldsymbol{I}} - \hat{\boldsymbol{r}}\hat{\boldsymbol{r}}) + O\left(\frac{1}{r^2}\right) \tag{2.150}$$

$$\begin{aligned}\nabla_{\boldsymbol{r}'}\widetilde{\boldsymbol{\Gamma}}(\boldsymbol{r},\boldsymbol{r}') = {} & \frac{k_p^2}{\lambda^+ + 2\mu^+} e^{-ik_p\hat{\boldsymbol{r}}\cdot\boldsymbol{r}'} h(k_p r)\hat{\boldsymbol{r}}\hat{\boldsymbol{r}}\hat{\boldsymbol{r}} \\ & + \frac{k_s^2}{\mu^+} e^{-ik_s\hat{\boldsymbol{r}}\cdot\boldsymbol{r}'} h(k_s r)\hat{\boldsymbol{r}}(\widetilde{\boldsymbol{I}} - \hat{\boldsymbol{r}}\hat{\boldsymbol{r}}) + O\left(\frac{1}{r^2}\right)\end{aligned} \tag{2.151}$$

where, recall (1.77),

$$h(x) = \frac{e^{ix}}{ix} \tag{2.152}$$

From the expression (1.230) an asymptotic formula for the surface traction of the fundamental dyadic is obtained in the form

$$\begin{aligned}\boldsymbol{T}(\partial_{\boldsymbol{r}'},\hat{\boldsymbol{n}}')\widetilde{\boldsymbol{\Gamma}}^+(\boldsymbol{r},\boldsymbol{r}') = {} & k_p^2 e^{-ik_p\hat{\boldsymbol{r}}\cdot\boldsymbol{r}'} h(k_p r)\hat{\boldsymbol{n}}' \cdot \frac{\lambda^+\widetilde{\boldsymbol{I}} + 2\mu^+\hat{\boldsymbol{r}}\hat{\boldsymbol{r}}}{\lambda^+ + 2\mu^+}\hat{\boldsymbol{r}} \\ & + k_s^2 e^{-ik_s\hat{\boldsymbol{r}}\cdot\boldsymbol{r}'} h(k_s r)[2(\hat{\boldsymbol{n}}' \cdot \hat{\boldsymbol{r}})\hat{\boldsymbol{\theta}} + (\hat{\boldsymbol{n}}' \times \hat{\boldsymbol{\varphi}})]\hat{\boldsymbol{\theta}} \\ & + k_s^2 e^{-ik_s\hat{\boldsymbol{r}}\cdot\boldsymbol{r}'} h(k_s r)[2(\hat{\boldsymbol{n}}' \cdot \hat{\boldsymbol{r}})\hat{\boldsymbol{\varphi}} - (\hat{\boldsymbol{n}}' \times \hat{\boldsymbol{\theta}})]\hat{\boldsymbol{\varphi}} \\ & + O\left(\frac{1}{r^2}\right), \quad r \to \infty\end{aligned} \tag{2.153}$$

These asymptotic relations can then be used in the integral representation (2.143) to obtain the far field form

$$\boldsymbol{u}(\boldsymbol{r}) = g_r(\hat{\boldsymbol{r}})\hat{\boldsymbol{r}}h(k_p r) + [g_\theta(\hat{\boldsymbol{r}})\hat{\boldsymbol{\theta}} + g_\varphi(\hat{\boldsymbol{r}})\hat{\boldsymbol{\varphi}}]h(k_s r) + O\left(\frac{1}{r^2}\right), \quad r \to +\infty \tag{2.154}$$

where the **radial** or **longitudinal scattering amplitude** g_r and the **tangential** or **transverse scattering amplitudes** g_θ and g_φ are given by

$$g_r(\hat{\boldsymbol{r}}) = k_p^2\left[\widetilde{\boldsymbol{H}}_p^+(\hat{\boldsymbol{r}}) : \frac{\lambda^+\widetilde{\boldsymbol{I}}\hat{\boldsymbol{r}} + 2\mu^+\hat{\boldsymbol{r}}\widetilde{\boldsymbol{I}}}{\lambda^+ + 2\mu^+} - \frac{ik_p}{\omega^2\rho^+}\boldsymbol{K}_p^+(\hat{\boldsymbol{r}})\right] \cdot \hat{\boldsymbol{r}} \tag{2.155}$$

$$g_\theta(\hat{\boldsymbol{r}}) = k_s^2\left[\widetilde{\boldsymbol{H}}_s^+(\hat{\boldsymbol{r}}) : (\widetilde{\boldsymbol{I}} \times \widetilde{\boldsymbol{I}} \times \hat{\boldsymbol{r}} + 2\hat{\boldsymbol{r}}\widetilde{\boldsymbol{I}}) - \frac{ik_s}{\omega^2\rho^+}\boldsymbol{K}_s^+(\hat{\boldsymbol{r}})\right] \cdot \hat{\boldsymbol{\theta}} \tag{2.156}$$

$$g_\varphi(\hat{\boldsymbol{r}}) = k_s^2\left[\widetilde{\boldsymbol{H}}_s^+(\hat{\boldsymbol{r}}) : (\widetilde{\boldsymbol{I}} \times \widetilde{\boldsymbol{I}} \times \hat{\boldsymbol{r}} + 2\hat{\boldsymbol{r}}\widetilde{\boldsymbol{I}}) - \frac{ik_s}{\omega^2\boldsymbol{r}^+}\boldsymbol{K}_s^+(\hat{\boldsymbol{r}})\right] \cdot \hat{\boldsymbol{\varphi}} \tag{2.157}$$

with

$$\widetilde{\boldsymbol{H}}_p^+(\hat{\boldsymbol{r}}) = \frac{1}{4\pi}\int_S \boldsymbol{u}^+(\boldsymbol{r}')\hat{\boldsymbol{n}}'e^{-ik_p\hat{\boldsymbol{r}}\cdot\boldsymbol{r}'}\,ds(\boldsymbol{r}') \tag{2.158}$$

$$\widetilde{\boldsymbol{H}}_s^+(\hat{\boldsymbol{r}}) = \frac{1}{4\pi}\int_S \boldsymbol{u}^+(\boldsymbol{r}')\hat{\boldsymbol{n}}'e^{-ik_s\hat{\boldsymbol{r}}\cdot\boldsymbol{r}'}\,ds(\boldsymbol{r}') \tag{2.159}$$

$$\boldsymbol{K}_p^+(\hat{\boldsymbol{r}}) = \frac{1}{4\pi}\int_S \boldsymbol{T}^+(\partial_{\boldsymbol{r}'}, \hat{\boldsymbol{n}}')\boldsymbol{u}^+(\boldsymbol{r}')e^{-ik_p\hat{\boldsymbol{r}}\cdot\boldsymbol{r}'}\,ds(\boldsymbol{r}') \tag{2.160}$$

$$\boldsymbol{K}_s^+(\hat{\boldsymbol{r}}) = \frac{1}{4\pi}\int_S \boldsymbol{T}^+(\partial_{\boldsymbol{r}'}, \hat{\boldsymbol{n}}')\boldsymbol{u}^+(\boldsymbol{r}')e^{-ik_s\hat{\boldsymbol{r}}\cdot\boldsymbol{r}'}\,ds(\boldsymbol{r}') \tag{2.161}$$

and the double inner product is defined as

$$\boldsymbol{ab} : \boldsymbol{cd} = (\boldsymbol{a} \cdot \boldsymbol{d})(\boldsymbol{b} \cdot \boldsymbol{c}) \tag{2.162}$$

All three scattering amplititudes, g_r, g_θ and g_φ, have dimensions of length. Observe that the radial amplitude g_r is a purely P wave characterized by k_p, while the tangential amplitudes g_θ and g_φ are purely S waves characterized by k_s.

The dyadics $\widetilde{\boldsymbol{H}}_{p,s}^+$ contain the boundary displacement field, while the vectors $\boldsymbol{K}_{p,s}^+$ involve the value of the traction field on the boundary. Thus, $\widetilde{\boldsymbol{H}}_{p,s}^+$ vanish for the rigid body problem, while $\boldsymbol{K}_{p,s}^+$ vanish in the case of a cavity. For the transmission problem, $\widetilde{\boldsymbol{H}}_{p,s}^+$ and $\boldsymbol{K}_{p,s}^+$ are the integrals that 'transmit' the deformation characteristics from the exterior to the interior region.

2.E.3 *The far field expansion theorem*

The asymptotic form (2.154) provides the first terms of the two convergent series representation for the scattered field in the exterior of a sphere circumscribing the scatterer (Dassios 1988*a*). This far field expansion, analogous to the Atkinson–Wilcox theorems in acoustics and electromagnetics, states that the scattered field has the representation

$$\boldsymbol{u}(\boldsymbol{r}) = h(k_p r)\sum_{n=0}^{\infty} \boldsymbol{F}_n^p(\hat{\boldsymbol{r}})r^{-n} + h(k_s r)\sum_{n=0}^{\infty} \boldsymbol{F}_n^s(\hat{\boldsymbol{r}})r^{-n} \tag{2.163}$$

where the convergence is absolute and uniform for $r > a$, a being the radius of the smallest sphere containing the scatterer.

In particular

$$\boldsymbol{F}_0^p(\hat{\boldsymbol{r}}) = g_r(\hat{\boldsymbol{r}})\hat{\boldsymbol{r}} \tag{2.164}$$

and

$$\boldsymbol{F}_0^s(\hat{\boldsymbol{r}}) = g_\theta(\hat{\boldsymbol{r}})\hat{\boldsymbol{\theta}} + g_\varphi(\hat{\boldsymbol{r}})\hat{\boldsymbol{\varphi}} \tag{2.165}$$

It has been shown (Dassios 1988*a*) that the entire series for the P wave can be recovered once g_r is given and similarly the series for the S wave can be obtained from a knowledge of g_θ and g_φ. This process requires the decomposition of all coefficients into radial and tangential components. Then, the determination of the coefficients becomes possible via the recurrence relations

$$2ik_p(\hat{\boldsymbol{r}}\cdot\boldsymbol{F}_1^p) = (\mathbb{B}-2)(\hat{\boldsymbol{r}}\cdot\boldsymbol{F}_0^p) \tag{2.166}$$

$$ik_p(\widetilde{\boldsymbol{I}}-\hat{\boldsymbol{r}}\hat{\boldsymbol{r}})\cdot\boldsymbol{F}_1^p = \boldsymbol{D}(\hat{\boldsymbol{r}}\cdot\boldsymbol{F}_0^p) \tag{2.167}$$

and for $n = 2, 3, \ldots$

$$\begin{aligned} 2n\frac{\lambda^+ + 2\mu^+}{\lambda^+ + \mu^+}k_p^2\,\hat{\boldsymbol{r}}\cdot\boldsymbol{F}_n^p &= (\boldsymbol{D}-2\hat{\boldsymbol{r}})\cdot\left[2ik_p(n-1)\frac{\mu^+}{\lambda^+ + \mu^+}\boldsymbol{F}_{n-1}^p - \widetilde{\boldsymbol{L}}_n\cdot\boldsymbol{F}_{n-2}^p\right] \\ &\quad -ik_p[\mathbb{B}(\hat{\boldsymbol{r}}\cdot\boldsymbol{F}_{n-1}^p) + \hat{\boldsymbol{r}}\cdot\widetilde{\boldsymbol{L}}_{n+1}\cdot\boldsymbol{F}_{n-1}^p] \end{aligned} \tag{2.168}$$

$$\begin{aligned} k_p^2(\widetilde{\boldsymbol{I}}-\hat{\boldsymbol{r}}\hat{\boldsymbol{r}})\cdot\boldsymbol{F}_n^p &= 2ik_p(n-1)\frac{\mu^+}{\lambda^+ + \mu^+}(\boldsymbol{I}-\hat{\boldsymbol{r}}\hat{\boldsymbol{r}})\cdot\boldsymbol{F}_{n-1}^p \\ &\quad -(\widetilde{\boldsymbol{I}}-\hat{\boldsymbol{r}}\hat{\boldsymbol{r}})\cdot\widetilde{\boldsymbol{L}}_n\cdot\boldsymbol{F}_{n-2}^p - ik_p\boldsymbol{D}(\hat{\boldsymbol{r}}\cdot\boldsymbol{F}_{n-1}^p) \end{aligned} \tag{2.169}$$

where $\boldsymbol{D}$ is the angular part of the gradient

$$\boldsymbol{D} = \hat{\boldsymbol{\theta}}\frac{\partial}{\partial\theta} + \frac{\hat{\boldsymbol{\varphi}}}{\sin\theta}\frac{\partial}{\partial\varphi} \tag{2.170}$$

$\mathbb{B}$ is the Beltrami operator

$$\mathbb{B} = \boldsymbol{D} \cdot \boldsymbol{D} \tag{2.171}$$

and

$$\begin{aligned}\widetilde{\boldsymbol{L}}_n = &\frac{\mu^+}{\lambda^+ + \mu^+}[\mathbb{B} + (n-1)(n-2)]\widetilde{\boldsymbol{I}} \\ &+(\boldsymbol{D} - n\hat{\boldsymbol{r}})(\boldsymbol{D} - (n-1)\hat{\boldsymbol{r}})\end{aligned} \tag{2.172}$$

Relations (2.166)–(2.169) recover the P wave. The corresponding relations for the S wave are as follows

$$ik_s(\hat{\boldsymbol{r}} \cdot \boldsymbol{F}_1^s) = -\boldsymbol{D} \cdot \boldsymbol{F}_0^s \tag{2.173}$$

$$2ik_s(\widetilde{\boldsymbol{I}} - \hat{\boldsymbol{r}}\hat{\boldsymbol{r}}) \cdot \boldsymbol{F}_1^s = (\widetilde{\boldsymbol{I}} - \hat{\boldsymbol{r}}\hat{\boldsymbol{r}}) \cdot \mathbb{B}\boldsymbol{F}_0^s \tag{2.174}$$

and for $n = 2, 3, 4, \ldots$

$$k_s^2(\hat{\boldsymbol{r}} \cdot \boldsymbol{F}_n^s) = ik_s(\boldsymbol{M}_n \cdot \boldsymbol{F}_{n-1}^s) + \hat{\boldsymbol{r}} \cdot \widetilde{\boldsymbol{L}}_n \cdot \boldsymbol{F}_{n-2}^s \tag{2.175}$$

$$\begin{aligned}&2n\frac{\mu^+}{\lambda^+ + \mu^+}k_s^2(\widetilde{\boldsymbol{I}} - \hat{\boldsymbol{r}}\hat{\boldsymbol{r}}) \cdot \boldsymbol{F}_n^s \\ &= -ik_s(\widetilde{\boldsymbol{I}} - \hat{\boldsymbol{r}}\hat{\boldsymbol{r}}) \cdot \widetilde{\boldsymbol{L}}_{n+1} \cdot \boldsymbol{F}_{n-1}^s + \boldsymbol{D}[ik_s(\boldsymbol{M}_n \cdot \boldsymbol{F}_{n-1}^s) + \hat{\boldsymbol{r}} \cdot \widetilde{\boldsymbol{L}}_n \cdot \boldsymbol{F}_{n-2}^s]\end{aligned} \tag{2.176}$$

where

$$\boldsymbol{M}_n = \boldsymbol{D} + 2\frac{\mu^+ - n(\lambda^+ + 2\mu^+)}{\lambda^+ + \mu^+}\hat{\boldsymbol{r}} \tag{2.177}$$

2.E.4 *Cross sections*

Recall that the power flux vector for time harmonic elastic fields was given in (1.251) and (1.252) as

$$\begin{aligned}\boldsymbol{P} &= \omega\,\mathrm{Im}\{\widetilde{\boldsymbol{\tau}} \cdot \boldsymbol{u}^*\} \\ &= \omega\,\mathrm{Im}\{2\mu\boldsymbol{u}^* \cdot \nabla\boldsymbol{u} + \lambda\boldsymbol{u}^*\nabla \cdot \boldsymbol{u} + \mu\boldsymbol{u}^* \times (\nabla \times \boldsymbol{u})\}\end{aligned} \tag{2.178}$$

For the incident field

$$\boldsymbol{u}^i(\boldsymbol{r}) = \alpha_p\hat{\boldsymbol{k}}e^{ik_p\hat{\boldsymbol{k}}\cdot\boldsymbol{r}} + \alpha_s\hat{\boldsymbol{b}}e^{ik_s\hat{\boldsymbol{k}}\cdot\boldsymbol{r}} \tag{2.179}$$

where α_p and α_s are the real amplitudes of the incident pressure and shear waves respectively, we have

$$\begin{aligned}\widetilde{\boldsymbol{\tau}}^i(\boldsymbol{r}) = &\lambda^+ik_p\alpha_pe^{ik_p\hat{\boldsymbol{k}}\cdot\boldsymbol{r}}\widetilde{\boldsymbol{I}} \\ &+\mu^+(\alpha_pik_pe^{ik_p\hat{\boldsymbol{k}}\cdot\boldsymbol{r}}\,\hat{\boldsymbol{k}}\hat{\boldsymbol{k}} + \alpha_sik_se^{ik_s\hat{\boldsymbol{k}}\cdot\boldsymbol{r}}\,\hat{\boldsymbol{k}}\hat{\boldsymbol{b}}) \\ &+\mu^+(\alpha_pik_pe^{ik_p\hat{\boldsymbol{k}}\cdot\boldsymbol{r}}\,\hat{\boldsymbol{k}}\hat{\boldsymbol{k}} + \alpha_sik_se^{ik_s\hat{\boldsymbol{k}}\cdot\boldsymbol{r}}\,\hat{\boldsymbol{b}}\hat{\boldsymbol{k}})\end{aligned} \tag{2.180}$$

Hence

$$\begin{aligned} \boldsymbol{u}^{i^*}(\boldsymbol{r}) \cdot \widetilde{\boldsymbol{\tau}}^i(\boldsymbol{r}) &= [\alpha_p e^{-ik_p \hat{\boldsymbol{k}}\cdot\boldsymbol{r}}\, \hat{\boldsymbol{k}} + \alpha_s e^{-ik_s \hat{\boldsymbol{k}}\cdot\boldsymbol{r}}\, \hat{\boldsymbol{b}}] \cdot [ik_p \lambda^+ \alpha_p e^{ik_p \hat{\boldsymbol{k}}\cdot\boldsymbol{r}} \widetilde{\boldsymbol{I}} \\ &\quad + 2ik_p \mu^+ \alpha_p e^{ik_p \hat{\boldsymbol{k}}\cdot\boldsymbol{r}}\, \hat{\boldsymbol{k}}\hat{\boldsymbol{k}} + ik_s \mu^+ \alpha_s e^{ik_s \hat{\boldsymbol{k}}\cdot\boldsymbol{r}}\, (\hat{\boldsymbol{k}}\hat{\boldsymbol{b}} + \hat{\boldsymbol{b}}\hat{\boldsymbol{k}})] \\ &= ik_p \lambda^+ \alpha_p^2\, \hat{\boldsymbol{k}} + 2ik_p \mu^+ \alpha_p^2\, \hat{\boldsymbol{k}} + ik_s \mu^+ \alpha_p \alpha_s e^{i(k_s - k_p)\hat{\boldsymbol{k}}\cdot\boldsymbol{r}}\, \hat{\boldsymbol{b}} \\ &\quad + ik_p \lambda^+ \alpha_p \alpha_s e^{-i(k_s - k_p)\hat{\boldsymbol{k}}\cdot\boldsymbol{r}}\, \hat{\boldsymbol{b}} + ik_s \mu^+ \alpha_s^2\, \hat{\boldsymbol{k}} \end{aligned} \tag{2.181}$$

In particular, in the direction of the propagation of the incident plane wave we obtain

$$\begin{aligned} \hat{\boldsymbol{k}} \cdot \boldsymbol{P}^i(\boldsymbol{r}) &= \omega \,\mathrm{Im}\, \{\hat{\boldsymbol{k}} \cdot \widetilde{\boldsymbol{\tau}}^i(\boldsymbol{r}) \cdot \boldsymbol{u}^{i^*}(\boldsymbol{r})\} \\ &= \omega \,\mathrm{Im}\, \{ik_p(\lambda^+ + 2\mu^+)\alpha_p^2 + ik_s \mu^+ \alpha_s^2\} \\ &= \omega^2 \rho^+ [c_p^+ \alpha_p^2 + c_s^+ \alpha_s^2] \end{aligned} \tag{2.182}$$

For the scattered field, in the radiation zone or far field, we have

$$\boldsymbol{u}(\boldsymbol{r}) = g_r \hat{\boldsymbol{r}} h(k_p r) + (g_\theta \hat{\boldsymbol{\theta}} + g_\varphi \hat{\boldsymbol{\varphi}}) h(k_s r) + O\left(\frac{1}{r^2}\right), \quad r \to \infty \tag{2.183}$$

and it can be shown (Dassios and Kiriaki 1984) that

$$\begin{aligned} \hat{\boldsymbol{r}} \cdot \widetilde{\boldsymbol{\tau}}(\boldsymbol{r}) &= i\omega\rho^+ c_p^+ g_r \hat{\boldsymbol{r}} h(k_p r) \\ &\quad + i\omega\rho^+ c_s^+ (g_\theta \hat{\boldsymbol{\theta}} + g_\varphi \hat{\boldsymbol{\varphi}}) h(k_s r) + O\left(\frac{1}{r^2}\right), \quad r \to \infty \end{aligned} \tag{2.184}$$

Hence, the outward power flux in the radiation zone is given by

$$\begin{aligned} \hat{\boldsymbol{r}} \cdot \boldsymbol{P}(\boldsymbol{r}) &= \omega \mathrm{Im}\{\hat{\boldsymbol{r}} \cdot \widetilde{\boldsymbol{\tau}}(\boldsymbol{r}) \cdot \boldsymbol{u}^*(\boldsymbol{r})\} \\ &= \frac{\rho^+ (c_p^+)^3}{r^2} |g_r|^2 + \frac{\rho^+ (c_s^+)^3}{r^2} (|g_\theta|^2 + |g_\varphi|^2) + O\left(\frac{1}{r^3}\right), \quad r \to \infty \end{aligned} \tag{2.185}$$

which is positive.

The **differential scattering cross section** is then defined as

$$\begin{aligned} \sigma(\hat{\boldsymbol{r}}) &= \lim_{r \to \infty} \frac{4\pi r^2 \hat{\boldsymbol{r}} \cdot \boldsymbol{P}(\boldsymbol{r})}{\hat{\boldsymbol{k}} \cdot \boldsymbol{P}^i(\boldsymbol{r})} \\ &= \frac{4\pi}{\omega^2} \frac{(c_p^+)^3 |g_r(\hat{\boldsymbol{r}})|^2 + (c_s^+)^3 (|g_\theta(\hat{\boldsymbol{r}})|^2 + |g_\varphi(\hat{\boldsymbol{r}})|^2)}{c_p^+ \alpha_p^2 + c_s^+ \alpha_s^2} \end{aligned} \tag{2.186}$$

Note that the acoustical case (2.34) is recovered from (2.186) if we assume that $\alpha_p = 1$ and $\alpha_s = 0$ and restrict attention to the longitudinal wave alone.

It is important to observe that because we deal with two different phase velocities in elasticity we **cannot** define the differential scattering cross section, as in the acoustical and the electromagnetic case, via the formula

$$\sigma(\hat{\boldsymbol{r}}) = \lim_{r\to\infty} 4\pi r^2 \frac{|\boldsymbol{u}(\boldsymbol{r})|^2}{|\boldsymbol{u}^i(\boldsymbol{r})|^2} \tag{2.187}$$

since this would imply that

$$\sigma(\hat{\boldsymbol{r}}) = \frac{4\pi}{\omega^2} \frac{(c_p^+)^2|g_r(\hat{\boldsymbol{r}})|^2 + (c_s^+)^2(|g_\theta(\hat{\boldsymbol{r}})|^2 + |g_\varphi(\hat{\boldsymbol{r}})|^2)}{\alpha_p^2 + \alpha_s^2} \tag{2.188}$$

which is **incorrect**. Expression (2.186) is based on energy arguments and represents the physics of the wave–obstacle interaction whereas (2.188) does not come from physical principles.

Averaging the differential scattering cross section over the full solid angle S^2 we obtain the **scattering cross section** or **total cross section**

$$\begin{aligned}\sigma_s &= \frac{1}{4\pi}\int_{S^2} \sigma(\hat{\boldsymbol{r}})\, ds(\hat{\boldsymbol{r}}) \\ &= \frac{1}{\omega^2(c_p^+\alpha_p^2 + c_s^+\alpha_s^2)}\int_{S^2} [(c_p^+)^3|g_r|^2 + (c_s^+)^3(|g_\theta|^2 + |g_\varphi|^2)]\, ds \\ &= \frac{1}{\omega\rho^+(c_p^+\alpha_p^2 + c_s^+\alpha_s^2)} \operatorname{Im} \int_{\partial B(\boldsymbol{0},\infty)} \hat{\boldsymbol{r}}\cdot\widetilde{\boldsymbol{\tau}}(\boldsymbol{r})\cdot\boldsymbol{u}^*(\boldsymbol{r})\, ds(\boldsymbol{r})\end{aligned} \tag{2.189}$$

This last expression for σ_s can be used to transfer the integral from the sphere at infinity to the surface integral over the scatterer, with the help of Betti's third identity (2.142), i.e.

$$\begin{aligned}\sigma_s &= \frac{1}{\omega\rho^+(c_p^+\alpha_p^2 + c_s^+\alpha_s^2)} \operatorname{Im}\int_S \hat{\boldsymbol{n}}\cdot\widetilde{\boldsymbol{\tau}}(\boldsymbol{r})\cdot\boldsymbol{u}^*(\boldsymbol{r})\, ds(\boldsymbol{r}) \\ &= \frac{1}{\omega\rho^+(c_p^+\alpha_p^2 + c_s^+\alpha_s^2)} \operatorname{Im}\int_S \boldsymbol{u}^*(\boldsymbol{r})\cdot\boldsymbol{T}(\partial_{\boldsymbol{r}},\hat{\boldsymbol{n}})\boldsymbol{u}(\boldsymbol{r})\, ds(\boldsymbol{r})\end{aligned} \tag{2.190}$$

The **absorption cross section** is defined by a similar integral involving the total rather than the scattered fields

$$\begin{aligned}\sigma_a &= -\frac{1}{\omega\rho^+(c_p^+\alpha_p^2 + c_s^+\alpha_s^2)} \operatorname{Im}\int_S \boldsymbol{u}^{+^*}(\boldsymbol{r})\cdot\boldsymbol{T}(\partial_{\boldsymbol{r}},\hat{\boldsymbol{n}})\boldsymbol{u}^+(\boldsymbol{r})\, ds(\boldsymbol{r}) \\ &= \frac{1}{\omega\rho^+(c_p^+\alpha_p^2 + c_s^+\alpha_s^2)} \operatorname{Im}\int_S \boldsymbol{u}^+(\boldsymbol{r})\cdot\boldsymbol{T}(\partial_{\boldsymbol{r}},\hat{\boldsymbol{n}})\boldsymbol{u}^{+^*}(\boldsymbol{r})\, ds(\boldsymbol{r})\end{aligned} \tag{2.191}$$

where the minus sign indicates the inward flux produced by the total field on the surface of the scatterer.

Finally, the **extinction cross section** is defined as in acoustics and electromagnetics by

$$\sigma_e = \sigma_s + \sigma_a \tag{2.192}$$

Equation (2.191) shows that in the case of a rigid body or a cavity the boundary conditions cause the absorption cross section to vanish. In fact, the same is true for any penetrable scatterer whenever λ^- and μ^- are real. This can be seen from the following

$$\begin{aligned}\sigma_a &= \frac{1}{2i\omega\rho^+(c_p^+\alpha_p^2 + c_s^+\alpha_s^2)} \int_S [\boldsymbol{u}^+ \cdot \boldsymbol{T}^+\boldsymbol{u}^{+^*} - \boldsymbol{u}^{+^*} \cdot \boldsymbol{T}^+\boldsymbol{u}^+]\, ds \\ &= \frac{1}{2i\omega\rho^+(c_p^+\alpha_p^2 + c_s^+\alpha_s^2)} \int_S [\boldsymbol{u}^- \cdot \boldsymbol{T}^-\boldsymbol{u}^{-^*} - \boldsymbol{u}^{-^*} \cdot \boldsymbol{T}^-\boldsymbol{u}^-]\, ds \\ &= \frac{1}{2i\omega\rho^+(c_p^+\alpha_p^2 + c_s^+\alpha_s^2)} \int_{V^-} [\boldsymbol{u}^- \cdot \Delta^{e-}\boldsymbol{u}^{-^*} - \boldsymbol{u}^{-^*} \cdot \Delta^{e-}\boldsymbol{u}^-]\, dv \\ &= 0 \end{aligned} \tag{2.193}$$

where the last equality is a consequence of the Navier equation in V^-.

Therefore, there is no absorption in any of the three problems we consider here. Absorption occurs only when λ^- and μ^- are complex, to account for viscosity effects in the interior of the scatterer (Achenbach 1973).

The lack of absorption implies that

$$\sigma_e = \sigma_s \tag{2.194}$$

in all three scattering problems considered here.

2.E.5 *The basic scattering theorems*

Different possible combinations of excitation in the case of elastic plane waves cause the reciprocity theorem to assume different forms, which are given below. Although the basic ideas in proving the reciprocity theorems in elasticity are the same as in acoustic and electromagnetic waves, the actual steps in the proofs are much more complicated. For this reason we will state the results without the proofs, which can be found in Dassios *et al.* (1987, 1995). For related theorems see also Tan (1976, 1977) and De Hoop (1995).

The reciprocity theorems: Let both $\boldsymbol{u}_1^+$ and $\boldsymbol{u}_2^+$ be solutions of any of the three scattering problems in elasticity corresponding to two incident plane waves propagating in the directions $\hat{\boldsymbol{k}}_1$ and $\hat{\boldsymbol{k}}_2$. The reciprocity theorems are all based on the following relation

$$\int_S [\boldsymbol{u}_1^+ \cdot \boldsymbol{T}^+\boldsymbol{u}_2^+ - \boldsymbol{u}_2^+ \cdot \boldsymbol{T}^+\boldsymbol{u}_1^+]\, ds = 0 \tag{2.195}$$

which can be proved with the help of Betti's third identity and the corresponding boundary or transmission conditions.

Depending on the type of incident wave (i.e. whether α_p or α_s vanishes) the following three versions of the reciprocity theorem have been obtained.

(i) If both incident waves $\boldsymbol{u}_1^i$ and $\boldsymbol{u}_2^i$ are longitudinal, i.e. $\alpha_s = 0$ in both cases, then

$$g_r(-\hat{\boldsymbol{k}}_1, \hat{\boldsymbol{k}}_2) = g_r(-\hat{\boldsymbol{k}}_2, \hat{\boldsymbol{k}}_1) \tag{2.196}$$

(ii) If both incident waves $\boldsymbol{u}_1^i$ and $\boldsymbol{u}_2^i$ are transverse, i.e. $\alpha_p = 0$ in both cases, and polarized along $\hat{\boldsymbol{b}}_1$ and $\hat{\boldsymbol{b}}_2$, respectively, then

$$\begin{aligned}(\hat{\boldsymbol{b}}_1 \cdot \hat{\boldsymbol{\theta}}) g_\theta(-\hat{\boldsymbol{k}}_1, \hat{\boldsymbol{k}}_2) + (\hat{\boldsymbol{b}}_1 \cdot \hat{\boldsymbol{\varphi}}) g_\varphi(-\hat{\boldsymbol{k}}_1, \hat{\boldsymbol{k}}_2) \\ = (\hat{\boldsymbol{b}}_2 \cdot \hat{\boldsymbol{\theta}}) g_\theta(-\hat{\boldsymbol{k}}_2, \hat{\boldsymbol{k}}_1) + (\hat{\boldsymbol{b}}_2 \cdot \hat{\boldsymbol{\varphi}}) g_\varphi(-\hat{\boldsymbol{k}}_2, \hat{\boldsymbol{k}}_1)\end{aligned} \tag{2.197}$$

(iii) If $\boldsymbol{u}_1^i$ is longitudinal and $\boldsymbol{u}_2^i$ is transverse and polarized along $\hat{\boldsymbol{b}}_2$, then

$$\begin{aligned}-\frac{1}{k_p^3} g_r(-\hat{\boldsymbol{k}}_1, \hat{\boldsymbol{k}}_2) = \frac{1}{k_s^3} (\hat{\boldsymbol{b}}_2 \cdot \hat{\boldsymbol{\theta}}) g_\theta(-\hat{\boldsymbol{k}}_2, \hat{\boldsymbol{k}}_1) \\ + \frac{1}{k_s^3} (\hat{\boldsymbol{b}}_2 \cdot \hat{\boldsymbol{\varphi}}) g_\varphi(-\hat{\boldsymbol{k}}_2, \hat{\boldsymbol{k}}_1)\end{aligned} \tag{2.198}$$

If we set $\hat{\boldsymbol{k}}_1 = -\hat{\boldsymbol{r}}$ and $\hat{\boldsymbol{k}}_2 = \hat{\boldsymbol{k}}$ in (2.196)–(2.198) we obtain the more familiar forms

(i) for longitudinal–longitudinal incidence

$$g_r(\hat{\boldsymbol{r}}, \hat{\boldsymbol{k}}) = g_r(-\hat{\boldsymbol{k}}, -\hat{\boldsymbol{r}}) \tag{2.199}$$

(ii) for transverse–transverse incidence

$$\hat{\boldsymbol{b}}_1 \cdot [g_\theta(\hat{\boldsymbol{r}}, \hat{\boldsymbol{k}})\hat{\boldsymbol{\theta}} + g_\varphi(\hat{\boldsymbol{r}}, \hat{\boldsymbol{k}})\hat{\boldsymbol{\varphi}}] = \hat{\boldsymbol{b}}_2 \cdot [g_\theta(-\hat{\boldsymbol{k}}, -\hat{\boldsymbol{r}})\hat{\boldsymbol{\theta}} + g_\varphi(-\hat{\boldsymbol{k}}, -\hat{\boldsymbol{r}})\hat{\boldsymbol{\varphi}}] \tag{2.200}$$

(iii) for longitudinal–transverse incidence

$$-\frac{1}{k_p^3} g_r(\hat{\boldsymbol{r}}, \hat{\boldsymbol{k}}) = \frac{1}{k_s^3} \hat{\boldsymbol{b}}_2 \cdot [g_\theta(-\hat{\boldsymbol{k}}, -\hat{\boldsymbol{r}})\hat{\boldsymbol{\theta}} + g_\varphi(-\hat{\boldsymbol{k}}, -\hat{\boldsymbol{r}})\hat{\boldsymbol{\varphi}}] \tag{2.201}$$

In acoustics and electromagnetics we presented a general scattering theorem from which the optical or forward scattering theorem followed. In elasticity such a general scattering theorem has also been found. However, it is a very complicated relation involving triadics and we content ourselves here with presenting only two very special cases which are the elastic versions of the **Optical Theorem:**

(i) for longitudinal incidence ($\alpha_s = 0$)

$$\sigma_e = -\frac{4\pi}{k_p^2} \operatorname{Re} g_r(\hat{\boldsymbol{k}}, \hat{\boldsymbol{k}}) \tag{2.202}$$

(ii) for transverse incidence ($\alpha_p = 0$)

$$\sigma_e = -\frac{4\pi}{k_s^2} \operatorname{Re}\{\, \hat{\boldsymbol{b}} \cdot [g_\theta(\hat{\boldsymbol{k}}, \hat{\boldsymbol{k}})\hat{\boldsymbol{\theta}} + g_\varphi(\hat{\boldsymbol{k}}, \hat{\boldsymbol{k}})\hat{\boldsymbol{\varphi}}]\} \tag{2.203}$$

Dassios *et al.* (1995) have presented an exhaustive analysis of reciprocity and scattering theorems and derived a general scattering theorem in the context of scattering by complete dyadic fields from which all known theorems in acoustics, electromagnetics and elasticity can be obtained as special cases. Moreover, any other reciprocity or scattering theorem involving scalar, vector or dyadic fields may also be deduced from this theorem.

2.E.6 *Problems in elasticity*

1. Prove Betti's third identity (2.142).
2. Derive the integral representation (2.144) for the interior domain V^- and the integral representation (2.146) for the exterior domain V^+.
3. Develop the asymptotic forms (2.149)–(2.151) as $r \to \infty$.
4. Use the asymptotic forms (2.149), (2.150) and (2.151) in the integral representation (2.143) for the scattered wave to derive formulae (2.154)–(2.161)
5. Prove the identity

$$[\widetilde{\boldsymbol{I}} \times \widetilde{\boldsymbol{I}} \times \hat{\boldsymbol{r}} + 2\hat{\boldsymbol{r}}\widetilde{\boldsymbol{I}}] \cdot \hat{\boldsymbol{a}} = \hat{\boldsymbol{r}}\hat{\boldsymbol{a}} + \hat{\boldsymbol{a}}\hat{\boldsymbol{r}}$$

and use it to show that

$$\begin{aligned}
\boldsymbol{u}(\boldsymbol{r}) = & \frac{k_p^2}{\lambda^+ + 2\mu^+}\widetilde{\boldsymbol{H}}_p^+ : (\lambda^+\widetilde{\boldsymbol{I}} + 2\mu^+\hat{\boldsymbol{r}}\hat{\boldsymbol{r}})\hat{\boldsymbol{r}}h(k_p r) \\
& + k_s^2\widetilde{\boldsymbol{H}}_s^+ : [\hat{\boldsymbol{r}}(\widetilde{\boldsymbol{I}} - \hat{\boldsymbol{r}}\hat{\boldsymbol{r}}) + \hat{\boldsymbol{\theta}}\hat{\boldsymbol{r}}\hat{\boldsymbol{\theta}} + \hat{\boldsymbol{\varphi}}\hat{\boldsymbol{r}}\hat{\boldsymbol{\varphi}}]h(k_s r) \\
& - \frac{ik_p}{\lambda^+ + 2\mu^+}\boldsymbol{K}_p^+ \cdot \hat{\boldsymbol{r}}\hat{\boldsymbol{r}}h(k_p r) \\
& - \frac{ik_s}{\mu^+}\boldsymbol{K}_s^+ \cdot (\widetilde{\boldsymbol{I}} - \hat{\boldsymbol{r}}\hat{\boldsymbol{r}})h(k_s r) \\
& + O\left(\frac{1}{r^2}\right), \qquad r \to \infty
\end{aligned}$$

6. Use the far field expansion theorem (2.163) to obtain the asymptotic term of order $1/r^2$, as $r \to \infty$, for the scattered wave in terms of the spherical scattering amplitudes g_r, g_θ and g_φ.
7. Derive the asymptotic form (2.185) for the outward power flux and the formula (2.186) for the differential scattering cross section.
8. Use Betti's identity (2.142) and the definition

$$\sigma_s = \frac{1}{4\pi}\int_{S^2} \sigma(\hat{\boldsymbol{r}})ds(\hat{\boldsymbol{r}})$$

of the scattering cross section to obtain the formula

$$\sigma_s = \frac{1}{\omega\rho^+(c_p^+\alpha_p^2 + c_s^+\alpha_s^2)} \operatorname{Im}\int_S \boldsymbol{u}^*(\boldsymbol{r}) \cdot \boldsymbol{T}(\partial_{\boldsymbol{r}}, \hat{\boldsymbol{n}})\boldsymbol{u}(\boldsymbol{r})\, ds(\boldsymbol{r})$$

where S^2 is the unit sphere in $\mathbb{R}^3$ and S is the surface of the scatterer.

3

LOW FREQUENCY EXPANSIONS

The motivation for considering **low frequency expansions** is to replace the general scattering problems described in Chapter 1 by sequences of simpler problems. If one is interested only in scattering at extremely low frequencies, then, in the spirit of Rayleigh (1897*a*), it is possible to replace the scattering problem by a potential problem. But, if one wants to improve the approximation, then additional terms in a power series expansion may be found (Stevenson 1953*a*,*b*) by solving boundary value or transmission problems for the Laplace equation in acoustics and electromagnetics and for the elastostatic equation in elasticity, each of which is simpler than the original scattering problem. In this chapter we describe this process in detail for each of the problems discussed in Chapter 1.

3.A Acoustics

3.A.1 *General low frequency expansions in acoustics*

First, we recall that we are restricting attention to those incident acoustic waves which are plane waves; point sources in V^+ or superpositions of such sources. Furthermore, all incident fields admit power series expansions of the form

$$u^i(\boldsymbol{r}) = \sum_{n=0}^{\infty} \frac{(ik)^n}{n!} u_n^i(\boldsymbol{r}) \tag{3.1}$$

These expansions converge for all points $\boldsymbol{r}$ except for singularities in the case of point sources, which are described in more detail in Chapter 1. Recall that we use $k = k^+$ to denote the wave number in V^+.

In all four scattering problems introduced in Chapter 1, we obtain formal low frequency expansions of the solutions by assuming that

$$u^+(\boldsymbol{r}) = \sum_{n=0}^{\infty} \frac{(ik)^n}{n!} u_n^+(\boldsymbol{r}), \quad \boldsymbol{r} \in V^+ \cup S \tag{3.2}$$

with the understanding that source points are excluded. Henceforth, we will not specifically mention this understanding. In addition, for the transmission problem we also assume that

$$u^-(\boldsymbol{r}) = \sum_{n=0}^{\infty} \frac{(ik)^n}{n!} u_n^-(\boldsymbol{r}), \quad \boldsymbol{r} \in V^- \cup S \tag{3.3}$$

That is, the fields in V^+ and V^- are analytic functions of the complex variable k in a neighborhood of $k = 0$. The validity of these expansions is established in (Kleinman

1965*a*, 1966). One might suppose that $u^-(\boldsymbol{r})$ should be expanded in powers of ik^- rather than ik and in fact such an expansion does exist. However, with the introduction of the index of refraction by the relation $k^- = \eta k$ such an expansion is equivalent to that given in (3.3) where the powers of η are absorbed in the coefficients $u_n^-(\boldsymbol{r})$. Observe that the **low frequency coefficients** $u_n^\pm$ have dimensions that differ from those of $u^\pm$ by a factor of length to the nth power because the expansion parameter k has dimensions of inverse length.

Recall that in all the acoustic scattering problems considered, $u^+(\boldsymbol{r})$ represents the total field, incident plus scattered, in V^+. That is

$$u^+(\boldsymbol{r}) = u^i(\boldsymbol{r}) + u(\boldsymbol{r}) \tag{3.4}$$

where

$$(\nabla^2 + k^2)u(\boldsymbol{r}) = 0, \quad \boldsymbol{r} \in V^+ \tag{3.5}$$

and u satisfies the radiation condition

$$\lim_{r\to\infty} r\left(\frac{\partial}{\partial r}u(\boldsymbol{r}) - iku(\boldsymbol{r})\right) = 0 \tag{3.6}$$

uniformly over all directions. In addition, the boundary conditions considered specify

The Dirichlet problem:

$$u^+(\boldsymbol{r}) = 0, \quad \boldsymbol{r} \in S \tag{3.7}$$

The Neumann problem:

$$\frac{\partial}{\partial n}u^+(\boldsymbol{r}) = 0, \quad \boldsymbol{r} \in S \tag{3.8}$$

The Robin problem:

$$\frac{\partial}{\partial n}u^+(\boldsymbol{r}) + ik\nu u^+(\boldsymbol{r}) = 0, \quad \boldsymbol{r} \in S \tag{3.9}$$

The transmission problem:

$$\left.\begin{aligned} u^+(\boldsymbol{r}) &= u^-(\boldsymbol{r}) \\ \frac{\partial}{\partial n}u^+(\boldsymbol{r}) &= \beta\frac{\partial}{\partial n}u^-(\boldsymbol{r}) \end{aligned}\right\} \quad \boldsymbol{r} \in S \tag{3.10}$$

and

$$(\nabla^2 + k^2\eta^2)u^-(\boldsymbol{r}) = 0, \quad \boldsymbol{r} \in V^- \tag{3.11}$$

Substitution of the expansion (3.2) in (3.5) leads to the relation

$$\nabla^2 u_n^+(\boldsymbol{r}) = n(n-1)u_{n-2}^+(\boldsymbol{r}), \quad \boldsymbol{r} \in V^+ \tag{3.12}$$

Similarly, the expansion (3.3) together with (3.11) implies that

$$\nabla^2 u_n^-(\boldsymbol{r}) = n(n-1)\eta^2 u_{n-2}^-(\boldsymbol{r}), \quad \boldsymbol{r} \in V^- \tag{3.13}$$

In this chapter we show how the coefficients u_n^+ and u_n^- may be characterized in terms of the previous coefficients, $u_m^\pm$, $m = 0, \ldots, n-1$ and the solution of potential problems.

Moreover, expressions of the scattering amplitude in terms of those coefficients are also presented.

To facilitate this development we need to utilize low frequency expansions of the integral representations given in Chapter 2. In particular, if we employ the low frequency expansions of the incident field (3.1), the total field (3.2), and the expansion of the fundamental solution

$$ikG^+(\boldsymbol{r},\boldsymbol{r}') = \sum_{n=0}^{\infty} \frac{(ik)^n}{n!}|\boldsymbol{r}-\boldsymbol{r}'|^{n-1} \tag{3.14}$$

in the integral representation (2.7) we obtain

$$\alpha(\boldsymbol{r})\sum_{n=0}^{\infty}\frac{(ik)^n}{n!}u_n^+(\boldsymbol{r}) = \sum_{n=0}^{\infty}\frac{(ik)^n}{n!}u_n^i(\boldsymbol{r}) + \frac{1}{4\pi}\sum_{n=0}^{\infty}\frac{(ik)^n}{n!}\sum_{m=0}^{n}\binom{n}{m}$$
$$\times\int_S\left[u_{n-m}^+(\boldsymbol{r}')\frac{\partial}{\partial n'}|\boldsymbol{r}-\boldsymbol{r}'|^{m-1} - |\boldsymbol{r}-\boldsymbol{r}'|^{m-1}\frac{\partial}{\partial n'}u_{n-m}^+(\boldsymbol{r}')\right]ds(\boldsymbol{r}') \tag{3.15}$$

In this equation the Cauchy formula has been used to rearrange the terms in the product of two series. Equating like powers of ik we obtain

$$\alpha(\boldsymbol{r})u_n^+(\boldsymbol{r}) = f_n^+(\boldsymbol{r})$$
$$+\frac{1}{4\pi}\int_S\left[u_n^+(\boldsymbol{r}')\frac{\partial}{\partial n'}\frac{1}{|\boldsymbol{r}-\boldsymbol{r}'|} - \frac{1}{|\boldsymbol{r}-\boldsymbol{r}'|}\frac{\partial}{\partial n'}u_n^+(\boldsymbol{r}')\right]ds(\boldsymbol{r}') \tag{3.16}$$

where $f_0^+(\boldsymbol{r}) = u_0^i(\boldsymbol{r})$ and

$$f_n^+(\boldsymbol{r}) = u_n^i(\boldsymbol{r}) + \frac{1}{4\pi}\sum_{m=1}^{n}\binom{n}{m}\int_S\left[u_{n-m}^+(\boldsymbol{r}')\frac{\partial}{\partial n'}|\boldsymbol{r}-\boldsymbol{r}'|^{m-1}\right.$$
$$\left.-|\boldsymbol{r}-\boldsymbol{r}'|^{m-1}\frac{\partial}{\partial n'}u_{n-m}^+(\boldsymbol{r}')\right]ds(\boldsymbol{r}'), \quad n \geq 1 \tag{3.17}$$

for all $\boldsymbol{r}$. Note that the sum in (3.17) begins with $m = 1$ so that $f_n^+(\boldsymbol{r})$ depends on the coefficients u_m^+ for $m \leq n-1$ and is independent of u_n^+. For $\boldsymbol{r} \in V^+$ we may rewrite (3.16) as

$$u_n^+(\boldsymbol{r}) = f_n^+(\boldsymbol{r}) + v_n^+(\boldsymbol{r}), \quad \boldsymbol{r} \in V^+ \tag{3.18}$$

where

$$v_n^+(\boldsymbol{r}) = \frac{1}{4\pi}\int_S\left[u_n^+(\boldsymbol{r}')\frac{\partial}{\partial n'}\frac{1}{|\boldsymbol{r}-\boldsymbol{r}'|}\right.$$
$$\left.-\frac{1}{|\boldsymbol{r}-\boldsymbol{r}'|}\frac{\partial}{\partial n'}u_n^+(\boldsymbol{r}')\right]ds(\boldsymbol{r}'), \quad \boldsymbol{r} \in V^+ \tag{3.19}$$

The function $v_n^+(\boldsymbol{r})$ is seen to be a combination of a single and a double layer distribution; that is, integrals of the fundamental solution of Laplace's equation and its normal derivative with unknown densities. As such, v_n^+ is a solution of Laplace's equation in V^+ and decays as r^{-1} as $r \to \infty$. If u_m^+ are known for $m = 0, \ldots, n-1$, then $f_n^+(\boldsymbol{r})$ is known and the specification of $u_n^+(\boldsymbol{r})$ is completed with the determination of the potential function $v_n^+(\boldsymbol{r})$

Observe that the fact that v_n^+ is harmonic in V^+ implies that f_n^+ is a particular solution of (3.12). Moreover, since v_n^+ is regular at infinity it has the integral representation

$$\alpha(\boldsymbol{r})v_n^+(\boldsymbol{r}) = \frac{1}{4\pi}\int_S \left[v_n^+(\boldsymbol{r}')\frac{\partial}{\partial n'}\frac{1}{|\boldsymbol{r}-\boldsymbol{r}'|} - \frac{1}{|\boldsymbol{r}-\boldsymbol{r}'|}\frac{\partial}{\partial n'}v_n^+(\boldsymbol{r}')\right] ds(\boldsymbol{r}'), \quad \boldsymbol{r} \in \mathbb{R}^3 \tag{3.20}$$

which can be obtained from (3.7) by letting $k \to 0$. Subtracting this from the representation for u_n^+ (3.16) and making use of the decomposition (3.18) leads to the representation

$$(\alpha(\boldsymbol{r})-1)f_n^+(\boldsymbol{r}) = \frac{1}{4\pi}\int_S \left[f_n^+(\boldsymbol{r}')\frac{\partial}{\partial n'}\frac{1}{|\boldsymbol{r}-\boldsymbol{r}'|} - \frac{1}{|\boldsymbol{r}-\boldsymbol{r}'|}\frac{\partial}{\partial n'}f_n^+(\boldsymbol{r}')\right] ds(\boldsymbol{r}') \tag{3.21}$$

for $\boldsymbol{r} \in \mathbb{R}^3$. This, coupled with the observation above, leads to the remarkable fact that f_n^+, defined by (3.17), satisfies

$$\left.\begin{aligned} \nabla^2 f_n^+ &= n(n-1)u_{n-2}^+ \quad &&\text{in } V^+ \\ \nabla^2 f_n^+ &= 0 &&\text{in } V^- \end{aligned}\right\} \tag{3.22}$$

Similarly, the low frequency expansion of $u^-(\boldsymbol{r})$, (3.3), together with the expansion of the fundamental solution

$$ik^-G^-(\boldsymbol{r},\boldsymbol{r}') = \sum_{n=0}^{\infty}\frac{(ik)^n}{n!}\eta^n|\boldsymbol{r}-\boldsymbol{r}'|^{n-1} \tag{3.23}$$

may be substituted in (2.5) yielding, after rearranging terms and equating like powers of ik

$$\begin{aligned}(\alpha(\boldsymbol{r})-1)u_n^-(\boldsymbol{r}) &= f_n^-(\boldsymbol{r}) \\ &+\frac{1}{4\pi}\int_S \left[u_n^-(\boldsymbol{r})\frac{\partial}{\partial n'}\frac{1}{|\boldsymbol{r}-\boldsymbol{r}'|} - \frac{1}{|\boldsymbol{r}-\boldsymbol{r}'|}\frac{\partial}{\partial n'}u_n^-(\boldsymbol{r}')\right] ds(\boldsymbol{r}')\end{aligned} \tag{3.24}$$

where $f_0^- = 0$ and

$$\begin{aligned} f_n^-(\boldsymbol{r}) = \frac{1}{4\pi}\sum_{m=1}^{n}\binom{n}{m}\eta^m \int_S \Bigg[u_{n-m}^-(\boldsymbol{r}')\frac{\partial}{\partial n'}|\boldsymbol{r}-\boldsymbol{r}'|^{m-1} \\ - |\boldsymbol{r}-\boldsymbol{r}'|^{m-1}\frac{\partial}{\partial n'}u_{n-m}^-(\boldsymbol{r}')\Bigg] ds(\boldsymbol{r}'), \quad n \geq 1 \end{aligned} \tag{3.25}$$

It is useful to note that not only is $f_0^- = 0$, but also that $f_1^- = 0$, since

$$\begin{aligned} f_1^-(\boldsymbol{r}) &= -\frac{\eta}{4\pi}\int_S \frac{\partial}{\partial n'} u_0^-(\boldsymbol{r}')\, ds'(\boldsymbol{r}') \\ &= -\frac{\eta}{4\pi}\int_{V^-} \nabla^2 u_0^-(\boldsymbol{r}')\, d\boldsymbol{r}' \\ &= 0 \end{aligned} \tag{3.26}$$

where the divergence theorem and the fact that u_0^- is a potential function in V^- (see (3.13)) has been used.

For $\boldsymbol{r} \in V^-$, (3.24) may be rewritten as

$$u_n^-(\boldsymbol{r}) = -f_n^-(\boldsymbol{r}) + v_n^-(\boldsymbol{r}), \quad \boldsymbol{r} \in V^- \tag{3.27}$$

where

$$\begin{aligned} v_n^-(\boldsymbol{r}) = \frac{1}{4\pi}\int_S \bigg[&\frac{1}{|\boldsymbol{r}-\boldsymbol{r}'|}\frac{\partial}{\partial n'} u_n^-(\boldsymbol{r}') \\ &-u_n^-(\boldsymbol{r}')\frac{\partial}{\partial n'}\frac{1}{|\boldsymbol{r}-\boldsymbol{r}'|}\bigg]\, ds(\boldsymbol{r}'), \quad \boldsymbol{r} \in V^- \end{aligned} \tag{3.28}$$

Clearly, $v_n^-(\boldsymbol{r})$ is a solution of the Laplace equation in V^-. If $u_m^-(\boldsymbol{r})$ are known for $m = 0, 1, \ldots, n-1$, then $f_n^-(\boldsymbol{r})$ is known and the complete specification of $u_n^-(\boldsymbol{r})$ depends on finding the potential function $v_n^-(\boldsymbol{r})$.

Observe that the fact that v_n^- is harmonic in V^- implies that $-f_n^-$ is a particular solution of (3.13). Moreover, v_n^- satisfies the identity

$$(\alpha(\boldsymbol{r}) - 1) v_n^-(\boldsymbol{r}) = \frac{1}{4\pi}\int_S \left[v_n^-(\boldsymbol{r}')\frac{\partial}{\partial n'}\frac{1}{|\boldsymbol{r}-\boldsymbol{r}'|} - \frac{1}{|\boldsymbol{r}-\boldsymbol{r}'|}\frac{\partial}{\partial n'} v_n^-(\boldsymbol{r}')\right] ds(\boldsymbol{r}') \tag{3.29}$$

for $\boldsymbol{r} \in \mathbb{R}^3$ which can be obtained from (2.5) by letting $k \to 0$. Subtracting this from the representation for u_n^- (3.24) and making use of the decomposition (3.27) leads to the representation

$$\alpha(\boldsymbol{r}) f_n^-(\boldsymbol{r}) = \frac{1}{4\pi}\int_S \left[f_n^-(\boldsymbol{r}')\frac{\partial}{\partial n'}\frac{1}{|\boldsymbol{r}-\boldsymbol{r}'|} - \frac{1}{|\boldsymbol{r}-\boldsymbol{r}'|}\frac{\partial}{\partial n'} f_n^-(\boldsymbol{r}')\right] ds(\boldsymbol{r}') \tag{3.30}$$

for $\boldsymbol{r} \in \mathbb{R}^3$. This, coupled with the observation above, leads to the fact that f_n^-, defined by (3.25), satisfies

$$\left.\begin{aligned} \nabla^2 f_n^- &= 0 && \text{in } V^+ \\ \nabla^2 f_n^- &= -n(n-1)\eta^2 u_{n-2}^- && \text{in } V^- \end{aligned}\right\} \tag{3.31}$$

A low frequency expression for the far field coefficient is obtained by substituting (3.2) and the expansion

$$e^{-ik\hat{\boldsymbol{r}}\cdot\boldsymbol{r}'} = \sum_{n=0}^{\infty} \frac{(ik)^n}{n!}(-\hat{\boldsymbol{r}}\cdot\boldsymbol{r}')^n \tag{3.32}$$

in equation (2.18) and using the Cauchy product formula yielding

$$\begin{aligned} g(\hat{\boldsymbol{r}}) = \frac{1}{4\pi}\sum_{n=0}^{\infty}\frac{(ik)^{n+1}}{n!}\sum_{m=0}^{n}\binom{n}{m}(-1)^{m+1} \\ \times\int_S\left[\frac{\partial}{\partial n'}u_{n-m}^{+}(\boldsymbol{r}') + ik(\hat{\boldsymbol{r}}\cdot\hat{\boldsymbol{n}}')u_{n-m}^{+}(\boldsymbol{r}')\right](\hat{\boldsymbol{r}}\cdot\boldsymbol{r}')^m\,ds(\boldsymbol{r}') \end{aligned} \tag{3.33}$$

These low frequency expressions simplify considerably in each of the four scattering problems.

3.A.2 *The Dirichlet problem*

The Dirichlet boundary condition implies that each coefficient in the expansion (3.2) also satisfies a Dirichlet condition; that is,

$$u_n^{+}(\boldsymbol{r}) = 0, \quad \boldsymbol{r}\in S \tag{3.34}$$

With this condition, the expressions for f_n^+ and v_n^+, (3.17) and (3.19), become $f_0^+ = u_0^i$ and

$$f_n^{+}(\boldsymbol{r}) = u_n^i(\boldsymbol{r}) - \frac{1}{4\pi}\sum_{m=1}^{n}\binom{n}{m}\int_S|\boldsymbol{r}-\boldsymbol{r}'|^{m-1}\frac{\partial}{\partial n'}u_{n-m}^{+}(\boldsymbol{r}')\,ds(\boldsymbol{r}'), \quad n\geq 1 \tag{3.35}$$

for $\boldsymbol{r}$ in $\mathbb{R}^3$ and

$$v_n^{+}(\boldsymbol{r}) = -\frac{1}{4\pi}\int_S\frac{1}{|\boldsymbol{r}-\boldsymbol{r}'|}\frac{\partial}{\partial n'}u_n^{+}(\boldsymbol{r}')\,ds(\boldsymbol{r}'), \quad \boldsymbol{r}\in V^+ \tag{3.36}$$

The coefficient $u_n^+(\boldsymbol{r})$ is given by (3.18) and $v_n^+(\boldsymbol{r})$ is determined as a solution of the following boundary value problem

$$\left.\begin{aligned} \nabla^2 v_n^{+} &= 0 && \text{in } V^+ \\ v_n^{+} &= -f_n^{+} && \text{on } S \\ v_n^{+} &= O\left(\frac{1}{r}\right) && \text{as } r\to\infty \end{aligned}\right\} \tag{3.37}$$

The scattering amplitude (3.27) simplifies in this case to

$$g(\hat{\boldsymbol{r}}) = \frac{1}{4\pi}\sum_{n=0}^{\infty}\frac{(ik)^{n+1}}{n!}\sum_{m=0}^{n}\binom{n}{m}(-1)^{m+1} \times \int_S (\hat{\boldsymbol{r}}\cdot\boldsymbol{r}')^m \frac{\partial}{\partial n'}u_{n-m}^{+}(\boldsymbol{r}')\,ds(\boldsymbol{r}') \tag{3.38}$$

from which we infer that

$$\left.\begin{aligned} \operatorname{Re} g(\hat{\boldsymbol{r}}) &= \sum_{n=1}^{\infty} k^{2n} A_{2n}(\hat{\boldsymbol{r}}) \\ \operatorname{Im} g(\hat{\boldsymbol{r}}) &= \sum_{n=0}^{\infty} k^{2n+1} A_{2n+1}(\hat{\boldsymbol{r}}) \end{aligned}\right\} \tag{3.39}$$

where

$$A_{2n}(\hat{\boldsymbol{r}}) = \frac{1}{4\pi}\sum_{m=0}^{2n-1}\frac{(-1)^{n+m+1}}{m!(2n-1-m)!}\int_S (\hat{\boldsymbol{r}}\cdot\boldsymbol{r}')^m \frac{\partial}{\partial n'}u_{2n-1-m}^{+}(\boldsymbol{r}')\,ds(\boldsymbol{r}') \tag{3.40}$$

and

$$A_{2n+1}(\hat{\boldsymbol{r}}) = \frac{1}{4\pi}\sum_{m=0}^{2n}\frac{(-1)^{n+m+1}}{m!(2n-m)!}\int_S (\hat{\boldsymbol{r}}\cdot\boldsymbol{r}')^m \frac{\partial}{\partial n'}u_{2n-m}^{+}(\boldsymbol{r}')\,ds(\boldsymbol{r}') \tag{3.41}$$

Examination of these expressions would seem to imply that in order to evaluate A_{n+1} it is necessary to have the coefficients u_i^+ for $i = 0, 1, \ldots, n$. Actually, as shown originally by Van Bladel (1968a), this is not the case and it suffices to know u_i^+ only for $i = 0, 1, \ldots, n-1$. This may be seen with the help of the conductor potential $\phi^c(\boldsymbol{r})$, which is the exterior potential function which solves the problem

$$\left.\begin{aligned} \nabla^2\phi^c &= 0 && \text{in } V^+ \\ \phi^c &= 1 && \text{on } S \\ \phi^c &= O\left(\frac{1}{r}\right) && \text{as } r\to\infty \end{aligned}\right\} \tag{3.42}$$

With the help of this function we see that the only term in A_{n+1} which involves u_n^+ has the integral

$$\int_S \frac{\partial}{\partial n'}u_n^{+}(\boldsymbol{r}')\,ds(\boldsymbol{r}') = \int_S \frac{\partial}{\partial n'}u_n^{+}(\boldsymbol{r}')\phi^c(\boldsymbol{r}')\,ds(\boldsymbol{r}') \tag{3.43}$$

as a factor. Using Green's theorem, the decomposition (3.18) and the boundary condition (3.37) this may be rewritten as

$$\int_S \frac{\partial}{\partial n'}u_n^{+}(\boldsymbol{r}')\,ds(\boldsymbol{r}') = \int_S \frac{\partial}{\partial n'}f_n^{+}(\boldsymbol{r}')\,ds(\boldsymbol{r}') - \int_S f_n^{+}(\boldsymbol{r}')\frac{\partial}{\partial n'}\phi^c(\boldsymbol{r}')\,ds(\boldsymbol{r}') \tag{3.44}$$

in which form it is seen to involve u_i^+ for $i = 0, 1, \ldots, n-1$. Obviously, what we have succeeded in doing is to replace the need for the solution of the boundary value problem for u_n^+ in calculating A_{n+1} with the solution of (3.42) which is not dependent on n. Hence, once (3.42) is solved, we can use this same solution to improve the approximation for the scattering amplitude by one order of magnitude for every n. We remark that subsequently, in Chapter 5, it is shown that for plane wave incidence

$$\phi^c(\boldsymbol{r}) = 1 - u_0^+(\boldsymbol{r}) \tag{3.45}$$

so that, in fact, problem (3.42) is not an 'extra' problem but one that we must solve in order to obtain the leading term in the low frequency expansion. For plane wave incidence in the direction $\hat{\boldsymbol{k}}$, the scattering amplitude depends on both $\hat{\boldsymbol{r}}$ and $\hat{\boldsymbol{k}}$ and the differential scattering cross section (2.33) may be expressed as

$$\begin{aligned}\sigma(\hat{\boldsymbol{r}}) &= \frac{4\pi}{k^2}|g(\hat{\boldsymbol{r}},\hat{\boldsymbol{k}})|^2\\ &= 4\pi\sum_{n=0}^{\infty}k^{2n}\sum_{m=0}^{n}A_{2n-2m+1}(\hat{\boldsymbol{r}})A_{2m+1}(\hat{\boldsymbol{r}})\\ &\quad+4\pi\sum_{n=0}^{\infty}k^{2n+2}\sum_{m=0}^{n}A_{2n-2m+2}(\hat{\boldsymbol{r}})A_{2m+2}(\hat{\boldsymbol{r}})\end{aligned} \tag{3.46}$$

where the A's are functions of $\hat{\boldsymbol{r}}$ and $\hat{\boldsymbol{k}}$ as is $\sigma(\hat{\boldsymbol{r}})$; however, it is common not to exhibit the $\hat{\boldsymbol{k}}$ dependence explicitly.

The absorption cross section σ_a, (2.39), vanishes in this case because of the boundary condition on S. Through the use of the optical, or forward scattering theorem, (2.59) we find that σ_s the (total) scattering cross section (2.35) which is equal to the extinction cross section σ_e (2.43) in this case is expressible as

$$\begin{aligned}\sigma_e = \sigma_s &= -\frac{4\pi}{k^2}\operatorname{Re}\, g(\hat{\boldsymbol{k}},\hat{\boldsymbol{k}})\\ &= -4\pi\sum_{n=0}^{\infty}k^{2n}A_{2n+2}(\hat{\boldsymbol{k}})\end{aligned} \tag{3.47}$$

where (3.39) has been used. Alternatively, with (3.46)

$$\begin{aligned}\sigma_e = \sigma_s = &\sum_{n=0}^{\infty}k^{2n}\sum_{m=0}^{n}\int_{S^2}A_{2n-2m+1}(\hat{\boldsymbol{r}})A_{2m+1}(\hat{\boldsymbol{r}})\,ds(\hat{\boldsymbol{r}})\\ &+\sum_{n=0}^{\infty}k^{2n+2}\sum_{m=0}^{n}\int_{S^2}A_{2n-2m+2}(\hat{\boldsymbol{r}})A_{2m+2}(\hat{\boldsymbol{r}})\,ds(\hat{\boldsymbol{r}})\end{aligned} \tag{3.48}$$

where all the coefficients A_n are functions of $\hat{\boldsymbol{r}}$ and $\hat{\boldsymbol{k}}$. While this last formula appears more complicated than (3.47), it requires fewer coefficients for a given order of k. For example, to determine the leading (k^0) term, one would require the knowledge of A_2, were equation (3.47) employed, whereas with equation (3.48) it is only necessary to know A_1. This disadvantage of equation (3.47) may be removed for a large class of scattering surfaces using the following technique introduced by Twersky (1953, 1954).

When the scatterer has symmetry with respect to the origin, by which is meant that if $\boldsymbol{r} \in S$, then $-\boldsymbol{r} \in S$, additional results are available for plane wave incidence. In particular,

$$A_{2n}(\hat{\boldsymbol{r}}, \hat{\boldsymbol{k}}) = -\frac{1}{4\pi} \sum_{m=1}^{2n-1} \int_{S^2} A_m(\hat{\boldsymbol{p}}, \hat{\boldsymbol{k}}) A_{2n-m}(\hat{\boldsymbol{p}}, \hat{\boldsymbol{r}})\, ds(\hat{\boldsymbol{p}}) \tag{3.49}$$

where for present purposes the dependence of A_n on the direction of propagation is explicitly indicated. For example, A_2 is given in terms of A_1. Twersky called this symmetry property **'inversion symmetry'**. Additional results for the first few terms are given in Chapter 5 for general surfaces, and in Chapters 7 and 8 for spheres and ellipsoids.

If a finite number of coefficients u_n^+ are found in the manner described above, then a low frequency approximation of $u^+(\boldsymbol{r})$ is determined by truncating the series (3.2). In like manner, approximation of the far field is found by truncating the series (3.38). It must be noted, however, that this process requires first the determination of the functions $u_n^+(\boldsymbol{r})$ in V^+ and then calculating their normal derivatives on S. A method for avoiding this two-step procedure by finding the normal derivative of u_n^+ directly is available through the use of integral equations as described in Chapter 6.

3.A.3 *The Neumann problem*

The Neumann condition (3.8) imposes a similar condition on the coefficients in the expansion (3.2); namely

$$\frac{\partial}{\partial n} u_n^+(\boldsymbol{r}) = 0, \qquad \boldsymbol{r} \in S \tag{3.50}$$

With this condition, equations (3.17) and (3.19) become $f_0^+ = u_0^i$ and $f_1^+ = u_1^i$

$$f_n^+(\boldsymbol{r}) = u_n^i(\boldsymbol{r}) + \frac{1}{4\pi} \sum_{m=2}^{n} \binom{n}{m} \int_S u_{n-m}^+(\boldsymbol{r}') \frac{\partial}{\partial n'} |\boldsymbol{r} - \boldsymbol{r}'|^{m-1}\, ds(\boldsymbol{r}'), \quad n \geq 2 \tag{3.51}$$

for every $\boldsymbol{r}$, and

$$v_n^+(\boldsymbol{r}) = \frac{1}{4\pi} \int_S u_n^+(\boldsymbol{r}') \frac{\partial}{\partial n'} \frac{1}{|\boldsymbol{r} - \boldsymbol{r}'|}\, ds(\boldsymbol{r}'), \quad \boldsymbol{r} \in V^+ \tag{3.52}$$

The coefficients $u_n^+(\boldsymbol{r}')$ are again given by (3.18) where now $f_n^+(\boldsymbol{r})$ assume the form (3.51). With the Neumann boundary condition (3.34) and the representation (3.36), v_n^+ may be characterized as the solution of the potential problem

$$\left.\begin{aligned} \nabla^2 v_n^+ &= 0 && \text{in } V^+ \\ \frac{\partial}{\partial n} v_n^+ &= -\frac{\partial}{\partial n} f_n^+ && \text{on } S \\ v_n^+ &= O\left(\frac{1}{r^2}\right) && \text{as } r \to \infty \end{aligned}\right\} \tag{3.53}$$

The scattering amplitude (3.32) becomes, in this case

$$g(\hat{\boldsymbol{r}}) = \frac{1}{4\pi}\sum_{n=0}^{\infty}\frac{(ik)^{n+2}}{n!}\sum_{m=0}^{n}\binom{n}{m}(-1)^{m+1} \times \int_S (\hat{\boldsymbol{r}}\cdot\hat{\boldsymbol{n}}')(\hat{\boldsymbol{r}}\cdot\boldsymbol{r}')^m u_{n-m}^{+}(\boldsymbol{r}')\,ds(\boldsymbol{r}') \tag{3.54}$$

Comparing this expression with the corresponding scattering amplitude for the Dirichlet problem reveals that g begins with $(ik)^2$ in the Neumann case compared with (ik) in the Dirichlet case. Actually, for plane wave incidence the Neumann scattering amplitude begins with $(ik)^3$ since the $n = 0$ term vanishes. That is,

$$\int_S (\hat{\boldsymbol{r}}\cdot\hat{\boldsymbol{n}}')u_0^{+}(\boldsymbol{r}')\,ds(\boldsymbol{r}') = 0 \tag{3.55}$$

This may be seen by noting that for plane wave incidence the normal derivative of f_0^+ vanishes since u_0^i is constant. Hence, v_0^+ is a regular solution of the homogeneous Neumann problem (3.53) for $n = 0$ and therefore vanishes due to uniqueness.

Then the divergence theorem ensures that

$$\int_S (\hat{\boldsymbol{r}}\cdot\hat{\boldsymbol{n}}')u_0^{+}(\boldsymbol{r}')\,ds(\boldsymbol{r}') = u_0^i\int_S (\hat{\boldsymbol{r}}\cdot\hat{\boldsymbol{n}}')\,ds(\boldsymbol{r}') = 0 \tag{3.56}$$

From these observations we infer that for plane wave incidence

$$\begin{aligned}\text{Re } g(\hat{\boldsymbol{r}}) &= \sum_{n=2}^{\infty}k^{2n}A_{2n}(\hat{\boldsymbol{r}})\\ \text{Im } g(\hat{\boldsymbol{r}}) &= \sum_{n=1}^{\infty}k^{2n+1}A_{2n+1}(\hat{\boldsymbol{r}})\end{aligned} \tag{3.57}$$

where

$$A_{2n}(\hat{\boldsymbol{r}}) = \frac{1}{4\pi}\sum_{m=0}^{2n-2}\frac{(-1)^{m+n+1}}{m!(2n-2-m)!}\int_S (\hat{\boldsymbol{n}}'\cdot\hat{\boldsymbol{r}})(\hat{\boldsymbol{r}}\cdot\hat{\boldsymbol{r}}')^m u_{2n-2-m}^{+}(\boldsymbol{r}')\,ds(\boldsymbol{r}') \tag{3.58}$$

and

$$A_{2n+1}(\hat{\boldsymbol{r}}) = \frac{1}{4\pi}\sum_{m=0}^{2n-1}\frac{(-1)^{n+m+1}}{m!(2n-1-m)!}\int_S (\hat{\boldsymbol{n}}'\cdot\hat{\boldsymbol{r}})(\hat{\boldsymbol{r}}\cdot\boldsymbol{r}')^m u_{2n-1-m}^{+}(\boldsymbol{r}')\,ds(\boldsymbol{r}') \tag{3.59}$$

To these we may add $A_1 = A_2 = 0$.

It would appear that it is necessary to know u_i^+ for $i \leq n-2$ in order to determine A_n but in fact it suffices to know u_i^+ for $i \leq n-3$. This may be seen as follows. The only term in A_n which involves u_{n-2}^+ has the integral

$$\int_S (\hat{\boldsymbol{n}}' \cdot \hat{\boldsymbol{r}}) u_{n-2}^+ \, ds = \hat{\boldsymbol{r}} \cdot \int_S \hat{\boldsymbol{n}}' (f_{n-2}^+ + v_{n-2}^+) \, ds \tag{3.60}$$

as a factor. Introducing the vector potential $\boldsymbol{\Psi}(\boldsymbol{r})$, which is used later in Chapter 4 to define the virtual mass and magnetic polarizability tensors, as the unique solution of the problem

$$\left.\begin{aligned} \nabla^2 \boldsymbol{\Psi} &= \boldsymbol{0} && \text{in } V^+ \\ \frac{\partial}{\partial n} \boldsymbol{\Psi} &= \hat{\boldsymbol{n}} && \text{on } S \\ \boldsymbol{\Psi} &= O\left(\frac{1}{r^2}\right) && \text{as } r \to \infty \end{aligned}\right\} \tag{3.61}$$

(3.60) may be rewritten, using the boundary condition on v_{n-2}^+ (3.53) as

$$\begin{aligned} &\int_S (\hat{\boldsymbol{n}}' \cdot \hat{\boldsymbol{r}}) u_{n-2}^+(\boldsymbol{r}') \, ds(\boldsymbol{r}') \\ &= \int_S \left[(\hat{\boldsymbol{n}}' \cdot \hat{\boldsymbol{r}}) f_{n-2}^+(\boldsymbol{r}') - (\hat{\boldsymbol{r}} \cdot \boldsymbol{\Psi}(\boldsymbol{r}')) \frac{\partial}{\partial n'} f_{n-2}^+(\boldsymbol{r}') \right] ds(\boldsymbol{r}') \end{aligned} \tag{3.62}$$

in which form, with the definition of f_{n-2}^+, (3.51), the right-hand side is seen to involve $u_i^+(\boldsymbol{r})$ for $i \leq n-3$ but not $i = n-2$. The comments after (3.44) apply in this case as well.

The differential scattering cross section (2.33) may be expressed as, using (3.57),

$$\begin{aligned} \sigma(\hat{\boldsymbol{r}}) &= 4\pi \sum_{n=2}^{\infty} k^{2n+2} \sum_{m=0}^{n-2} A_{2n-2m}(\hat{\boldsymbol{r}}) A_{2m+4}(\hat{\boldsymbol{r}}) \\ &\quad + 4\pi \sum_{n=1}^{\infty} k^{2n+2} \sum_{m=0}^{n-1} A_{2n-2m+1}(\hat{\boldsymbol{r}}) A_{2m+3}(\hat{\boldsymbol{r}}) \end{aligned} \tag{3.63}$$

where the functions A_n are given in (3.58) and (3.59).

As in the Dirichlet case the absorption cross section vanishes and the extinction cross section is, using the optical or forward scattering theorem

$$\sigma_e = -\frac{4\pi}{k^2} \operatorname{Re} g(\hat{\boldsymbol{k}}, \hat{\boldsymbol{k}}) = -4\pi \sum_{n=2}^{\infty} k^{2n-2} A_{2n}(\hat{\boldsymbol{k}}) \tag{3.64}$$

On the other hand, employing (3.63) in the definition of σ_s, (2.35), we find, since $\sigma_a = 0$

$$\sigma_e = \sum_{n=2}^{\infty} k^{2n+2} \sum_{m=0}^{n-2} \int_{S^2} A_{2n-2m}(\hat{\boldsymbol{r}}) A_{2m+4}(\hat{\boldsymbol{r}})\, ds(\hat{\boldsymbol{r}})$$

$$+ \sum_{n=1}^{\infty} k^{2n+2} \sum_{m=0}^{n-1} \int_{S^2} A_{2n-2m+1}(\hat{\boldsymbol{r}}) A_{2m+3}(\hat{\boldsymbol{r}})\, ds(\hat{\boldsymbol{r}}) \tag{3.65}$$

The second form of σ_e indicates that the leading term is $O(k^4)$, whereas the first form (3.64) indicates that the leading term is $O(k^2)$ unless $A_4(\hat{\boldsymbol{k}}, \hat{\boldsymbol{k}}) = 0$. No independent demonstration that $A_4(\hat{\boldsymbol{k}}, \hat{\boldsymbol{k}}) = 0$ is available for arbitrary shapes; however, the above analysis constitutes a proof that this must be the case. For surfaces with inversion symmetry, equation (3.49) relating $A_{2n}(\hat{\boldsymbol{r}}, \hat{\boldsymbol{k}})$ to previous coefficients remains valid. This ensures, since $A_1 = A_2 = 0$, that not only is $A_4 = 0$ when $\hat{\boldsymbol{r}} = \hat{\boldsymbol{k}}$, but $A_4(\hat{\boldsymbol{r}}) = 0$ for all $\hat{\boldsymbol{r}}$ for this class of surfaces. Whether this is true for arbitrary surfaces is an open question.

3.A.4 *The Robin problem*

The presence of the factor ik in the boundary condition makes the determination of the coefficients in the expansion of the solution of the Robin problem very similar to the corresponding process for the Neumann problem. Indeed, the Robin condition (3.9) implies, with (3.2), that

$$\frac{\partial}{\partial n} u_n^+ + n\nu u_{n-1}^+ = 0 \quad \text{on } S \tag{3.66}$$

So at the nth step, the normal derivative of u_n^+ is known in terms of previous coefficients but u_n^+ itself is not known, just as in the Neumann problem. With condition (3.66), equation (3.17) becomes $f_0^+ = u_0^i$, $f_1^+ = u_1^i$ and

$$f_n^+(\boldsymbol{r}) = u_n^i(\boldsymbol{r}) + \frac{1}{4\pi} \sum_{m=2}^{n} \binom{n}{m}$$

$$\times \int_S u_{n-m}^+(\boldsymbol{r}') \left[\frac{\partial}{\partial n'} |\boldsymbol{r} - \boldsymbol{r}'|^{m-1} + \nu m |\boldsymbol{r} - \boldsymbol{r}'|^{m-2} \right] ds(\boldsymbol{r}), \quad n \geq 2 \tag{3.67}$$

for all $\boldsymbol{r}$, while equation (3.19) remains a valid representation of $v_n^+(\boldsymbol{r})$. The coefficients $u_n^+(\boldsymbol{r})$ are given by (3.18) where the f_n^+ are defined above in (3.67) and the function v_n^+ is a solution of the following potential problem

$$\left.\begin{aligned} \nabla^2 v_n^+ &= 0 && \text{in } V^+ \\ \frac{\partial}{\partial n} v_n^+ &= -n\nu u_{n-1}^+ - \frac{\partial}{\partial n} f_n^+ && \text{on } S \\ v_n^+ &= O\left(\frac{1}{r}\right) && \text{as } r \to \infty \end{aligned}\right\} \tag{3.68}$$

If $u_m^+(\boldsymbol{r})$ are known for $m = 0, 1, \ldots, n-2$, then the normal derivative of f_n^+ is known and the determination of v_n^+ involves the solution of a Neumann problem, rather than a Robin problem, for the Laplacian. One might surmise that the solution of the Robin problem is merely a perturbation of the Neumann problem and indeed, as pointed out in Chapter 1, it is a regular perturbation in the sense that the difference between the solutions of the Neumann and Robin problems, for the same incident field, is $O(\nu)$ as $\nu \to 0$. However, the presence of the perturbation is much more evident in the far field where it changes the low frequency behavior. The complete low frequency expansion of the scattering amplitude or far field coefficient is obtained by substituting the boundary condition (3.66) in (3.33) obtaining

$$g(\hat{\boldsymbol{r}}) = \frac{1}{4\pi}\sum_{n=0}^{\infty}(ik)^{n+2}\sum_{m=0}^{n}\frac{(-1)^{m+1}}{m!(n-m)!} \times \int_S (\hat{\boldsymbol{n}}'\cdot\hat{\boldsymbol{r}}' - \nu)(\hat{\boldsymbol{r}}\cdot\boldsymbol{r}')^m u^+_{n-m}(\boldsymbol{r}')\,ds(\boldsymbol{r}') \tag{3.69}$$

For plane wave incidence, u_0^+ is the constant u_0^i, just as in the Neumann case. Then, we may rewrite (3.69) as

$$\begin{aligned} g(\hat{\boldsymbol{r}}) &= -\frac{(ik)^2}{4\pi}\int_S (\hat{\boldsymbol{n}}'\cdot\hat{\boldsymbol{r}} - \nu)u_0^+(\boldsymbol{r}')\,ds(\boldsymbol{r}') + O(k^3) \\ &= -\frac{(ik)^2}{4\pi}u_0^i\int_S (\hat{\boldsymbol{n}}'\cdot\hat{\boldsymbol{r}} - \nu)\,ds(\boldsymbol{r}') + O(k^3) \\ &= \nu\frac{(ik)^2}{4\pi}u_0^i\int_S ds(\boldsymbol{r}') + O(k^3) \\ &= \nu|S|\frac{(ik)^2}{4\pi}u_0^i + O(k^3) \end{aligned} \tag{3.70}$$

where $|S|$ is the surface area of S and equation (3.55) has been employed. This explicitly demonstrates the different low frequency behavior of g when $\nu \neq 0$, where $g = O(k^2)$, and when $\nu = 0$, which is the Neumann problem where $g = O(k^3)$.

With expression (3.69) for $g(\boldsymbol{r})$ it follows that

$$\left.\begin{aligned} \operatorname{Re} g(\hat{\boldsymbol{r}}) &= \sum_{n=1}^{\infty} k^{2n}A_{2n}(\hat{\boldsymbol{r}}) \\ \operatorname{Im} g(\hat{\boldsymbol{r}}) &= \sum_{n=1}^{\infty} k^{2n+1}A_{2n+1}(\hat{\boldsymbol{r}}) \end{aligned}\right\} \tag{3.71}$$

where

$$A_{2n}(\hat{\boldsymbol{r}}) = \frac{1}{4\pi}\sum_{m=0}^{2n-2}\frac{(-1)^{n+m+1}}{m!(2n-m-2)!}\int_S (\hat{\boldsymbol{n}}'\cdot\hat{\boldsymbol{r}} - \nu)(\hat{\boldsymbol{r}}\cdot\boldsymbol{r}')^m u^+_{2n-m-2}(\boldsymbol{r}')\,ds(\boldsymbol{r}') \tag{3.72}$$

$$A_{2n+1}(\hat{\boldsymbol{r}}) = \frac{1}{4\pi}\sum_{m=0}^{2n-1}\frac{(-1)^{n+m+1}}{m!(2n-m-1)!}\int_S (\hat{\boldsymbol{n}}'\cdot\hat{\boldsymbol{r}} - \nu)(\hat{\boldsymbol{r}}\cdot\boldsymbol{r}')^m u^+_{2n-m-1}(\boldsymbol{r}')\,ds(\boldsymbol{r}') \tag{3.73}$$

to which we may add $A_1 = 0$. This decomposition is correct only if the parameter ν is real because this ensures that the coefficients u_n^+ are also real.

It would appear from (3.69) that the coefficient of $(ik)^{n+2}$ involves u_i^+ for $i \leq n$. However, the term involving u_n^+ may be re-expressed in terms of the preceding coefficients, u_i^+ for $i \leq n-1$, as follows. The only term of order k^{n+2} involving u_n^+ contains the factor

$$\int_S (\nu - \hat{\boldsymbol{n}}' \cdot \hat{\boldsymbol{r}}) u_n^+(\boldsymbol{r}')\, ds(\boldsymbol{r}')$$

This may be rewritten using the potential function $\boldsymbol{\Psi}(\boldsymbol{r})$, introduced in the previous section as a solution of (3.61) and $\psi^c(\boldsymbol{r})$ which is defined as the solution of

$$\left.\begin{array}{lll} \nabla^2 \psi^c = 0 & & \text{in } V^+ \\ \dfrac{\partial}{\partial n}\psi^c = 1 & & \text{on } S \\ \psi^c = O\left(\dfrac{1}{r}\right) & & \text{as } r \to \infty \end{array}\right\} \tag{3.74}$$

Both of these potential functions are used in Chapter 4. With these functions, together with the decomposition of u_n^+, (3.18), the boundary condition in (3.68), and Green's theorem, we obtain

$$\begin{aligned} &\int_S (\nu - \hat{\boldsymbol{n}}' \cdot \hat{\boldsymbol{r}}) u_n^+(\boldsymbol{r}')\, ds(\boldsymbol{r}') \\ &\quad = \int_S (\nu - \hat{\boldsymbol{n}}' \cdot \hat{\boldsymbol{r}}) f_n^+(\hat{\boldsymbol{r}}')\, ds(\boldsymbol{r}') \\ &\quad\quad + \int_S \left(\nu \frac{\partial}{\partial n'}\psi^c(\boldsymbol{r}') - \hat{\boldsymbol{r}} \cdot \frac{\partial}{\partial n'}\,\boldsymbol{\Psi}(\boldsymbol{r}')\right) v_n^+(\boldsymbol{r}')\, ds(\boldsymbol{r}') \\ &\quad = \int_S (\nu - \hat{\boldsymbol{n}}' \cdot \hat{\boldsymbol{r}}) f_n^+(\boldsymbol{r}')\, ds(\boldsymbol{r}') \\ &\quad\quad - \int_S \left(\nu\psi^c(\boldsymbol{r}') - \hat{\boldsymbol{r}} \cdot\, \boldsymbol{\Psi}(\boldsymbol{r}')\right)\left(n\nu u_{n-1}^+(\boldsymbol{r}') + \frac{\partial}{\partial n'} f_n^+(\boldsymbol{r}')\right) ds(\boldsymbol{r}') \end{aligned} \tag{3.75}$$

In this form, with the definition, (3.67), of f_n^+ it is seen that only u_i^+ for $i \leq n-1$ are involved. The solution of (3.74) can be used to improve the approximation at any order in k.

The differential scattering cross section becomes

$$\begin{aligned} \sigma(\hat{\boldsymbol{r}}) &= 4\pi \sum_{n=1}^{\infty} k^{2n} \sum_{m=1}^{n} A_{2n-2m+2}(\hat{\boldsymbol{r}}) A_{2m}(\hat{\boldsymbol{r}}) \\ &\quad + 4\pi \sum_{n=2}^{\infty} k^{2n} \sum_{m=1}^{n-1} A_{2n-2m+1}(\hat{\boldsymbol{r}}) A_{2m+1}(\hat{\boldsymbol{r}}) \end{aligned} \tag{3.76}$$

and the scattering cross section is

$$\sigma_s = \sum_{n=1}^{\infty} k^{2n} \sum_{m=1}^{m} \int_{S^2} A_{2n-2m+2}(\hat{\boldsymbol{r}}) A_{2m}(\hat{\boldsymbol{r}})\, ds(\hat{\boldsymbol{r}})$$
$$+ \sum_{n=2}^{\infty} k^{2n} \sum_{m=1}^{n-1} \int_{S^2} A_{2n-2m+1}(\hat{\boldsymbol{r}}) A_{2m+1}(\hat{\boldsymbol{r}})\, ds(\hat{\boldsymbol{r}}) \tag{3.77}$$

In this case, the absorption cross section is nonzero and, from (2.42) and the low frequency expansion (3.2), is found to be

$$\sigma_a = \nu \int_S |u^+(\boldsymbol{r}')|^2\, ds(\boldsymbol{r}')$$
$$= \nu \sum_{n=0}^{\infty} k^{2n} \sum_{m=0}^{2n} \frac{(-1)^{m+n}}{m!(2n-m)!} \int_S u^+_{2n-m}(\boldsymbol{r}') u^+_m(\boldsymbol{r}')\, ds(\boldsymbol{r}') \tag{3.78}$$

This form of the expansion for σ_a is possible only because the coefficients u_n^+ are real.

We remark that an argument similar to that which showed that it is only required to know u_i^+ for $i \le n-1$ in order to determine A_{n+2} in the scattering amplitude, may be invoked to show that for plane wave incidence a knowledge of u_i^+ for $i \le 2n-1$ suffices to determine the coefficient of k^{2n} in the absorption cross section. This follows, since the only term that occurs in the coefficient of k^{2n} which involves u_{2n}^+ is, since $u_0^+ = 1$ for plane wave incidence

$$\int_S u^+_{2n}(\boldsymbol{r}')\, ds(\boldsymbol{r}')$$

Introducing the decomposition of u_{2n}^+, (3.18), the function ψ^c and the boundary condition on the normal derivative of v_n^+ we find, via Green's theorem, that

$$\int_S u^+_{2n}(\boldsymbol{r}')\, ds(\boldsymbol{r}') = \int_S \left[f^+_{2n}(\boldsymbol{r}') - \psi^c(\boldsymbol{r}') \frac{\partial}{\partial n'} f^+_{2n}(\boldsymbol{r}') - 2n\nu\psi^c(\boldsymbol{r}') u^+_{2n-1}(\boldsymbol{r}') \right] ds(\boldsymbol{r}') \tag{3.79}$$

which only involves u_i^+ for $i \le n-1$.

The extinction cross section is $\sigma_e = \sigma_s + \sigma_a$, and its low frequency expansion may be found by combining the expression for σ_s and σ_a.

3.A.5 *The transmission problem*

We treat the problems of lossless and lossy interior media separately in order to obtain true low frequency expansions. It should be kept in mind that if, after finding the coefficients in a low frequency expansion in terms of real β and η, we allow them to assume the complex values of the lossy media, each coefficient then becomes an analytic function of k. Then the series expansion is no longer a power series expansion in k since the coefficients are no longer independent of k. If we take the Nth partial sum of this series, expand each coefficient in an infinite series in powers of k and rearrange terms to obtain a power series again, the first N terms will coincide with the first N terms of the true low frequency expansion which we determine separately in the lossy case. Thus, the process of determining the low frequency expansion for real β and η and then allowing them to be functions of k, is in some sense a partial summation of the true low frequency expansion.

Here we present the low frequency expansion based on surface integrals. In Chapter 4 we give an alternative development based on volume integrals.

The lossless transmission problem: In this case, $\delta^- = 0$ in V^-, so that $\beta = \rho^+/\rho^-$ and $\beta\eta^2 = \gamma^-/\gamma^+$. The transmission conditions (3.10) together with the expansions (3.2) and (3.3) imply that the coefficients in the low frequency expansions also satisfy the transmission conditions

$$\left.\begin{aligned} u_n^+ &= u_n^- \\ \frac{\partial}{\partial n}u_n^+ &= \beta\frac{\partial}{\partial n}u_n^- \end{aligned}\right\} \quad \text{on } S \tag{3.80}$$

The coefficients $u_n^{\pm}(\boldsymbol{r})$ are given by (3.18) and (3.27) for $\boldsymbol{r}$ in V^+, and V^-, respectively, where $f_n^{\pm}$ are defined in (3.17) and (3.25). The full determination of $u_n^{\pm}$, assuming that $u_m^{\pm}$ are known for $m = 0, 1, \ldots, n-1$, is accomplished by solving the following transmission problem for $v_n^{\pm}(\boldsymbol{r})$

$$\left.\begin{aligned} \nabla^2 v_n^+ &= 0 && \text{in } V^+ \\ \nabla^2 v_n^- &= 0 && \text{in } V^- \\ v_n^+ &= v_n^- - (f_n^- + f_n^+) && \text{on } S \\ \frac{\partial}{\partial n}v_n^+ &= \beta\frac{\partial}{\partial n}v_n^- - \left(\beta\frac{\partial}{\partial n}f_n^- + \frac{\partial}{\partial n}f_n^+\right) && \text{on } S \\ v_n^+ &= O\left(\frac{1}{r}\right) && \text{as } r \to \infty \end{aligned}\right\} \tag{3.81}$$

With the transmission conditions, (3.80), the potentials $v_n^{\pm}$ have the representations

$$\begin{aligned} v_n^+(\boldsymbol{r}) &= \frac{1}{4\pi}\int_S \left[u_n^+(\boldsymbol{r}')\frac{\partial}{\partial n'}\frac{1}{|\boldsymbol{r}-\boldsymbol{r}'|} - \frac{1}{|\boldsymbol{r}-\boldsymbol{r}'|}\frac{\partial}{\partial n'}u_n^+(\boldsymbol{r}')\right] ds(\boldsymbol{r}') \\ &= \frac{1}{4\pi}\int_S \left[u_n^-(\boldsymbol{r}')\frac{\partial}{\partial n'}\frac{1}{|\boldsymbol{r}-\boldsymbol{r}'|} - \frac{\beta}{|\boldsymbol{r}-\boldsymbol{r}'|}\frac{\partial}{\partial n'}u_n^-(\boldsymbol{r}')\right] ds(\boldsymbol{r}') \\ &= \frac{1}{4\pi}\int_S \left[u_n^-(\boldsymbol{r}')\frac{\partial}{\partial n'}\frac{1}{|\boldsymbol{r}-\boldsymbol{r}'|} - \frac{1}{|\boldsymbol{r}-\boldsymbol{r}'|}\frac{\partial}{\partial n'}u_n^-(\boldsymbol{r}')\right] ds(\boldsymbol{r}') \\ &\quad + \frac{1-\beta}{4\pi}\int_S \frac{1}{|\boldsymbol{r}-\boldsymbol{r}'|}\frac{\partial}{\partial n'}u_n^-(\boldsymbol{r}')\, ds(\boldsymbol{r}') \\ &= -f_n^-(\boldsymbol{r}) + \frac{1-\beta}{4\pi}\int_S \frac{1}{|\boldsymbol{r}-\boldsymbol{r}'|}\frac{\partial}{\partial n'}u_n^-(\boldsymbol{r}')\, ds(\boldsymbol{r}'), \quad \boldsymbol{r} \in V^+, \end{aligned} \tag{3.82}$$

where (3.24) has been employed, and

$$
\begin{aligned}
v_n^-(\boldsymbol{r}) &= \frac{1}{4\pi}\int_S \left[\frac{1}{|\boldsymbol{r}-\boldsymbol{r}'|}\frac{\partial}{\partial n'}u_n^-(\boldsymbol{r}') - u_n^-(\boldsymbol{r}')\frac{\partial}{\partial n'}\frac{1}{|\boldsymbol{r}-\boldsymbol{r}'|}\right] ds(\boldsymbol{r}') \\
&= \frac{1}{4\pi}\int_S \left[\frac{1}{\beta}\frac{1}{|\boldsymbol{r}-\boldsymbol{r}'|}\frac{\partial}{\partial n'}u_n^+(\boldsymbol{r}') - u_n^+(\boldsymbol{r}')\frac{\partial}{\partial n'}\frac{1}{|\boldsymbol{r}-\boldsymbol{r}'|}\right] ds(\boldsymbol{r}') \\
&= \frac{1}{4\pi}\int_S \left[\frac{1}{|\boldsymbol{r}-\boldsymbol{r}'|}\frac{\partial}{\partial n'}u_n^+(\boldsymbol{r}') - u_n^+(\boldsymbol{r}')\frac{\partial}{\partial n'}\frac{1}{|\boldsymbol{r}-\boldsymbol{r}'|}\right] ds(\boldsymbol{r}') \\
&\quad + \frac{1-\beta}{4\pi\beta}\int_S \frac{1}{|\boldsymbol{r}-\boldsymbol{r}'|}\frac{\partial}{\partial n'}u_n^+(\boldsymbol{r}')\, ds(\boldsymbol{r}') \\
&= f_n^+(\boldsymbol{r}) + \frac{1-\beta}{4\pi}\int_S \frac{1}{|\boldsymbol{r}-\boldsymbol{r}'|}\frac{\partial}{\partial n'}u_n^-(\boldsymbol{r}')\, ds(\boldsymbol{r}'), \quad \boldsymbol{r}' \in V^- \qquad (3.83)
\end{aligned}
$$

where (3.16) has been used. Hence, the coefficients may be written as

$$
u_n^+(\boldsymbol{r}) = f_n^+(\boldsymbol{r}) - f_n^-(\boldsymbol{r}) + (1-\beta)w_n^+(\boldsymbol{r}), \quad \boldsymbol{r} \in V^+ \qquad (3.84)
$$

$$
u_n^-(\boldsymbol{r}) = f_n^+(\boldsymbol{r}) - f_n^-(\boldsymbol{r}) + (1-\beta)w_n^-(\boldsymbol{r}), \quad \boldsymbol{r} \in V^- \qquad (3.85)
$$

with

$$
w_n^\pm(\boldsymbol{r}) = \frac{1}{4\pi}\int_S \frac{1}{|\boldsymbol{r}-\boldsymbol{r}'|}\frac{\partial}{\partial n'}u_n^-(\boldsymbol{r}')\, ds', \quad \boldsymbol{r} \in V^\pm \qquad (3.86)
$$

and $f_n^\pm$ are given by (3.17) and (3.25). Note that $w_n^\pm$, being single layer potentials, coincide on S.

It is remarkable to observe that when $\beta = 1$ it is not necessary to solve any potential problem as (3.84) and (3.85) clearly show. In this case, the determination of u_n^+ and u_n^- is simply a matter of quadratures. Indeed, with (3.17), (3.25), (3.84) and (3.85) it follows that $u_0^\pm = u_0^i$, $u_1^\pm = u_1^i$ and

$$
\begin{aligned}
u_n^\pm(\boldsymbol{r}) &= u_n^i(\boldsymbol{r}) + \frac{1}{4\pi}\sum_{m=1}^{n}\binom{n}{m}(1-\eta^m)\int_S \left[u_{n-m}^-(\boldsymbol{r}')\frac{\partial}{\partial n'}|\boldsymbol{r}-\boldsymbol{r}'|^{m-1}\right. \\
&\quad \left. - |\boldsymbol{r}-\boldsymbol{r}'|^{m-1}\frac{\partial}{\partial n'}u_{n-m}^-(\boldsymbol{r}')\right] ds(\boldsymbol{r}'), \quad \boldsymbol{r} \in V^\pm, \ n \geq 2 \qquad (3.87)
\end{aligned}
$$

When $\beta \neq 1$ one may pose a transmission problem for the potentials $w_n^\pm$ as an alternative to (3.81); namely

$$\left.\begin{aligned} \nabla^2 w_n^+ &= 0 && \text{in } V^+ \\ \nabla^2 w_n^- &= 0 && \text{in } V^- \\ w_n^+ &= w_n^- && \text{on } S \\ \frac{\partial}{\partial n} w_n^+ + \frac{\partial}{\partial n}(f_n^+ - f_n^-) &= \beta \frac{\partial}{\partial n} w_n^- && \text{on } S \\ w_n^+ &= O\left(\frac{1}{r}\right) && \text{as } r \to \infty \end{aligned}\right\} \tag{3.88}$$

From this investigation of the cases for $\beta = 1$ and $\beta \neq 1$, it follows that in the lossless case with equal mass densities in V^+ and V^- it is possible to construct the entire low frequency expansion of the solution of the scattering problem generated by differences in compressibility in V^+ and V^- simply by quadratures. However, when the mass densities differ it is necessary to solve a potential problem at each step. Hence, the density has a more pronounced effect on low frequency scattering than the compressibility.

The scattering amplitude is given again by (3.33). However, a few simplifications are possible. First, we rewrite (3.33) as

$$\begin{aligned} g(\hat{\boldsymbol{r}}) = &-\frac{ik}{4\pi}\int_S \frac{\partial}{\partial n'} u_0^+(\boldsymbol{r}')\, ds(\boldsymbol{r}') - \frac{1}{4\pi}\sum_{n=0}^{\infty}\frac{(ik)^{n+2}}{(n+1)!}\int_S \frac{\partial}{\partial n'} u_{n+1}^+(\boldsymbol{r}')\, ds(\boldsymbol{r}') \\ &+ \frac{1}{4\pi}\sum_{n=0}^{\infty}\frac{(ik)^{n+2}}{n!}\sum_{m=0}^{n}\binom{n}{m}(-1)^m \int_S \left[\frac{(\hat{\boldsymbol{r}}\cdot\boldsymbol{r}')^{m+1}}{m+1}\frac{\partial}{\partial n'} u_{n-m}^+(\boldsymbol{r}') \right. \\ &\left. - (\hat{\boldsymbol{r}}\cdot\hat{\boldsymbol{n}}')(\hat{\boldsymbol{r}}\cdot\boldsymbol{r}')^m u_{n-m}^+(\boldsymbol{r}')\right] ds(\boldsymbol{r}') \end{aligned} \tag{3.89}$$

With the transmission conditions (3.80) and equation (3.13) we see that

$$\begin{aligned} \int_S \frac{\partial}{\partial n} u_n^+(\boldsymbol{r}')\, ds(\boldsymbol{r}') &= \beta \int_S \frac{\partial}{\partial n'} u_n^-(\boldsymbol{r}')\, ds(\boldsymbol{r}') \\ &= \beta \int_{V^-} \nabla^2 u_n^-(\boldsymbol{r}')\, dv(\boldsymbol{r}') \\ &= n(n-1)\beta\eta^2 \int_{V^-} u_{n-2}^-(\boldsymbol{r}')\, dv(\boldsymbol{r}'), \quad n \geq 0 \end{aligned} \tag{3.90}$$

Hence, we may rewrite the expression for $g(\hat{\boldsymbol{r}})$ as

$$g(\hat{\boldsymbol{r}}) = \sum_{n=0}^{\infty}[k^{2n+2}A_{2n+2}(\hat{\boldsymbol{r}}) + ik^{2n+3}A_{2n+3}(\hat{\boldsymbol{r}})] \tag{3.91}$$

where

$$A_{2n+2}(\hat{\boldsymbol r}) = \frac{(-1)^n}{4\pi(2n)!}\left\{2n\beta\eta^2\int\limits_{V^-} u^-_{2n-1}(\boldsymbol r')\,dv(\boldsymbol r')\right.$$
$$+\sum_{m=0}^{2n}\binom{2n}{m}(-1)^{m+1}\int\limits_S\left[\frac{(\hat{\boldsymbol r}\cdot\boldsymbol r')^{m+1}}{m+1}\frac{\partial}{\partial n'}u^+_{2n-m}(\boldsymbol r')\right.$$
$$\left.\left.-(\hat{\boldsymbol r}\cdot\hat{\boldsymbol n}')(\hat{\boldsymbol r}\cdot\boldsymbol r')^m u^+_{2n-m}(\boldsymbol r')\right]ds(\boldsymbol r')\right\} \tag{3.92}$$

$$A_{2n+3}(\hat{\boldsymbol r}) = \frac{(-1)^n}{4\pi(2n+1)!}\left\{(2n+1)\beta\eta^2\int\limits_{V^-} u^-_{2n}(\boldsymbol r')\,dv(\boldsymbol r')\right.$$
$$+\sum_{m=0}^{2n+1}\binom{2n+1}{m}(-1)^{m+1}\int\limits_S\left[\frac{(\hat{\boldsymbol r}\cdot\boldsymbol r')^{m+1}}{m+1}\frac{\partial}{\partial n'}u^+_{2n+1-m}(\boldsymbol r')\right.$$
$$\left.\left.-(\hat{\boldsymbol r}\cdot\hat{\boldsymbol n}')(\hat{\boldsymbol r}\cdot\boldsymbol r')^m u^+_{2n+1-m}(\boldsymbol r')\right]ds(\boldsymbol r')\right\} \tag{3.93}$$

One must bear in mind that the coefficients A_{n+2} are real, since we have assumed the scatterer to be lossless; that is, both β and η are real. It would appear from (3.89) and (3.90) that it is necessary to know the functions u^+_m for $m = 0, \ldots, n$ in order to determine the coefficient of k^{n+2}, but in fact one only needs to know u^+_m up to order $n-1$ if use is made of the vector field $\boldsymbol v^\pm$ associated with the solution of the following transmission potential problem

$$\left.\begin{aligned}
\nabla^2\boldsymbol v^+ &= \boldsymbol 0 && \text{in } V^+\\
\nabla^2\boldsymbol v^- &= \boldsymbol 0 && \text{in } V^-\\
\boldsymbol v^+ &= \boldsymbol v^- + \boldsymbol r && \text{on } S\\
\frac{\partial}{\partial n}\boldsymbol v^+ &= \beta\frac{\partial}{\partial n}\boldsymbol v^- + \hat{\boldsymbol n} && \text{on } S\\
\boldsymbol v^+ &= O\left(\frac{1}{r^2}\right) && \text{as } r\to\infty
\end{aligned}\right\} \tag{3.94}$$

This vector field $\boldsymbol v^\pm$ is used in Chapter 5 to define the generalized polarizability tensor. To see this one may find, after some manipulation, that the only factor in the term of order k^{n+2} which apparently involves u^+_n may be rewritten as

$$\int\limits_S\left[(\hat{\boldsymbol r}\cdot\hat{\boldsymbol n}')u^+_n(\boldsymbol r') - (\hat{\boldsymbol r}\cdot\boldsymbol r')\frac{\partial}{\partial n'}u^+_n(\boldsymbol r')\right]ds(\boldsymbol r')$$
$$= \int\limits_S\left[\hat{\boldsymbol r}\cdot\left(\frac{\partial}{\partial n'}\boldsymbol v^+(\boldsymbol r') - \beta\frac{\partial}{\partial n'}\boldsymbol v^-(\boldsymbol r')\right)u^+_n(\boldsymbol r')\right.$$
$$\left.-\hat{\boldsymbol r}\cdot(\boldsymbol v^+(\boldsymbol r') - \boldsymbol v^-(\boldsymbol r'))\frac{\partial}{\partial n'}u^+_n(\boldsymbol r')\right]ds(\boldsymbol r')$$

$$
\begin{aligned}
&= \hat{\boldsymbol{r}} \cdot \int_S \left[(f_n^+(\boldsymbol{r}') + v_n^+(\boldsymbol{r}')) \frac{\partial}{\partial n'} \boldsymbol{v}^+(\boldsymbol{r}') \right. \\
&\quad \left. - \boldsymbol{v}^+(\boldsymbol{r}') \frac{\partial}{\partial n'} (f_n^+(\boldsymbol{r}') + v_n^+(\boldsymbol{r}')) \right] ds(\boldsymbol{r}') \\
&\quad + \beta \hat{\boldsymbol{r}} \cdot \int_S \left[\boldsymbol{v}^-(\boldsymbol{r}') \frac{\partial}{\partial n'} u_n^-(\boldsymbol{r}') - u_n^-(\boldsymbol{r}') \frac{\partial}{\partial n'} \boldsymbol{v}^-(\boldsymbol{r}') \right] ds(\boldsymbol{r}') \\
&= \hat{\boldsymbol{r}} \cdot \int_S \left[f_n^+(\boldsymbol{r}') \frac{\partial}{\partial n'} \boldsymbol{v}^+(\boldsymbol{r}') - \boldsymbol{v}^+(\boldsymbol{r}') \frac{\partial}{\partial n'} f_n^+(\boldsymbol{r}') \right] ds(\boldsymbol{r}') \\
&\quad + \beta \hat{\boldsymbol{r}} \cdot \int_S \left[f_n^-(\boldsymbol{r}') \frac{\partial}{\partial n'} \boldsymbol{v}^-(\boldsymbol{r}') - \boldsymbol{v}^-(\boldsymbol{r}') \frac{\partial}{\partial n'} f_n^-(\boldsymbol{r}') \right] ds(\boldsymbol{r}')
\end{aligned}
\tag{3.95}
$$

where Green's theorem has been invoked twice to obtain the last equality. Relation (3.95) now shows that in order to calculate the k^{n+2} term of g we only need $u_n^{\pm}$ for $m = 0, 1, 2, \ldots, n-1$. Hence, as in the other cases of acoustic scattering, the solution of one potential problem can be used to improve the approximation of g.

The lossy transmission problem: When the scatterer causes dissipation, the compressional viscosity δ^- in V^- is a positive real number. In this case, β and η^2 are no longer real. In fact, they assume the values

$$
\beta = \frac{\rho^+}{\rho^-} \left(1 - ik \frac{\delta^- \gamma^-}{\sqrt{\rho^+ \gamma^+}} \right) \tag{3.96}
$$

and

$$
\eta^2 = \frac{\gamma^- \rho^-}{\gamma^+ \rho^+} \left(1 - ik \frac{\delta^- \gamma^-}{\sqrt{\rho^+ \gamma^+}} \right)^{-1} \tag{3.97}
$$

or, for

$$
k < \frac{\sqrt{\rho^+ \gamma^+}}{\delta^- \gamma^-} \tag{3.98}
$$

$$
\eta^2 = \frac{\rho^- \gamma^-}{\rho^+ \gamma^+} \sum_{n=0}^{\infty} (ik)^n \left(\frac{\delta^- \gamma^-}{\sqrt{\rho^+ \gamma^+}} \right)^n \tag{3.99}
$$

Therefore, for a lossy scatterer, β is a linear function of k, η^2 has a series expansion in powers of k and $\beta\eta^2$ is real.

Substituting (3.2), (3.3), (3.96) and (3.99) into the Helmholtz equations (3.12) and (3.13) in V^+ and V^-, respectively, as well as in the transmission conditions (3.80), and equating like powers of k we obtain the following equations for every $n \geq 0$

$$
\nabla^2 u_n^+ = n(n-1) u_{n-2}^+ \qquad \text{in } V^+ \tag{3.100}
$$

$$
\nabla^2 u_n^- = n(n-1) \frac{\rho^- \gamma^-}{\rho^+ \gamma^+} \sum_{m=0}^{n-2} \frac{(n-2)!}{m!} \left(\frac{\delta^- \gamma^-}{\sqrt{\rho^+ \gamma^+}} \right)^{n-m-2} u_m^- \qquad \text{in } V^- \tag{3.101}
$$

and the transmission conditions

$$u_n^+ = u_n^- \qquad \text{on } S \tag{3.102}$$

$$\frac{\partial}{\partial n}u_n^+ = \frac{\rho^+}{\rho^-}\left(\frac{\partial}{\partial n}u_n^- - n\frac{\delta^-\gamma^-}{\sqrt{\rho^+\gamma^+}}\frac{\partial}{\partial n}u_{n-1}^-\right) \qquad \text{on } S \tag{3.103}$$

The decomposition (3.18) of u_n^+, where f_n^+ and v_n^+ are given by (3.17) and (3.19), respectively, hold true for lossy scatters as well, but the decomposition (3.27), (3.25) and (3.28) no longer holds. Nevertheless, the relation

$$u_n^+(\boldsymbol{r}) = f_n^+(\boldsymbol{r}) + O\left(\frac{1}{r}\right), \qquad r \to \infty \tag{3.104}$$

where f_n^+ are given in (3.17) and are particular solutions of (3.100), ensures that the transmission problems (3.100)–(3.103) are well posed. Restricting attention to the case of plane wave incidence, from (3.100)–(3.104) it is easy to see that $u_0^\pm = 1$ and that the leading two low frequency approximations $u_0^\pm$ and $u_1^\pm$ coincide with the corresponding approximations for the lossless case. Therefore, dissipation in low frequency scattering is a second-order effect for the near field.

In fact, the first coefficient that involves the compressional viscosity parameter, δ^-, is the solution of the following transmission problem

$$\nabla^2 u_2^+ = 2 \qquad \text{in } V^+ \tag{3.105}$$

$$\nabla^2 u_2^- = 2\frac{\rho^-\gamma^-}{\rho^+\gamma^+} \qquad \text{in } V^- \tag{3.106}$$

$$u_2^+ = u_2^- \qquad \text{on } S \tag{3.107}$$

$$\frac{\partial}{\partial n}u_2^+ = \frac{\rho^+}{\rho^-}\left(\frac{\partial}{\partial n}u_2^- - 2\frac{\delta^-\gamma^-}{\sqrt{\rho^+\gamma^+}}\frac{\partial}{\partial n}u_1^-\right) \qquad \text{on } S \tag{3.108}$$

$$u_2^+ = u_2^i + \frac{1}{4\pi}\int_S \frac{\partial}{\partial n'}|\boldsymbol{r}-\boldsymbol{r}'|\,ds(\boldsymbol{r}') - \frac{1}{2\pi}\int_S \frac{\partial}{\partial n'}u_1^+(\boldsymbol{r}')\,ds(\boldsymbol{r}') + O\left(\frac{1}{r}\right), \qquad r \to \infty \tag{3.109}$$

where u_1^- is the corresponding solution for the lossless case, and $u_0^+ = u_0^- = 1$. Note that for plane wave incidence

$$u_2^i(\boldsymbol{r}) = (\hat{\boldsymbol{k}}\cdot\boldsymbol{r})^2, \qquad \boldsymbol{r} \in \mathbb{R}^3 \tag{3.110}$$

Hence

$$\nabla^2 u_2^i(\boldsymbol{r}) = 2, \qquad \boldsymbol{r} \in \mathbb{R}^3 \tag{3.111}$$

and the two particular solutions needed to solve (3.105) and (3.106) are given by

$$u_{p2}^{+}(\boldsymbol{r}) = (\hat{\boldsymbol{k}} \cdot \boldsymbol{r})^2, \quad \boldsymbol{r} \in V^{+} \tag{3.112}$$

and

$$u_{p2}^{-}(\boldsymbol{r}) = \frac{\rho^{-}\gamma^{-}}{\rho^{+}\gamma^{+}}(\hat{\boldsymbol{k}} \cdot \boldsymbol{r})^2, \quad \boldsymbol{r} \in V^{-} \tag{3.113}$$

respectively. In this case, we can reduce the solution of (3.105)–(3.109) to potential problems.

Higher order approximations for the lossy transmission problem are obtained similarly from (3.100)–(3.104). Obviously, in the limit as $\delta^{-} \to 0$ we recover the lossless transmission problem.

The scattering amplitude is given as before in terms of the u_n^{+} and through that the differential scattering cross section as well as the scattering cross section can be evaluated by means of (2.33) and (2.35), respectively.

The absorption cross section (2.39) is not zero for lossy scatterers and its low frequency expansion is given by

$$\begin{aligned}\sigma_a = {} & \frac{\gamma^{-}}{\gamma^{+}} \sum_{n=0}^{\infty} k^{2n+2} \sum_{m=0}^{n} \sum_{s=0}^{2n-2m} \frac{(-1)^{n-s}}{s!(2n-2m-s)!} \left(\frac{\delta^{-}\gamma^{-}}{\sqrt{\rho^{+}\gamma^{+}}}\right)^{2m+1} \\ & + \int_{V^{-}} u_s^{-}(\boldsymbol{r}')u_{2n-2m-s}^{-}(\boldsymbol{r}')\, dv(\boldsymbol{r}') \\ & + \frac{\rho^{+}}{\rho^{-}} \frac{\delta^{-}\gamma^{-}}{\sqrt{\rho^{+}\gamma^{+}}} \sum_{n=0}^{\infty} k^{2n+2} \sum_{m=0}^{2n+2} \frac{(-1)^{n-m+1}}{m!(2n-m+2)!} \\ & \times \int_{S} u_m^{-}(\boldsymbol{r}') \frac{\partial}{\partial n'} u_{2n-m-2}^{-}(\boldsymbol{r}')\, ds(\boldsymbol{r}')\end{aligned} \tag{3.114}$$

and, in particular, the leading approximation is

$$\sigma_a = k^2 \delta^{-} \frac{\gamma^{-}}{\rho^{-}} \sqrt{\frac{\rho^{+}}{\gamma^{+}}} \int_{V^{-}} |\nabla u_1^{-}(\boldsymbol{r})|^2\, dv(\boldsymbol{r}) + O(k^4), \quad k \to 0. \tag{3.115}$$

Note that the coefficients $u_0^{\pm}$ and $u_1^{\pm}$ are the same whether the medium is lossless or lossy. Thus, in a sense, these first terms cannot distinguish between lossy and lossless scatterers. Nevertheless, the first term in the absorption cross section for a lossy medium is expressed in terms of u_1^{-}, a coefficient which does not 'know' that the medium is lossy.

3.A.6 *Problems in acoustics*

1. Use the expansions (3.1), (3.2), (3.3), (3.14) and (3.23) in the representations (2.7) and (2.5) to generate the expressions (3.16) and (3.17), (3.24) and (3.25) for the exterior and interior coefficients of order n, respectively.

2. Show that expression (3.43) can be transformed into (3.44).
3. Derive formulae (3.46) and (3.48).
4. Provide a detailed proof of (3.56).
5. Provide a detailed proof of (3.62).
6. Justify expression (3.67).
7. Consider the lossless transmission problem for which $\rho^+ = \rho^-$ and let v_n^+ be the solution of (3.81) in V^+. Prove that

$$v_n^+(\boldsymbol{r}) = -\frac{n(n-1)\eta^2}{4\pi} \int_{V^-} \frac{u_{n-2}^-(\boldsymbol{r}')}{|\boldsymbol{r}-\boldsymbol{r}'|}\, dv(\boldsymbol{r}')$$

8. Show all details in the derivation of (3.95).
9. Derive the Poisson equation (3.101) and the transmission condition (3.103) for the lossy transmission problem.
10. Derive formula (3.114) for the absorption coefficient of a lossy penetrable scatterer.
11. Prove that the Rayleigh term (leading low frequency approximation) of the absorption cross section for a lossy penetrable scatterer is given by

$$\sigma_a = k^2 \delta^- \frac{\gamma^-}{\rho^-} \sqrt{\frac{\rho^+}{\gamma^+}} \int_{V^-} |\nabla u_1^-(\boldsymbol{r})|^2\, dv(\boldsymbol{r}) + O(k^4), \quad k \to 0$$

where u_1^- is the first-order interior approximation.
12. Show that an alternative form of (3.101) is given by

$$\nabla^2 u_n^-(\boldsymbol{r}) = n(n-1)\frac{\gamma^-\rho^-}{\gamma^+\rho^+} u_{n-2}^-(\boldsymbol{r}) + n\frac{\gamma^-\delta^-}{\sqrt{\gamma^+\rho^+}} \nabla^2 u_{n-1}^-(\boldsymbol{r})$$

3.EM Electromagnetics

3.EM.1 *General low frequency expansions in electromagnetics*

We recall that attention will be restricted to incident waves which are either plane waves, radiating dipoles in V^+, or superpositions of such sources. The fields from such sources admit of power series expansion of the form

$$\boldsymbol{E}^i(\boldsymbol{r}) = \sum_{n=0}^{\infty} \frac{(ik)^n}{n!} \boldsymbol{E}_n^i(\boldsymbol{r}), \quad \boldsymbol{H}^i(\boldsymbol{r}) = \sum_{n=0}^{\infty} \frac{(ik)^n}{n!} \boldsymbol{H}_n^i(\boldsymbol{r}), \quad \boldsymbol{r} \in \mathbb{R}^3 \tag{3.116}$$

except at singularities of the incident fields.

In all three electromagnetic scattering problems introduced in Chapter 1 we obtain formal low frequency expansions of the solutions by assuming that

$$\boldsymbol{E}^+(\boldsymbol{r}) = \sum_{n=0}^{\infty} \frac{(ik)^n}{n!} \boldsymbol{E}_n^+(\boldsymbol{r}), \quad \boldsymbol{H}^+(\boldsymbol{r}) = \sum_{n=0}^{\infty} \frac{(ik)^n}{n!} \boldsymbol{H}_n^+(\boldsymbol{r}), \quad \boldsymbol{r} \in V^+ \tag{3.117}$$

and, in the transmission problem

$$\boldsymbol{E}^-(\boldsymbol{r}) = \sum_{n=0}^{\infty} \frac{(ik)^n}{n!} \boldsymbol{E}_n^-(\boldsymbol{r}), \quad \boldsymbol{H}^-(\boldsymbol{r}) = \sum_{n=0}^{\infty} \frac{(ik)^n}{n!} \boldsymbol{H}_n^-(\boldsymbol{r}), \quad \boldsymbol{r} \in V^- \tag{3.118}$$

The expansions of $\boldsymbol{E}^-$ and $\boldsymbol{H}^-$ are in powers of k rather than k^-. Expansions in powers of k^- do exist, but they may be changed to expansions in powers of k via the relation $k^- = k\eta$ and absorbing the powers of η into the coefficients. Note that, just as in acoustics, the dimensions of the coefficients are a factor of length to the nth power higher than the dimensions of the corresponding fields.

In all the electromagnetic scattering problems considered, the terms $\boldsymbol{E}^+$ and $\boldsymbol{H}^+$ consist of sums of incident and scattered fields in V^+; that is

$$\boldsymbol{E}^+(\boldsymbol{r}) = \boldsymbol{E}^i(\boldsymbol{r}) + \boldsymbol{E}(\boldsymbol{r}), \quad \boldsymbol{H}^+(\boldsymbol{r}) = \boldsymbol{H}^i(\boldsymbol{r}) + \boldsymbol{H}(\boldsymbol{r}) \tag{3.119}$$

where

$$\left.\begin{aligned} \nabla \times \boldsymbol{E} &= ikZ^+\boldsymbol{H} \\ \nabla \times \boldsymbol{H} &= -ikY^+\boldsymbol{E} \end{aligned}\right\} \quad \text{in } V^+ \tag{3.120}$$

$$\nabla \cdot \boldsymbol{E}^+ = \nabla \cdot \boldsymbol{H}^+ = 0 \quad \text{in } V^+ \tag{3.121}$$

and $\boldsymbol{E}$ and $\boldsymbol{H}$ satisfy the radiation conditions (1.148) and (1.149).

The boundary and transmission conditions introduced in Chapter 1 define the following problems.

The perfect conductor problem:

$$\hat{\boldsymbol{n}} \times \boldsymbol{E}^+ = \boldsymbol{0}, \qquad \hat{\boldsymbol{n}} \cdot \boldsymbol{H}^+ = 0 \quad \text{on } S \tag{3.122}$$

The impedance problem:

$$\hat{\boldsymbol{n}} \times (\hat{\boldsymbol{n}} \times \boldsymbol{E}^+) = -Z_s Z^+ \hat{\boldsymbol{n}} \times \boldsymbol{H}^+ \quad \text{on } S \tag{3.123}$$

The transmission problem:

$$\left.\begin{aligned} \hat{\boldsymbol{n}} \times \boldsymbol{E}^+ &= \hat{\boldsymbol{n}} \times \boldsymbol{E}^- \\ \hat{\boldsymbol{n}} \times \boldsymbol{H}^+ &= \hat{\boldsymbol{n}} \times \boldsymbol{H}^- \end{aligned}\right\} \quad \text{on } S \tag{3.124}$$

$$\left.\begin{aligned} \epsilon^+ \hat{\boldsymbol{n}} \cdot \boldsymbol{E}^+ &= \epsilon^- \left(1 + \frac{i\sigma^-}{k} \frac{\epsilon^+}{\epsilon^-} Z^+\right) \hat{\boldsymbol{n}} \cdot \boldsymbol{E}^- \\ \mu^+ \hat{\boldsymbol{n}} \cdot \boldsymbol{H}^+ &= \mu^- \hat{\boldsymbol{n}} \cdot \boldsymbol{H}^- \end{aligned}\right\} \quad \text{on } S \tag{3.125}$$

where

$$\left.\begin{aligned} \nabla \times \boldsymbol{E}^- &= ik\frac{\mu^-}{\mu^+}Z^+\boldsymbol{H}^- \\ \nabla \times \boldsymbol{H}^- &= -ik\frac{\epsilon^-}{\epsilon^+}Y^+\left(1+\frac{i\sigma^-}{k}\frac{\epsilon^+}{\epsilon^-}Z^+\right)\boldsymbol{E}^- \\ \nabla \cdot \boldsymbol{E}^- &= \nabla \cdot \boldsymbol{H}^- = 0 \end{aligned}\right\} \quad \text{in } V^- \tag{3.126}$$

Substituting the expansions (3.117) and (3.118) into the Maxwell equations and equating like powers of ik leads to

$$\left.\begin{aligned} \nabla \times \boldsymbol{E}_n^+ &= nZ^+\boldsymbol{H}_{n-1}^+ \\ \nabla \times \boldsymbol{H}_n^+ &= -nY^+\boldsymbol{E}_{n-1}^+ \\ \nabla \cdot \boldsymbol{E}_n^+ &= \nabla \cdot \boldsymbol{H}_n^+ = 0 \end{aligned}\right\} \quad \text{in } V^+ \tag{3.127}$$

$$\left.\begin{aligned} \nabla \times \boldsymbol{E}_n^- &= n\frac{\mu^-}{\mu^+}Z^+\boldsymbol{H}_{n-1}^- \\ \nabla \times \boldsymbol{H}_n^- &= -n\frac{\epsilon^-}{\epsilon^+}Y^+\boldsymbol{E}_{n-1}^- + \sigma^-\boldsymbol{E}_n^- \\ \nabla \cdot \boldsymbol{E}_n^- &= \nabla \cdot \boldsymbol{H}_n^- = 0 \end{aligned}\right\} \quad \text{in } V^- \tag{3.128}$$

for $n = 0, 1, 2, \ldots$.

Observe that since S is closed, a straightforward application of Stokes' theorem together with (3.127) and (3.128) leads to the facts that

$$\int_S \hat{\boldsymbol{n}} \cdot \boldsymbol{H}_n^\pm \, ds = \int_S \hat{\boldsymbol{n}} \cdot \boldsymbol{E}_n^\pm \, ds = 0 \tag{3.129}$$

These relations will prove useful in what follows. Equations (3.129) not only imply that there is no net surface charge density on S (see the definition of ρ_s in Chapter 1) but also that there would be no net 'magnetic charge' on S ($\int_S \hat{\boldsymbol{n}} \cdot \boldsymbol{B}\, ds = 0$) if this concept made physical sense.

In this section we show how the coefficients $\boldsymbol{E}_n^\pm$ and $\boldsymbol{H}_n^\pm$ may be characterized in terms of the previous coefficients, $\boldsymbol{E}_m^\pm$ and $\boldsymbol{H}_m^\pm$, $m = 0, 1, \ldots, n-1$ and the solution of potential problems. In addition, expressions for the scattering amplitude in terms of these coefficients are also presented.

As in the acoustic case the key to this development lies in low frequency expansions of the integral representations given in Chapter 2. Any of the three forms given there may be used to obtain the low frequency expansion. All require a certain amount of manipulation, but perhaps the most convenient for the present purpose is the Stratton–Chu representation (2.69) and (2.70). Substituting the expansions of the fields (3.117) together with the expansions of G^+ (3.14) into the Stratton–Chu representations, reordering terms and equating coefficients of like powers of ik, we obtain

$$\left.\begin{aligned} \alpha(\boldsymbol{r})\boldsymbol{E}_n^+(\boldsymbol{r}) &= \boldsymbol{F}_{en}^+(\boldsymbol{r}) + \boldsymbol{U}_{en}^+(\boldsymbol{r}) \\ \alpha(\boldsymbol{r})\boldsymbol{H}_n^+(\boldsymbol{r}) &= \boldsymbol{F}_{mn}^+(\boldsymbol{r}) + \boldsymbol{U}_{mn}^+(\boldsymbol{r}) \end{aligned}\right\} \tag{3.130}$$

where

$$F^{+}_{en}(\boldsymbol{r}) = E^{i}_{n}(\boldsymbol{r}) + \frac{Z^{+}}{4\pi}\sum_{m=1}^{n}\binom{n}{m}m\int_{S}|\boldsymbol{r}-\boldsymbol{r}'|^{m-2}(\hat{\boldsymbol{n}}'\times \boldsymbol{H}^{+}_{n-m}(\boldsymbol{r}'))\,ds(\boldsymbol{r}')$$

$$+\frac{1}{4\pi}\sum_{m=2}^{n}\binom{n}{m}(m-1)\int_{S}|\boldsymbol{r}-\boldsymbol{r}'|^{m-3}[(\boldsymbol{r}-\boldsymbol{r}')\times(\hat{\boldsymbol{n}}\times \boldsymbol{E}^{+}_{n-m}(\boldsymbol{r}'))$$

$$-(\boldsymbol{r}-\boldsymbol{r}')(\hat{\boldsymbol{n}}'\cdot \boldsymbol{E}^{+}_{n-m}(\boldsymbol{r}'))]\,ds(\boldsymbol{r}') \tag{3.131}$$

$$F^{+}_{mn}(\boldsymbol{r}) = H^{i}_{n}(\boldsymbol{r}) - \frac{Y^{+}}{4\pi}\sum_{m=1}^{n}\binom{n}{m}m\int_{S}|\boldsymbol{r}-\boldsymbol{r}'|^{m-2}(\hat{\boldsymbol{n}}'\times \boldsymbol{E}^{+}_{n-m}(\boldsymbol{r}'))\,ds(\boldsymbol{r}')$$

$$+\frac{1}{4\pi}\sum_{m=2}^{n}\binom{n}{m}(m-1)\int_{S}|\boldsymbol{r}-\boldsymbol{r}'|^{m-3}[(\boldsymbol{r}-\boldsymbol{r}')\times(\hat{\boldsymbol{n}}\times \boldsymbol{H}^{+}_{n-m}(\boldsymbol{r}'))$$

$$-(\boldsymbol{r}-\boldsymbol{r}')(\hat{\boldsymbol{n}}'\cdot \boldsymbol{H}^{+}_{n-m}(\boldsymbol{r}'))]\,ds(\boldsymbol{r}') \tag{3.132}$$

$$U^{+}_{en}(\boldsymbol{r}) = \frac{1}{4\pi}\nabla\times\int_{S}\frac{\hat{\boldsymbol{n}}'\times \boldsymbol{E}^{+}_{n}(\boldsymbol{r}')}{|\boldsymbol{r}-\boldsymbol{r}'|}\,ds(\boldsymbol{r}') - \frac{1}{4\pi}\nabla\int_{S}\frac{\hat{\boldsymbol{n}}'\cdot \boldsymbol{E}^{+}_{n}(\boldsymbol{r}')}{|\boldsymbol{r}-\boldsymbol{r}'|}\,ds(\boldsymbol{r}') \tag{3.133}$$

and

$$U^{+}_{mn}(\boldsymbol{r}) = \frac{1}{4\pi}\nabla\times\int_{S}\frac{\hat{\boldsymbol{n}}'\times \boldsymbol{H}^{+}_{n}(\boldsymbol{r}')}{|\boldsymbol{r}-\boldsymbol{r}'|}\,ds(\boldsymbol{r}') - \frac{1}{4\pi}\nabla\int_{S}\frac{\hat{\boldsymbol{n}}'\cdot \boldsymbol{H}^{+}_{n}(\boldsymbol{r}')}{|\boldsymbol{r}-\boldsymbol{r}'|}\,ds(\boldsymbol{r}') \tag{3.134}$$

We use the convention that

$$\sum_{m=p}^{q}(\cdot) = 0 \qquad \text{if } q < p \tag{3.135}$$

Thus

$$F^{+}_{eo}(\boldsymbol{r}) = E^{i}_{0}(\boldsymbol{r}), \quad F^{+}_{m0}(\boldsymbol{r}) = H^{i}_{0}(\boldsymbol{r}) \tag{3.136}$$

Observe that F^{+}_{em} and F^{+}_{mn} are given in terms of the coefficients E^{+}_{m} and H^{+}_{m} for $m = 0, \ldots, n-1$. Thus, if the coefficients are known up to $n-1$, the determination of the nth coefficient depends on finding the vector potential functions U^{+}_{en} and U^{+}_{mn}. These problems will be further simplified to scalar potential problems when the boundary, or transmission, conditions are introduced.

Corresponding to equations (3.130)–(3.134) are similar expressions for the coefficients in expansions of the fields E^{-} and H^{-} obtained by substituting the expansions

(3.118) together with expansion for G^-, (3.23) in the Stratton–Chu representations (2.65) and (2.66). After reordering terms and equating like powers of ik we have

$$\left.\begin{aligned}(\alpha(\boldsymbol{r})-1)\boldsymbol{E}_n^-(\boldsymbol{r}) &= \boldsymbol{F}_{en}^-(\boldsymbol{r}) + \boldsymbol{U}_{en}^-(\boldsymbol{r})\\ (\alpha(\boldsymbol{r})-1)\boldsymbol{H}_n^-(\boldsymbol{r}) &= \boldsymbol{F}_{mn}^-(\boldsymbol{r}) + \boldsymbol{U}_{mn}^-(\boldsymbol{r})\end{aligned}\right\} \tag{3.137}$$

where

$$\begin{aligned}\boldsymbol{F}_{en}^-(\boldsymbol{r}) = {} & \frac{Z^-}{4\pi}\sum_{m=1}^{n}\binom{n}{m} m\eta^m \int_S |\boldsymbol{r}-\boldsymbol{r}'|^{m-2}(\hat{\boldsymbol{n}}' \times \boldsymbol{H}_{n-m}^-(\boldsymbol{r}'))\, ds(\boldsymbol{r}')\\ & + \frac{1}{4\pi}\sum_{m=2}^{n}\binom{n}{m}(m-1)\eta^m\\ & \times \int_S |\boldsymbol{r}-\boldsymbol{r}'|^{m-3}[(\boldsymbol{r}-\boldsymbol{r}') \times (\hat{\boldsymbol{n}}' \times \boldsymbol{E}_{n-m}^-(\boldsymbol{r}'))\\ & - (\boldsymbol{r}-\boldsymbol{r}')(\hat{\boldsymbol{n}}' \cdot \boldsymbol{E}_{n-m}^-(\boldsymbol{r}'))]\, ds(\boldsymbol{r}')\end{aligned} \tag{3.138}$$

$$\begin{aligned}\boldsymbol{F}_{mn}^-(\boldsymbol{r}) = {} & -\frac{Y^-}{4\pi}\sum_{m=1}^{n}\binom{n}{m} m\eta^m \int_S |\boldsymbol{r}-\boldsymbol{r}'|^{m-2}(\hat{\boldsymbol{n}}' \times \boldsymbol{E}_{n-m}^-(\boldsymbol{r}'))\, ds(\boldsymbol{r}')\\ & + \frac{1}{4\pi}\sum_{m=2}^{n}\binom{n}{m}(m-1)\eta^m\\ & \times \int_S |\boldsymbol{r}-\boldsymbol{r}'|^{m-3}[(\boldsymbol{r}-\boldsymbol{r}') \times (\hat{\boldsymbol{n}}' \times \boldsymbol{H}_{n-m}^-(\boldsymbol{r}'))\\ & - (\boldsymbol{r}-\boldsymbol{r}')(\hat{\boldsymbol{n}}' \cdot \boldsymbol{H}_{n-m}^-(\boldsymbol{r}'))]\, ds(\boldsymbol{r}')\end{aligned} \tag{3.139}$$

$$\boldsymbol{U}_{en}^-(\boldsymbol{r}) = \frac{1}{4\pi}\nabla \times \int_S \frac{\hat{\boldsymbol{n}}' \times \boldsymbol{E}_n^-(\boldsymbol{r}')}{|\boldsymbol{r}-\boldsymbol{r}'|}\, ds(\boldsymbol{r}') - \frac{1}{4\pi}\nabla \int_S \frac{\hat{\boldsymbol{n}}' \cdot \boldsymbol{E}_n^-(\boldsymbol{r}')}{|\boldsymbol{r}-\boldsymbol{r}'|}\, ds(\boldsymbol{r}') \tag{3.140}$$

and

$$\boldsymbol{U}_{mn}^-(\boldsymbol{r}) = \frac{1}{4\pi}\nabla \times \int_S \frac{\hat{\boldsymbol{n}}' \times \boldsymbol{H}_n^-(\boldsymbol{r}')}{|\boldsymbol{r}-\boldsymbol{r}'|}\, ds(\boldsymbol{r}') - \frac{1}{4\pi}\nabla \int_S \frac{\hat{\boldsymbol{n}}' \cdot \boldsymbol{H}_n^-(\boldsymbol{r}')}{|\boldsymbol{r}-\boldsymbol{r}'|}\, ds(\boldsymbol{r}') \tag{3.141}$$

Observe that

$$\boldsymbol{F}_{eo}^- = \boldsymbol{F}_{mo}^- = 0 \tag{3.142}$$

and we remark that the relations (3.130)–(3.142) hold for all $\boldsymbol{r}$ in $\mathbb{R}^3$, provided that the derivatives of the integrals occurring in the functions $\boldsymbol{U}$ are interpreted as direct values.

The scattering amplitudes are expressible in terms of the coefficients in the low frequency expansions by substituting the expansions (3.117) into one of the expressions for $\boldsymbol{g}_e$ and $\boldsymbol{g}_m$ given in Chapter 2, expanding the exponential that appears in the integrand and collecting like powers of ik. The most convenient forms of the scattering amplitudes for this purpose are (2.95) and (2.96) since these forms explicitly exhibit the fact that the lowest order terms in the scattering amplitudes are of order k^3, a fact that is obscured when other expressions, i.e. (2.88)–(2.91) are employed. Demonstration that they are equivalent is reserved for the problems. The expressions for the scattering amplitudes that result are

$$\begin{aligned}\boldsymbol{g}_e(\hat{\boldsymbol{r}}) = \frac{1}{4\pi}\sum_{n=0}^{\infty}\frac{(ik)^{n+3}}{n!}\sum_{m=0}^{n}\binom{n}{m}(-1)^m\Bigg\{\hat{\boldsymbol{r}}\times\Bigg[\hat{\boldsymbol{r}}\times\int_S[\hat{\boldsymbol{n}}'\cdot\boldsymbol{E}^+_{n-m}(\boldsymbol{r}')\\ - Z^+\hat{\boldsymbol{r}}\cdot(\hat{\boldsymbol{n}}'\times\boldsymbol{H}^+_{n-m}(\boldsymbol{r}'))]\boldsymbol{r}'(\hat{\boldsymbol{r}}\cdot\boldsymbol{r}')^m\,ds(\boldsymbol{r}')\Bigg]\\ +\hat{\boldsymbol{r}}\times\int_S[Z^+(\hat{\boldsymbol{n}}'\cdot\boldsymbol{H}^+_{n-m}(\boldsymbol{r}'))\\ +\hat{\boldsymbol{r}}\cdot(\hat{\boldsymbol{n}}'\times\boldsymbol{E}^+_{n-m}(\boldsymbol{r}'))]\boldsymbol{r}'(\hat{\boldsymbol{r}}\cdot\boldsymbol{r}')^m\,ds(\boldsymbol{r}')\Bigg\}\end{aligned} \tag{3.143}$$

$$\begin{aligned}\boldsymbol{g}_m(\hat{\boldsymbol{r}}) = \frac{1}{4\pi}\sum_{n=0}^{\infty}\frac{(ik)^{n+3}}{n!}\sum_{m=0}^{n}\binom{n}{m}(-1)^m\Bigg\{\hat{\boldsymbol{r}}\times\Bigg[\hat{\boldsymbol{r}}\times\int_S[(\hat{\boldsymbol{n}}'\cdot\boldsymbol{H}^+_{n-m}(\boldsymbol{r}'))\\ + Y^+\hat{\boldsymbol{r}}\cdot(\hat{\boldsymbol{n}}'\times\boldsymbol{E}^+_{n-m}(\boldsymbol{r}'))]\boldsymbol{r}'(\hat{\boldsymbol{r}}\cdot\boldsymbol{r}')^m\,ds(\boldsymbol{r}')\Bigg]\\ +\hat{\boldsymbol{r}}\times\int_S[-Y^+(\hat{\boldsymbol{n}}'\cdot\boldsymbol{E}^+_{n-m}(\boldsymbol{r}'))\\ +\hat{\boldsymbol{r}}\cdot(\hat{\boldsymbol{n}}'\times\boldsymbol{H}^+_{n-m}(\boldsymbol{r}'))]\boldsymbol{r}'(\hat{\boldsymbol{r}}\cdot\boldsymbol{r}')^m\,ds(\boldsymbol{r}')\Bigg\}\end{aligned} \tag{3.144}$$

These expressions simplify somewhat when boundary or transmission conditions for the problems considered are introduced.

3.EM.2 *The perfect conductor problem*

The boundary conditions for infinite conductivity (3.122) together with the low frequency expansions (3.117) imply that

$$\hat{\boldsymbol{n}}\times\boldsymbol{E}^+_n = \boldsymbol{0}, \qquad \hat{\boldsymbol{n}}\cdot\boldsymbol{H}^+_n = 0 \ \text{ on } S \tag{3.145}$$

for $n = 0, 1, 2, \ldots$. With these simplifications, the expressions for $\boldsymbol{F}^+_{en}$ and $\boldsymbol{F}^+_{mn}$, (3.131) and (3.132) become

$$\begin{aligned}
\boldsymbol{F}^+_{en}(\boldsymbol{r}) &= \boldsymbol{E}^i_n(\boldsymbol{r}) \\
&+ \frac{Z^+}{4\pi}\sum_{m=1}^{n}\binom{n}{m} m \int_S |\boldsymbol{r}-\boldsymbol{r}'|^{m-2}(\hat{\boldsymbol{n}}' \times \boldsymbol{H}^+_{n-m}(\boldsymbol{r}'))\, ds(\boldsymbol{r}') \\
&- \frac{1}{4\pi}\sum_{m=2}^{n}\binom{n}{m}(m-1)\int_S |\boldsymbol{r}-\boldsymbol{r}'|^{m-3}(\hat{\boldsymbol{n}}' \cdot \boldsymbol{E}^+_{n-m}(\boldsymbol{r}'))(\boldsymbol{r}-\boldsymbol{r}')\, ds(\boldsymbol{r}')
\end{aligned} \tag{3.146}$$

and

$$\begin{aligned}
\boldsymbol{F}^+_{mn}(\boldsymbol{r}) &= \boldsymbol{H}^i_n(\boldsymbol{r}) \\
&+ \frac{1}{4\pi}\sum_{m=2}^{n}\binom{n}{m}(m-1) \\
&\times \int_S |\boldsymbol{r}-\boldsymbol{r}'|^{m-3}(\boldsymbol{r}-\boldsymbol{r}') \times (\hat{\boldsymbol{n}}' \times \boldsymbol{H}^+_{n-m}(\boldsymbol{r}'))\, ds(\boldsymbol{r}')
\end{aligned} \tag{3.147}$$

while the potentials $\boldsymbol{U}^+_{en}$ and $\boldsymbol{U}^+_{mn}$, (3.133) and (3.134) become

$$\boldsymbol{U}^+_{en}(\boldsymbol{r}) = -\frac{1}{4\pi}\nabla \int_S \frac{\hat{\boldsymbol{n}}' \cdot \boldsymbol{E}^+_n(\boldsymbol{r}')}{|\boldsymbol{r}-\boldsymbol{r}'|}\, ds(\boldsymbol{r}') \tag{3.148}$$

$$\boldsymbol{U}^+_{mn}(\boldsymbol{r}) = \frac{1}{4\pi}\nabla \times \int_S \frac{\hat{\boldsymbol{n}}' \times \boldsymbol{H}^+_n(\boldsymbol{r}')}{|\boldsymbol{r}-\boldsymbol{r}'|}\, ds(\boldsymbol{r}') \tag{3.149}$$

If we assume that the coefficients $\boldsymbol{E}^+_m$ and $\boldsymbol{H}^+_m$ have been found for $m = 0, 1, \ldots, n-1$, then the quantities $\boldsymbol{F}^+_{en}$ and $\boldsymbol{H}^+_{mn}$ are known in terms of these previous coefficients. The problems of determining $\boldsymbol{E}^+_n$ and $\boldsymbol{H}^+_n$ are reduced to finding the vector valued functions $\boldsymbol{U}^+_{en}$ and $\boldsymbol{U}^+_{mn}$. These problems are further reduced to solving some standard scalar potential problems as follows.

Determination of $\boldsymbol{U}^+_{en}$: To begin, we define the scalar function $\phi^+_{en}(\boldsymbol{r})$ as

$$\phi^+_{en}(\boldsymbol{r}) = \frac{1}{4\pi}\int_S \frac{\hat{\boldsymbol{n}}' \cdot \boldsymbol{E}^+_n(\boldsymbol{r}')}{|\boldsymbol{r}-\boldsymbol{r}'|}\, ds(\boldsymbol{r}'), \quad \boldsymbol{r} \in \mathbb{R}^3 \tag{3.150}$$

Of course, since $\boldsymbol{E}^+_n$ is not known, ϕ^+_{en} is not known either. However, the definition (3.150) reveals some important properties of ϕ^+_{en}; namely, that it is a solution of Laplace's equation in V^+ and decays as $1/r^2$ as $r \to \infty$. This rapid decay is a consequence of (3.129) and the asymptotic form of $1/|\boldsymbol{r}-\boldsymbol{r}'|$ as $r \to \infty$ which ensures that

$$\int_S \frac{\partial}{\partial n'}\phi^+_{en}\, ds = 0 \tag{3.151}$$

Then in V^+ we have from (3.130) and (3.148) that

$$\boldsymbol{E}_n^+(\boldsymbol{r}) = \boldsymbol{F}_{en}^+(\boldsymbol{r}) - \nabla\phi_{en}^+(\boldsymbol{r}) \tag{3.152}$$

and furthermore, with relations (3.129) and (3.151), it follows that

$$\int_S \hat{\boldsymbol{n}}' \cdot \boldsymbol{F}_{en}^+(\boldsymbol{r}')\,ds(\boldsymbol{r}') = 0 \tag{3.153}$$

This, together with the boundary conditions (3.145), leads to the following problem for finding ϕ_{en}^+

$$\left.\begin{aligned} \nabla^2\phi_{en}^+ &= 0 && \text{in } V^+ \\ \hat{\boldsymbol{n}} \times \nabla\phi_{en}^+ &= \hat{\boldsymbol{n}} \times \boldsymbol{F}_{en}^+ && \text{on } S \\ \phi_{en}^+ &= O\left(\frac{1}{r^2}\right) && \text{as } r \to \infty \end{aligned}\right\} \tag{3.154}$$

This is almost a standard exterior potential problem with Dirichlet boundary conditions for ϕ_{en}^+. The boundary condition is slightly unusual in that the tangential derivatives, rather than the values of ϕ_{en}^+, are specified on S. This introduces an ambiguity since the boundary values of ϕ_{en}^+ are determined up to a constant. However, the ambiguity is removed by the strong decay as $r \to \infty$ which is equivalent to the additional condition (3.151). It will be shown subsequently, in Chapter 6, that an integral equation may be derived to determine $\hat{\boldsymbol{n}} \cdot \boldsymbol{E}_n^+$ and, with (3.148), $\boldsymbol{U}_{en}^+$.

Determination of $\boldsymbol{U}_{mn}^+$: The determination of $\boldsymbol{U}_{mn}^+$ is not quite so straightforward. The reason is that while $\boldsymbol{U}_{en}^+$ could easily be expressed as the gradient of a scalar potential function (in fact that was implicit in the definition (3.148)), $\boldsymbol{U}_{mn}^+$ is not expressible in this way. Indeed, except when $n = 0$, $\boldsymbol{U}_{mn}^+$ is not the gradient of a scalar potential function. To overcome this difficulty we use an idea of Stevenson (1953*a*). We add and subtract a known function $\boldsymbol{h}_n^+$, so that the difference $\boldsymbol{U}_{mn}^+ - \boldsymbol{h}_n^+$ is the gradient of a scalar potential function. To find the appropriate function $\boldsymbol{h}_n^+$, perhaps surprisingly, requires the solution of a scalar interior potential problem. To facilitate this process we pose a scalar potential problem for a function ϕ_{mn}^-; namely

$$\left.\begin{aligned} \nabla^2\phi_{mn}^- &= 0 && \text{in } V^- \\ \frac{\partial}{\partial n}\phi_{mn}^- &= -nY^+\hat{\boldsymbol{n}} \cdot \boldsymbol{E}_{n-1}^+ && \text{on } S \end{aligned}\right\} \tag{3.155}$$

The jump condition for the single layer potential (3.150) enables us to obtain the following relation

$$\frac{\partial}{\partial n^+}\phi_{e(n-1)}^+ - \frac{\partial}{\partial n^-}\phi_{e(n-1)}^+ = -\hat{\boldsymbol{n}} \cdot \boldsymbol{E}_{n-1}^+ \tag{3.156}$$

where $\partial/\partial n^\pm$ indicate normal differentiation from $V^\pm$. In view of (3.156), the boundary condition in (3.155) reads as

$$\frac{\partial}{\partial n}\phi_{mn}^- = nY^+\left(\frac{\partial}{\partial n^+}\phi_{e(n-1)}^+ - \frac{\partial}{\partial n^-}\phi_{e(n-1)}^+\right) \quad \text{on } S \tag{3.157}$$

This interior Neumann problem is solvable since the additional compatibility condition

$$\int_S \frac{\partial}{\partial n}\phi_{mn}^{-}\, ds = -nY^{+}\int_S \hat{\boldsymbol{n}}\cdot\boldsymbol{E}_{n-1}^{+}\, ds = 0 \tag{3.158}$$

is satisfied by virtue of (3.129). In order to proceed, it is necessary to solve this interior potential problem. There is no unique solution, since any constant satisfies all the conditions, but any solution will suffice.

With this helpful function, we may in turn define the auxiliary function

$$\boldsymbol{h}_n^{+}(\boldsymbol{r}) = \frac{1}{4\pi}\nabla\times\int_S \frac{\phi_{mn}^{-}(\boldsymbol{r}')\hat{\boldsymbol{n}}'}{|\boldsymbol{r}-\boldsymbol{r}'|}\, ds(\boldsymbol{r}'),\quad \boldsymbol{r}\in V^{+} \tag{3.159}$$

and write

$$\boldsymbol{H}_n^{+}(\boldsymbol{r}) = \boldsymbol{F}_{mn}^{+}(\boldsymbol{r}) + \boldsymbol{h}_n^{+}(\boldsymbol{r}) + \boldsymbol{U}_{mn}^{+}(\boldsymbol{r}) - \boldsymbol{h}_n^{+}(\boldsymbol{r}) \tag{3.160}$$

The utility of this representation lies in the following Lemma.

Lemma: *There exists a potential function $\phi_{mn}^{+}(\boldsymbol{r})$ in V^{+} which decays as r^{-1} as $r\to\infty$, such that*

$$\boldsymbol{U}_{mn}^{+} - \boldsymbol{h}_n^{+} = -\nabla\phi_{mn}^{+}\quad \text{in } V^{+} \tag{3.161}$$

Proof: First, we show that the difference $\boldsymbol{U}_{mn}^{+} - \boldsymbol{h}_n^{+}$ is the gradient of a scalar function. For this, it is sufficient to show that

$$\nabla\times\boldsymbol{U}_{mn}^{+} - \nabla\times\boldsymbol{h}_n^{+} = \boldsymbol{0}\quad \text{in } V^{+} \tag{3.162}$$

or with (3.149) and (3.159)

$$\nabla\times\left[\nabla\times\int_S \frac{\hat{\boldsymbol{n}}'\times\boldsymbol{H}_n^{+}(\boldsymbol{r}')}{|\boldsymbol{r}-\boldsymbol{r}'|}\, ds(\boldsymbol{r}')\right]$$

$$-\nabla\times\left[\nabla\times\int_S \frac{\phi_{mn}^{-}(\boldsymbol{r}')\hat{\boldsymbol{n}}'}{|\boldsymbol{r}-\boldsymbol{r}'|}\, ds(\boldsymbol{r}')\right] = \boldsymbol{0},\quad \boldsymbol{r}\in V^{+} \tag{3.163}$$

Since both integrals are solutions of Laplace's equation, we may rewrite the left-hand side as

$$\nabla\int_S \left(\nabla_{\boldsymbol{r}}\frac{1}{|\boldsymbol{r}-\boldsymbol{r}'|}\right)\cdot(\hat{\boldsymbol{n}}'\times\boldsymbol{H}_n^{+}(\boldsymbol{r}'))\, ds(\boldsymbol{r}')$$

$$-\nabla\int_S \left(\nabla_{\boldsymbol{r}}\frac{1}{|\boldsymbol{r}-\boldsymbol{r}'|}\right)\cdot\hat{\boldsymbol{n}}'\phi_{mn}^{-}(\boldsymbol{r}')\, ds(\boldsymbol{r}') = \boldsymbol{0},\quad \boldsymbol{r}\in V^{+} \tag{3.164}$$

Then, invoking the relation

$$\nabla_{\boldsymbol{r}} \frac{1}{|\boldsymbol{r}-\boldsymbol{r}'|} = -\nabla_{\boldsymbol{r}'} \frac{1}{|\boldsymbol{r}-\boldsymbol{r}'|} \tag{3.165}$$

familiar vector identities and Green's theorem, together with the fact that $\boldsymbol{r} \in V^+$, (3.164) becomes

$$\begin{aligned} &\nabla \int_S \hat{\boldsymbol{n}}' \cdot \left[\left(\nabla_{\boldsymbol{r}'} \frac{1}{|\boldsymbol{r}-\boldsymbol{r}'|}\right) \times \boldsymbol{H}_n^+(\boldsymbol{r}')\right] ds(\boldsymbol{r}') \\ &+ \nabla \int_S \frac{1}{|\boldsymbol{r}-\boldsymbol{r}'|} \frac{\partial}{\partial n'} \phi_{mn}^-(\boldsymbol{r}')\, ds(\boldsymbol{r}') = \boldsymbol{0}, \quad \boldsymbol{r} \in V^+ \end{aligned} \tag{3.166}$$

Now using Stokes' theorem for the first term and the boundary condition in (3.155) for the second, we obtain

$$\begin{aligned} &\nabla \int_S \hat{\boldsymbol{n}}' \cdot \frac{\nabla_{\boldsymbol{r}'} \times \boldsymbol{H}_n^+(\boldsymbol{r}')}{|\boldsymbol{r}-\boldsymbol{r}'|} ds(\boldsymbol{r}') \\ &+ nY^+ \nabla \int_S \frac{\hat{\boldsymbol{n}}' \cdot \boldsymbol{E}_{n-1}^+(\boldsymbol{r}')}{|\boldsymbol{r}-\boldsymbol{r}'|} ds(\boldsymbol{r}') = \boldsymbol{0}, \quad \boldsymbol{r} \in V^+ \end{aligned} \tag{3.167}$$

That this vanishes follows using the second equation in (3.127).

Hence

$$\boldsymbol{U}_{mn}^+(\boldsymbol{r}) - \boldsymbol{h}_n^+(\boldsymbol{r}) = -\nabla \phi_{mn}^+(\boldsymbol{r}), \quad \boldsymbol{r} \in V^+ \tag{3.168}$$

for some ϕ_{mn}^+, since the left-hand side is an irrotational field. That $\phi_{mn}^+(\boldsymbol{r})$ is a potential function follows from the fact that $\boldsymbol{U}_{mn}^+$ and $\boldsymbol{h}_n^+$ are defined as curls, hence their divergence vanishes. That $\phi_{mn}^+(\boldsymbol{r})$ is of order r^{-1} as $r \to \infty$, follows from (3.149) and (3.159) which ensure that

$$\begin{aligned} -\nabla \phi_{mn}^+(\boldsymbol{r}) &= \boldsymbol{U}_{mn}^+(\boldsymbol{r}) - \boldsymbol{h}_n^+(\boldsymbol{r}) \\ &= \frac{1}{4\pi} \nabla \times \int_S \frac{\hat{\boldsymbol{n}}' \times \boldsymbol{H}_n^+(\boldsymbol{r}') - \phi_{mn}^-(\boldsymbol{r}')\boldsymbol{n}'}{|\boldsymbol{r}-\boldsymbol{r}'|} ds(\boldsymbol{r}') \\ &= O\left(\frac{1}{r^2}\right), \quad r \to \infty \end{aligned} \tag{3.169}$$

Hence

$$\phi_{mn}^+(\boldsymbol{r}) = c + O\left(\frac{1}{r}\right) \quad \text{as } r \to \infty \tag{3.170}$$

and we restrict attention to that ϕ_{mn}^+ for which $c = 0$. This completes the proof of the lemma.

To complete the determination of $\boldsymbol{H}_n^+(\boldsymbol{r})$, we must actually find $\phi_{mn}^+(\boldsymbol{r})$. This is accomplished by solving yet another scalar potential problem. Since we may now write (3.160) as

$$\boldsymbol{H}_n^+ = \boldsymbol{F}_{mn}^+ + \boldsymbol{h}_n^+ - \nabla\phi_{mn}^+ \quad \text{in } V^+ \tag{3.171}$$

we use the second boundary condition (3.145) to pose the following problem for ϕ_{mn}^+

$$\left.\begin{aligned} \nabla^2\phi_{mn}^+(\boldsymbol{r}) &= 0 && \text{in } V^+ \\ \frac{\partial}{\partial n}\phi_{mn}^+ &= \hat{\boldsymbol{n}}\cdot\boldsymbol{F}_{mn}^+ + \hat{\boldsymbol{n}}\cdot\boldsymbol{h}_n^+ && \text{on } S \\ \phi_{mn}^+ &= O\left(\frac{1}{r}\right) && \text{as } r\to\infty \end{aligned}\right\} \tag{3.172}$$

This completes the process of finding $\boldsymbol{H}_n^+$.

Thus, the determination of the nth coefficients $\boldsymbol{E}_n^+$ and $\boldsymbol{H}_n^+$ in the expansion require the solution of three scalar potential problems, the exterior problem (3.154) for ϕ_{en}^+ to determine $\boldsymbol{E}_n^+$ and both the interior problem (3.155) for ϕ_{mn}^- and the exterior problem (3.172) for ϕ_{mn}^+ to determine $\boldsymbol{H}_n^+$.

With the boundary conditions for perfect conductivity, the expression for the electric scattering amplitude (3.143) becomes

$$\begin{aligned} \boldsymbol{g}_e(\hat{\boldsymbol{r}}) = \frac{1}{4\pi}\sum_{n=0}^{\infty}\frac{(ik)^{n+3}}{n!}\sum_{m=0}^{n}\binom{n}{m}(-1)^m\hat{\boldsymbol{r}}\times\Bigg[\hat{\boldsymbol{r}}\times\int_S[\hat{\boldsymbol{n}}'\cdot\boldsymbol{E}_{n-m}^+(\boldsymbol{r}') \\ - Z^+\hat{\boldsymbol{r}}\cdot(\hat{\boldsymbol{n}}'\times\boldsymbol{H}_{n-m}^+(\boldsymbol{r}'))]\boldsymbol{r}'(\hat{\boldsymbol{r}}\cdot\boldsymbol{r}')^m\,ds(\boldsymbol{r}')\Bigg] \end{aligned} \tag{3.173}$$

and through (2.93)

$$\boldsymbol{g}_m(\hat{\boldsymbol{r}}) = Y^+\hat{\boldsymbol{r}}\times\boldsymbol{g}_e(\hat{\boldsymbol{r}}) \tag{3.174}$$

Since $\boldsymbol{E}_n^+$ and $\boldsymbol{H}_n^+$ are real valued, these coefficients may be decomposed into real and imaginary parts.

$$\left.\begin{aligned} \operatorname{Re}\boldsymbol{g}_e(\hat{\boldsymbol{r}}) &= k^3\sum_{n=0}^{\infty}(-1)^n k^{2n+1}\,\boldsymbol{A}_{2n+1}(\hat{\boldsymbol{r}}) \\ \operatorname{Im}\boldsymbol{g}_e(\hat{\boldsymbol{r}}) &= k^3\sum_{n=0}^{\infty}(-1)^{n+1}k^{2n}\,\boldsymbol{A}_{2n}(\hat{\boldsymbol{r}}) \end{aligned}\right\} \tag{3.175}$$

where

$$\begin{aligned} \boldsymbol{A}_n(\hat{\boldsymbol{r}}) = \frac{1}{4\pi n!}\sum_{m=0}^{n}\binom{n}{m}(-1)^m\hat{\boldsymbol{r}}\times\Bigg[\hat{\boldsymbol{r}}\times\int_S[\hat{\boldsymbol{n}}'\cdot\boldsymbol{E}_{n-m}^+(\boldsymbol{r}') \\ - Z^+\hat{\boldsymbol{r}}\cdot(\hat{\boldsymbol{n}}'\times\boldsymbol{H}_{n-m}^+(\boldsymbol{r}'))]\boldsymbol{r}'(\hat{\boldsymbol{r}}\cdot\boldsymbol{r}')^m\,ds(\boldsymbol{r}')\Bigg] \end{aligned} \tag{3.176}$$

for $n = 0, 1, 2, \ldots$. The differential scattering cross section (2.113) is given by

$$\sigma(\hat{\boldsymbol{r}}) = 4\pi k^4 \sum_{n=0}^{\infty} k^{2n}(-1)^n \sum_{m=0}^{n} \boldsymbol{A}_{2n-2m}(\hat{\boldsymbol{r}}) \cdot \boldsymbol{A}_{2m}(\hat{\boldsymbol{r}})$$
$$+ 4\pi k^6 \sum_{n=0}^{\infty} k^{2n}(-1)^n \sum_{m=0}^{n} \boldsymbol{A}_{2n-2m+1}(\hat{\boldsymbol{r}}) \cdot \boldsymbol{A}_{2m+1}(\hat{\boldsymbol{r}}) \qquad (3.177)$$

The scattering cross section is given by

$$\sigma_s = \frac{1}{4\pi} \int_{S^2} \sigma(\hat{\boldsymbol{r}})\, ds(\hat{\boldsymbol{r}}) \qquad (3.178)$$

and no general simplifications are possible. For the first term, see Chapter 5. Since by (2.117), $\sigma_a = 0$ in this case, it follows that $\sigma_e = \sigma_s$.

3.EM.3 *The impedance problem*

The impedance boundary conditions (3.123) together with the low frequency expansions (3.117) imply that

$$\hat{\boldsymbol{n}} \times (\hat{\boldsymbol{n}} \times \boldsymbol{E}_n^+) = -Z_s Z^+ \hat{\boldsymbol{n}} \times \boldsymbol{H}_n^+ \quad \text{on } S \qquad (3.179)$$

Unlike the case of perfect conductivity, the problem of determining the potential functions $\boldsymbol{U}_{en}^+$ and $\boldsymbol{U}_{mn}^+$ has not been reduced to the solution of scalar potential problems. This remains an open problem, even though integral equations for the full problem have been derived (Colton and Kress 1983).

3.EM.4 *The transmission problem*

In view of the absence of low frequency results for the impedance boundary condition, it is perhaps surprising that in the more complicated case of transmission conditions there are complete results for the reduction of the determination of the coefficients in the low frequency expansion to scalar potential problems. We distinguish between lossy and lossless scatterers. The introduction of losses into the interior medium presents a different picture in electromagnetics than in acoustics at low frequencies. In acoustics, the compressional viscosity δ^- always appears as $k\delta^-$ so that small wave number and small loss coincide and the lossless case is easily recovered from the lossy case by letting $\delta^- \to 0$. In electromagnetics, however, the conductivity enters in the form σ^-/k so that the low frequency expansion in powers of k is not uniformly convergent with respect to σ^-. The power series expansion in k for the lossless case is not recovered from that for the lossy case merely by letting $\sigma^- = 0$. In fact, the zeroth order coefficient $\boldsymbol{E}_0^+$ for $\sigma^- \neq 0$ is independent of σ^- and is identical to that for a perfect conductor, which is very different from $\boldsymbol{E}_0^+$ for the lossless case. It is tempting to follow a strategy similar to that in acoustics of determining the low frequency expansion for the lossless case and then replacing ϵ^- by the complex dielectric constant $\epsilon^- + i(\sigma^-/k)\sqrt{\epsilon^+\mu^+}$ in each coefficient to obtain results for the lossy case. Indeed, this strategy has been followed, for example, by Stevenson (1953*a*) but it is difficult to assess when the resulting series converge to the actual fields.

Here we present a complete development of the low frequency expansion in the lossless case, $\sigma^- = 0$, based on boundary integral equations. The comparable lossy case was treated by Kleinman (1978). In Chapter 4 we present an alternative development based on volume integrals for both the lossy and lossless cases.

The low frequency expansions (3.117) and (3.118) together with the transmission conditions (3.124) and (3.125) lead to the following transmission conditions for the coefficients

$$\left.\begin{aligned} \hat{\boldsymbol{n}} \times \boldsymbol{E}_n^+ &= \hat{\boldsymbol{n}} \times \boldsymbol{E}_n^- \\ \hat{\boldsymbol{n}} \times \boldsymbol{H}_n^+ &= \hat{\boldsymbol{n}} \times \boldsymbol{H}_n^- \end{aligned}\right\} \quad \text{on } S \tag{3.180}$$

$$\left.\begin{aligned} Y^+\hat{\boldsymbol{n}} \cdot \boldsymbol{E}_n^+ &= \eta Y^- \hat{\boldsymbol{n}} \cdot \boldsymbol{E}_n^- \\ Z^+\hat{\boldsymbol{n}} \cdot \boldsymbol{H}_n^+ &= \eta Z^- \hat{\boldsymbol{n}} \cdot \boldsymbol{H}_n^- \end{aligned}\right\} \quad \text{on } S \tag{3.181}$$

It may be easily verified that for $\sigma^- = 0$

$$\eta Y^- = \frac{\epsilon^-}{\epsilon^+} Y^+ \quad \text{and} \quad \eta Z^- = \frac{\mu^-}{\mu^+} Z^+ \tag{3.182}$$

so that the transmission conditions (3.181) for the normal components may also be written as

$$\left.\begin{aligned} \epsilon^+\hat{\boldsymbol{n}} \cdot \boldsymbol{E}_n^+ &= \epsilon^- \hat{\boldsymbol{n}} \cdot \boldsymbol{E}_n^- \\ \mu^+\hat{\boldsymbol{n}} \cdot \boldsymbol{H}_n^+ &= \mu^- \hat{\boldsymbol{n}} \cdot \boldsymbol{H}_n^- \end{aligned}\right\} \quad \text{on } S \tag{3.183}$$

The unknown field coefficients $\boldsymbol{E}_n^\pm$ and $\boldsymbol{H}_n^\pm$ are represented in terms of the preceding coefficients of order $\leq n-1$ and unknown vector valued potential functions as given in (3.130)–(3.141). However, the transmission conditions permit some simplification. Examination of the structure of the coefficients in the low frequency expansions (3.130)–(3.141) shows that each *n*th coefficient is expressible in terms of a function which is completely specified in terms of coefficients up to order $n-1$ and a vector potential which in general consists of both solenoidal and irrotational terms. Just as was done for the magnetic field in the perfectly conducting case, it is possible to re-express the solenoidal term as an irrotational term, thus representing the unknown part of the coefficient as the gradient of a scalar potential.

Consider first the solenoidal part of $\boldsymbol{U}_{en}^-$ given in (3.140)

$$\boldsymbol{U}_{en}^{s-}(\boldsymbol{r}) = \frac{1}{4\pi} \nabla \times \int_S \frac{\hat{\boldsymbol{n}}' \times \boldsymbol{E}_n^-(\boldsymbol{r}')}{|\boldsymbol{r} - \boldsymbol{r}'|} \, ds(\boldsymbol{r}') \tag{3.184}$$

First, we prove that, except when $n = 0$, $\nabla \times \boldsymbol{U}_{en}^{s-} \neq \boldsymbol{0}$. Explicitly, for $\boldsymbol{r} \notin S$ it may be shown, using standard vector identities and the fact that the integral is a potential function, that

$$\begin{aligned} \nabla \times \boldsymbol{U}_{en}^{s-}(\boldsymbol{r}) &= \frac{1}{4\pi} \nabla \times \left[\nabla \times \int_S \frac{\hat{\boldsymbol{n}}' \times \boldsymbol{E}_n^-(\boldsymbol{r}')}{|\boldsymbol{r} - \boldsymbol{r}'|} \, ds(\boldsymbol{r}') \right] \\ &= \frac{1}{4\pi} \nabla\nabla \cdot \int_S \frac{\hat{\boldsymbol{n}}' \times \boldsymbol{E}_n^-(\boldsymbol{r}')}{|\boldsymbol{r} - \boldsymbol{r}'|} \, ds(\boldsymbol{r}') \end{aligned} \tag{3.185}$$

Thus, (3.185) may be rewritten, with Stokes' theorem, as

$$\begin{aligned}\nabla \times \boldsymbol{U}^{s-}_{en}(\boldsymbol{r}) &= -\frac{1}{4\pi}\nabla \int_S \left(\nabla_{\boldsymbol{r}'}\frac{1}{|\boldsymbol{r}-\boldsymbol{r}'|}\right)\cdot(\hat{\boldsymbol{n}}'\cdot \boldsymbol{E}^-_n(\boldsymbol{r}'))\,ds(\boldsymbol{r}') \\ &= -\frac{1}{4\pi}\nabla \int_S \frac{\hat{\boldsymbol{n}}'\cdot(\nabla_{\boldsymbol{r}'}\times \boldsymbol{E}^-_n(\boldsymbol{r}'))}{|\boldsymbol{r}-\boldsymbol{r}'|}\,ds(\boldsymbol{r}')\end{aligned} \tag{3.186}$$

Finally, using the reduced Maxwell equations, this becomes

$$\nabla \times \boldsymbol{U}^{s-}_{en}(\boldsymbol{r}) = -\frac{n\eta Z^-}{4\pi}\nabla \int_S \frac{\hat{\boldsymbol{n}}'\cdot \boldsymbol{H}^-_{n-1}(\boldsymbol{r}')}{|\boldsymbol{r}-\boldsymbol{r}'|}\,ds(\boldsymbol{r}') \tag{3.187}$$

in which form it is clear that $\boldsymbol{U}^{s-}_{en}$ is irrotational when $n = 0$, but not necessarily otherwise. However, this result contains the hint as to how to add a function which will cause the curl to vanish. To this end, define an exterior scalar potential ψ^+_{en} to be the solution of the problem

$$\left.\begin{aligned} \nabla^2\psi^+_{en} &= 0 && \text{in } V^+ \\ \frac{\partial}{\partial n}\psi^+_{en} &= -\frac{n\eta Z^-}{4\pi}\hat{\boldsymbol{n}}\cdot \boldsymbol{H}^-_{n-1} && \text{on } S \\ \psi^+_{en} &= O\left(\frac{1}{r^2}\right) && \text{as } r\to\infty \end{aligned}\right\} \tag{3.188}$$

This is a well-posed scalar potential problem with data on S which is assumed to be known. In terms of this function

$$\nabla \times \boldsymbol{U}^{s-}_{en}(\boldsymbol{r}) = \nabla \int_S \frac{1}{|\boldsymbol{r}-\boldsymbol{r}'|}\frac{\partial}{\partial n'}\psi^+_{en}(\boldsymbol{r}')\,ds(\boldsymbol{r}'), \quad \boldsymbol{r}\in V^- \tag{3.189}$$

Using Green's theorem, the fact that the integral is harmonic and some vector manipulation we find, for $\boldsymbol{r}\in V^-$

$$\begin{aligned}\nabla \times \boldsymbol{U}^{s-}_{en}(\boldsymbol{r}) &= \nabla \int_S \psi^+_{en}(\boldsymbol{r}')\frac{\partial}{\partial n'}\frac{1}{|\boldsymbol{r}-\boldsymbol{r}'|}\,ds(\boldsymbol{r}') \\ &= -\nabla\times\left[\nabla\times \int_S \frac{\hat{\boldsymbol{n}}'\psi^+_{en}(\boldsymbol{r}')}{|\boldsymbol{r}-\boldsymbol{r}'|}\,ds(\boldsymbol{r}')\right]\end{aligned} \tag{3.190}$$

Defining

$$\boldsymbol{V}^-_{en}(\boldsymbol{r}) = \nabla\times \int_S \frac{\hat{\boldsymbol{n}}'\psi^+_{en}(\boldsymbol{r}')}{|\boldsymbol{r}-\boldsymbol{r}'|}\,ds(\boldsymbol{r}') \tag{3.191}$$

the relation (3.190) may be written as

$$\nabla \times (\boldsymbol{U}^{s-}_{en} + \boldsymbol{V}^{-}_{en}) = \mathbf{0} \quad \text{in } V^{-} \tag{3.192}$$

so that $\boldsymbol{U}^{s-}_{en} + \boldsymbol{V}^{-}_{en}$ is the gradient of a scalar function. Moreover $\boldsymbol{U}^{s-}_{en}$ is solenoidal and by virtue of (3.191), $\boldsymbol{V}^{-}_{en}$ is also solenoidal. Therefore, this scalar function must satisfy Laplace's equation. Then with (3.137) and (3.140)

$$-\boldsymbol{E}^{-}_{n} = \boldsymbol{F}^{-}_{en} - \boldsymbol{V}^{-}_{en} + \nabla\phi^{-}_{en} \quad \text{in } V^{-} \tag{3.193}$$

where ϕ^{-}_{en} is an interior scalar potential to be determined.

Now we consider the exterior coefficients. To rewrite $\boldsymbol{E}^{+}_{n}$ as a known function plus the gradient of a scalar potential does not require the introduction of an auxiliary potential function comparable to ψ^{+}_{en}. It is possible to utilize the transmission conditions to rewrite the solenoidal part (3.133) of $\boldsymbol{U}^{+}_{en}$ as

$$\begin{aligned}\boldsymbol{U}^{s+}_{en}(\boldsymbol{r}) &= \frac{1}{4\pi}\nabla \times \int_{S} \frac{\hat{\boldsymbol{n}}' \times \boldsymbol{E}^{+}_{n}(\boldsymbol{r}')}{|\boldsymbol{r}-\boldsymbol{r}'|}\, ds(\boldsymbol{r}') \\ &= \frac{1}{4\pi}\nabla \times \int_{S} \frac{\hat{\boldsymbol{n}}' \times \boldsymbol{E}^{-}_{n}(\boldsymbol{r}')}{|\boldsymbol{r}-\boldsymbol{r}'|}\, ds(\boldsymbol{r}'), \quad \boldsymbol{r} \in V^{+}\end{aligned} \tag{3.194}$$

Although this has the same form as $\boldsymbol{U}^{s-}_{en}$, it is evaluated in V^{+} rather than in V^{-} and so the previous analysis cannot be carried over directly. However, through the use of Gauss' theorem, the reduced Maxwell equations (3.128) and standard vector manipulation it may be shown that

$$\begin{aligned}\boldsymbol{U}^{s+}_{en}(\boldsymbol{r}) &= \frac{1}{4\pi}\nabla \times \int_{V^{-}} \nabla_{\boldsymbol{r}'} \times \left(\frac{\boldsymbol{E}^{-}_{n}(\boldsymbol{r}')}{|\boldsymbol{r}-\boldsymbol{r}'|}\right) dv(\boldsymbol{r}') \\ &= \frac{1}{4\pi}\nabla \times \int_{V^{-}} \left(\nabla_{\boldsymbol{r}'} \frac{1}{|\boldsymbol{r}-\boldsymbol{r}'|}\right) \times \boldsymbol{E}^{-}_{n}(\boldsymbol{r}')\, dv(\boldsymbol{r}') \\ &\quad + \frac{n\eta}{4\pi} Z^{-} \nabla \times \int_{V^{-}} \frac{\boldsymbol{H}^{-}_{n-1}(\boldsymbol{r}')}{|\boldsymbol{r}-\boldsymbol{r}'|}\, dv(\boldsymbol{r}') \\ &= \frac{n\eta}{4\pi} Z^{-} \nabla \times \int_{V^{-}} \frac{\boldsymbol{H}^{-}_{n-1}(\boldsymbol{r}')}{|\boldsymbol{r}-\boldsymbol{r}'|}\, dv(\boldsymbol{r}') \\ &\quad + \frac{1}{4\pi}\nabla \int_{S} \frac{\hat{\boldsymbol{n}}' \cdot \boldsymbol{E}^{-}_{n}(\boldsymbol{r}')}{|\boldsymbol{r}-\boldsymbol{r}'|}\, ds(\boldsymbol{r}'), \quad \boldsymbol{r} \in V^{+}\end{aligned} \tag{3.195}$$

Defining

$$\boldsymbol{V}_{en}^{+}(\boldsymbol{r}) = \frac{n\eta}{4\pi} Z^{-} \nabla \times \int_{V^{-}} \frac{\boldsymbol{H}_{n-1}^{-}(\boldsymbol{r}')}{|\boldsymbol{r}-\boldsymbol{r}'|}\, dv(\boldsymbol{r}'), \quad \boldsymbol{r} \in V^{+} \tag{3.196}$$

and again employing the transmission conditions (3.183) leads to

$$\boldsymbol{U}_{en}^{s+}(\boldsymbol{r}) = \boldsymbol{V}_{en}^{+} + \frac{1}{4\pi}\frac{\epsilon^{+}}{\epsilon^{-}} \nabla \int_{S} \frac{\hat{\boldsymbol{n}}' \cdot \boldsymbol{E}_{n}^{+}(\boldsymbol{r}')}{|\boldsymbol{r}-\boldsymbol{r}'|}\, ds(\boldsymbol{r}'), \quad \boldsymbol{r} \in V^{+} \tag{3.197}$$

Combining this result with the irrotational part of $\boldsymbol{U}_{en}^{+}$ we find that

$$\boldsymbol{E}_{n}^{+}(\boldsymbol{r}) = \boldsymbol{F}_{en}^{+}(\boldsymbol{r}) + \boldsymbol{V}_{en}^{+}(\boldsymbol{r}) + \frac{1}{4\pi}\left(\frac{\epsilon^{+}}{\epsilon^{-}} - 1\right) \nabla \int_{S} \frac{\hat{\boldsymbol{n}}' \cdot \boldsymbol{E}_{n}^{+}(\boldsymbol{r}')}{|\boldsymbol{r}-\boldsymbol{r}'|}\, ds(\boldsymbol{r}') \tag{3.198}$$

or defining

$$\phi_{en}^{+}(\boldsymbol{r}) = \frac{1}{4\pi}\left(1 - \frac{\epsilon^{+}}{\epsilon^{-}}\right) \int_{S} \frac{\hat{\boldsymbol{n}}' \cdot \boldsymbol{E}_{n}^{+}(\boldsymbol{r}')}{|\boldsymbol{r}-\boldsymbol{r}'|}\, ds(\boldsymbol{r}') \tag{3.199}$$

we obtain

$$\boldsymbol{E}_{n}^{+} = \boldsymbol{F}_{en}^{+} + \boldsymbol{V}_{en}^{+} - \nabla \phi_{en}^{+} \quad \text{in } V^{+} \tag{3.200}$$

where ϕ_{en}^{+} is an exterior potential function which decays as $O\left(1/r^{2}\right)$ since

$$\int_{S} \hat{\boldsymbol{n}} \cdot \boldsymbol{E}_{n}^{+}\, ds = 0 \tag{3.201}$$

If ϵ^{+} and ϵ^{-} are equal, then ϕ_{en}^{+} vanishes as can be seen from (3.199). Then (3.200) provides a representation of $\boldsymbol{E}_{n}^{+}$ in terms of coefficients of order lower than n. However, it is still necessary to find ϕ_{en}^{-} in order to determine $\boldsymbol{E}_{n}^{-}$. The determination of $\phi_{en}^{\pm}$, which are needed in order to calculate $\boldsymbol{E}_{n}^{\pm}$, requires the solution of the following transmission potential problem

$$\left.\begin{aligned}
\nabla^{2}\phi_{en}^{+} &= 0 && \text{in } V^{+} \\
\nabla^{2}\phi_{en}^{-} &= 0 && \text{in } V^{-} \\
\hat{\boldsymbol{n}} \times \nabla(\phi_{en}^{+} - \phi_{en}^{-}) &= \hat{\boldsymbol{n}} \times (\boldsymbol{F}_{en}^{-} + \boldsymbol{F}_{en}^{+} + \boldsymbol{V}_{en}^{+} - \boldsymbol{V}_{en}^{-}) && \text{on } S \\
\epsilon^{+}\frac{\partial}{\partial n}\phi_{en}^{+} - \epsilon^{-}\frac{\partial}{\partial n}\phi_{en}^{-} &= \epsilon^{+}\hat{\boldsymbol{n}} \cdot (\boldsymbol{F}_{en}^{+} + \boldsymbol{V}_{en}^{+}) \\
&\quad + \epsilon^{-}\hat{\boldsymbol{n}} \cdot (\boldsymbol{F}_{en}^{-} - \boldsymbol{V}_{en}^{-}) && \text{on } S \\
\phi_{en}^{+} &= O\left(\frac{1}{r^{2}}\right) && \text{as } r \to \infty
\end{aligned}\right\} \tag{3.202}$$

as can be seen from the transmission conditions (3.180) and (3.181) and the expressions (3.193) and (3.200). To these conditions we add the requirements that

$$\int_{S} \hat{\boldsymbol{n}} \cdot \boldsymbol{E}_{n}^{\pm}\, ds = 0 \tag{3.203}$$

The problem of determining $\boldsymbol{H}_n^+$ and $\boldsymbol{H}_n^-$, assuming all coefficients are known up to and including order $n-1$, is resolved in a similar fashion. For the solenoidal part of $\boldsymbol{U}_{mn}^-$ we have from (3.141)

$$\boldsymbol{U}_{mn}^{s-}(\boldsymbol{r}) = \frac{1}{4\pi}\nabla \times \int_S \frac{\hat{\boldsymbol{n}}' \times \boldsymbol{H}_n^-(\boldsymbol{r}')}{|\boldsymbol{r}-\boldsymbol{r}'|}\, ds(\boldsymbol{r}') \tag{3.204}$$

and a straightforward calculation similar to that for $\boldsymbol{U}_{en}^{s-}$ in (3.185)–(3.187) leads to

$$\nabla \times \boldsymbol{U}_{mn}^{s-}(\boldsymbol{r}) = \frac{n\eta}{4\pi} Y^- \nabla \int_S \frac{\hat{\boldsymbol{n}}' \cdot \boldsymbol{E}_{n-1}^-(\boldsymbol{r}')}{|\boldsymbol{r}-\boldsymbol{r}'|}\, ds(\boldsymbol{r}') \tag{3.205}$$

Defining the exterior potential function ψ_{mn}^+ to be the unique solution of the problem

$$\left.\begin{aligned} \nabla^2 \psi_{mn}^+ &= 0 && \text{in } V^+ \\ \frac{\partial}{\partial n}\psi_{mn}^+ &= \frac{n\eta}{4\pi} Y^- \hat{\boldsymbol{n}} \cdot \boldsymbol{E}_{n-1}^- && \text{on } S \\ \psi_{mn}^+ &= O\left(\frac{1}{r}\right) && \text{as } r \to \infty \end{aligned}\right\} \tag{3.206}$$

we see that

$$\nabla \times \boldsymbol{U}_{mn}^{s-}(\boldsymbol{r}) = \nabla \int_S \frac{1}{|\boldsymbol{r}-\boldsymbol{r}'|} \frac{\partial}{\partial n'} \psi_{mn}^+(\boldsymbol{r}')\, ds(\boldsymbol{r}') \tag{3.207}$$

and defining

$$\boldsymbol{V}_{mn}^-(\boldsymbol{r}) = \nabla \times \int_S \frac{\psi_{mn}^+(\boldsymbol{r}')}{|\boldsymbol{r}-\boldsymbol{r}'|} \hat{\boldsymbol{n}}'\, ds(\boldsymbol{r}'), \quad \boldsymbol{r} \in V^- \tag{3.208}$$

it follows, with Green's theorem and standard vector operations, that

$$\nabla \times (\boldsymbol{U}_{mn}^{s-}(\boldsymbol{r}) + \boldsymbol{V}_{mn}^-(\boldsymbol{r})) = \boldsymbol{0}, \quad \boldsymbol{r} \in V^- \tag{3.209}$$

Hence, $\boldsymbol{U}_{mn}^{s-} + \boldsymbol{V}_{mn}^-$ is irrotational and therefore it is the gradient of a scalar function, which is also harmonic because both $\boldsymbol{U}_{mn}^{s-}$ and $\boldsymbol{V}_{mn}^-$ are defined as curls. It thus follows, with (3.137) and (3.141), that

$$-\boldsymbol{H}_n^-(\boldsymbol{r}) = \boldsymbol{F}_{mn}^-(\boldsymbol{r}) - \boldsymbol{V}_{mn}^-(\boldsymbol{r}) + \nabla\phi_{mn}^-(\boldsymbol{r}), \quad \boldsymbol{r} \in V^- \tag{3.210}$$

where ϕ_{mn}^- is an unknown scalar potential function.

On the other hand, defining

$$V^+_{mn}(\boldsymbol{r}) = -\frac{n\eta}{4\pi} Y^- \nabla \times \int\limits_{V^-} \frac{\boldsymbol{E}^-_{n-1}(\boldsymbol{r}')}{|\boldsymbol{r}-\boldsymbol{r}'|}\, dv(\boldsymbol{r}') \tag{3.211}$$

it may be shown, using (3.130), (3.134) and (3.180) and analysis similar to that used in deriving the corresponding result (3.198) for $\boldsymbol{E}^+_n$, that

$$\boldsymbol{H}^+_n(\boldsymbol{r}) = \boldsymbol{F}^+_{mn}(\boldsymbol{r}) + \boldsymbol{V}^+_{mn}(\boldsymbol{r}) + \frac{1}{4\pi}\left(\frac{\mu^+}{\mu^-} - 1\right) \nabla \int\limits_S \frac{\hat{\boldsymbol{n}}' \cdot \boldsymbol{H}^+_n(\boldsymbol{r}')}{|\boldsymbol{r}-\boldsymbol{r}'|}\, ds(\boldsymbol{r}') \tag{3.212}$$

Defining

$$\phi^+_{mn}(\boldsymbol{r}) = \frac{1}{4\pi}\left(1 - \frac{\mu^+}{\mu^-}\right) \int\limits_S \frac{\hat{\boldsymbol{n}}' \cdot \boldsymbol{H}^+_n(\boldsymbol{r}')}{|\boldsymbol{r}-\boldsymbol{r}'|}\, ds(\boldsymbol{r}'), \qquad \boldsymbol{r} \in V^+ \tag{3.213}$$

the representation (3.212) becomes

$$\boldsymbol{H}^+_n(\boldsymbol{r}) = \boldsymbol{F}^+_{mn}(\boldsymbol{r}) + \boldsymbol{V}^+_{mn}(\boldsymbol{r}) - \nabla\phi^+_{mn}(\boldsymbol{r}), \qquad \boldsymbol{r} \in V^+ \tag{3.214}$$

where ϕ^+_{mn} is an exterior scalar potential function which, in view of (3.129), decays as $1/r^2$ as $r \to \infty$. Note that in the case where $\mu^+ = \mu^-$ we obtain $\phi^+_{mn} = 0$, which in turn implies that $\boldsymbol{H}^+_n$ is completely determined by coefficients of order lower than n. Then the problem of determining $\boldsymbol{H}^\pm_n$ is reduced to the solution of the interior potential problem for ϕ^-_{mn}.

In general, however, ϕ^+_{mn} and ϕ^-_{mn} must be determined simultaneously as the solution of the following scalar potential transmission problem

$$\left.\begin{aligned}
\nabla^2\phi^+_{mn} &= 0 && \text{in } V^+\\
\nabla^2\phi^-_{mn} &= 0 && \text{in } V^-\\
\hat{\boldsymbol{n}} \times \nabla(\phi^+_{mn} - \phi^-_{mn}) &= \hat{\boldsymbol{n}} \times (\boldsymbol{F}^+_{mn} + \boldsymbol{F}^-_{mn} \\
&\quad + \boldsymbol{V}^+_{mn} - \boldsymbol{V}^-_{mn}) && \text{on } S\\
\mu^+\frac{\partial}{\partial n}\phi^+_{mn} - \mu^-\frac{\partial}{\partial n}\phi^-_{mn} &= \mu^+\hat{\boldsymbol{n}} \cdot (\boldsymbol{F}^+_{mn} + \boldsymbol{V}^+_{mn}) \\
&\quad + \mu^-\hat{\boldsymbol{n}} \cdot (\boldsymbol{F}^-_{mn} - \boldsymbol{V}^-_{mn}) && \text{on } S\\
\phi^+_{mn} &= O\!\left(\frac{1}{r^2}\right) && \text{as } r \to \infty
\end{aligned}\right\} \tag{3.215}$$

which incorporates relations (3.130), (3.137), (3.210) and (3.214) and the transmission conditions (3.180) and (3.181). To these may be added the auxiliary conditions

$$\int\limits_S \hat{\boldsymbol{n}} \cdot \boldsymbol{H}^\pm_n \, ds = 0 \tag{3.216}$$

It is clear that in the case when $\mu^+ = \mu^-$, where $\phi^+_{mn} = 0$, the second condition provides a Neumann boundary condition for the interior potential ϕ^-_{mn}, from which ϕ^-_{mn} may be determined to within a constant, provided that the data satisfy the compatibility condition that the integral of the normal derivative of ϕ^-_{mn} over S vanishes. It must then be true that the first condition on the tangential components is automatically satisfied (otherwise the problem would be over-determined). However, a proof of this fact is as yet unavailable.

To summarize the steps that one must follow to determine the nth coefficients once all lower order coefficients have been found, we provide the following checklist.

For $\boldsymbol{E}^{\pm}_n$:

Step 1: Evaluate $\boldsymbol{F}^+_{en}$ and $\boldsymbol{F}^-_{en}$ from (3.131) and (3.138).
Step 2: Solve the exterior Neumann problem (3.188) for ψ^+_{en}.
Step 3: Use ψ^+_{en} in (3.191) to calculate $\boldsymbol{V}^-_{en}$.
Step 4: Calculate $\boldsymbol{V}^+_{en}$ from (3.196).
Step 5: Solve the transmission problem (3.202) for $\phi^{\pm}_{en}$.
Step 6: Substitute $\boldsymbol{F}^+_{en}$, $\boldsymbol{V}^+_{en}$ and ϕ^+_{en} into (3.200) to determine $\boldsymbol{E}^+_n$ and $\boldsymbol{F}^-_{en}$, $\boldsymbol{V}^-_{en}$ and ϕ^-_{en} into (3.193) to determine $\boldsymbol{E}^-_n$.

For $\boldsymbol{H}^{\pm}_n$:

Step 1: Evaluate $\boldsymbol{F}^+_{mn}$ and $\boldsymbol{F}^-_{mn}$ from (3.132) and (3.139).
Step 2: Solve the exterior Neumann problem (3.206) for ψ^+_{mn}.
Step 3: Use ψ^+_{mn} in (3.208) to calculate $\boldsymbol{V}^-_{mn}$.
Step 4: Calculate $\boldsymbol{V}^+_{mn}$ from (3.211).
Step 5: Solve the transmission problem (3.215) for $\phi^{\pm}_{mn}$.
Step 6: Substitute $\boldsymbol{F}^+_{mn}$, $\boldsymbol{V}^+_{mn}$ and ϕ^+_{mn} into (3.214) to determine $\boldsymbol{H}^+_n$ and $\boldsymbol{F}^-_{mn}$, $\boldsymbol{V}^-_{mn}$ and ϕ^-_{mn} into (3.210) to determine $\boldsymbol{H}^-_n$.

Thus, the determination of the coefficients $\boldsymbol{E}^{\pm}_n$ and $\boldsymbol{H}^{\pm}_n$, each of which is a coupled system of six functions, has been reduced to the determination of two independent exterior Neumann and two independent transmission problems for potential functions.

The scattering amplitudes $\boldsymbol{g}_e$ and $\boldsymbol{g}_m$ are given by (3.143) and (3.144) or (2.94b)

$$\boldsymbol{g}_e(\hat{\boldsymbol{r}}) = \operatorname{Re} \boldsymbol{g}_e(\hat{\boldsymbol{r}}) + i \operatorname{Im} \boldsymbol{g}_e(\hat{\boldsymbol{r}}) \tag{3.217}$$

$$\boldsymbol{g}_m(\hat{\boldsymbol{r}}) = Y^+ \hat{\boldsymbol{r}} \times \boldsymbol{g}_e(\hat{\boldsymbol{r}}) \tag{3.218}$$

where

$$\operatorname{Re} \boldsymbol{g}_e(\hat{\boldsymbol{r}}) = \sum_{n=0}^{\infty} k^{2n+3}(-1)^n A_{2n}(\hat{\boldsymbol{r}}) \tag{3.219}$$

$$\operatorname{Im} \boldsymbol{g}_e(\hat{\boldsymbol{r}}) = \sum_{n=0}^{\infty} k^{2n+4}(-1)^n A_{2n+1}(\hat{\boldsymbol{r}}) \tag{3.220}$$

and

$$\begin{aligned}
\boldsymbol{A}_n(\hat{\boldsymbol{r}}) = \frac{1}{4\pi n!} \sum_{m=0}^{n} \binom{n}{m} (-1)^m \int_S \{ & \hat{\boldsymbol{r}} \times (\hat{\boldsymbol{r}} \times \boldsymbol{r}')[\hat{\boldsymbol{n}}' \cdot \boldsymbol{E}^+_{n-m}(\boldsymbol{r}') \\
& - Z^+ \hat{\boldsymbol{r}} \cdot (\hat{\boldsymbol{n}}' \times \boldsymbol{H}^+_{n-m}(\boldsymbol{r}'))] \\
& + (\hat{\boldsymbol{r}} \times \boldsymbol{r}')[Z^+(\hat{\boldsymbol{n}}' \cdot \boldsymbol{H}^+_{n-m}(\boldsymbol{r}')) \\
& + \hat{\boldsymbol{r}} \cdot (\hat{\boldsymbol{n}}' \times \boldsymbol{E}^+_{n-m}(\boldsymbol{r}'))]\} (\hat{\boldsymbol{r}} \cdot \boldsymbol{r}')^m \, ds(\boldsymbol{r}')
\end{aligned} \tag{3.221}$$

for $n = 0, 1, 2, \ldots$. The differential scattering cross section is given by (2.113) as

$$\begin{aligned}
\sigma(\hat{\boldsymbol{r}}) &= 4\pi \sum_{n=0}^{\infty} k^{2n+4} (-1)^n \sum_{m=0}^{n} \boldsymbol{A}_{2n-2m}(\hat{\boldsymbol{r}}) \cdot \boldsymbol{A}_{2m}(\hat{\boldsymbol{r}}) \\
&\quad + 4\pi \sum_{n=0}^{\infty} k^{2n+6} (-1)^n \sum_{m=0}^{n} \boldsymbol{A}_{2n-2m+1}(\hat{\boldsymbol{r}}) \cdot \boldsymbol{A}_{2m+1}(\hat{\boldsymbol{r}})
\end{aligned} \tag{3.222}$$

and the scattering cross section by

$$\begin{aligned}
\sigma_s &= \sum_{n=0}^{\infty} k^{2n+4} (-1)^n \sum_{m=0}^{n} \int_S \boldsymbol{A}_{2n-2m}(\hat{\boldsymbol{r}}) \cdot \boldsymbol{A}_{2m}(\hat{\boldsymbol{r}}) \, ds(\hat{\boldsymbol{r}}) \\
&\quad + \sum_{n=0}^{\infty} k^{2n+6} (-1)^n \sum_{m=0}^{n} \int_S \boldsymbol{A}_{2n-2m+1}(\hat{\boldsymbol{r}}) \cdot \boldsymbol{A}_{2m+1}(\hat{\boldsymbol{r}}) \, ds(\hat{\boldsymbol{r}})
\end{aligned} \tag{3.223}$$

The absorption cross section vanishes, $\sigma_a = 0$, as can be seen from (2.123) for $\sigma^- = 0$, which implies that the extinction cross section σ_e is equal to σ_s.

3.EM.5 *Problems in electromagnetics*

1. Use the expansions (3.116), (3.117), (3.118), (3.14) and (3.23) in the Stratton–Chu representations (2.69) and (2.70), (2.65) and (2.66) to derive the relations (3.130)–(3.134) and (3.137)–(3.141).
2. Derive the low frequency expansions for the electric and the magnetic scattering amplitudes based on the representations (2.88) and (2.89). Repeat the derivations for the representations (2.90) and (2.91). Finally, prove that both expansions obtained above are equivalent to the expansions (3.143) and (3.144) which are based on the representations (2.95) and (2.96).
3. Derive the jump condition (3.156).
4. Provide the details that lead to the expressions (3.166) and (3.167).
5. Justify the expansions (3.175) and (3.176) for the electric scattering amplitude $\boldsymbol{g}_e$.
6. Use (3.143) to obtain the low frequency expansion of the magnetic scattering amplitude $\boldsymbol{g}_m$. Then provide formulae for the $\operatorname{Re}\boldsymbol{g}_m$ and the $\operatorname{Im}\boldsymbol{g}_m$ analogous to those obtained in (3.175) and (3.176) for $\boldsymbol{g}_e$. Finally, use the connection formulae (2.93)

   $$\boldsymbol{g}_e = -Z^+ \hat{\boldsymbol{r}} \times \boldsymbol{g}_m$$

   to recover (3.175) and (3.176) from the expansion of $\boldsymbol{g}_m$.

7. Expand the fundamental dyadics $ik^{\pm}\widetilde{G}^{\pm}(\boldsymbol{r}, \boldsymbol{r}')$ given by (1.166) in $V^{\pm}$ in powers of $k^{+} = k$.
8. Use the expansions obtained in Problem 7 and (3.117) and (3.118) to obtain low frequency expansions of $\boldsymbol{E}^{\pm}$ and $\boldsymbol{H}^{\pm}$ based on the integral representations (2.72) and (2.73).
9. The expansion for $\boldsymbol{E}^{+}$ obtained in Problem 8 will involve the following singular integral

$$\boldsymbol{I}(\boldsymbol{r}) = \int_S [\nabla \times \boldsymbol{E}_0^{+}(\boldsymbol{r}') + ik\nabla \times \boldsymbol{E}_1^{+}(\boldsymbol{r}')] \cdot \left[\frac{1}{(ik)^2}\hat{\boldsymbol{n}}' \times \nabla_{\boldsymbol{r}'}\left(\nabla_{\boldsymbol{r}'}\frac{1}{|\boldsymbol{r}-\boldsymbol{r}'|}\right)\right] ds(\boldsymbol{r}')$$

Prove that $\boldsymbol{I} = \boldsymbol{0}$. Then show that similar arguments can be used to prove that all singular integrals in the expansions for $\boldsymbol{E}^{-}$, $\boldsymbol{H}^{+}$ and $\boldsymbol{H}^{-}$ vanish.
10. Provide the details in deriving expressions (3.186), (3.190), (3.195) and (3.212) for the lossless transmission problem.

3.E Elasticity

3.E.1 *General low frequency expansions in elasticity*

In elastodynamic scattering there are in general four wave numbers, $k_p^{\pm}$ for the longitudinal and $k_s^{\pm}$ for the transverse waves. From our requirements on the Lamé parameters (1.195), (1.199) and (1.211), we find that

$$\frac{k_p^{\pm}}{k_s^{\pm}} = \frac{c_s^{\pm}}{c_p^{\pm}} = \sqrt{\frac{\mu^{\pm}}{\lambda^{\pm} + 2\mu^{\pm}}} < 1 \tag{3.224}$$

where the longitudinal, or P wave, has a larger velocity of propagation then the transverse, or S wave.

We use k_p^{+} as the expansion parameter for all the low frequency expansions and introduce the notation

$$\tau^{\pm} = \frac{k_p^{\pm}}{k_s^{\pm}} < 1 \tag{3.225}$$

For convenience, we will omit the superscript plus sign on the exterior wave numbers k_p and k_s.

Then a general incident field has the expansion

$$\boldsymbol{u}^i(\boldsymbol{r}) = \sum_{n=0}^{\infty} \frac{(ik_p)^n}{n!} \boldsymbol{u}_n^i(\boldsymbol{r}) \tag{3.226}$$

while a plane wave is expressed as

$$\alpha_p \hat{\boldsymbol{k}} e^{ik_p \hat{\boldsymbol{k}}\cdot \boldsymbol{r}} + \alpha_s \hat{\boldsymbol{b}} e^{ik_s \hat{\boldsymbol{k}}\cdot \boldsymbol{r}}$$

$$= \sum_{n=0}^{\infty} \frac{(ik_p)^n}{n!} \left[\alpha_p \hat{\boldsymbol{k}} + \left(\frac{1}{\tau^+}\right)^n \alpha_s \hat{\boldsymbol{b}} \right] (\hat{\boldsymbol{k}} \cdot \boldsymbol{r})^n, \quad \boldsymbol{r} \in \mathbb{R}^3 \tag{3.227}$$

where $\hat{\boldsymbol{k}} \cdot \hat{\boldsymbol{b}} = 0$. The scattered field $\boldsymbol{u}(\boldsymbol{r})$ and the total field $\boldsymbol{u}^+(\boldsymbol{r})$ for $\boldsymbol{r} \in V^+$ have expansions

$$\boldsymbol{u}(\boldsymbol{r}) = \sum_{n=0}^{\infty} \frac{(ik_p)^n}{n!} \boldsymbol{u}_n(\boldsymbol{r}) \tag{3.228}$$

and

$$\boldsymbol{u}^+(\boldsymbol{r}) = \sum_{n=0}^{\infty} \frac{(ik_p)^n}{n!} \boldsymbol{u}_n^+(\boldsymbol{r}) \tag{3.229}$$

where

$$\boldsymbol{u}_n^+(\boldsymbol{r}) = \boldsymbol{u}_n^i(\boldsymbol{r}) + \boldsymbol{u}_n(\boldsymbol{r}) \tag{3.230}$$

Moreover, the displacement field in V^- is also expanded in powers of ik_p as

$$\boldsymbol{u}^-(\boldsymbol{r}) = \sum_{n=0}^{\infty} \frac{(ik_p)^n}{n!} \boldsymbol{u}_n^-(\boldsymbol{r}) \tag{3.231}$$

The fundamental dyadic (1.224), (1.228) and (1.229) has the expansion

$$\widetilde{\boldsymbol{\Gamma}}^{\pm}(\boldsymbol{r}, \boldsymbol{r}') = \frac{1}{\mu^{\pm}} \sum_{n=0}^{\infty} \frac{(ik_p^{\pm})^n}{n!} \widetilde{\gamma}_n^{\pm}(\boldsymbol{r}, \boldsymbol{r}') \tag{3.232}$$

where, for $n = 0, 1, 2, \ldots$

$$\widetilde{\gamma}_n^{\pm}(\boldsymbol{r}, \boldsymbol{r}') = \frac{|\boldsymbol{r} - \boldsymbol{r}'|^{n-1}}{(\tau^{\pm})^n} \left[\left(1 + \frac{((\tau^{\pm})^{n+2} - 1)}{n+2} \right) \widetilde{\boldsymbol{I}} + (n-1) \frac{((\tau^{\pm})^{n+2} - 1)}{n+2} \frac{(\boldsymbol{r} - \boldsymbol{r}')(\boldsymbol{r} - \boldsymbol{r}')}{|\boldsymbol{r} - \boldsymbol{r}'|^2} \right] \tag{3.233}$$

In particular, $\widetilde{\gamma}_1^{\pm}$ are constant multiples of the identity, i.e.

$$\widetilde{\gamma}_1^{\pm}(\boldsymbol{r}, \boldsymbol{r}') = \frac{(\tau^{\pm})^3 + 2}{3\tau^{\pm}} \widetilde{\boldsymbol{I}} \tag{3.234}$$

Recall (1.218) that the relative index of refraction for the longitudinal wave is

$$\eta_p = \frac{k_p^-}{k_p^+} \tag{3.235}$$

Then, the fundamental dyadic for V^- may be expanded in powers of ik_p as

$$\widetilde{\boldsymbol{\Gamma}}^-(\boldsymbol{r},\boldsymbol{r}') = \frac{1}{\mu^-}\sum_{n=0}^{\infty}\frac{(ik_p)^n}{n!}\eta_p^n\,\widetilde{\gamma}_n^-(\boldsymbol{r},\boldsymbol{r}') \tag{3.236}$$

Similarly, the surface traction operating on the fundamental dyadic (1.230) has the following expansion in powers of ik_p

$$\boldsymbol{T}^+(\partial_{\boldsymbol{r}'},\hat{\boldsymbol{n}}')\,\widetilde{\boldsymbol{\Gamma}}^+(\boldsymbol{r},\boldsymbol{r}') = \frac{1}{\mu^+}\sum_{n=0}^{\infty}\frac{(ik_p)^n}{n!}\boldsymbol{T}^+(\partial_{\boldsymbol{r}'},\hat{\boldsymbol{n}}')\,\widetilde{\gamma}_n^+(\boldsymbol{r},\boldsymbol{r}') \tag{3.237}$$

and

$$\boldsymbol{T}^-(\partial_{\boldsymbol{r}'},\hat{\boldsymbol{n}}')\,\widetilde{\boldsymbol{\Gamma}}^-(\boldsymbol{r},\boldsymbol{r}') = \frac{1}{\mu^-}\sum_{n=0}^{\infty}\frac{(ik_p)^n}{n!}\eta_p^n\boldsymbol{T}^-(\partial_{\boldsymbol{r}'},\hat{\boldsymbol{n}}')\,\widetilde{\gamma}_n^-(\boldsymbol{r},\boldsymbol{r}') \tag{3.238}$$

where, for $n = 0, 1, 2, \ldots$

$$\begin{aligned}\boldsymbol{T}^\pm(\partial_{\boldsymbol{r}'},\hat{\boldsymbol{n}}')\,\widetilde{\gamma}_n^\pm(\boldsymbol{r},\boldsymbol{r}') &= -\frac{n-1}{(\tau^\pm)^n(n+2)}|\boldsymbol{r}-\boldsymbol{r}'|^{n-2}\\ &\times\Bigg\{\mu^\pm(2(\tau^\pm)^{n+2}+n)\left[\frac{\hat{\boldsymbol{n}}'\cdot(\boldsymbol{r}-\boldsymbol{r}')}{|\boldsymbol{r}-\boldsymbol{r}'|}\widetilde{\boldsymbol{I}}+\frac{(\boldsymbol{r}-\boldsymbol{r}')\hat{\boldsymbol{n}}'}{|\boldsymbol{r}-\boldsymbol{r}'|}\right]\\ &+2\mu^\pm(n-3)((\tau^\pm)^{n+2}-1)\frac{\hat{\boldsymbol{n}}'\cdot(\boldsymbol{r}-\boldsymbol{r}')(\boldsymbol{r}-\boldsymbol{r}')(\boldsymbol{r}-\boldsymbol{r}')}{|\boldsymbol{r}-\boldsymbol{r}'|^3}\\ &+[\lambda^\pm(n+2)(\tau^\pm)^{n+2}+2\mu^\pm((\tau^\pm)^{n+2}-1)]\frac{\hat{\boldsymbol{n}}'(\boldsymbol{r}-\boldsymbol{r}')}{|\boldsymbol{r}-\boldsymbol{r}'|}\Bigg\}\end{aligned} \tag{3.239}$$

In particular, since $\widetilde{\gamma}_1^\pm$ are constant, see (3.234),

$$\boldsymbol{T}^\pm(\partial_{\boldsymbol{r}'},\hat{\boldsymbol{n}}')\,\widetilde{\gamma}_1^\pm(\boldsymbol{r},\boldsymbol{r}') = \widetilde{\boldsymbol{0}} \tag{3.240}$$

Note that the Lamé operators

$$\Delta^{e\pm} = [(c_p^\pm)^2 - (c_s^\pm)^2]\nabla(\nabla\cdot) + (c_s^\pm)^2\nabla^2 \tag{3.241}$$

differ in V^+ and in V^-. Note also that in all the low frequency expansions the dimensions of the coefficients have an extra factor of length to the nth power to compensate for the dimensions of k_p^n.

Substituting the low frequency expansions (3.229) and (3.231) into the Navier equations (1.201) and (3.241) and equating like powers of ik_p leads to the following differential equations for the coefficients

$$\Delta^{e+}\boldsymbol{u}_n^+(\boldsymbol{r}) = n(n-1)(c_p^+)^2\boldsymbol{u}_{n-2}^+(\boldsymbol{r}), \quad \boldsymbol{r}\in V^+ \tag{3.242}$$

$$\Delta^{e-}\boldsymbol{u}_n^-(\boldsymbol{r}) = n(n-1)(c_p^+)^2\boldsymbol{u}_{n-2}^-(\boldsymbol{r}), \quad \boldsymbol{r}\in V^- \tag{3.243}$$

Remark: Note that since the angular frequency is the same in V^+ and in V^-, equation (1.201) with $\Delta^{e\pm}$ and $w = k_p c_p^+$ will produce the same factor $(c_p^+)^2$ in the right-hand sides of equations (3.242) and (3.243).

Similarly, substituting the low frequency expansions into the boundary and transmission conditions (1.214), (1.215), (1.216) and (1.217) yields

The rigid problem:

$$\boldsymbol{u}_n^+(\boldsymbol{r}) = \mathbf{0}, \quad \boldsymbol{r} \in S \tag{3.244}$$

The cavity problem:

$$\boldsymbol{T}^+(\partial_{\boldsymbol{r}}, \hat{\boldsymbol{n}})\boldsymbol{u}_n^+(\boldsymbol{r}) = \mathbf{0}, \quad \boldsymbol{r} \in S \tag{3.245}$$

The transmission problem:

$$\boldsymbol{u}_n^+(\boldsymbol{r}) = \boldsymbol{u}_n^-(\boldsymbol{r}), \quad \boldsymbol{r} \in S \tag{3.246}$$

$$\boldsymbol{T}^+(\partial_{\boldsymbol{r}}, \hat{\boldsymbol{n}})\boldsymbol{u}_n^+(\boldsymbol{r}) = \boldsymbol{T}^-(\partial_{\boldsymbol{r}}, \hat{\boldsymbol{n}})\boldsymbol{u}_n^-(\boldsymbol{r}), \quad \boldsymbol{r} \in S \tag{3.247}$$

for $n = 0, 1, 2, \ldots$.

On the other hand, substituting the low frequency expansions for the displacement fields (3.229) and (3.231) and the fundamental dyadic (3.232) into the integral representations (2.146), utilizing the Cauchy product formula and equating like coefficients of $(ik_p)^n$ yields

$$\alpha(\boldsymbol{r})\boldsymbol{u}_n^+(\boldsymbol{r}) = \boldsymbol{f}_n^+(\boldsymbol{r}) + \boldsymbol{v}_n^+(\boldsymbol{r}), \quad \boldsymbol{r} \in \mathbb{R}^3 \tag{3.248}$$

where

$$\begin{aligned}\boldsymbol{f}_n^+(\boldsymbol{r}) &= \boldsymbol{u}_n^i(\boldsymbol{r}) \\ &\quad + \frac{1}{4\pi\mu^+}\sum_{m=1}^{n}\binom{n}{m}\int_S [\boldsymbol{u}_{n-m}^+(\boldsymbol{r}') \cdot \boldsymbol{T}^+(\partial_{\boldsymbol{r}'}, \hat{\boldsymbol{n}}')\,\widetilde{\boldsymbol{\gamma}}_m^+(\boldsymbol{r}, \boldsymbol{r}') \\ &\quad - \widetilde{\boldsymbol{\gamma}}_m^+(\boldsymbol{r}, \boldsymbol{r}') \cdot \boldsymbol{T}^+(\partial_{\boldsymbol{r}'}, \hat{\boldsymbol{n}}')\boldsymbol{u}_{n-m}^+(\boldsymbol{r}')]\, ds(\boldsymbol{r}')\end{aligned} \tag{3.249}$$

and

$$\begin{aligned}\boldsymbol{v}_n^+(\boldsymbol{r}) &= \frac{1}{4\pi\mu^+}\int_S [\boldsymbol{u}_n^+(\boldsymbol{r}') \cdot \boldsymbol{T}^+(\partial_{\boldsymbol{r}'}, \hat{\boldsymbol{n}}')\,\widetilde{\boldsymbol{\gamma}}_0^+(\boldsymbol{r}, \boldsymbol{r}') \\ &\quad - \widetilde{\boldsymbol{\gamma}}_0^+(\boldsymbol{r}, \boldsymbol{r}') \cdot \boldsymbol{T}^+(\partial_{\boldsymbol{r}'}, \hat{\boldsymbol{n}}')\boldsymbol{u}_n^+(\boldsymbol{r}')]\, ds(\boldsymbol{r}') \\ &= O\left(\frac{1}{r}\right), \quad r \to \infty\end{aligned} \tag{3.250}$$

A similar process involving the expansions of $\boldsymbol{u}^-$ and $\widetilde{\boldsymbol{\Gamma}}^-$ and the integral representation (2.144) leads to

$$(\alpha(\boldsymbol{r}) - 1)\boldsymbol{u}_n^-(\boldsymbol{r}) = \boldsymbol{f}_n^-(\boldsymbol{r}) + \boldsymbol{v}_n^-(\boldsymbol{r}), \quad \boldsymbol{r} \in \mathbb{R}^3 \tag{3.251}$$

where

$$\begin{aligned}\boldsymbol{f}_n^-(\boldsymbol{r}) &= \frac{1}{4\pi\mu^-}\sum_{m=1}^{n}\binom{n}{m}\eta_p^m\int_S [\boldsymbol{u}_{n-m}^-(\boldsymbol{r}') \cdot \boldsymbol{T}^-(\partial_{\boldsymbol{r}'}, \hat{\boldsymbol{n}}')\,\widetilde{\boldsymbol{\gamma}}_m^-(\boldsymbol{r}, \boldsymbol{r}') \\ &\quad - \widetilde{\boldsymbol{\gamma}}_m^-(\boldsymbol{r}, \boldsymbol{r}') \cdot \boldsymbol{T}^-(\partial_{\boldsymbol{r}'}, \hat{\boldsymbol{n}}')\boldsymbol{u}_{n-m}^-(\boldsymbol{r}')]\, ds(\boldsymbol{r}')\end{aligned} \tag{3.252}$$

and

$$\boldsymbol{v}_n^-(\boldsymbol{r}) = \frac{1}{4\pi\mu^-}\int_S [\boldsymbol{u}_n^-(\boldsymbol{r}')\cdot \boldsymbol{T}^-(\partial_{\boldsymbol{r}'},\hat{\boldsymbol{n}}')\,\widetilde{\gamma}_0^-(\boldsymbol{r},\boldsymbol{r}') - \widetilde{\gamma}_0^-(\boldsymbol{r},\boldsymbol{r}')\cdot \boldsymbol{T}^-(\partial_{\boldsymbol{r}'},\hat{\boldsymbol{n}}')\boldsymbol{u}_n^-(\boldsymbol{r}')]\,ds(\boldsymbol{r}') \tag{3.253}$$

Note that the representations (3.248) and (3.251), for $\boldsymbol{u}_n^{\pm}$, are valid for all $\boldsymbol{r}$. While we will confine attention to $\boldsymbol{r}\in V^+$ for $\boldsymbol{u}_n^+$, and $\boldsymbol{r}\in V^-$ for $\boldsymbol{u}_n^-$, the more general representations may prove useful if the null field method were applied in low frequency analysis. As yet this has not been done.

Observe that since

$$\Delta_{\boldsymbol{r}}^{e+}\,\widetilde{\gamma}_0^+(\boldsymbol{r},\boldsymbol{r}') = \widetilde{\boldsymbol{0}}, \quad \boldsymbol{r}\in V^+, \quad \boldsymbol{r}'\in S \tag{3.254}$$

and

$$\Delta_{\boldsymbol{r}}^{e-}\,\widetilde{\gamma}_0^-(\boldsymbol{r},\boldsymbol{r}') = \widetilde{\boldsymbol{0}}, \quad \boldsymbol{r}\in V^-, \quad \boldsymbol{r}'\in S \tag{3.255}$$

it follows, from (3.250) and (3.253), that

$$\Delta^{e\pm}\boldsymbol{v}_n^{\pm}(\boldsymbol{r}) = \boldsymbol{0}, \quad \boldsymbol{r}\in V^{\pm} \tag{3.256}$$

Thus, $\boldsymbol{f}_n^{\pm}(\boldsymbol{r})$ furnish particular solutions of the inhomogeneous differential equations (3.242) and (3.243), respectively. In fact

$$\left.\begin{aligned} \Delta^{e+}\boldsymbol{f}_n^+ &= n(n-1)(c_p^+)^2\boldsymbol{u}_{n-2}^+ && \text{in } V^+\\ \Delta^{e+}\boldsymbol{f}_n^+ &= \boldsymbol{0} && \text{in } V^- \end{aligned}\right\} \tag{3.257}$$

$$\left.\begin{aligned} \Delta^{e-}\boldsymbol{f}_n^- &= \boldsymbol{0} && \text{in } V^+\\ \Delta^{e-}\boldsymbol{f}_n^- &= -n(n-1)(c_p^+)^2\boldsymbol{u}_{n-2}^- && \text{in } V^- \end{aligned}\right\} \tag{3.258}$$

The determination of the nth coefficient in the low frequency expansion of the displacement field, assuming the coefficients of order $0, 1, \ldots, n-1$ are known, is thus reduced to finding the elastostatic displacement fields $\boldsymbol{v}_n^{\pm}(\boldsymbol{r})$ where $\boldsymbol{v}_n^-$ is regular in V^- and

$$\boldsymbol{v}_n^+(\boldsymbol{r}) = O\left(\frac{1}{r}\right), \quad r\to+\infty \tag{3.259}$$

As described in Chapter 2, the scattering amplitude g_r given in (2.155), associated with the P wave, is expressed in terms of the dyadic $\widetilde{\boldsymbol{H}}_p^+$ in (2.158) and the vector $\boldsymbol{K}_p^+$ in (2.160), while the scattering amplitudes g_θ and $g_{\mathscr{L}}$ given in (2.156) and (2.157) associated with the S wave, are expressed in terms of $\widetilde{\boldsymbol{H}}_s^+$ and $\boldsymbol{K}_s^+$ in (2.159) and (2.161), respectively. Expanding the exponential in the integrands of (2.158)–(2.161) and using the low frequency expansion of $\boldsymbol{u}^+$, (3.229), in the expressions for $\widetilde{\boldsymbol{H}}_p^+$, $\widetilde{\boldsymbol{H}}_s^+$, $\boldsymbol{K}_p^+$ and $\boldsymbol{K}_s^+$, yields the following low frequency expansions

$$\widetilde{\boldsymbol{H}}_p^+(\hat{\boldsymbol{r}}) = \frac{1}{4\pi}\sum_{n=0}^{\infty}\frac{(ik_p)^n}{n!}\sum_{m=0}^{n}\binom{n}{m}(-1)^m \times \int_S \boldsymbol{u}_{n-m}^+(\boldsymbol{r}')\hat{\boldsymbol{n}}'(\hat{\boldsymbol{r}}\cdot\boldsymbol{r}')^m\,ds(\boldsymbol{r}') \tag{3.260}$$

$$\widetilde{\boldsymbol{H}}_s^+(\hat{\boldsymbol{r}}) = \frac{1}{4\pi}\sum_{n=0}^{\infty}\frac{(ik_p)^n}{n!}\sum_{m=0}^{n}\binom{n}{m}\left(-\frac{1}{\tau^+}\right)^m \times \int_S \boldsymbol{u}_{n-m}^+(\boldsymbol{r}')\hat{\boldsymbol{n}}'(\hat{\boldsymbol{r}}\cdot\boldsymbol{r}')^m\,ds(\boldsymbol{r}') \tag{3.261}$$

$$\boldsymbol{K}_p^+(\hat{\boldsymbol{r}}) = \frac{1}{4\pi}\sum_{n=0}^{\infty}\frac{(ik_p)^n}{n!}\sum_{m=0}^{n}\binom{n}{m}(-1)^m \times \int_S (\hat{\boldsymbol{r}}\cdot\boldsymbol{r}')^m\boldsymbol{T}^+(\partial_{\boldsymbol{r}'},\hat{\boldsymbol{n}}')\boldsymbol{u}_{n-m}^+(\boldsymbol{r}')\,ds(\boldsymbol{r}') \tag{3.262}$$

$$\boldsymbol{K}_s^+(\hat{\boldsymbol{r}}) = \frac{1}{4\pi}\sum_{n=0}^{\infty}\frac{(ik_p)^n}{n!}\sum_{m=0}^{n}\binom{n}{m}\left(-\frac{1}{\tau^+}\right)^m \times \int_S (\hat{\boldsymbol{r}}\cdot\boldsymbol{r}')^m\boldsymbol{T}^+(\partial_{\boldsymbol{r}'},\hat{\boldsymbol{n}}')\boldsymbol{u}_{n-m}^+(\boldsymbol{r}')\,ds(\boldsymbol{r}') \tag{3.263}$$

With these expansions, the spherical scattering amplitudes (2.155)–(2.157) assume the form

$$\begin{aligned} g_r(\hat{\boldsymbol{r}}) = {} & \frac{1}{4\pi(\lambda^+ + 2\mu^+)}\sum_{n=0}^{\infty}\frac{(ik_p)^{n+1}}{n!}\sum_{m=0}^{n}\binom{n}{m}(-1)^{m+1} \\ & \times \hat{\boldsymbol{r}}\cdot\int_S (\hat{\boldsymbol{r}}\cdot\boldsymbol{r}')^m\boldsymbol{T}^+(\partial_{\boldsymbol{r}'},\hat{\boldsymbol{n}}')\boldsymbol{u}_{n-m}^+(\boldsymbol{r}')\,ds(\boldsymbol{r}') \\ & + \frac{1}{4\pi(\lambda^+ + 2\mu^+)}\sum_{n=0}^{\infty}\frac{(ik_p)^{n+2}}{n!}\sum_{m=0}^{n}\binom{n}{m}(-1)^{m+1} \\ & \times [\lambda^+\widetilde{\boldsymbol{I}} + 2\mu^+\hat{\boldsymbol{r}}\hat{\boldsymbol{r}}] : \int_S \boldsymbol{u}_{n-m}^+(\boldsymbol{r}')\hat{\boldsymbol{n}}'(\hat{\boldsymbol{r}}\cdot\boldsymbol{r}')^m\,ds(\boldsymbol{r}') \end{aligned} \tag{3.264}$$

and

$$\left.\begin{aligned} g_\theta(\hat{\boldsymbol{r}}) &= \hat{\boldsymbol{\theta}}\cdot\boldsymbol{g}_t(\hat{\boldsymbol{r}}) \\ g_\varphi(\hat{\boldsymbol{r}}) &= \hat{\boldsymbol{\varphi}}\cdot\boldsymbol{g}_t(\hat{\boldsymbol{r}}) \end{aligned}\right\} \tag{3.265}$$

where

$$g_t(\hat{r}) = \frac{1}{4\pi\mu^+}\sum_{n=0}^{\infty}\frac{(ik_p)^{n+1}}{n!}\sum_{m=0}^{n}\binom{n}{m}\left(-\frac{1}{\tau^+}\right)^{m+1}$$
$$\times\int_S (\hat{r}\cdot r')^m \boldsymbol{T}^+(\partial_{r'},\hat{n}')\boldsymbol{u}^+_{n-m}(r')\,ds(r')$$
$$+\frac{1}{4\pi\tau^+}\sum_{n=0}^{\infty}\frac{(ik_p)^{n+2}}{n!}\sum_{m=0}^{n}\binom{n}{m}\left(-\frac{1}{\tau^+}\right)^{m+1}$$
$$\times\left[\int_S \boldsymbol{u}^+_{n-m}(r')\hat{n}'(\hat{r}\cdot r')^m\,ds(r')\right]:[\tilde{\boldsymbol{I}}\times\tilde{\boldsymbol{I}}\times\hat{r}+2\hat{r}\tilde{\boldsymbol{I}}] \tag{3.266}$$

The double contraction in (3.264) and (3.266) is to be interpreted as

$$(\boldsymbol{ab}):(\boldsymbol{cd}) = (\boldsymbol{a}\cdot\boldsymbol{d})(\boldsymbol{b}\cdot\boldsymbol{c}) \tag{3.267}$$

and the identity

$$\boldsymbol{ab}:[\tilde{\boldsymbol{I}}\times\tilde{\boldsymbol{I}}\times\hat{r}+2\hat{r}\tilde{\boldsymbol{I}}] = (\boldsymbol{ab}+\boldsymbol{ba})\cdot\hat{r}$$
$$= \boldsymbol{a}(\boldsymbol{b}\cdot\hat{r})+\boldsymbol{b}(\boldsymbol{a}\cdot\hat{r}) \tag{3.268}$$

may be used. These low frequency expressions for the scattering amplitudes are simplified when specific boundary conditions are introduced.

3.E.2 *The rigid problem*

The boundary condition on a rigid surface implies that

$$\boldsymbol{u}^+_n(\boldsymbol{r}) = \boldsymbol{0}, \quad \boldsymbol{r}\in S \tag{3.269}$$

for every $n = 0, 1, 2, \ldots$.

Substituting these conditions into the expressions (3.248), (3.249) and (3.250) we obtain

$$\boldsymbol{v}^+_n(\boldsymbol{r}) = -\boldsymbol{f}^+_n(\boldsymbol{r}), \quad \boldsymbol{r}\in S \tag{3.270}$$

where $\boldsymbol{f}^+_n$ and $\boldsymbol{v}^+_n$ are defined for all $\boldsymbol{r}$ in $\mathbb{R}^3$ as

$$\boldsymbol{f}^+_n(\boldsymbol{r}) = \boldsymbol{u}^i_n(\boldsymbol{r}) - \frac{1}{4\pi\mu^+}\sum_{m=1}^{n}\binom{n}{m}\int_S \tilde{\gamma}^+_m(\boldsymbol{r},\boldsymbol{r}')\cdot\boldsymbol{T}^+(\partial_{r'},\hat{n}')\boldsymbol{u}^+_{n-m}(\boldsymbol{r}')\,ds(\boldsymbol{r}') \tag{3.271}$$

and

$$\boldsymbol{v}^+_n(\boldsymbol{r}) = -\frac{1}{4\pi\mu^+}\int_S \tilde{\gamma}^+_0(\boldsymbol{r},\boldsymbol{r}')\cdot\boldsymbol{T}^+(\partial_{r'},\hat{n}')\boldsymbol{u}^+_n(\boldsymbol{r}')\,ds(\boldsymbol{r}') \tag{3.272}$$

Since $\boldsymbol{f}^+_n$ is a particular solution of the inhomogeneous equation (3.242) expressed in terms of $\boldsymbol{u}^+_m$ for $m = 0, 1, \ldots, n-1$, it follows that the nth order approximation $\boldsymbol{u}^+_n$ is given by (3.248) where $\boldsymbol{f}_n^+$ is as in (3.271) and $\boldsymbol{v}_n^+$ is the solution of the following problem of static elasticity

$$\left.\begin{aligned} \Delta^{+e}\boldsymbol{v}_n^+ &= \mathbf{0} && \text{in } V^+ \\ \boldsymbol{v}_n^+ &= -\boldsymbol{f}_n^+ && \text{on } S \\ \boldsymbol{v}_n^+ &= O\left(\frac{1}{r}\right) && \text{as } r \to \infty \end{aligned}\right\} \tag{3.273}$$

Hence, we see that the evaluation of the coefficients $\boldsymbol{u}_n^+$ is reduced to the solution of homogeneous elastostatic problems with prescribed surface displacement and regular behavior at infinity.

For simply connected domains V^+ these elastostatic problems admit the **Papkovich representation** in terms of a scalar and a vector harmonic function, known as **Papkovich potentials**. Specifically, the solution $\boldsymbol{v}_n^+$ of (3.273) can be represented as

$$\boldsymbol{v}_n^+(\boldsymbol{r}) = \boldsymbol{A}_n^+(\boldsymbol{r}) + \frac{(c_s^+)^2 - (c_p^+)^2}{2(c_p^+)^2}\nabla[\boldsymbol{r}\cdot \boldsymbol{A}_n^+(\boldsymbol{r}) + B_n^+(\boldsymbol{r})], \quad \boldsymbol{r} \in V^+ \tag{3.274}$$

where

$$\left.\begin{aligned} \nabla^2 \boldsymbol{A}_n^+ &= \mathbf{0} && \text{in } V^+ \\ \nabla^2 B_n^+ &= 0 && \text{in } V^+ \end{aligned}\right\} \tag{3.275}$$

Papkovich potentials are complete, in the sense that any solution of the elastostatic equation may be represented in terms of these potentials (Eubanks and Sternberg 1956). However, the representation is not unique. In fact, B_n^+ and the three scalar components of $\boldsymbol{A}_n^+$ are not linearly independent; any one may be expressed in terms of the other three (Gurtin 1972). This suggests that B_n^+ could be taken as zero thus simplifying the representation of $\boldsymbol{v}_n^+$. Although this in fact may be done, it has proved to be useful to keep all four potentials and use the extra freedom to satisfy the boundary or transmission conditions. This technique was used to obtain low frequency approximations of elastic scattering problems for spheres and ellipsoids (Dassios and Kiriaki 1984, 1986, 1987; Kiriaki 1982, 1989; Kiriaki and Polyzos 1988). It enabled replacing an infinite algebraic system with two finite algebraic systems.

In terms of Papkovich potentials the coefficient $\boldsymbol{u}_n^+$ is obtained from the representation (2.248) where $\boldsymbol{f}_n^+$ and $\boldsymbol{v}_n^+$ are given by (3.249) and (3.274), respectively and the potential functions $\boldsymbol{A}_n^+$ and B_n^+ which are solutions of

$$\left.\begin{aligned} \nabla^2 \boldsymbol{A}_n^+ &= \mathbf{0} && \text{in } V^+ \\ \nabla^2 B_n^+ &= 0 && \text{in } V^+ \\ \boldsymbol{A}_n^+ + \frac{(c_s^+)^2 - (c_p^+)^2}{2(c_p^+)^2}\nabla(\boldsymbol{r}\cdot \boldsymbol{A}_n^+ + B_n^+) &= -\boldsymbol{f}_n^+ && \text{on } S \\ \boldsymbol{A}_n^+ &= O\left(\frac{1}{r}\right) && \text{as } r \to \infty \\ B_n^+ &= O(1) && \text{as } r \to \infty \end{aligned}\right\} \tag{3.276}$$

For the scattering amplitudes we obtain from (3.264)–(3.266) that

$$\operatorname{Re} g_r(\hat{\boldsymbol{r}}) = \sum_{n=0}^{\infty} k_p^{2n+2}(-1)^{n+1} A_{r(2n+2)}(\hat{\boldsymbol{r}}) \tag{3.277}$$

$$\operatorname{Im} g_r(\hat{\boldsymbol{r}}) = \sum_{n=0}^{\infty} k_p^{2n+1}(-1)^{n} A_{r(2n+1)}(\hat{\boldsymbol{r}}) \tag{3.278}$$

$$\operatorname{Re} g_\theta(\hat{\boldsymbol{r}}) = \sum_{n=0}^{\infty} k_p^{2n+2}(-1)^{n+1} A_{\theta(2n+2)}(\hat{\boldsymbol{r}}) \tag{3.279}$$

$$\operatorname{Im} g_\theta(\hat{\boldsymbol{r}}) = \sum_{n=0}^{\infty} k_p^{2n+1}(-1)^{n} A_{\theta(2n+1)}(\hat{\boldsymbol{r}}) \tag{3.280}$$

$$\operatorname{Re} g_\varphi(\hat{\boldsymbol{r}}) = \sum_{n=0}^{\infty} k_p^{2n+2}(-1)^{n+1} A_{\varphi(2n+2)}(\hat{\boldsymbol{r}}) \tag{3.281}$$

$$\operatorname{Im} g_\varphi(\hat{\boldsymbol{r}}) = \sum_{n=0}^{\infty} k_p^{2n+1}(-1)^{n} A_{\varphi(2n+1)}(\hat{\boldsymbol{r}}) \tag{3.282}$$

where A_{rn}, $A_{\theta n}$ and $A_{\varphi n}$ are given by

$$\begin{aligned} A_{r(n+1)}(\hat{\boldsymbol{r}}) = {} & \frac{1}{4\pi(\lambda^+ + 2\mu^+)n!} \sum_{m=0}^{n} \binom{n}{m} (-1)^{m+1} \\ & \times \hat{\boldsymbol{r}} \cdot \int_S (\hat{\boldsymbol{r}} \cdot \boldsymbol{r}')^m \boldsymbol{T}^+(\partial_{\boldsymbol{r}'}, \hat{\boldsymbol{n}}') \boldsymbol{u}_{n-m}^+(\boldsymbol{r}') \, ds(\boldsymbol{r}') \end{aligned} \tag{3.283}$$

$$\begin{aligned} A_{\theta(n+1)}(\hat{\boldsymbol{r}}) = {} & \frac{1}{4\pi \mu^+ n!} \sum_{m=0}^{n} \binom{n}{m} \left(-\frac{1}{\tau^+}\right)^{m+1} \\ & \times \hat{\boldsymbol{\theta}} \cdot \int_S (\hat{\boldsymbol{r}} \cdot \boldsymbol{r}')^m \boldsymbol{T}^+(\partial_{\boldsymbol{r}'}, \hat{\boldsymbol{n}}') \boldsymbol{u}_{n-m}^+(\boldsymbol{r}') \, ds(\boldsymbol{r}') \end{aligned} \tag{3.284}$$

$$\begin{aligned} A_{\varphi(n+1)}(\hat{\boldsymbol{r}}) = {} & \frac{1}{4\pi \mu^+ n!} \sum_{m=0}^{n} \binom{n}{m} \left(-\frac{1}{\tau^+}\right)^{m+1} \\ & \times \hat{\boldsymbol{\varphi}} \cdot \int_S (\hat{\boldsymbol{r}} \cdot \boldsymbol{r}')^m \boldsymbol{T}^+(\partial_{\boldsymbol{r}'}, \hat{\boldsymbol{n}}') \boldsymbol{u}_{n-m}^+(\boldsymbol{r}') \, ds(\boldsymbol{r}') \end{aligned} \tag{3.285}$$

for $n = 0, 1, 2, \ldots$.

For general plane wave incidence (3.227), the differential scattering cross section (2.186) may then be written as

$$\sigma(\hat{\boldsymbol{r}}) = \frac{4\pi[|g_r|^2 + (\tau^+)^3(|g_\theta|^2 + |g_\varphi|^2)]}{k_p^2(\alpha_p^2 + \tau^+\alpha_s^2)}$$

$$= \frac{4\pi}{\alpha_p^2 + \tau^+\alpha_s^2}\{A_{r1}^2(\hat{\boldsymbol{r}}) + (\tau^+)^3[A_{\theta 1}^2(\hat{\boldsymbol{r}}) + A_{\varphi 1}^2(\hat{\boldsymbol{r}})]\}$$
$$+ \frac{4\pi}{\alpha_p^2 + \tau^+\alpha_s^2}\sum_{n=0}^{\infty} k_p^{2n+2}(-1)^n \left\{\sum_{m=0}^{n}[A_{r(2n-2m+2)}(\hat{\boldsymbol{r}})A_{r(2m+2)}(\hat{\boldsymbol{r}})\right.$$
$$+ (\tau^+)^3(A_{\theta(2n-2m+2)}(\hat{\boldsymbol{r}})A_{\theta(2m+2)}(\hat{\boldsymbol{r}}) + A_{\varphi(2n-2m+2)}(\hat{\boldsymbol{r}})A_{\varphi(2m+2)}(\hat{\boldsymbol{r}}))]$$
$$- \sum_{m=0}^{n+1}[A_{r(2n-2m+3)}(\hat{\boldsymbol{r}})A_{r(2m+1)}(\hat{\boldsymbol{r}})$$
$$\left. + (\tau^+)^3(A_{\theta(2n-2m+3)}(\hat{\boldsymbol{r}})A_{\theta(2m+1)}(\hat{\boldsymbol{r}}) + A_{\varphi(2n-2m+3)}(\hat{\boldsymbol{r}})A_{\varphi(2m+1)}(\hat{\boldsymbol{r}}))]\right\} \tag{3.286}$$

For a general incident plane wave ($\alpha_s \neq 0, \alpha_p \neq 0$) the scattering cross section is computed from the definition

$$\sigma_s = \frac{1}{4\pi}\int_{S^2} \sigma(\hat{\boldsymbol{r}})\, ds(\hat{\boldsymbol{r}}) \tag{3.287}$$

However, in special cases there are simplifications. For longitudinal incidence ($\alpha_s = 0$) the elastic forward scattering theorem (2.202) yields, in view of (3.277), the following result for the extinction cross section

$$\sigma_e = 4\pi \sum_{n=0}^{\infty} k_p^{2n}(-1)^n A_{r(2n+2)}(\hat{\boldsymbol{k}}) \tag{3.288}$$

whereas for transverse incidence ($\alpha_p = 0$, polarized in the direction $\hat{\boldsymbol{b}}$), theorem (2.203) and expansions (3.279) and (3.281) yield

$$\sigma_e = 4\pi(\tau^+)^2 \sum_{n=0}^{\infty} k_p^{2n}(-1)^n[(\hat{\boldsymbol{b}} \cdot \hat{\boldsymbol{\theta}})A_{\theta(2n+2)}(\hat{\boldsymbol{k}}) + (\hat{\boldsymbol{b}} \cdot \hat{\boldsymbol{\varphi}})A_{\varphi(2n+2)}(\hat{\boldsymbol{k}})] \tag{3.289}$$

and since the absorption cross section vanishes we have

$$\sigma_s = \sigma_e \tag{3.290}$$

3.E.3 *The cavity problem*

The boundary condition (3.245) for a traction-free surface reads as

$$T^+(\partial_{\boldsymbol{r}}, \hat{\boldsymbol{n}})\boldsymbol{u}_n^+(\boldsymbol{r}) = \boldsymbol{0}, \quad \boldsymbol{r} \in S \tag{3.291}$$

for $n = 0, 1, 2, \ldots$. Then, from equations (3.248)–(3.250) we obtain

$$\boldsymbol{T}^+(\partial_{\boldsymbol{r}}, \hat{\boldsymbol{n}})\boldsymbol{v}_n^+(\boldsymbol{r}) = -\boldsymbol{T}^+(\partial_{\boldsymbol{r}}, \hat{\boldsymbol{n}})\boldsymbol{f}_n^+(\boldsymbol{r}), \quad \boldsymbol{r} \in S \tag{3.292}$$

where for all $\boldsymbol{r} \in \mathbb{R}^3$

$$\boldsymbol{f}_n^+(\boldsymbol{r}) = \boldsymbol{u}_n^i(\boldsymbol{r}) + \frac{1}{4\pi\mu^+}\sum_{m=1}^{n}\binom{n}{m}\int_S \boldsymbol{u}_{n-m}^+(\boldsymbol{r}') \cdot \boldsymbol{T}^+(\partial_{\boldsymbol{r}'}, \hat{\boldsymbol{n}}')\, \widetilde{\boldsymbol{\gamma}}_m^+(\boldsymbol{r}, \boldsymbol{r}')\, ds(\boldsymbol{r}') \tag{3.293}$$

and

$$\boldsymbol{v}_n^+(\boldsymbol{r}) = \frac{1}{4\pi\mu^+}\int_S \boldsymbol{u}_n^+(\boldsymbol{r}') \cdot \boldsymbol{T}^+(\partial_{\boldsymbol{r}'}, \hat{\boldsymbol{n}}')\, \widetilde{\boldsymbol{\gamma}}_0^+(\boldsymbol{r}, \boldsymbol{r}')\, ds(\boldsymbol{r}') \tag{3.294}$$

which implies that in fact

$$\boldsymbol{v}_n^+(\boldsymbol{r}) = O\left(\frac{1}{r^2}\right) \qquad \text{as } r \to \infty \tag{3.295}$$

From (3.248) and (3.257) we see that the evaluation of the coefficient $\boldsymbol{u}_n^+$ is reduced to the following problem in elastostatics where the surface traction field is prescribed on S

$$\left.\begin{aligned} \Delta^{e+}\boldsymbol{v}_n^+ &= \boldsymbol{0} && \text{in } V^+ \\ \boldsymbol{T}^+\boldsymbol{v}_n^+ &= -\boldsymbol{T}^+\boldsymbol{f}_n^+ && \text{on } S \\ \boldsymbol{v}_n^+ &= O\left(\frac{1}{r^2}\right) && \text{as } r \to \infty \end{aligned}\right\} \tag{3.296}$$

where the function $\boldsymbol{f}_n^+$ is given by (3.293) in terms of the coefficients $\boldsymbol{u}_m^+, m = 0, 1, 2, \ldots, n-1$.

The exterior elastostatic problem (3.296) is further reduced to a potential problem if the Papkovich representation (3.274), (3.275) is used. In fact, if $\boldsymbol{A}_n^+$, B_n^+ is the solution of

$$\left.\begin{aligned} &\nabla^2 \boldsymbol{A}_n^+ = \boldsymbol{0} && \text{in } V^+ \\ &\nabla^2 B_n^+ = 0 && \text{in } V^+ \\ &\boldsymbol{T}^+\left[\boldsymbol{A}_n^+ + \frac{(c_s^+)^2 - (c_p^+)^2}{2(c_p^+)^2}\nabla\left(\boldsymbol{r}\cdot \boldsymbol{A}_n^+ + B_n^+\right)\right] = -\boldsymbol{T}^+\boldsymbol{f}_n^+ && \text{on } S \\ &\boldsymbol{A}_n^+ = O\left(\frac{1}{r^2}\right) && \text{as } r \to \infty \\ &B_n^+ = O\left(\frac{1}{r}\right) && \text{as } r \to \infty. \end{aligned}\right\} \tag{3.297}$$

the coefficient $\boldsymbol{u}_n^+$ is obtained from (3.248), via (3.249) and (3.274). It is not hard to see that, if the incident field is the plane wave given in (3.227), then the coefficient $\boldsymbol{u}_0^+$ is equal to the leading term in the expansion (3.227) which is a constant.

The scattering amplitudes can be obtained from (3.264)–(3.266) and they are given by

$$\operatorname{Re} g_r(\hat{\boldsymbol{r}}) = \sum_{n=0}^{\infty} k_p^{2n+2}(-1)^n A_{r(2n+2)}(\hat{\boldsymbol{r}}) \tag{3.298}$$

$$\operatorname{Im} g_r(\hat{\boldsymbol{r}}) = \sum_{n=0}^{\infty} k_p^{2n+3}(-1)^n A_{r(2n+3)}(\hat{\boldsymbol{r}}) \tag{3.299}$$

$$\text{Re } g_\theta(\hat{\boldsymbol{r}}) = \sum_{n=0}^{\infty} k_p^{2n+2}(-1)^{n+1} A_{\theta(2n+2)}(\hat{\boldsymbol{r}}) \tag{3.300}$$

$$\text{Im } g_\theta(\hat{\boldsymbol{r}}) = \sum_{n=0}^{\infty} k_p^{2n+3}(-1)^{n+1} A_{\theta(2n+3)}(\hat{\boldsymbol{r}}) \tag{3.301}$$

$$\text{Re } g_\varphi(\hat{\boldsymbol{r}}) = \sum_{n=0}^{\infty} k_p^{2n+2}(-1)^{n+1} A_{\varphi(2n+2)}(\hat{\boldsymbol{r}}) \tag{3.302}$$

$$\text{Im } g_\varphi(\hat{\boldsymbol{r}}) = \sum_{n=0}^{\infty} k_p^{2n+3}(-1)^{n+1} A_{\varphi(2n+3)}(\hat{\boldsymbol{r}}) \tag{3.303}$$

where

$$A_{r(n+2)}(\hat{\boldsymbol{r}}) = \frac{1}{4\pi(\lambda^+ + 2\mu^+)n!} \sum_{m=0}^{n} \binom{n}{m} (-1)^m \int_S [\lambda^+(\hat{\boldsymbol{n}}' \cdot \boldsymbol{u}_{n-m}^+(\boldsymbol{r}')) + 2\mu^+(\hat{\boldsymbol{r}} \cdot \boldsymbol{u}_{n-m}^+(\boldsymbol{r}'))(\hat{\boldsymbol{r}} \cdot \hat{\boldsymbol{n}}')](\hat{\boldsymbol{r}} \cdot \boldsymbol{r}')^m \, ds(\boldsymbol{r}') \tag{3.304}$$

$$A_{\theta(n+2)}(\hat{\boldsymbol{r}}) = -\frac{1}{4\pi n!} \sum_{m=0}^{n} \binom{n}{m} \left(-\frac{1}{\tau^+}\right)^{m+2} \hat{\boldsymbol{\theta}} \cdot \int_S [\hat{\boldsymbol{n}}'(\hat{\boldsymbol{r}} \cdot \boldsymbol{u}_{n-m}^+(\boldsymbol{r}')) + (\hat{\boldsymbol{n}}' \cdot \hat{\boldsymbol{r}})\boldsymbol{u}_{n-m}^+(\boldsymbol{r}')](\hat{\boldsymbol{r}} \cdot \boldsymbol{r}')^m \, ds(\boldsymbol{r}') \tag{3.305}$$

$$A_{\varphi(n+2)}(\hat{\boldsymbol{r}}) = -\frac{1}{4\pi n!} \sum_{n=0}^{n} \binom{n}{m} \left(-\frac{1}{\tau^+}\right)^{m+2} \hat{\boldsymbol{\varphi}} \cdot \int_S [\hat{\boldsymbol{n}}'(\hat{\boldsymbol{r}} \cdot \boldsymbol{u}_{n-m}^+(\boldsymbol{r}')) + (\hat{\boldsymbol{n}}' \cdot \hat{\boldsymbol{r}})\boldsymbol{u}_{n-m}^+(\boldsymbol{r}')](\hat{\boldsymbol{r}} \cdot \boldsymbol{r}')^m \, ds(\boldsymbol{r}') \tag{3.306}$$

for $n = 0, 1, 2, \ldots$.

For plane wave incidence it may be shown (see Chapter 5) that

$$A_{r2}(\hat{\boldsymbol{r}}) = A_{\theta 2}(\hat{\boldsymbol{r}}) = A_{\varphi 2}(\hat{\boldsymbol{r}}) = 0 \tag{3.307}$$

For either plane waves or point sources, the differential scattering cross section is given by (2.186) and takes the following form

$$\begin{aligned}
\sigma(\hat{\boldsymbol{r}}) &= \frac{4\pi}{\alpha_p^2 + \tau^+\alpha_s^2} \frac{|g_r|^2 + (\tau^+)^3(|g_\theta|^2 + |g_\varphi|^2)}{k_p^2} \\
&= \frac{4\pi}{\alpha_p^2 + \tau^+\alpha_s^2} \sum_{n=0}^{\infty} k_p^{2n+2}(-1)^n \sum_{m=0}^{n} [A_{r(2n-2m+2)}(\hat{\boldsymbol{r}}) A_{r(2m+2)}(\hat{\boldsymbol{r}}) \\
&\quad + (\tau^+)^3 A_{\theta(2n-2m+2)}(\hat{\boldsymbol{r}}) A_{\theta(2m+2)}(\hat{\boldsymbol{r}}) \\
&\quad + (\tau^+)^3 A_{\varphi(2n-2m+2)}(\hat{\boldsymbol{r}}) A_{\varphi(2m+2)}(\hat{\boldsymbol{r}})] \\
&\quad + \frac{4\pi}{\alpha_p^2 + \tau^+\alpha_s^2} \sum_{n=0}^{\infty} k_p^{2n+4}(-1)^n \sum_{m=0}^{n} [A_{r(2n-2m+3)}(\hat{\boldsymbol{r}}) A_{r(2m+3)}(\hat{\boldsymbol{r}}) \\
&\quad + (\tau^+)^3 A_{\theta(2n-2m+3)}(\hat{\boldsymbol{r}}) A_{\theta(2m+3)}(\hat{\boldsymbol{r}}) \\
&\quad + (\tau^+)^3 A_{\varphi(2n-2m+3)}(\hat{\boldsymbol{r}}) A_{\varphi(2m+3)}(\hat{\boldsymbol{r}})]
\end{aligned} \tag{3.308}$$

For plane wave incidence the $n = 0$ term in the first sum vanishes and the lowest order term in $\sigma(\hat{\boldsymbol{r}})$ is $O(k_p^4)$. Substituting (3.308) into the formula

$$\sigma_s = \frac{1}{4\pi} \int_{S^2} \sigma(\hat{\boldsymbol{r}})\, ds(\hat{\boldsymbol{r}}) \tag{3.309}$$

we obtain the scattering cross section which coincides with the extinction cross section σ_e since the absorption cross section for a cavity vanishes. Finally, for plane wave incidence the forward scattering theorem (2.202) for a longitudinal incident wave gives

$$\sigma_e = \sigma_s = 4\pi \sum_{n=0}^{\infty} k_p^{2n+2} (-1)^n A_{r(2n+4)}(\hat{\boldsymbol{k}}) \tag{3.310}$$

and for a transverse incident wave, polarized along $\hat{\boldsymbol{b}}$, (2.203) gives

$$\begin{aligned} \sigma_e = \sigma_s = 4\pi(\tau^+)^2 \sum_{n=0}^{\infty} k_p^{2n+2} (-1)^{n+1} \\ \times [(\hat{\boldsymbol{b}} \cdot \hat{\boldsymbol{\theta}}) A_{\theta(2n+4)}(\hat{\boldsymbol{k}}) + (\hat{\boldsymbol{b}} \cdot \hat{\boldsymbol{\varphi}}) A_{\varphi(2n+4)}(\hat{\boldsymbol{k}})] \end{aligned} \tag{3.311}$$

3.E.4 *The transmission problem*

A penetrable elastic scatterer satisfies the transmission conditions (1.216) and (1.217) which imply that the low frequency coefficients satisfy

$$\boldsymbol{u}_n^+(\boldsymbol{r}) = \boldsymbol{u}_n^-(\boldsymbol{r}), \quad \boldsymbol{r} \in S \tag{3.312}$$

$$\boldsymbol{T}^+(\partial_{\boldsymbol{r}}, \hat{\boldsymbol{n}})\boldsymbol{u}_n^+(\boldsymbol{r}) = \boldsymbol{T}^-(\partial_{\boldsymbol{r}}, \hat{\boldsymbol{n}})\boldsymbol{u}_n^-(\boldsymbol{r}), \quad \boldsymbol{r} \in S \tag{3.313}$$

for $n = 0, 1, 2, \ldots$.

Then, in order to find the coefficient $\boldsymbol{u}_n^\pm$ from the representations (3.248)–(3.253), we need to solve equations (3.242) and (3.243) together with conditions (3.312) and (3.313). For plane wave incidence it is easily seen that $\boldsymbol{u}_0^\pm$ is a constant equal to the leading term in the expansion (3.227) of the plane wave. Relations (3.248)–(3.253) can be used to obtain

$$\boldsymbol{f}_n^+(\boldsymbol{r}) + \boldsymbol{v}_n^+(\boldsymbol{r}) = -\boldsymbol{f}_n^-(\boldsymbol{r}) - \boldsymbol{v}_n^-(\boldsymbol{r}), \quad \boldsymbol{r} \in S \tag{3.314}$$

$$\boldsymbol{T}^+(\partial_{\boldsymbol{r}}, \hat{\boldsymbol{n}})(\boldsymbol{f}_n^+(\boldsymbol{r}) + \boldsymbol{v}_n^+(\boldsymbol{r})) = -\boldsymbol{T}^-(\partial_{\boldsymbol{r}}, \hat{\boldsymbol{n}})(\boldsymbol{f}_n^-(\boldsymbol{r}) + \boldsymbol{v}_n^-(\boldsymbol{r})), \quad \boldsymbol{r} \in S \tag{3.315}$$

where $\boldsymbol{f}_n^+$ is given by (3.249), $\boldsymbol{v}_n^+$ by (3.250), $\boldsymbol{f}_n^-$ by (3.252) and $\boldsymbol{v}_n^-$ by (3.253). Furthermore

$$\boldsymbol{v}_n^+(\boldsymbol{r}) = O\left(\frac{1}{r}\right), \quad r \to \infty \tag{3.316}$$

Since $\boldsymbol{f}_n^+$ and $\boldsymbol{f}_n^-$ are particular solutions of equations (3.242) and (3.243), respectively, it follows that the coefficients $\boldsymbol{u}_n^\pm$ are obtained from (3.248) and (3.251) where $\boldsymbol{f}_n^\pm$ are dependent on $\boldsymbol{u}_m^\pm$, $m = 0, 1, \ldots, n-1$ and $\boldsymbol{v}_n^\pm$ solves the following transmission problem

$$\left.\begin{aligned} \Delta^{e+}\boldsymbol{v}_n^+ &= \mathbf{0} && \text{in } V^+ \\ \Delta^{e-}\boldsymbol{v}_n^- &= \mathbf{0} && \text{in } V^- \\ \boldsymbol{v}_n^+ + \boldsymbol{v}_n^- &= -\boldsymbol{f}_n^+ - \boldsymbol{f}_n^- && \text{on } S \\ \boldsymbol{T}^+\boldsymbol{v}_n^+ + \boldsymbol{T}^-\boldsymbol{v}_n^- &= -\boldsymbol{T}^+\boldsymbol{f}_n^+ - \boldsymbol{T}^-\boldsymbol{f}_n^- && \text{on } S \\ \boldsymbol{v}_n^+ &= O\left(\frac{1}{r}\right) && \text{as } r \to \infty \end{aligned}\right\} \tag{3.317}$$

This transmission problem can be reduced to the following transmission problem for the Papkovich potentials $\boldsymbol{A}_n^\pm$ and $B_n^\pm$

$$\left.\begin{aligned} \nabla^2 \boldsymbol{A}_n^+ &= \mathbf{0} && \text{in } V^+ \\ \nabla^2 B_n^+ &= 0 && \text{in } V^+ \\ \nabla^2 \boldsymbol{A}_n^- &= \mathbf{0} && \text{in } V^- \\ \nabla^2 B_n^- &= 0 && \text{in } V^- \end{aligned}\right\} \tag{3.318a}$$

$$\left.\begin{aligned} &\boldsymbol{A}_n^+ + \frac{(c_s^+)^2 - (c_p^+)^2}{2(c_p^+)^2}\nabla(\boldsymbol{r}\cdot\boldsymbol{A}_n^+ + B_n^+) \\ &\quad + \boldsymbol{A}_n^- + \frac{(c_s^-)^2 - (c_p^-)^2}{2(c_p^-)^2}\nabla(\boldsymbol{r}\cdot\boldsymbol{A}_n^- + B_n^-) = -\boldsymbol{f}_n^+ - \boldsymbol{f}_n^- && \text{on } S \\ &\boldsymbol{T}^+\left[\boldsymbol{A}_n^+ + \frac{(c_s^+)^2 - (c_p^+)^2}{2(c_p^+)^2}\nabla(\boldsymbol{r}\cdot\boldsymbol{A}_n^+ + B_n^+)\right] \\ &\quad + \boldsymbol{T}^-\left[\boldsymbol{A}_n^- + \frac{(c_s^-)^2 - (c_p^-)^2}{2(c_p^-)^2}\nabla(\boldsymbol{r}\cdot\boldsymbol{A}_n^- + B_n^-)\right] = -\boldsymbol{T}^+\boldsymbol{f}_n^+ - \boldsymbol{T}^-\boldsymbol{f}_n^- && \text{on } S \\ &\boldsymbol{A}_n^+ = O\left(\frac{1}{r}\right) && \text{as } r \to \infty \\ &B_n = O(1) && \text{as } r \to \infty \end{aligned}\right\} \tag{3.318b}$$

Once $\boldsymbol{A}_n^\pm$ and $B_n^\pm$ are obtained from the solution of (3.318), $\boldsymbol{v}_n^\pm$ can be evaluated from

$$\boldsymbol{v}_n^\pm = \boldsymbol{A}_n^\pm + \frac{(c_s^\pm)^2 - (c_p^\pm)^2}{2(c_p^\pm)^2}\nabla(\boldsymbol{r}\cdot\boldsymbol{A}_n^\pm + B_n^\pm) \quad \text{in } V^\pm \tag{3.319}$$

and the coefficient $\boldsymbol{u}_n^\pm$ is derived from (3.248) and (3.251).

For the transmission conditions (3.312) and (3.313), Betti's third identity (2.142) and equation (3.243) imply that

$$\begin{aligned} \int_S \boldsymbol{T}^+(\partial_{\boldsymbol{r}}, \hat{\boldsymbol{n}})\boldsymbol{u}_n^+(\boldsymbol{r})\, ds(\boldsymbol{r}) &= \int_S \boldsymbol{T}^-(\partial_{\boldsymbol{r}}, \hat{\boldsymbol{n}})\boldsymbol{u}_n^-(\boldsymbol{r})\, ds(\boldsymbol{r}) \\ &= \rho^- \int_{V^-} \Delta^{e-}\boldsymbol{u}_n^-(\boldsymbol{r})\, dv(\boldsymbol{r}) \\ &= n(n-1)\rho^-(c_p^+)^2 \int_{V^-} \boldsymbol{u}_{n-2}^-(\boldsymbol{r})\, dv(\boldsymbol{r}) \end{aligned} \tag{3.320}$$

which shows that the total traction on the penetrable surface S generated by the first two coefficients $\boldsymbol{u}_0^\pm$ and $\boldsymbol{u}_1^\pm$ is equal to zero.

Then, from the general expressions (3.264)–(3.266) for the scattering amplitudes we see that for a penetrable elastic body

$$\left.\begin{aligned} g_r(\hat{\boldsymbol{r}}) &= O(k_p^2) \\ \boldsymbol{g}_t(\hat{\boldsymbol{r}}) &= O(k_p^2) \end{aligned}\right\} \quad \text{as } k_p \to 0 \tag{3.321}$$

For plane wave incidence (3.321) comes from the trivial observation that $\boldsymbol{u}_0^\pm$ is a constant.

Because of (3.321) the scattering amplitudes are given by the same formulae as for the cavity problem, i.e. by (3.298)–(3.303) but the coefficients are expressed as follows

$$\begin{aligned} A_{r(n+2)}(\hat{\boldsymbol{r}}) = {} & \frac{1}{4\pi(\lambda^+ + 2\mu^+)n!}\sum_{m=0}^{n}\binom{n}{m}(-1)^m \\ & \times\left[\frac{\hat{\boldsymbol{r}}}{n+1-m}\cdot\int_S \boldsymbol{T}^+(\partial_{\boldsymbol{r}'},\hat{\boldsymbol{n}}')\boldsymbol{u}^+_{n+1-m}(\boldsymbol{r}')(\hat{\boldsymbol{r}}\cdot\boldsymbol{r}')^m\,ds(\boldsymbol{r}')\right. \\ & \left. + \;(\lambda^+\widetilde{\boldsymbol{I}} + 2\mu^+\hat{\boldsymbol{r}}\hat{\boldsymbol{r}}) : \int_S \boldsymbol{u}^+_{n-m}(\boldsymbol{r}')\hat{\boldsymbol{n}}'(\hat{\boldsymbol{r}}\cdot\boldsymbol{r}')^m\,ds(\boldsymbol{r}')\right] \end{aligned} \tag{3.322}$$

$$\begin{aligned} A_{\theta(n+2)}(\hat{\boldsymbol{r}}) = {} & \frac{1}{4\pi\mu^+(n+1)!}\sum_{m=0}^{n+1}\binom{n+1}{m}\left(-\frac{1}{\tau^+}\right)^{m+1} \\ & \times\hat{\boldsymbol{\theta}}\cdot\int_S \boldsymbol{T}^+(\partial_{\boldsymbol{r}'},\hat{\boldsymbol{n}}')\boldsymbol{u}^+_{n+1-m}(\boldsymbol{r}')(\hat{\boldsymbol{r}}\cdot\boldsymbol{r}')^m\,ds(\boldsymbol{r}') \\ & + \frac{1}{4\pi\tau^+ n!}\sum_{m=0}^{n}\binom{n}{m}\left(-\frac{1}{\tau^+}\right)^{m+1} \\ & \times\hat{\boldsymbol{\theta}}\cdot\int_S \boldsymbol{u}^+_{n-m}(\boldsymbol{r}')\hat{\boldsymbol{n}}'(\hat{\boldsymbol{r}}\cdot\boldsymbol{r}')^m\,ds(\boldsymbol{r}') : [\widetilde{\boldsymbol{I}}\times\widetilde{\boldsymbol{I}}\times\hat{\boldsymbol{r}} + 2\hat{\boldsymbol{r}}\widetilde{\boldsymbol{I}}] \end{aligned} \tag{3.323}$$

$$\begin{aligned} A_{\varphi(n+2)}(\hat{\boldsymbol{r}}) = {} & \frac{1}{4\pi\mu^+(n+1)!}\sum_{m=0}^{n+1}\binom{n+1}{m}\left(-\frac{1}{\tau^+}\right)^{m+1} \\ & \times\hat{\boldsymbol{\varphi}}\cdot\int_S \boldsymbol{T}^+(\partial_{\boldsymbol{r}'},\hat{\boldsymbol{n}}')\boldsymbol{u}^+_{n+1-m}(\boldsymbol{r}')(\hat{\boldsymbol{r}}\cdot\boldsymbol{r}')^m\,ds(\boldsymbol{r}') \\ & + \frac{1}{4\pi\tau^+ n!}\sum_{m=0}^{n}\binom{n}{m}\left(-\frac{1}{\tau^+}\right)^{m+1} \\ & \times\hat{\boldsymbol{\varphi}}\cdot\int_S \boldsymbol{u}^+_{n-m}(\boldsymbol{r}')\hat{\boldsymbol{n}}'(\hat{\boldsymbol{r}}\cdot\boldsymbol{r}')^m\,ds(\boldsymbol{r}') : [\widetilde{\boldsymbol{I}}\times\widetilde{\boldsymbol{I}}\times\hat{\boldsymbol{r}} + 2\hat{\boldsymbol{r}}\widetilde{\boldsymbol{I}}] \end{aligned} \tag{3.324}$$

For the case of plane wave incidence these coefficients can be written as

$$
\begin{aligned}
A_{r(n+2)}(\hat{\boldsymbol{r}}) = & \frac{1}{4\pi(\lambda^+ + 2\mu^+)n!}\sum_{m=0}^{n}\binom{n}{m}(-1)^{m+1} \\
& \times \int_S \left[\frac{\hat{\boldsymbol{r}}\cdot\boldsymbol{r}'}{m+1}\hat{\boldsymbol{r}}\cdot\boldsymbol{T}^+(\partial_{\boldsymbol{r}'},\hat{\boldsymbol{n}}')\boldsymbol{u}^+_{n-m}(\boldsymbol{r}')\right. \\
& \left. -\lambda^+\hat{\boldsymbol{n}}'\cdot\boldsymbol{u}^+_{n-m}(\boldsymbol{r}') - 2\mu^+(\hat{\boldsymbol{r}}\cdot\boldsymbol{u}^+_{n-m}(\boldsymbol{r}'))(\hat{\boldsymbol{r}}\cdot\hat{\boldsymbol{n}}')\right](\hat{\boldsymbol{r}}\cdot\boldsymbol{r}')^m\,ds(\boldsymbol{r}') \\
& - \frac{n\rho^-}{4\pi\rho^+ n!}\hat{\boldsymbol{r}}\cdot\int_{V^-}\boldsymbol{u}^-_{n-1}(\boldsymbol{r}')\,dv(\boldsymbol{r}')
\end{aligned} \tag{3.325}
$$

$$
\begin{aligned}
A_{\theta(n+2)}(\hat{\boldsymbol{r}}) = & \frac{1}{4\pi n!}\sum_{m=0}^{n}\binom{n}{m}\left(-\frac{1}{\tau^+}\right)^{m+2} \\
& \times \hat{\boldsymbol{\theta}}\cdot\int_S \left[\frac{\hat{\boldsymbol{r}}\cdot\boldsymbol{r}'}{(m+1)\mu^+}\boldsymbol{T}^+(\partial_{\boldsymbol{r}'},\hat{\boldsymbol{n}}')\boldsymbol{u}^+_{n-m}(\boldsymbol{r}')\right. \\
& \left. -\hat{\boldsymbol{n}}'(\hat{\boldsymbol{r}}\cdot\boldsymbol{u}^+_{n-m}(\boldsymbol{r}')) - (\hat{\boldsymbol{n}}'\cdot\hat{\boldsymbol{r}})\boldsymbol{u}^+_{n-m}(\boldsymbol{r}')\right](\hat{\boldsymbol{r}}\cdot\boldsymbol{r}')^m\,ds(\boldsymbol{r}') \\
& - \frac{n}{4\pi(\tau^+)^3 n!}\frac{\rho^-}{\rho^+}\hat{\boldsymbol{\theta}}\cdot\int_{V^-}\boldsymbol{u}^-_{n-1}(\boldsymbol{r}')\,dv(\boldsymbol{r}')
\end{aligned} \tag{3.326}
$$

$$
\begin{aligned}
A_{\varphi(n+2)}(\hat{\boldsymbol{r}}) = & \frac{1}{4\pi n!}\sum_{m=0}^{n}\binom{n}{m}\left(-\frac{1}{\tau^+}\right)^{m+2} \\
& \times \hat{\boldsymbol{\varphi}}\cdot\int_S \left[\frac{\hat{\boldsymbol{r}}\cdot\boldsymbol{r}'}{(m+1)\mu^+}\boldsymbol{T}^+(\partial_{\boldsymbol{r}'},\hat{\boldsymbol{n}}')\boldsymbol{u}^+_{n-m}(\boldsymbol{r}')\right. \\
& \left. -\hat{\boldsymbol{n}}'(\hat{\boldsymbol{r}}\cdot\boldsymbol{u}^+_{n-m}(\boldsymbol{r}')) - (\hat{\boldsymbol{n}}'\cdot\hat{\boldsymbol{r}})\boldsymbol{u}^+_{n-m}(\boldsymbol{r}')\right](\hat{\boldsymbol{r}}\cdot\boldsymbol{r}')^m\,ds(\boldsymbol{r}') \\
& - \frac{n}{4\pi(\tau^+)^3 n!}\frac{\rho^-}{\rho^+}\hat{\boldsymbol{\varphi}}\cdot\int_{V^-}\boldsymbol{u}^-_{n-1}(\boldsymbol{r}')\,dv(\boldsymbol{r}')
\end{aligned} \tag{3.327}
$$

The scattering cross section is given by (3.309) and the special cases of longitudinal and transverse incident plane waves can be obtained from (3.310) and (3.311), respectively, where A_{rn}, $A_{\theta n}$ and $A_{\varphi n}$ are as in (3.325)–(3.327).

3.E.5 *Problems in elasticity*

1. Derive the low frequency expansion (3.232) and (3.233) for the fundamental dyadic, and (3.237)–(3.239) for the surface traction it generates.

2. Establish the validity of equations (3.242) in V^+ and (3.243) in V^-.
3. Use the exterior integral representation (2.146) to obtain the representation (3.248)–(3.250), and the interior integral representation (2.144) to obtain (3.251)–(3.253).
4. Show by direct calculations that f_n^+, as given by (3.249), is a particular solution of equation (3.242), and that f_n^-, as given by (3.252), is a particular solution of (3.243).
5. Derive the general low frequency expansions (3.260)–(3.266) for the scattering amplitudes.
6. Prove that if $\boldsymbol{A}$ and B are harmonic functions, then the function $\boldsymbol{u}$ defined as

$$\boldsymbol{u}(\boldsymbol{r}) = \boldsymbol{A}(\boldsymbol{r}) + \frac{c_s^2 - c_p^2}{2c_p^2}\nabla(\boldsymbol{r} \cdot \boldsymbol{A}(\boldsymbol{r}) + B(\boldsymbol{r}))$$

solves the static Navier equation

$$c_s^2\nabla^2\boldsymbol{u}(\boldsymbol{r}) + (c_p^2 - c_s^2)\nabla\nabla \cdot \boldsymbol{u}(\boldsymbol{r}) = \boldsymbol{0}$$

7. Derive the form (3.286) for the differential scattering cross section of a rigid body.
8. For the elastic transmission problem prove that for every $n = 0, 1, 2, \ldots$

$$\int_S \boldsymbol{T}^+(\partial_{\boldsymbol{r}}, \hat{\boldsymbol{n}})\boldsymbol{v}_n^+(\boldsymbol{r})\, ds(\boldsymbol{r}) = \boldsymbol{0}$$

where $\boldsymbol{v}_n^+$ are given by (3.250).
9. Obtain formulae (3.322)–(3.324) for the expansion coefficients of the scattering amplitudes and show that for the case of plane wave incidence they can be written as in (3.325)–(3.327).

4

THE TRANSMISSION PROBLEM REVISITED

In this chapter we present another derivation of the low frequency expansions for transmission problems. In contrast to the problems involving impenetrable scatters, the existence of fields in V^- in transmission problems enables an alternate derivation of low frequency expansions employing **volume integrals** over V^-. In many respects the procedure for determining the coefficients in low frequency expansions is simpler than that based entirely on **boundary integrals**. In acoustics the lossless case was treated in Dassios (1977) and the lossy case was considered in Dassios (1989*b*) but not in the explicit form presented here. The development in acoustics and electromagnetics has not appeared in the literature, while the material on elasticity is based on the work of Kiriaki and Polyzos (1988).

4.A Acoustics

4.A.1 *The volume integral formulation*

Recall that the excess acoustic pressure satisfies the Helmholtz equation with wave number k in V^+ (1.82) and wave number $k\eta$ in V^- (1.87) and transmission conditions

$$\left.\begin{aligned} u^+ &= u^- \\ \frac{\partial}{\partial n}u^+ &= \beta\frac{\partial}{\partial n}u^- \end{aligned}\right\} \quad \text{on } S \tag{4.1}$$

where β and η are given in terms of material parameters in (3.96) and (3.97) which for convenience we repeat here.

$$\beta = \frac{\rho^+}{\rho^-}\left(1 - ik\frac{\delta^-\gamma^-}{\sqrt{\gamma^+\rho^+}}\right) \tag{4.2}$$

$$\eta^2 = \frac{\gamma^-\rho^-}{\gamma^+\rho^+}\left(1 - ik\frac{\delta^-\gamma^-}{\sqrt{\gamma^+\rho^+}}\right)^{-1} \tag{4.3}$$

and

$$\beta\eta^2 = \frac{\gamma^-}{\gamma^+} \tag{4.4}$$

The low frequency analysis in Chapter 3 began with the integral representations (2.5) and (2.7). The representation (2.5) involved G^-, the Green's function with wave number k^-. Here, we start with (2.7) and base the analysis entirely upon the exterior Green's function G^+ which will result in the appearance of volume integrals over V^- as follows. Introducing the transmission conditions (4.1) into (2.7) produces the representation

$$\alpha(\boldsymbol{r})u^+(\boldsymbol{r}) = u^i(\boldsymbol{r}) + \frac{ik}{4\pi}\int_S \left[u^-(\boldsymbol{r}')\frac{\partial}{\partial n'}G^+(\boldsymbol{r},\boldsymbol{r}') - \beta G^+(\boldsymbol{r},\boldsymbol{r}')\frac{\partial}{\partial n'}u^-(\boldsymbol{r}')\right] ds(\boldsymbol{r}') \tag{4.5}$$

which we rewrite by adding and subtracting the term

$$\frac{ik}{4\pi}\int_S G^+(\boldsymbol{r},\boldsymbol{r}')\frac{\partial}{\partial n'}u^-(\boldsymbol{r}')\,ds(\boldsymbol{r}') \tag{4.6}$$

as

$$\begin{aligned}\alpha(\boldsymbol{r})u^+(\boldsymbol{r}) &= u^i(\boldsymbol{r}) \\ &\quad + \frac{ik}{4\pi}\int_S \left[u^-(\boldsymbol{r}')\frac{\partial}{\partial n'}G^+(\boldsymbol{r},\boldsymbol{r}') - G^+(\boldsymbol{r},\boldsymbol{r}')\frac{\partial}{\partial n'}u^-(\boldsymbol{r}')\right] ds(\boldsymbol{r}') \\ &\quad + \frac{ik}{4\pi}(1-\beta)\int_S G^+(\boldsymbol{r},\boldsymbol{r}')\frac{\partial}{\partial n'}u^-(\boldsymbol{r}')\,ds(\boldsymbol{r}')\end{aligned} \tag{4.7}$$

The reason for rewriting (4.5) is to obtain a term which invites the use of Green's second identity. This could also have been achieved by adding and subtracting

$$\frac{ik\beta}{4\pi}\int_S u^-(\boldsymbol{r}')\frac{\partial}{\partial n'}G^+(\boldsymbol{r},\boldsymbol{r}')\,ds(\boldsymbol{r}')$$

but we will not follow that avenue. Now, we accept the invitation to apply Green's second identity in (4.7), taking into account the facts that G^+ and u^- satisfy Helmholtz equations with wave numbers k and $k\eta$, respectively, and that G^+ has a singularity at $\boldsymbol{r} = \boldsymbol{r}'$. This results in

$$\begin{aligned}\alpha(\boldsymbol{r})u^+(\boldsymbol{r}) &+ (1-\alpha(\boldsymbol{r}))u^-(\boldsymbol{r}) \\ &= u^i(\boldsymbol{r}) - \frac{(ik)^3}{4\pi}(\eta^2-1)\int_{V^-} u^-(\boldsymbol{r}')G^+(\boldsymbol{r},\boldsymbol{r}')\,dv(\boldsymbol{r}') \\ &\quad + \frac{ik}{4\pi}(1-\beta)\int_S G^+(\boldsymbol{r},\boldsymbol{r}')\frac{\partial}{\partial n'}u^-(\boldsymbol{r}')\,ds(\boldsymbol{r}')\end{aligned} \tag{4.8}$$

This representation is valid for all $\boldsymbol{r}$ in $\mathbb{R}^3$ and provides the basis for the low frequency expansions. We will treat the lossless and lossy cases separately although, as we will see, we could confine attention to the lossy case and obtain the lossless results by letting $\delta^- \to 0$.

In both cases, it will prove useful to have the following expansions which may be obtained using the low frequency expansions for u^-, (3.3), and G^+, (3.14), together with the Cauchy product rule and some rearrangement of terms

$$ikG^+(\boldsymbol{r},\boldsymbol{r}')u^-(\boldsymbol{r}') = \sum_{n=0}^{\infty} \frac{(ik)^n}{n!} \sum_{m=0}^{n} \binom{n}{m} |\boldsymbol{r}-\boldsymbol{r}'|^{m-1} u_{n-m}^-(\boldsymbol{r}') \tag{4.9}$$

$$(ik)^2 G^+(\boldsymbol{r},\boldsymbol{r}')u^-(\boldsymbol{r}') = \sum_{n=1}^{\infty} \frac{(ik)^n}{n!} \sum_{m=1}^{n} \binom{n}{m} m|\boldsymbol{r}-\boldsymbol{r}'|^{m-2} u_{n-m}^-(\boldsymbol{r}') \tag{4.10}$$

$$(ik)^3 G^+(\boldsymbol{r},\boldsymbol{r}')u^-(\boldsymbol{r}') = \sum_{n=2}^{\infty} \frac{(ik)^n}{n!} \sum_{m=2}^{n} \binom{n}{m} m(m-1)|\boldsymbol{r}-\boldsymbol{r}'|^{m-3} u_{n-m}^-(\boldsymbol{r}') \tag{4.11}$$

and

$$(ik)^4 G^+(\boldsymbol{r},\boldsymbol{r}')u^-(\boldsymbol{r}') = \sum_{n=3}^{\infty} \frac{(ik)^n}{n!} \times \sum_{m=3}^{n} \binom{n}{m} m(m-1)(m-2)|\boldsymbol{r}-\boldsymbol{r}'|^{m-4} u_{n-m}^-(\boldsymbol{r}') \tag{4.12}$$

The lossless transmission problem: $\delta^- = 0$. In this case, η and β are real and we rewrite (4.8) exhibiting η^2 and β explicitly in terms of the material parameters

$$\begin{aligned}\alpha(\boldsymbol{r})u^+(\boldsymbol{r}) + (1-\alpha(\boldsymbol{r}))u^-(\boldsymbol{r}) &= u^i(\boldsymbol{r}) \\ &+ \frac{(ik)^3}{4\pi}\left(1-\frac{\gamma^-\rho^-}{\gamma^+\rho^+}\right)\int_{V^-} u^-(\boldsymbol{r}')G^+(\boldsymbol{r},\boldsymbol{r}')\,dv(\boldsymbol{r}') \\ &+ \frac{ik}{4\pi}\left(1-\frac{\rho^+}{\rho^-}\right)\int_S G^+(\boldsymbol{r},\boldsymbol{r}')\frac{\partial}{\partial n'}u^-(\boldsymbol{r}')\,ds(\boldsymbol{r}')\end{aligned} \tag{4.13}$$

We now introduce the low frequency expansions for u^+, u^- and u^i, (3.1)–(3.3), as well as the relations (4.9) and (4.11) into (4.13) and equate like powers of ik to obtain

$$\begin{aligned}\alpha u_n^+(\boldsymbol{r}) &+ (1-\alpha(\boldsymbol{r}))u_n^-(\boldsymbol{r}) = u_n^i(\boldsymbol{r}) \\ &+ \frac{1}{4\pi}\left(1-\frac{\gamma^-\rho^-}{\gamma^+\rho^+}\right)\sum_{m=2}^{n}\binom{n}{m} m(m-1)\int_{V^-} |\boldsymbol{r}-\boldsymbol{r}'|^{m-3} u_{n-m}^-(\boldsymbol{r}')\,dv(\boldsymbol{r}') \\ &+ \frac{1}{4\pi}\left(1-\frac{\rho^+}{\rho^-}\right)\sum_{m=0}^{n}\binom{n}{m}\int_S |\boldsymbol{r}-\boldsymbol{r}'|^{m-1}\frac{\partial}{\partial n'}u_{n-m}^-(\boldsymbol{r}')\,ds(\boldsymbol{r}')\end{aligned} \tag{4.14}$$

The right-hand side of (4.14) consists of terms which depend on the coefficients u_m^- for $m \le n-1$ as well as one term which depends on u_n^-. Thus, we may rewrite (4.14) as

$$\alpha(\boldsymbol{r})u_n^+(\boldsymbol{r}) + (1-\alpha(\boldsymbol{r}))u_n^-(\boldsymbol{r}) = f_n(\boldsymbol{r}) + v_n(\boldsymbol{r}) \tag{4.15}$$

where f_n depends only on previous coefficients and is given by

$$\begin{aligned} f_n(\boldsymbol{r}) = u_n^i(\boldsymbol{r}) &+ \frac{1}{4\pi}\left(1-\frac{\gamma^-\rho^-}{\gamma^+\rho^+}\right)\sum_{m=2}^{n}\binom{n}{m} m(m-1) \\ &\times \int_{V^-} |\boldsymbol{r}-\boldsymbol{r}'|^{m-3} u_{n-m}^-(\boldsymbol{r}')\, dv(\boldsymbol{r}') \\ &+ \frac{1}{4\pi}\left(1-\frac{\rho^+}{\rho^-}\right)\sum_{m=1}^{n}\binom{n}{m}\int_S |\boldsymbol{r}-\boldsymbol{r}'|^{m-1}\frac{\partial}{\partial n'}u_{n-m}^-(\boldsymbol{r}')\, ds(\boldsymbol{r}') \end{aligned} \tag{4.16}$$

while v_n depends on u_n^- and is given by

$$v_n(\boldsymbol{r}) = \frac{1}{4\pi}\left(1-\frac{\rho^+}{\rho^-}\right)\int_S \frac{1}{|\boldsymbol{r}-\boldsymbol{r}'|}\frac{\partial}{\partial n'}u_n^-(\boldsymbol{r}')\, ds(\boldsymbol{r}') \tag{4.17}$$

The term f_n may be considered as known once the coefficients up to and including order $n-1$ have been determined. Because the singularities in the integrands in f_n are of sufficiently low order, the functions f_n are continuously differentiable in $\mathbb{R}^3$.

On the other hand, v_n is not known in terms of lower order coefficients. However, its definition in (4.17) reveals that v_n is a single layer distribution which means that it is a solution of Laplace's equation in V^+ and V^- and decays as $1/r$ as $r \to \infty$. If we affix superscript $+$ and $-$ to v_n to denote that $\boldsymbol{r}$ is in V^+ or V^-, respectively, and recall the definition of α, (2.4), then (4.15) may be rewritten as the two equations

$$u_n^+ = f_n^+ + v_n^+ \quad \text{in } V^+ \tag{4.18}$$

and

$$u_n^- = f_n^- + v_n^- \quad \text{in } V^- \tag{4.19}$$

The problem of determining these coefficients is thus reduced to solving a transmission problem for the potential functions $v_n^\pm$ which, with the transmission conditions (4.1), may be stated as follows.

$$\left.\begin{aligned} \nabla^2 v_n^+ &= 0 && \text{in } V^+ \\ \nabla^2 v_n^- &= 0 && \text{in } V^- \\ v_n^+ &= v_n^- && \text{on } S \\ \frac{\partial}{\partial n}v_n^+ - \frac{\rho^+}{\rho^-}\frac{\partial}{\partial n}v_n^- &= -\left(1-\frac{\rho^+}{\rho^-}\right)\frac{\partial}{\partial n}f_n && \text{on } S \\ v_n^+ &= O\left(\frac{1}{r}\right) && \text{as } r \to \infty \end{aligned}\right\} \tag{4.20}$$

Since f_n is continuously differentiable, we omit the superscript when $\boldsymbol{r}$ is on S. This potential problem is very similar to that in Chapter 3, (3.88), using the boundary integral approach. Moreover, the expressions for scattering amplitude (3.89), (3.91), (3.92) and (3.93) are still valid with $u_n^{\pm}$ determined as described above.

The lossy transmission problem: $\delta^- > 0$. The introduction of losses does not present a major difficulty. With η^2 and β now given by (4.2) and (4.3), we may rewrite the basic integral representation (4.8) explicitly exhibiting the dependence on k and δ^- as

$$\begin{aligned}\alpha(\boldsymbol{r})u^+(\boldsymbol{r}) + (1-\alpha(\boldsymbol{r}))u^-(\boldsymbol{r}) &= u^i(\boldsymbol{r})\\ &+ \frac{(ik)^3}{4\pi}\left[1 - \frac{\gamma^-\rho^-}{\gamma^+\rho^+}\left(1 - \frac{ik\delta^-\gamma^-}{\sqrt{\rho^+\gamma^+}}\right)^{-1}\right]\int_{V^-} u^-(\boldsymbol{r}')G^+(\boldsymbol{r},\boldsymbol{r}')\,dv(\boldsymbol{r}')\\ &+ \frac{ik}{4\pi}\left[1 - \frac{\rho^+}{\rho^-}\left(1 - \frac{ik\delta^-\gamma^-}{\sqrt{\rho^+\gamma^+}}\right)\right]\int_S G^+(\boldsymbol{r},\boldsymbol{r}')\frac{\partial}{\partial n'}u^-(\boldsymbol{r}')\,ds(\boldsymbol{r}')\end{aligned} \tag{4.21}$$

Multiplying by $\left(1 - \dfrac{ik\delta^-\gamma^-}{\sqrt{\rho^+\gamma^+}}\right)$ and rearranging terms, this becomes

$$\begin{aligned}\alpha(\boldsymbol{r})u^+(\boldsymbol{r}) + (1-\alpha(\boldsymbol{r}))u^-(\boldsymbol{r}') &= u^i(\boldsymbol{r})\\ &+ \frac{(ik)^3}{4\pi}\left(1 - \frac{\gamma^-\rho^-}{\gamma^+\rho^+}\right)\int_{V^-} u^-(\boldsymbol{r}')G^+(\boldsymbol{r},\boldsymbol{r}')\,dv(\boldsymbol{r}')\\ &+ \frac{ik}{4\pi}\left(1 - \frac{\rho^+}{\rho^-}\right)\int_S G^+(\boldsymbol{r},\boldsymbol{r}')\frac{\partial}{\partial n'}u^-(\boldsymbol{r}')\,ds(\boldsymbol{r}')\\ &+ \frac{\delta^-\gamma^-}{\sqrt{\rho^+\gamma^+}}\Bigg\{ik[\alpha(\boldsymbol{r})u^+(\boldsymbol{r}) + (1-\alpha(\boldsymbol{r}))u^-(\boldsymbol{r}) - u^i(\boldsymbol{r})]\\ &- \frac{(ik)^2}{4\pi}\left(1 - 2\frac{\rho^+}{\rho^-}\right)\int_S G^+(\boldsymbol{r},\boldsymbol{r}')\frac{\partial}{\partial n'}u^-(\boldsymbol{r}')\,ds(\boldsymbol{r}')\\ &- \frac{(ik)^3}{4\pi}\frac{\delta^-\gamma^-}{\sqrt{\rho^+\gamma^+}}\frac{\rho^+}{\rho^-}\int_S G^+(\boldsymbol{r},\boldsymbol{r}')\frac{\partial}{\partial n'}u^-(\boldsymbol{r}')\,ds(\boldsymbol{r}')\\ &- \frac{(ik)^4}{4\pi}\int_{V^-} u^-(\boldsymbol{r}')G^+(\boldsymbol{r},\boldsymbol{r}')\,dv(\boldsymbol{r}')\Bigg\}\end{aligned} \tag{4.22}$$

This form is convenient for the introduction of the low frequency expansions for u^i, u^+, and u^-, (3.1)–(3.3), and the application of the relations (4.9)–(4.12). After doing so and equating like powers of ik we obtain

$$\begin{aligned}\alpha(\boldsymbol{r})u_n^+(\boldsymbol{r}) + (1-\alpha(\boldsymbol{r}))u_n^-(\boldsymbol{r}) &= f_n(\boldsymbol{r}) + v_n(\boldsymbol{r})\\ &+ \frac{\delta^-\gamma^-}{\sqrt{\rho^+\gamma^+}}[n\alpha(\boldsymbol{r})u_{n-1}^+(\boldsymbol{r}) + n(1-\alpha(\boldsymbol{r}))u_{n-1}^-(\boldsymbol{r}) + g_n(\boldsymbol{r})]\end{aligned} \tag{4.23}$$

where f_n and v_n are defined exactly as in the lossless case by (4.16) and (4.17), respectively, and

$$
\begin{aligned}
g_n(\boldsymbol{r}) = & -nu^i_{n-1}(\boldsymbol{r}) \\
& -\frac{1}{4\pi}\sum_{m=1}^{n}\binom{n}{m}\left\{m\left(1-2\frac{\rho^+}{\rho^-}\right)\int_S |\boldsymbol{r}-\boldsymbol{r}'|^{m-2}\frac{\partial}{\partial n'}u^-_{n-m}(\boldsymbol{r}')\,ds(\boldsymbol{r}')\right. \\
& +\frac{\delta^-\gamma^-}{\sqrt{\rho^+\gamma^+}}\frac{\rho^+}{\rho^-}m(m-1)\int_S |\boldsymbol{r}-\boldsymbol{r}'|^{m-3}\frac{\partial}{\partial n'}u^-_{n-m}(\boldsymbol{r}')\,ds(\boldsymbol{r}') \\
& \left. +\, m(m-1)(m-2)\int_{V^-}|\boldsymbol{r}-\boldsymbol{r}'|^{m-4}u^-_{n-m}(\boldsymbol{r}')\,ds(\boldsymbol{r}')\right\}
\end{aligned}
\tag{4.24}
$$

It should be noted that unlike f_n, g_n is continuous but not continuously differentiable in $\mathbb{R}^3$ because of the occurrence of a single layer potential whose derivatives are not continuous across S. Now we affix superscripts to denote whether $\boldsymbol{r}$ is in V^+ or V^- and rewrite (4.23) as the two equations

$$
u^+_n = f^+_n + \frac{\delta^-\gamma^-}{\sqrt{\rho^+\gamma^+}}[nu^+_{n-1} + g^+_n] + v^+_n \quad \text{in } V^+ \tag{4.25}
$$

$$
u^-_n = f^-_n + \frac{\delta^-\gamma^-}{\sqrt{\rho^+\gamma^+}}[nu^-_{n-1} + g^-_n] + v^-_n \quad \text{in } V^- \tag{4.26}
$$

All the terms on the right are known or involve coefficients of order less than n, except for v^+_n and v^-_n. These functions may be determined by solving a transmission problem for Laplace's equation where the transmission conditions are given by (3.102) and (3.103). Using the representations (4.25) and (4.26) and the continuity of f_n and g_n at S the transmission problem for v^+_n and v^-_n is

$$
\left.
\begin{aligned}
\nabla^2 v^+_n &= 0 && \text{in } V^+ \\
\nabla^2 v^-_n &= 0 && \text{in } V^- \\
v^+_n &= v^-_n && \text{on } S \\
\frac{\partial}{\partial n}v^+_n - \frac{\rho^+}{\rho^-}\frac{\partial}{\partial n}v^-_n &= -\left(1-\frac{\rho^+}{\rho^-}\right)\frac{\partial}{\partial n}f_n && \\
&\quad +\frac{\delta^-\gamma^-}{\sqrt{\rho^+\gamma^+}}\left(-\frac{\partial}{\partial n}g^+_n + \frac{\rho^+}{\rho^-}\frac{\partial}{\partial n}g^-_n - n\frac{\partial}{\partial n}u^+_{n-1}\right) && \text{on } S \\
v^+_n &= O\left(\frac{1}{r}\right) && \text{as } r\to\infty
\end{aligned}
\right\}
\tag{4.27}
$$

Comparison with (4.20) reveals that the same transmission problem arises in both the lossless and lossy cases, the only difference appearing in the inhomogeneous term in the second transmission condition. It is clear that when δ^- vanishes the problem becomes exactly the same as (4.20) in the lossless case. Once u_n^+ have been found, the same expressions used before for the scattering amplitude (3.89), (3.91), (3.92) and (3.93) may be employed. Of course, if $\delta^- \neq 0$ the absorption cross section no longer vanishes. The low frequency expansion of σ_a may be obtained from (2.40) as

$$\sigma_a = \delta^- \frac{\gamma^-}{\rho^-}\sqrt{\frac{\rho^+}{\gamma^+}} \sum_{n=0}^{\infty} \frac{(ik)^n}{n!} \sum_{m=0}^{n} \binom{n}{m} (-1)^m \times \int_{V^-} \nabla u_{n-m}^-(\boldsymbol{r}') \cdot \nabla u_m^{-*}(\boldsymbol{r}')\, dv(\boldsymbol{r}') \tag{4.28}$$

Since for plane wave incidence $u_0^- = 1$, this may be rewritten as

$$\sigma_a = \delta^- \frac{\gamma^-}{\rho^-}\sqrt{\frac{\rho^+}{\gamma^+}} \sum_{n=2}^{\infty} \frac{(ik)^n}{n!} \sum_{m=1}^{n-1} \binom{n}{m} (-1)^m \times \int_{V^-} \nabla u_{n-m}^-(\boldsymbol{r}') \cdot \nabla u_m^{-*}(\boldsymbol{r}')\, dv(\boldsymbol{r}') \tag{4.29}$$

which clearly shows that the effects of loss first appear in the k^2 term. An alternative form for σ_a is given in (3.114)

4.A.2 *Problems in acoustics*

1. Derive equation (4.14) for the low frequency coefficients.
2. Show that (4.20) defines a well-posed transmission problem.
3. Derive equations (4.23) and (4.24).
4. Show that the transmission problem (4.27) is well posed.

4.EM Electromagnetics

4.EM.1 *The volume integral formulation*

While in acoustics the volume integral approach simplified the analysis of the low frequency expansion somewhat, the transmission potential problem that must be solved to determine the nth coefficient was essentially the same as in the development in Chapter 3 using boundary integrals only. This was true in both lossless and lossy cases. In electromagnetics, however, we will show in this section that the volume integral approach considerably simplifies the analysis in the lossless case, reducing the number of potential problems to be solved from four (two exterior Neumann problems and two transmission problems) to two (both transmission problems). For the lossy case in electromagnetics the volume integral approach will be seen to require the solution of one exterior Neumann and one interior Dirichlet problem to determine the nth coefficients in the expansions of the electric fields in V^+ and V^- and the solution of one transmission problem to determine the nth coefficients in the expansions of the magnetic fields.

The transmission conditions (1.142)–(1.147) are reproduced here for convenience

$$\left.\begin{aligned} \hat{\boldsymbol{n}} \times \boldsymbol{E}^+ &= \hat{\boldsymbol{n}} \times \boldsymbol{E}^- \\ \hat{\boldsymbol{n}} \times \boldsymbol{H}^+ &= \hat{\boldsymbol{n}} \times \boldsymbol{H}^- \end{aligned}\right\} \quad \text{on } S \tag{4.30}$$

$$\left.\begin{aligned} Z^+\hat{\boldsymbol{n}} \cdot \boldsymbol{H}^+ &= \eta Z^-\hat{\boldsymbol{n}} \cdot \boldsymbol{H}^- \\ Y^+\hat{\boldsymbol{n}} \cdot \boldsymbol{E}^+ &= \eta Y^-\hat{\boldsymbol{n}} \cdot \boldsymbol{E}^- \end{aligned}\right\} \quad \text{on } S \tag{4.31}$$

where

$$\eta Z^- = \frac{\mu^-}{\mu^+} Z^+ \tag{4.32}$$

$$\eta Y^- = \frac{\epsilon^-}{\epsilon^+} Y^+ \left(1 - \frac{\sigma^-}{ik} \frac{\epsilon^+}{\epsilon^-} Z^+\right) \tag{4.33}$$

and

$$\eta^2 = \frac{\mu^-\epsilon^-}{\mu^+\epsilon^+} \left(1 - \frac{\sigma^-}{ik} \frac{\epsilon^+}{\epsilon^-} Z^+\right) \tag{4.34}$$

The implicit presence of the factor σ^-/ik in condition (4.31) for the normal components of the electric field is made evident in (4.33) and this factor plays a determining role in the development of low frequency expansions in the lossy case as will become evident subsequently.

We started the low frequency analysis in Chapter 3 with the integral representations for the exterior fields, (2.69) and (2.70), which involve the exterior Green's function G^+. Rather than use (2.65) and (2.66) to represent the fields in V^- in terms of the interior Green's function G^-, we base our present analysis entirely upon the exterior Green's function, which will result in the appearance of volume integrals.

Introducing the transmission conditions (4.30) and (4.31) into (2.69) and (2.70) yields

$$\begin{aligned} \alpha(\boldsymbol{r})\boldsymbol{E}^+(\boldsymbol{r}) = \boldsymbol{E}^i(\boldsymbol{r}) + \frac{ik}{4\pi} \int_S & \big[ikZ^+G^+(\boldsymbol{r},\boldsymbol{r}')(\hat{\boldsymbol{n}}' \times \boldsymbol{H}^-(\boldsymbol{r}')) \\ & + \eta Y^- Z^+ (\nabla_{\boldsymbol{r}'} G^+(\boldsymbol{r},\boldsymbol{r}'))(\hat{\boldsymbol{n}}' \cdot \boldsymbol{E}^-(\boldsymbol{r}')) \\ & - (\nabla_{\boldsymbol{r}'} G^+(\boldsymbol{r},\boldsymbol{r}')) \times (\hat{\boldsymbol{n}}' \times \boldsymbol{E}^-(\boldsymbol{r}'))\big]\, ds(r') \end{aligned} \tag{4.35}$$

and

$$\begin{aligned} \alpha(\boldsymbol{r})\boldsymbol{H}^+(\boldsymbol{r}) = \boldsymbol{H}^i(\boldsymbol{r}) + \frac{ik}{4\pi} \int_S & \big[-ikY^+G^+(\boldsymbol{r},\boldsymbol{r}')(\hat{\boldsymbol{n}}' \times \boldsymbol{E}^-(\boldsymbol{r}')) \\ & + \eta Z^- Y^+ (\nabla_{\boldsymbol{r}'} G^+(\boldsymbol{r},\boldsymbol{r}'))(\hat{\boldsymbol{n}}' \cdot \boldsymbol{H}^-(\boldsymbol{r}')) \\ & - (\nabla_{\boldsymbol{r}'} G^+(\boldsymbol{r},\boldsymbol{r}')) \times (\hat{\boldsymbol{n}}' \times \boldsymbol{H}^-(\boldsymbol{r}'))\big]\, ds(\boldsymbol{r}') \end{aligned} \tag{4.36}$$

Next, we add and subtract terms to rewrite (4.35) and (4.36) in the entirely equivalent forms

$$
\begin{aligned}
\alpha(\boldsymbol{r})\boldsymbol{E}^{+}(\boldsymbol{r}) = \boldsymbol{E}^{i}(\boldsymbol{r}) + \frac{ik}{4\pi}\int_{S} & \big[ik^{-}Z^{-}G^{+}(\boldsymbol{r},\boldsymbol{r}')(\hat{\boldsymbol{n}}'\times\boldsymbol{H}^{-}(\boldsymbol{r}')) \\
& + (\nabla_{\boldsymbol{r}'}G^{+}(\boldsymbol{r},\boldsymbol{r}'))(\boldsymbol{n}'\cdot\boldsymbol{E}^{-}(\boldsymbol{r}')) \\
& - (\nabla_{\boldsymbol{r}'}G^{+}(\boldsymbol{r},\boldsymbol{r}'))\times(\hat{\boldsymbol{n}}'\times\boldsymbol{E}^{-}(\boldsymbol{r}'))\big]\,ds(\boldsymbol{r}') \\
& + \frac{ik}{4\pi}(ikZ^{+} - ik^{-}Z^{-})\int_{S} G^{+}(\boldsymbol{r},\boldsymbol{r}')(\hat{\boldsymbol{n}}'\times\boldsymbol{H}^{-}(\boldsymbol{r}'))\,ds(\boldsymbol{r}') \\
& + \frac{ik}{4\pi}(\eta Y^{-}Z^{+} - 1)\int_{S}(\nabla_{\boldsymbol{r}'}G^{+}(\boldsymbol{r},\boldsymbol{r}'))(\hat{\boldsymbol{n}}'\cdot\boldsymbol{E}^{-}(\boldsymbol{r}'))\,ds(\boldsymbol{r}')
\end{aligned} \tag{4.37}
$$

$$
\begin{aligned}
\alpha(\boldsymbol{r})\boldsymbol{H}^{+}(\boldsymbol{r}) = \boldsymbol{H}^{i}(\boldsymbol{r}) + \frac{ik}{4\pi}\int_{S} & \big[-ik^{-}Y^{-}G^{+}(\boldsymbol{r},\boldsymbol{r}')(\hat{\boldsymbol{n}}'\times\boldsymbol{E}^{-}(\boldsymbol{r}')) \\
& + (\nabla_{\boldsymbol{r}'}G^{+}(\boldsymbol{r},\boldsymbol{r}'))(\hat{\boldsymbol{n}}'\cdot\boldsymbol{H}^{-}(\boldsymbol{r}')) \\
& - (\nabla_{\boldsymbol{r}'}G^{+}(\boldsymbol{r},\boldsymbol{r}'))\times(\hat{\boldsymbol{n}}'\times\boldsymbol{H}^{-}(\boldsymbol{r}'))\big]\,ds(\boldsymbol{r}') \\
& + \frac{ik}{4\pi}(ik^{-}Y^{-} - ikY^{+})\int_{S} G^{+}(\boldsymbol{r},\boldsymbol{r}')(\hat{\boldsymbol{n}}'\times\boldsymbol{E}^{-}(\boldsymbol{r}'))\,ds(\boldsymbol{r}') \\
& + \frac{ik}{4\pi}(\eta Z^{-}Y^{+} - 1)\int_{S}(\nabla_{\boldsymbol{r}'}G^{+}(\boldsymbol{r},\boldsymbol{r}'))(\hat{\boldsymbol{n}}'\cdot\boldsymbol{H}^{-}(\boldsymbol{r}'))\,ds(\boldsymbol{r}')
\end{aligned} \tag{4.38}
$$

We now introduce a vector identity which may be derived in exactly the same way as (2.62) was used to derive (2.65) and (2.66), which we state as follows. If

$$
\left.\begin{aligned}
\nabla\times(\nabla\times\boldsymbol{f}) &= k^{2}\eta^{2}\boldsymbol{f} \\
\nabla\cdot\boldsymbol{f} &= 0
\end{aligned}\right\}\quad \text{in } V^{-} \tag{4.39}
$$

then

$$
\begin{aligned}
\frac{ik}{4\pi}\int_{S} & \big[G^{+}(\boldsymbol{r},\boldsymbol{r}')(\boldsymbol{n}'\times(\nabla_{\boldsymbol{r}'}\times\boldsymbol{f}(\boldsymbol{r}')) + (\nabla_{\boldsymbol{r}'}G^{+}(\boldsymbol{r},\boldsymbol{r}'))(\hat{\boldsymbol{n}}'\cdot\boldsymbol{f}(\boldsymbol{r}')) \\
& - (\nabla_{\boldsymbol{r}'}G^{+}(\boldsymbol{r},\boldsymbol{r}'))\times(\hat{\boldsymbol{n}}'\times\boldsymbol{f}(\boldsymbol{r}'))\big]\,ds(\boldsymbol{r}') \\
& = \frac{(ik)^{3}}{4\pi}(1-\eta^{2})\int_{V^{-}}\boldsymbol{f}(\boldsymbol{r}')G^{+}(\boldsymbol{r},\boldsymbol{r}')\,dv(\boldsymbol{r}') + (\alpha(\boldsymbol{r})-1)\boldsymbol{f}(\boldsymbol{r})
\end{aligned} \tag{4.40}
$$

Identifying $\boldsymbol{f}$ first with $\boldsymbol{E}^{-}$ and then with $\boldsymbol{H}^{-}$ and employing the Maxwell equations as well as (4.35) and (4.36) allows us to obtain the representations

$$
\begin{aligned}
\alpha(\boldsymbol{r})\boldsymbol{E}^{+}(\boldsymbol{r}) + (1-\alpha(\boldsymbol{r}))\boldsymbol{E}^{-}(\boldsymbol{r}) &= \boldsymbol{E}^{i}(\boldsymbol{r}) \\
&+ \frac{(ik)^3}{4\pi}(1-\eta^2)\int_{V^-} \boldsymbol{E}^{-}(\boldsymbol{r}')G^{+}(\boldsymbol{r},\boldsymbol{r}')\,dv(\boldsymbol{r}') \\
&+ \frac{ik}{4\pi}(ikZ^{+} - ik^{-}Z^{-})\int_{S} G^{+}(\boldsymbol{r},\boldsymbol{r}')(\hat{\boldsymbol{n}}'\times\boldsymbol{H}^{-}(\boldsymbol{r}'))\,ds(\boldsymbol{r}') \\
&+ \frac{ik}{4\pi}(\eta Y^{-}Z^{+} - 1)\int_{S} (\nabla_{\boldsymbol{r}'}G^{+}(\boldsymbol{r},\boldsymbol{r}'))(\hat{\boldsymbol{n}}'\cdot\boldsymbol{E}^{-}(\boldsymbol{r}'))\,ds(\boldsymbol{r}') \qquad (4.41)
\end{aligned}
$$

and

$$
\begin{aligned}
\alpha(\boldsymbol{r})\boldsymbol{H}^{+}(\boldsymbol{r}) + (1-\alpha(\boldsymbol{r}))\boldsymbol{H}^{-}(\boldsymbol{r}) &= \boldsymbol{H}^{i}(\boldsymbol{r}) \\
&+ \frac{(ik)^3}{4\pi}(1-\eta^2)\int_{V^-} \boldsymbol{H}^{-}(\boldsymbol{r}')G^{+}(\boldsymbol{r},\boldsymbol{r}')\,dv(\boldsymbol{r}') \\
&+ \frac{ik}{4\pi}(ik^{-}Y^{-} - ikY^{+})\int_{S} G^{+}(\boldsymbol{r},\boldsymbol{r}')(\hat{\boldsymbol{n}}'\times\boldsymbol{E}^{-}(\boldsymbol{r}'))\,ds(\boldsymbol{r}') \\
&+ \frac{ik}{4\pi}(\eta Z^{-}Y^{+} - 1)\int_{S} (\nabla_{\boldsymbol{r}'}G^{+}(\boldsymbol{r},\boldsymbol{r}'))(\hat{\boldsymbol{n}}'\cdot\boldsymbol{H}^{-}(\boldsymbol{r}'))\,ds(\boldsymbol{r}') \qquad (4.42)
\end{aligned}
$$

These representations are valid for all $\boldsymbol{r}$ in $\mathbb{R}^3$ and provide the basis for the low frequency expansions. We treat the lossless and lossy cases separately since the construction of the low frequency expansion is quite different in the two cases.

The lossless transmission problem: $\sigma^{-} = 0$. In this case, η and Y^{-} are real and we rewrite (4.41) and (4.42) exhibiting the constitutive parameters explicitly.

$$
\begin{aligned}
\alpha(\boldsymbol{r})\boldsymbol{E}^{+}(\boldsymbol{r}) + (1-\alpha(\boldsymbol{r}))\boldsymbol{E}^{-}(\boldsymbol{r}) &= \boldsymbol{E}^{i}(\boldsymbol{r}) \\
&+ \frac{(ik)^3}{4\pi}\left(1-\frac{\mu^{-}\epsilon^{-}}{\mu^{+}\epsilon^{+}}\right)\int_{V^-} \boldsymbol{E}^{-}(\boldsymbol{r}')G^{+}(\boldsymbol{r},\boldsymbol{r}')\,dv(\boldsymbol{r}') \\
&+ \frac{(ik)^2}{4\pi}\left(1-\frac{\mu^{-}}{\mu^{+}}\right)Z^{+}\int_{S} G^{+}(\boldsymbol{r},\boldsymbol{r}')(\hat{\boldsymbol{n}}'\times\boldsymbol{H}^{-}(\boldsymbol{r}'))\,ds(\boldsymbol{r}') \\
&- \frac{ik}{4\pi}\left(1-\frac{\epsilon^{-}}{\epsilon^{+}}\right)\int_{S} (\nabla_{\boldsymbol{r}'}G^{+}(\boldsymbol{r},\boldsymbol{r}'))(\hat{\boldsymbol{n}}'\cdot\boldsymbol{E}^{-}(\boldsymbol{r}'))\,ds(\boldsymbol{r}') \qquad (4.43)
\end{aligned}
$$

and

$$
\begin{aligned}
\alpha(\boldsymbol{r})\boldsymbol{H}^{+}(\boldsymbol{r}) &+ (1-\alpha(\boldsymbol{r}))\boldsymbol{H}^{-}(\boldsymbol{r}) = \boldsymbol{H}^{i}(\boldsymbol{r}) \\
&+ \frac{(ik)^3}{4\pi}\left(1-\frac{\mu^{-}\epsilon^{-}}{\mu^{+}\epsilon^{+}}\right)\int_{V^{-}} \boldsymbol{H}^{-}(\boldsymbol{r}')G^{+}(\boldsymbol{r},\boldsymbol{r}')\,dv(\boldsymbol{r}') \\
&- \frac{(ik)^2}{4\pi}\left(1-\frac{\epsilon^{-}}{\epsilon^{+}}\right)Y^{+}\int_{S} G^{+}(\boldsymbol{r},\boldsymbol{r}')(\hat{\boldsymbol{n}}'\times\boldsymbol{E}^{-}(\boldsymbol{r}'))\,ds(\boldsymbol{r}') \\
&- \frac{ik}{4\pi}\left(1-\frac{\mu^{-}}{\mu^{+}}\right)\int_{S} (\nabla_{\boldsymbol{r}'}G^{+}(\boldsymbol{r},\boldsymbol{r}'))(\hat{\boldsymbol{n}}'\cdot\boldsymbol{H}^{-}(\boldsymbol{r}'))\,ds(\boldsymbol{r}')
\end{aligned}
\tag{4.44}
$$

Now we introduce the low frequency expansions for $\boldsymbol{E}^{\pm}, \boldsymbol{H}^{\pm}, \boldsymbol{E}^{i}$ and $\boldsymbol{H}^{i}$, (3.116)–(3.118), as well as the relations (4.9)–(4.11) with u^{-}_{n-m} replaced by $\boldsymbol{E}^{-}_{n-m}$ or $\boldsymbol{H}^{-}_{n-m}$ as appropriate and equate like powers of ik to obtain

$$
\begin{aligned}
\alpha(\boldsymbol{r})\boldsymbol{E}^{+}_{n}(\boldsymbol{r}) &+ (1-\alpha(\boldsymbol{r}))\boldsymbol{E}^{-}_{n}(\boldsymbol{r}) = \boldsymbol{E}^{i}_{n}(\boldsymbol{r}) \\
&+ \frac{1}{4\pi}\left(1-\frac{\mu^{-}\epsilon^{-}}{\mu^{+}\epsilon^{+}}\right)\sum_{m=2}^{n}\binom{n}{m} m(m-1)\int_{V^{-}} |\boldsymbol{r}-\boldsymbol{r}'|^{m-3}\boldsymbol{E}^{-}_{n-m}(\boldsymbol{r}')\,dv(\boldsymbol{r}') \\
&+ \frac{1}{4\pi}\left(1-\frac{\mu^{-}}{\mu^{+}}\right)Z^{+}\sum_{m=1}^{n}\binom{n}{m} m\int_{S} |\boldsymbol{r}-\boldsymbol{r}'|^{m-2}(\hat{\boldsymbol{n}}'\times\boldsymbol{H}^{-}_{n-m}(\boldsymbol{r}'))\,ds(\boldsymbol{r}') \\
&- \frac{1}{4\pi}\left(1-\frac{\epsilon^{-}}{\epsilon^{+}}\right)\sum_{m=0}^{n}\binom{n}{m}\int_{S} (\nabla_{\boldsymbol{r}'}|\boldsymbol{r}-\boldsymbol{r}'|^{m-1})(\hat{\boldsymbol{n}}'\cdot\boldsymbol{E}^{-}_{n-m}(\boldsymbol{r}'))\,ds(\boldsymbol{r}')
\end{aligned}
\tag{4.45}
$$

and

$$
\begin{aligned}
\alpha(\boldsymbol{r})\boldsymbol{H}^{+}_{n}(\boldsymbol{r}) &+ (1-\alpha(\boldsymbol{r}))\boldsymbol{H}^{-}_{n}(\boldsymbol{r}) = \boldsymbol{H}^{i}_{n}(\boldsymbol{r}) \\
&+ \frac{1}{4\pi}\left(1-\frac{\mu^{-}\epsilon^{-}}{\mu^{+}\epsilon^{+}}\right)\sum_{m=2}^{n}\binom{n}{m} m(m-1)\int_{V^{-}} |\boldsymbol{r}-\boldsymbol{r}'|^{m-3}\boldsymbol{H}^{-}_{n-m}(\boldsymbol{r}')\,dv(\boldsymbol{r}') \\
&- \frac{1}{4\pi}\left(1-\frac{\epsilon^{-}}{\epsilon^{+}}\right)Y^{+}\sum_{m=1}^{n}\binom{n}{m} m\int_{S} |\boldsymbol{r}-\boldsymbol{r}'|^{m-2}(\hat{\boldsymbol{n}}'\times\boldsymbol{E}^{-}_{n-m}(\boldsymbol{r}'))\,ds(\boldsymbol{r}') \\
&- \frac{1}{4\pi}\left(1-\frac{\mu^{-}}{\mu^{+}}\right)\sum_{m=0}^{n}\binom{n}{m}\int_{S} (\nabla_{\boldsymbol{r}'}|\boldsymbol{r}-\boldsymbol{r}'|^{m-1})(\hat{\boldsymbol{n}}'\cdot\boldsymbol{H}^{-}_{n-m}(\boldsymbol{r}'))\,ds(\boldsymbol{r}')
\end{aligned}
\tag{4.46}
$$

The right-hand sides of (4.45) and (4.46) consist of terms which depend on $\boldsymbol{E}^{-}_{m}$ and $\boldsymbol{H}^{-}_{m}$ for $m \le n-1$ as well as one term which involves the nth term. Thus, we rewrite (4.45) and (4.46) as

$$
\alpha(\boldsymbol{r})\boldsymbol{E}^{+}_{n}(\boldsymbol{r}) + (1-\alpha(\boldsymbol{r}))\boldsymbol{E}^{-}_{n}(\boldsymbol{r}) = \boldsymbol{F}_{en}(\boldsymbol{r}) + \nabla\phi_{n}(\boldsymbol{r})
\tag{4.47}
$$

and

$$\alpha(\boldsymbol{r})\boldsymbol{H}_n^+(\boldsymbol{r}) + (1-\alpha(\boldsymbol{r}))\boldsymbol{H}_n^-(\boldsymbol{r}) = \boldsymbol{F}_{mn}(\boldsymbol{r}) + \nabla\psi_n(\boldsymbol{r}) \tag{4.48}$$

where

$$\begin{aligned}\boldsymbol{F}_{en}(\boldsymbol{r}) &= \boldsymbol{E}_n^i(\boldsymbol{r})\\ &+\frac{1}{4\pi}\left(1-\frac{\mu^-\epsilon^-}{\mu^+\epsilon^+}\right)\sum_{m=2}^{n}\binom{n}{m}m(m-1)\int_{V^-}|\boldsymbol{r}-\boldsymbol{r}'|^{m-3}\boldsymbol{E}_{n-m}^-(\boldsymbol{r}')\,dv(\boldsymbol{r}')\\ &+\frac{1}{4\pi}\left(1-\frac{\mu^-}{\mu^+}\right)Z^+\sum_{m=1}^{n}\binom{n}{m}m\int_{S}|\boldsymbol{r}-\boldsymbol{r}'|^{m-2}(\hat{\boldsymbol{n}}'\times\boldsymbol{H}_{n-m}^-(\boldsymbol{r}'))\,ds(\boldsymbol{r}')\\ &-\frac{1}{4\pi}\left(1-\frac{\epsilon^-}{\epsilon^+}\right)\sum_{m=1}^{n}\binom{n}{m}\int_{S}(\nabla_{\boldsymbol{r}'}|\boldsymbol{r}-\boldsymbol{r}'|^{m-1})(\hat{\boldsymbol{n}}'\cdot\boldsymbol{E}_{n-m}^-(\boldsymbol{r}'))\,ds(\boldsymbol{r}')\end{aligned} \tag{4.49}$$

$$\phi_n(\boldsymbol{r}) = \frac{1}{4\pi}\left(1-\frac{\epsilon^-}{\epsilon^+}\right)\int_S \frac{1}{|\boldsymbol{r}-\boldsymbol{r}'|}\hat{\boldsymbol{n}}'\cdot\boldsymbol{E}_n^-(\boldsymbol{r}')\,ds(\boldsymbol{r}') \tag{4.50}$$

$$\begin{aligned}\boldsymbol{F}_{mn}(\boldsymbol{r}) &= \boldsymbol{H}_n^i(\boldsymbol{r})\\ &+\frac{1}{4\pi}\left(1-\frac{\mu^-\epsilon^-}{\mu^+\epsilon^+}\right)\sum_{m=2}^{n}\binom{n}{m}m(m-1)\int_{V^-}|\boldsymbol{r}-\boldsymbol{r}'|^{m-3}\boldsymbol{H}_{n-m}^-(\boldsymbol{r}')\,dv(\boldsymbol{r}')\\ &-\frac{1}{4\pi}\left(1-\frac{\epsilon^-}{\epsilon^+}\right)Y^+\sum_{m=1}^{n}\binom{n}{m}m\int_{S}|\boldsymbol{r}-\boldsymbol{r}'|^{m-2}(\hat{\boldsymbol{n}}'\times\boldsymbol{E}_{n-m}^-(\boldsymbol{r}'))\,ds(\boldsymbol{r}')\\ &-\frac{1}{4\pi}\left(1-\frac{\mu^-}{\mu^+}\right)\sum_{m=1}^{n}\binom{n}{m}\int_{S}(\nabla_{\boldsymbol{r}'}|\boldsymbol{r}-\boldsymbol{r}'|^{m-1})(\hat{\boldsymbol{n}}'\cdot\boldsymbol{H}_{n-m}^-(\boldsymbol{r}'))\,ds(\boldsymbol{r}')\end{aligned} \tag{4.51}$$

$$\psi_n(\boldsymbol{r}) = \frac{1}{4\pi}\left(1-\frac{\mu^-}{\mu^+}\right)\int_S \frac{1}{|\boldsymbol{r}-\boldsymbol{r}'|}\hat{\boldsymbol{n}}'\cdot\boldsymbol{H}_n^-(\boldsymbol{r}')\,ds(\boldsymbol{r}') \tag{4.52}$$

The functions $\boldsymbol{F}_{en}$ and $\boldsymbol{F}_{mn}$ may be considered as known once the coefficients $\boldsymbol{E}_m^-$ and $\boldsymbol{H}_m^-$ are known for $m \leq n-1$. Observe that $\boldsymbol{F}_{en}$ and $\boldsymbol{F}_{mn}$ are continuous in $\mathbb{R}^3$ but not continuously differentiable on S. On the other hand, at the nth step ϕ_n and ψ_n are not known since they involve the nth coefficients. However, the forms (4.50) and (4.52) reveal that ϕ_n and ψ_n are single layer potentials hence are solutions of Laplace's equation and are of order $1/r^2$ as r goes to infinity. The high asymptotic decay is a consequence of (3.129) and the vanishing of the integrals of $\hat{\boldsymbol{n}}\cdot\boldsymbol{E}^-$ and $\hat{\boldsymbol{n}}\cdot\boldsymbol{H}_n^-$ on S. To facilitate the determination of ϕ_n and ψ_n we affix superscripts to indicate whether $\boldsymbol{r}$ is in V^+ or V^- and, recalling the definition of α, rewrite (4.47) and (4.48) as

$$\boldsymbol{E}_n^+ = \boldsymbol{F}_{en}^+ + \nabla\phi_n^+ \quad \text{in } V^+ \tag{4.53}$$

$$\boldsymbol{E}_n^- = \boldsymbol{F}_{en}^- + \nabla\phi_n^- \quad \text{in } V^- \tag{4.54}$$

$$\boldsymbol{H}_n^+ = \boldsymbol{F}_{mn}^+ + \nabla\psi_n^+ \quad \text{in } V^+ \tag{4.55}$$

$$\boldsymbol{H}_n^- = \boldsymbol{F}_{mn}^- + \nabla\psi_n^- \quad \text{in } V^- \tag{4.56}$$

The problem of determining the coefficients $\boldsymbol{E}_n^\pm$ and $\boldsymbol{H}_n^\pm$ is reduced, using the transmission conditions (3.180) and (3.183), to the following transmission potential problems for ϕ_n and ψ_n

$$\left.\begin{aligned}
\nabla^2\phi_n^+ &= 0 && \text{in } V^+\\
\nabla^2\phi_n^- &= 0 && \text{in } V^-\\
\hat{\boldsymbol{n}} \times \nabla(\phi_n^+ - \phi_n^-) &= \boldsymbol{0} && \text{on } S\\
\frac{\partial}{\partial n}\phi_n^+ - \frac{\epsilon^-}{\epsilon^+}\frac{\partial}{\partial n}\phi_n^- &= -\left(1 - \frac{\epsilon^-}{\epsilon^+}\right)\hat{\boldsymbol{n}} \cdot \boldsymbol{F}_{en} && \text{on } S\\
\phi_n^+ &= O\left(\frac{1}{r^2}\right) && \text{as } r \to \infty
\end{aligned}\right\} \tag{4.57}$$

$$\left.\begin{aligned}
\nabla^2\psi_n^+ &= 0 && \text{in } V^+\\
\nabla^2\psi_n^- &= 0 && \text{in } V^-\\
\hat{\boldsymbol{n}} \times \nabla(\psi_n^+ - \psi_n^-) &= \boldsymbol{0} && \text{on } S\\
\frac{\partial}{\partial n}\psi_n^+ - \frac{\mu^-}{\mu^+}\frac{\partial}{\partial n}\psi^- &= -\left(1 - \frac{\mu^-}{\mu^+}\right)\hat{\boldsymbol{n}} \cdot \boldsymbol{F}_{mn} && \text{on } S\\
\psi_n^+ &= O\left(\frac{1}{r^2}\right) && \text{as } r \to \infty
\end{aligned}\right\} \tag{4.58}$$

We have omitted the superscripts on $\boldsymbol{F}_{em}$ and $\boldsymbol{F}_{mn}$ for $\boldsymbol{r}$ on S since the functions are continuous across S. The potential problems are seen to be of exactly the same form for ϕ_n and ψ_n. Note that $\boldsymbol{E}_n^\pm$ and $\boldsymbol{H}_n^\pm$ are decoupled in the sense that they each depend on terms of previous order but at the nth step the electric and magnetic coefficients must be determined independently. The expressions for the scattering amplitudes and cross sections (3.217)–(3.223) are all still valid with $\boldsymbol{E}_n^\pm$ and $\boldsymbol{H}_n^\pm$ determined as described above.

The lossy transmission problem: $\sigma^- > 0$. Unlike the acoustic case, where the transmission potential problem to be solved at each step was essentially the same in both the lossy and lossless cases, the presence of losses in the electromagnetic case significantly changes the process of determining the coefficients in the low frequency expansions from that in the lossless case. The electric and magnetic fields are coupled at every step but in a nonsymmetric way because conductivity appears in the Maxwell equations in a nonsymmetric way. That is, $\boldsymbol{E}_n^\pm$ is independent of $\boldsymbol{H}_n^\pm$ but $\boldsymbol{H}_n^\pm$ depends on $\boldsymbol{E}_n^\pm$. We begin the development of the lossy case by rewriting the integral representations (4.41) and (4.42) specifically exhibiting the dependence on k and σ^-.

$$\begin{aligned}\alpha(\boldsymbol{r})\boldsymbol{E}^+(\boldsymbol{r}) + (1-\alpha(\boldsymbol{r}))\boldsymbol{E}^-(\boldsymbol{r}) &= \boldsymbol{E}^i(\boldsymbol{r})\\ &+ \frac{(ik)^3}{4\pi}\left(1-\frac{\mu^-\epsilon^-}{\mu^+\epsilon^+}+\frac{\sigma^-}{ik}\frac{\mu^-}{\mu^+}Z^+\right)\int_{V^-}\boldsymbol{E}^-(\boldsymbol{r}')G^+(\boldsymbol{r},\boldsymbol{r}')\,dv(\boldsymbol{r}')\\ &+ \frac{(ik)^2}{4\pi}\left(1-\frac{\mu^-}{\mu^+}\right)Z^+\int_S G^+(\boldsymbol{r},\boldsymbol{r}')(\hat{\boldsymbol{n}}'\times\boldsymbol{H}^-(\boldsymbol{r}'))\,ds(\boldsymbol{r}')\\ &- \frac{ik}{4\pi}\left(1-\frac{\epsilon^-}{\epsilon^+}+\frac{\sigma^-}{ik}Z^+\right)\int_S(\nabla_{\boldsymbol{r}'}G^+(\boldsymbol{r},\boldsymbol{r}'))(\hat{\boldsymbol{n}}'\cdot\boldsymbol{E}^-(\boldsymbol{r}'))\,ds(\boldsymbol{r}')\end{aligned} \tag{4.59}$$

and

$$\begin{aligned}\alpha(\boldsymbol{r})\boldsymbol{H}^+(\boldsymbol{r}) + (1-\alpha(\boldsymbol{r}))\boldsymbol{H}^-(\boldsymbol{r}) &= \boldsymbol{H}^i(\boldsymbol{r})\\ &+ \frac{(ik)^3}{4\pi}\left(1-\frac{\mu^-\epsilon^-}{\mu^+\epsilon^+}+\frac{\sigma^-}{ik}\frac{\mu^-}{\mu^+}Z^+\right)\int_{V^-}\boldsymbol{H}^-(\boldsymbol{r}')G^+(\boldsymbol{r},\boldsymbol{r}')\,dv(\boldsymbol{r}')\\ &- \frac{(ik)^2}{4\pi}\left(1-\frac{\epsilon^-}{\epsilon^+}+\frac{\sigma^-}{ik}Z^+\right)Y^+\int_S G^+(\boldsymbol{r},\boldsymbol{r}')(\hat{\boldsymbol{n}}'\times\boldsymbol{E}^-(\boldsymbol{r}'))\,ds(\boldsymbol{r}')\\ &- \frac{ik}{4\pi}\left(1-\frac{\mu^-}{\mu^+}\right)\int_S(\nabla_{\boldsymbol{r}'}G^+(\boldsymbol{r},\boldsymbol{r}'))(\hat{\boldsymbol{n}}'\cdot\boldsymbol{H}^-(\boldsymbol{r}'))\,ds(\boldsymbol{r}')\end{aligned} \tag{4.60}$$

We rewrite these equations to show more clearly both the k dependence and the effect of the losses.

$$\begin{aligned}\alpha(\boldsymbol{r})\boldsymbol{E}^+(\boldsymbol{r}) + (1-\alpha(\boldsymbol{r}))\boldsymbol{E}^-(\boldsymbol{r}) &= \boldsymbol{E}^i(\boldsymbol{r})\\ &+ \frac{(ik)^3}{4\pi}\left(1-\frac{\mu^-\epsilon^-}{\mu^+\epsilon^+}\right)\int_{V^-}\boldsymbol{E}^-(\boldsymbol{r}')G^+(\boldsymbol{r},\boldsymbol{r}')\,dv(\boldsymbol{r}')\\ &+ \frac{(ik)^2}{4\pi}\left(1-\frac{\mu^-}{\mu^+}\right)Z^+\int_S G^+(\boldsymbol{r},\boldsymbol{r}')(\hat{\boldsymbol{n}}'\times\boldsymbol{H}^-(\boldsymbol{r}'))\,ds(\boldsymbol{r}')\\ &- \frac{ik}{4\pi}\left(1-\frac{\epsilon^-}{\epsilon^+}\right)\int_S(\nabla_{\boldsymbol{r}'}G^+(\boldsymbol{r},\boldsymbol{r}'))(\hat{\boldsymbol{n}}'\cdot\boldsymbol{E}^-(\boldsymbol{r}'))\,ds(\boldsymbol{r}')\\ &+\sigma^-Z^+\left[\frac{(ik)^2}{4\pi}\frac{\mu^-}{\mu^+}\int_{V^-}\boldsymbol{E}^-(\boldsymbol{r}')G^+(\boldsymbol{r},\boldsymbol{r}')\,dv(\boldsymbol{r}')\right.\\ &\left.-\frac{1}{4\pi}\int_S(\nabla_{\boldsymbol{r}'}G^+(\boldsymbol{r},\boldsymbol{r}'))(\hat{\boldsymbol{n}}'\cdot\boldsymbol{E}^-(\boldsymbol{r}'))\,ds(\boldsymbol{r}')\right]\end{aligned} \tag{4.61}$$

and

$$\begin{aligned}
\alpha(\boldsymbol{r})\boldsymbol{H}^{+}(\boldsymbol{r}) + (1-\alpha(\boldsymbol{r}))(\boldsymbol{H}^{-}(\boldsymbol{r}) &= \boldsymbol{H}^{i}(\boldsymbol{r}) \\
&+ \frac{(ik)^3}{4\pi}\left(1-\frac{\mu^{-}\epsilon^{-}}{\mu^{+}\epsilon^{+}}\right)\int\limits_{V^-} \boldsymbol{H}^{-}(\boldsymbol{r}')G^{+}(\boldsymbol{r},\boldsymbol{r}')\,dv(\boldsymbol{r}') \\
&- \frac{(ik)^2}{4\pi}\left(1-\frac{\epsilon^{-}}{\epsilon^{+}}\right)Y^{+}\int\limits_{S} G^{+}(\boldsymbol{r},\boldsymbol{r}')(\hat{\boldsymbol{n}}'\times\boldsymbol{E}^{-}(\boldsymbol{r}'))\,ds(\boldsymbol{r}') \\
&- \frac{ik}{4\pi}\left(1-\frac{\mu^{-}}{\mu^{+}}\right)\int\limits_{S}(\nabla_{\boldsymbol{r}'}G^{+}(\boldsymbol{r},\boldsymbol{r}'))(\hat{\boldsymbol{n}}'\cdot\boldsymbol{H}^{-}(\boldsymbol{r}'))\,ds(\boldsymbol{r}') \\
&+ \sigma^{-}\left[\frac{(ik)^2}{4\pi}\frac{\mu^{-}}{\mu^{+}}Z^{+}\int\limits_{V^-}\boldsymbol{H}^{-}(\boldsymbol{r})G^{+}(\boldsymbol{r},\boldsymbol{r}')\,dv(\boldsymbol{r}')\right. \\
&\left. - \frac{ik}{4\pi}\int\limits_{S} G^{+}(\boldsymbol{r},\boldsymbol{r}')(\hat{\boldsymbol{n}}'\times\boldsymbol{E}^{-}(\boldsymbol{r}'))\,ds(\boldsymbol{r}')\right]
\end{aligned} \tag{4.62}$$

Observe that the last term on the right-hand side of (4.61) has a lower order dependence on ik than any of the other terms in (4.61) or (4.62). This requires us to do some preliminary analysis before introducing the low frequency expansions (3.116)–(3.118) into (4.61) and (4.62).

Recall that (3.128), the low frequency expansions, together with the Maxwell equations, imply that

$$\nabla\times\boldsymbol{E}_n^{-} = n\frac{\mu^{-}}{\mu^{+}}Z^{+}\boldsymbol{H}_{n-1}^{-} \quad \text{in } V^{-} \tag{4.63}$$

and

$$\nabla\times\boldsymbol{H}_n^{-} = -n\frac{\epsilon^{-}}{\epsilon^{+}}Y^{+}\boldsymbol{E}_{n-1}^{-} + \sigma^{-}\boldsymbol{E}_n^{-} \quad \text{in } V^{-} \tag{4.64}$$

while the transmission conditions (3.124) and (3.125) imply that

$$\left.\begin{aligned}\hat{\boldsymbol{n}}\times\boldsymbol{E}_n^{+} &= \hat{\boldsymbol{n}}\times\boldsymbol{E}_n^{-}\\ \hat{\boldsymbol{n}}\times\boldsymbol{H}_n^{+} &= \hat{\boldsymbol{n}}\times\boldsymbol{H}_n^{-}\end{aligned}\right\} \quad \text{on } S \tag{4.65}$$

$$\mu^{+}\hat{\boldsymbol{n}}\cdot\boldsymbol{H}_n^{+} = \mu^{-}\hat{\boldsymbol{n}}\cdot\boldsymbol{H}_n^{-} \quad \text{on } S \tag{4.66}$$

and

$$\sigma^{-}\hat{\boldsymbol{n}}\cdot\boldsymbol{E}_n^{-} = nY^{+}\left(\frac{\epsilon^{-}}{\epsilon^{+}}\hat{\boldsymbol{n}}\cdot\boldsymbol{E}_{n-1}^{-} - \hat{\boldsymbol{n}}\cdot\boldsymbol{E}_{n-1}^{+}\right) \quad \text{on } S \tag{4.67}$$

Equation (4.67) is particularly important as it shows that when $\sigma^- > 0, \hat{\boldsymbol{n}} \cdot \boldsymbol{E}_n^-$ is fully determined by lower order terms. Moreover, for $n = 0$, (4.63) and (4.67) imply that

$$\nabla \times \boldsymbol{E}_0^- = \boldsymbol{0} \quad \text{on } V^- \tag{4.68}$$

and

$$\hat{\boldsymbol{n}} \cdot \boldsymbol{E}_0^- = 0 \quad \text{on } S \tag{4.69}$$

These equations may be used to show that $\boldsymbol{E}_0^-$ vanishes, since with (4.68) we infer that $\boldsymbol{E}_0^-$ is the gradient of a scalar function which, because of (4.69) and the divergence theorem, has vanishing divergence. Hence, the scalar function is a potential function whose normal derivative vanishes on the boundary by virtue of (4.69). This means that the function is constant and therefore its gradient vanishes. The fact that $\boldsymbol{E}_0^-$ vanishes together with (4.65) implies that $\hat{\boldsymbol{n}} \times \boldsymbol{E}_0^+$ vanishes. Thus, for $\sigma^- > 0$, the lowest order term in the electric field in V^+ is exactly the same as if S were perfectly conducting.

The fact that $\boldsymbol{E}_0^-$ vanishes allows us to obtain the following low frequency expansion

$$\int_S (\nabla_{\boldsymbol{r}'} G^+(\boldsymbol{r}, \boldsymbol{r}'))(\hat{\boldsymbol{n}}' \cdot \boldsymbol{E}^-(\boldsymbol{r}'))\, ds(\boldsymbol{r}') = \sum_{n=0}^{\infty} \frac{(ik)^n}{n!} \sum_{m=0}^{n} \binom{n}{m} \frac{1}{(n+1-m)}$$
$$\times \int_S (\nabla_{\boldsymbol{r}'} |\boldsymbol{r} - \boldsymbol{r}'|^{m-1})(\hat{\boldsymbol{n}}' \cdot \boldsymbol{E}_{n+1-m}^-(\boldsymbol{r}'))\, ds(\boldsymbol{r}') \tag{4.70}$$

With this important result we may now introduce the low frequency expansions into (4.61) and (4.62) and equate like powers of ik to obtain

$$\alpha(\boldsymbol{r})\boldsymbol{E}_n^+(\boldsymbol{r}) + (1 - \alpha(\boldsymbol{r}))\boldsymbol{E}_n^-(\boldsymbol{r}) = \boldsymbol{E}_n^i(\boldsymbol{r})$$
$$+ \frac{1}{4\pi}\left(1 - \frac{\mu^-\epsilon^-}{\mu^+\epsilon^+}\right) \sum_{m=2}^{n-1} \binom{n}{m} m(m-1) \int_{V^-} |\boldsymbol{r} - \boldsymbol{r}'|^{m-3} \boldsymbol{E}_{n-m}^-(\boldsymbol{r}')\, dv(\boldsymbol{r}')$$
$$+ \frac{1}{4\pi}\left(1 - \frac{\mu^-}{\mu^+}\right) Z^+ \sum_{m=1}^{n} \binom{n}{m} m \int_S |\boldsymbol{r} - \boldsymbol{r}'|^{m-2} (\hat{\boldsymbol{n}}' \times \boldsymbol{H}_{n-m}^-(\boldsymbol{r}'))\, ds(\boldsymbol{r}')$$
$$- \frac{1}{4\pi}\left(1 - \frac{\epsilon^-}{\epsilon^+}\right) \sum_{m=0}^{n-1} \binom{n}{m} \int_S (\nabla_{\boldsymbol{r}'} |\boldsymbol{r} - \boldsymbol{r}'|^{m-1})(\hat{\boldsymbol{n}}' \cdot \boldsymbol{E}_{n-m}^-(\boldsymbol{r}'))\, ds(\boldsymbol{r}')$$
$$+ \sigma^- Z^+ \left\{ \frac{1}{4\pi} \frac{\mu^-}{\mu^+} \sum_{m=1}^{n-1} \binom{n}{m} m \int_{V^-} |\boldsymbol{r} - \boldsymbol{r}'|^{m-2} \boldsymbol{E}_{n-m}^-(\boldsymbol{r}')\, dv(\boldsymbol{r}') \right.$$
$$\left. - \frac{1}{4\pi} \sum_{m=0}^{n-1} \binom{n}{m} \frac{1}{(n+1-m)} \int_S (\nabla_{\boldsymbol{r}'} |\boldsymbol{r} - \boldsymbol{r}'|^{m-1})(\hat{\boldsymbol{n}}' \cdot \boldsymbol{E}_{n+1-m}^-(\boldsymbol{r}'))\, ds(\boldsymbol{r}') \right\} \tag{4.71}$$

and

$$
\begin{aligned}
\alpha(\boldsymbol{r})\boldsymbol{H}_n^+(\boldsymbol{r}) + (1-\alpha(\boldsymbol{r}))\boldsymbol{H}_n^-(\boldsymbol{r}) &= \boldsymbol{H}_n^i(\boldsymbol{r}) \\
&+ \frac{1}{4\pi}\left(1-\frac{\mu^-\epsilon^-}{\mu^+\epsilon^+}\right)\sum_{m=2}^{n}\binom{n}{m} m(m-1)\int_{V^-} |\boldsymbol{r}-\boldsymbol{r}'|^{m-3}\boldsymbol{H}_{n-m}^-(\boldsymbol{r}')\,dv(\boldsymbol{r}') \\
&- \frac{1}{4\pi}\left(1-\frac{\epsilon^-}{\epsilon^+}\right)Y^+\sum_{m=1}^{n-1}\binom{n}{m} m\int_S |\boldsymbol{r}-\boldsymbol{r}'|^{m-2}(\hat{\boldsymbol{n}}'\times\boldsymbol{E}_{n-m}^-(\boldsymbol{r}'))\,ds(\boldsymbol{r}') \\
&- \frac{1}{4\pi}\left(1-\frac{\mu^-}{\mu^+}\right)\sum_{m=0}^{n}\binom{n}{m}\int_S (\nabla_{r'}|\boldsymbol{r}-\boldsymbol{r}'|^{m-1})(\hat{\boldsymbol{n}}'\cdot\boldsymbol{H}_{n-m}^-(\boldsymbol{r}'))\,ds(\boldsymbol{r}') \\
&+ \sigma^-\left\{\frac{1}{4\pi}\frac{\mu^-}{\mu^+}Z^+\sum_{m=1}^{n}\binom{n}{m} m\int_{V^-} |\boldsymbol{r}-\boldsymbol{r}'|^{m-2}\boldsymbol{H}_{n-m}^-(\boldsymbol{r}')\,dv(\boldsymbol{r}')\right. \\
&\left. - \frac{1}{4\pi}\sum_{m=0}^{n-1}\binom{n}{m}\int_S |\boldsymbol{r}-\boldsymbol{r}'|^{m-1}(\hat{\boldsymbol{n}}'\times\boldsymbol{E}_{n-m}^-(\boldsymbol{r}'))\,ds(\boldsymbol{r}')\right\}
\end{aligned}
\tag{4.72}
$$

Now the presence of loss makes its effect known by the different way in which we treat the electric and magnetic coefficients. First, we rewrite (4.71) as

$$
\alpha(\boldsymbol{r})\boldsymbol{E}_n^+(\boldsymbol{r}) + (1-\alpha(\boldsymbol{r}))\boldsymbol{E}_n^-(\boldsymbol{r}) = \boldsymbol{F}_{en}(\boldsymbol{r}) + \sigma^-\boldsymbol{G}_{en}(\boldsymbol{r}) + \nabla\phi_n(\boldsymbol{r}) \tag{4.73}
$$

where $\boldsymbol{F}_{en}$ is defined exactly as in the lossless case by (4.49) while

$$
\begin{aligned}
\boldsymbol{G}_{en}(\boldsymbol{r}) = {} & \frac{1}{4\pi}\frac{\mu^-}{\mu^+}Z^+\sum_{m=1}^{n-1}\binom{n}{m} m\int_{V^-} |\boldsymbol{r}-\boldsymbol{r}'|^{m-2}\boldsymbol{E}_{n-m}^-(\boldsymbol{r}')\,dv(\boldsymbol{r}') \\
& - \frac{1}{4\pi}Z^+\sum_{m=1}^{n}\binom{n}{m}\frac{1}{n+1-m}\int_S (\nabla_{r'}|\boldsymbol{r}-\boldsymbol{r}'|^{m-1})(\hat{\boldsymbol{n}}'\cdot\boldsymbol{E}_{n+1-m}^-(\boldsymbol{r}'))\,ds(\boldsymbol{r}')
\end{aligned}
\tag{4.74}
$$

and

$$
\begin{aligned}
\phi_n(\boldsymbol{r}) = {} & \frac{1}{4\pi}\left(1-\frac{\epsilon^-}{\epsilon^+}\right)\int_S \frac{1}{|\boldsymbol{r}-\boldsymbol{r}'|}\hat{\boldsymbol{n}}'\cdot\boldsymbol{E}_n^-(\boldsymbol{r}')\,ds(\boldsymbol{r}') \\
& + \frac{\sigma^-}{4\pi}\frac{Z^+}{(n+1)}\int_S \frac{1}{|\boldsymbol{r}-\boldsymbol{r}'|}\hat{\boldsymbol{n}}'\cdot\boldsymbol{E}_{n+1}^-(\boldsymbol{r}')\,ds(\boldsymbol{r}')
\end{aligned}
\tag{4.75}
$$

Note that $\boldsymbol{G}_{eo}$ is zero. If we assume that all coefficients up to order $(n-1)$ have been determined, then $\boldsymbol{F}_{en}$ and $\boldsymbol{G}_{en}$ may be considered as known. Note that unlike $\boldsymbol{F}_{en}$, which is continuous but not continuously differentiable in $\mathbb{R}^3$, $\boldsymbol{G}_{en}$ is continuously differentiable. Observe also that with (4.67) we could express $\hat{\boldsymbol{n}}\cdot\boldsymbol{E}_n^-$ and $\hat{\boldsymbol{n}}\cdot\boldsymbol{E}_{n+1}^-$ in terms of coefficients of lower order. However, it is just as convenient to consider ϕ_n, given by (4.75), as an unknown function which, because it is expressible as a single layer potential, is a solution of Laplace's equation in V^+ and V^- and which, by virtue of (3.129), decays as $1/r^2$ as r goes to infinity. As before, we may affix superscripts to indicate whether $\boldsymbol{r}$ is in V^+ or V^- and rewrite (4.73) as

$$\boldsymbol{E}_n^+ = \boldsymbol{F}_{en}^+ + \sigma^-\boldsymbol{G}_{en}^+ + \nabla\phi_n^+ \quad \text{in } V^+ \tag{4.76}$$

$$\boldsymbol{E}_n^- = \boldsymbol{F}_{en}^- + \sigma^-\boldsymbol{G}_{en}^- + \nabla\phi_n^- \quad \text{in } V^- \tag{4.77}$$

Unlike the lossless case, the determination of ϕ_n^+ and ϕ_n^- does not require the solution of a transmission problem but rather two boundary value problems. First, we determine ϕ_n^- as the solution to the following Neumann problem.

$$\left.\begin{aligned} \nabla^2\phi_n^- &= 0 && \text{in } V^- \\ \frac{\partial}{\partial n}\phi_n^- &= -\hat{\boldsymbol{n}}\cdot\boldsymbol{F}_{en}^- - \sigma^-\hat{\boldsymbol{n}}\cdot\boldsymbol{G}_{en}^- & & \\ &\quad + \frac{n}{\sigma^-}Y^+\left(\frac{\epsilon^-}{\epsilon^+}\hat{\boldsymbol{n}}\cdot\boldsymbol{E}_{n-1}^- - \hat{\boldsymbol{n}}\cdot\boldsymbol{E}_{n-1}^+\right) && \text{on } S \end{aligned}\right\} \tag{4.78}$$

As it stands, problem (4.78) is not well posed unless the compatibility relation

$$\int_S \left[\hat{\boldsymbol{n}}\cdot\boldsymbol{F}_{en}^-(\boldsymbol{r}) + \sigma^-\hat{\boldsymbol{n}}\cdot\boldsymbol{G}_{en}^-(\boldsymbol{r}) - \frac{n}{\sigma^-}Y^+\left(\frac{\epsilon^-}{\epsilon^+}\hat{\boldsymbol{n}}\cdot\boldsymbol{E}_{n-1}^-(\boldsymbol{r}) - \hat{\boldsymbol{n}}\cdot\boldsymbol{E}_{n-1}^+(\boldsymbol{r})\right)\right] ds(\boldsymbol{r}) = 0 \tag{4.79}$$

is satisfied. Using (3.129) which ensures that integrals of $\hat{\boldsymbol{n}}\cdot\boldsymbol{E}_m^-$ over S vanish for all m, this condition becomes

$$\int_S [\hat{\boldsymbol{n}}\cdot\boldsymbol{F}_{en}^-(\boldsymbol{r}) + \sigma^-\hat{\boldsymbol{n}}\cdot\boldsymbol{G}_{en}^-(\boldsymbol{r})]\, ds(\boldsymbol{r}) = 0 \tag{4.80}$$

To see that this holds we take the normal components of (4.77), integrate over S and invoke (3.129) to obtain

$$\int_S \left[\hat{\boldsymbol{n}}\cdot\boldsymbol{F}_{en}^-(\boldsymbol{r}) + \sigma^-\hat{\boldsymbol{n}}\cdot\boldsymbol{G}_{en}^-(\boldsymbol{r}) + \frac{\partial}{\partial n}\phi_n^-(\boldsymbol{r})\right] ds(\boldsymbol{r}) = 0 \tag{4.81}$$

Since ϕ_n^- satisfies Laplace's equation in V^-, Gauss' theorem ensures that

$$\int_S \frac{\partial}{\partial n}\phi_n^-(\boldsymbol{r})\, ds(\boldsymbol{r}) = \int_{V^-} \nabla^2\phi_n^-(\boldsymbol{r})\, dr(\boldsymbol{r}) = 0 \tag{4.82}$$

Hence, condition (4.80) is fulfilled and problem (4.78) is guaranteed to have a solution, although not a unique solution since any constant may be added. However, $\boldsymbol{E}_n^-$ will be uniquely determined.

Once this problem is solved we may use the solution as data for the determination of ϕ_n^+. That is, using the continuity of the tangential components of $\boldsymbol{E}_n$ on S we may formulate the following exterior Dirichlet problem for ϕ_n^+

$$\left.\begin{aligned} \nabla^2\phi_n^+ &= 0 && \text{in } V^+ \\ \phi_n^+ &= \phi_n^- + c && \text{on } S \\ \phi_n^+ &= O\left(\frac{1}{r^2}\right) && \text{as } r \to \infty \end{aligned}\right\} \tag{4.83}$$

where the boundary condition in (4.83) follows from the continuity of $\boldsymbol{F}_{en}$ and $\boldsymbol{G}_{en}$ on S so that

$$\hat{\boldsymbol{n}} \times \boldsymbol{E}^+ - \hat{\boldsymbol{n}} \times \boldsymbol{E}^- = \hat{\boldsymbol{n}} \times \nabla(\phi_n^+ - \phi_n^-) = \mathbf{0} \quad \text{on } S \tag{4.84}$$

The constant c is determined by requiring the rapid decay of ϕ_n^+ as r approaches infinity, which is equivalent to requiring that

$$\int_S \frac{\partial}{\partial n}\phi_n^+(\boldsymbol{r})\,ds(\boldsymbol{r}) = 0 \tag{4.85}$$

This completes the determination of $\boldsymbol{E}_n^\pm$ and we have seen that we had to solve the interior Neumann problem (4.78) and the exterior Dirichlet problem (4.83) in sequence. Having solved for the nth order electric field coefficients we now determine the magnetic field coefficients. Rewriting (4.72) as

$$\alpha(\boldsymbol{r})\boldsymbol{H}_n^+(\boldsymbol{r}) + (1 - \alpha(\boldsymbol{r}))\boldsymbol{H}_n^-(\boldsymbol{r}) = \boldsymbol{F}_{mn}(\boldsymbol{r}) + \sigma^-\boldsymbol{G}_{mn}(\boldsymbol{r}) + \nabla\psi_n(\boldsymbol{r}) \tag{4.86}$$

where $\boldsymbol{F}_{mn}$ and ψ_n are defined exactly as in the lossless case by (4.51) and (4.52), respectively, and

$$\begin{aligned} \boldsymbol{G}_{mn}(\boldsymbol{r}) = {}& \frac{1}{4\pi}\frac{\mu^-}{\mu^+}Z^+\sum_{m=1}^{n}\binom{n}{m} m \int_{V^-} |\boldsymbol{r} - \boldsymbol{r}'|^{m-2}\boldsymbol{H}_{n-m}^-(\boldsymbol{r}')\,dv(\boldsymbol{r}') \\ & - \frac{1}{4\pi}\sum_{m=0}^{n-1}\binom{n}{m}\int_S |\boldsymbol{r} - \boldsymbol{r}'|^{m-1}(\hat{\boldsymbol{n}}' \times \boldsymbol{E}_{n-m}^-(\boldsymbol{r}'))\,ds(\boldsymbol{r}') \end{aligned} \tag{4.87}$$

observe that $\boldsymbol{G}_{mn}$ contains a term involving $\hat{\boldsymbol{n}}' \times \boldsymbol{E}_n^-$ but this is now known. Observe also that $\boldsymbol{G}_{mn}$, like $\boldsymbol{F}_{mn}$, is continuous in $\mathbb{R}^3$ but not continuously differentiable across S. To determine the potential function ψ_n, and hence the coefficients $\boldsymbol{H}_n^+$ and $\boldsymbol{H}_n^-$, we rewrite the relation (4.86), affixing superscripts to indicate whether $\boldsymbol{r}$ is in V^+ or V^-, as

$$\boldsymbol{H}_n^+ = \boldsymbol{F}_{mn}^+ + \sigma^-\boldsymbol{G}_{mn}^+ + \nabla\psi_n^+ \quad \text{in } V^+ \tag{4.88}$$

and

$$\boldsymbol{H}_n^- = \boldsymbol{F}_{mn}^- + \sigma^- \boldsymbol{G}_{mn}^- + \nabla \psi_n^- \quad \text{in } V^- \tag{4.89}$$

and use the transmission conditions (4.65) and (4.66) to pose the following transmission potential problem for ψ_n^+ and ψ_n^-

$$\left.\begin{aligned} \nabla^2 \psi_n^+ &= 0 && \text{in } V^+ \\ \nabla^2 \psi_n^- &= 0 && \text{in } V^- \\ \hat{\boldsymbol{n}} \times \nabla(\psi_n^+ - \psi_n^-) &= \boldsymbol{0} && \text{on } S \\ \mu^+ \frac{\partial}{\partial n} \psi_n^+ - \mu^- \frac{\partial}{\partial n} \psi_n^- &= -(\mu^+ - \mu^-) \hat{\boldsymbol{n}} \cdot (\boldsymbol{F}_{mn} + \sigma^- \boldsymbol{G}_{mn}) && \text{on } S \\ \psi_n^+ &= O\left(\frac{1}{r^2}\right) && \text{as } r \to \infty \end{aligned}\right\} \tag{4.90}$$

The superscripts on $\boldsymbol{F}_{mn}$ and $\boldsymbol{G}_{mn}$ have been omitted in (4.90) because these functions are continuous on S. If the permeabilities are equal in V^+ and V^-, then ψ_n^+ and ψ_n^- vanish and the coefficients $\boldsymbol{H}_n^+$ and $\boldsymbol{H}_n^-$ are determined from (4.88) and (4.89) without the necessity of solving any potential problems.

The expressions for the scattering amplitudes and cross sections (3.217)–(3.223) are all still valid with $\boldsymbol{E}_n^{\pm}$ and $\boldsymbol{H}_n^{\pm}$ determined as described above.

4.EM.2 *Problems in electromagnetics*

1. Derive equations (4.37) and (4.38).
2. Use formulae (2.62) and (4.39) to establish the identity (4.40).
3. Use expressions (4.40) together with (4.35) and (4.36) and the Maxwell equations to obtain (4.41) and (4.42).
4. Prove that the transmission problems (4.57) and (4.58) are well posed.
5. Prove that the asymptotic relation $\phi_n^+ = O(1/r^2)$, as $r \to \infty$, in (4.83), is equivalent to the condition (4.85).
6. Show that when the magnetic permeabilities in V^+ and V^- are equal the only solution of (4.90) vanishes identically.

4.E Elasticity

4.E.1 *The volume integral formulation*

As in the cases of acoustics and electromagnetics, it is possible to derive an alternative form for the low frequency expansion involving volume integrals. However, because of the appearance of the Lamé parameters in the Navier and traction operators, we cannot obtain a different formulation of the problems of determining the low frequency coefficients but this development does produce a more effective form for the scattering amplitude and cross sections. Following the work of Kiriaki and Polyzos (1988) we introduce the transmission conditions

$$\left.\begin{aligned} \boldsymbol{u}^+(\boldsymbol{r}) &= \boldsymbol{u}^-(\boldsymbol{r}) \\ \boldsymbol{T}^+(\partial_{\boldsymbol{r}}, \hat{\boldsymbol{n}})\boldsymbol{u}^+(\boldsymbol{r}) &= \boldsymbol{T}^-(\partial_{\boldsymbol{r}}, \hat{\boldsymbol{n}})\boldsymbol{u}^-(\boldsymbol{r}) \end{aligned}\right\} \quad \boldsymbol{r} \in S \tag{4.91}$$

into the integral representation (2.146) and obtain, for $\boldsymbol{r} \in V^{+}$

$$\begin{aligned}
\boldsymbol{u}^{+}(\boldsymbol{r}) &= \boldsymbol{u}^{i}(\boldsymbol{r}) + \frac{1}{4\pi}\int_{S}[\boldsymbol{u}^{-}(\boldsymbol{r}')\cdot \boldsymbol{T}^{+}(\partial_{\boldsymbol{r}'},\hat{\boldsymbol{n}}')\,\widetilde{\boldsymbol{\Gamma}}^{+}(\boldsymbol{r},\boldsymbol{r}') \\
&\quad - \widetilde{\boldsymbol{\Gamma}}^{+}(\boldsymbol{r},\boldsymbol{r}')\cdot \boldsymbol{T}^{-}(\partial_{\boldsymbol{r}'},\hat{\boldsymbol{n}}')\boldsymbol{u}^{-}(\boldsymbol{r}')]\,ds(\boldsymbol{r}') \\
&= \boldsymbol{u}^{i}(\boldsymbol{r}) + \frac{1}{4\pi}\int_{S}\boldsymbol{u}^{-}(\boldsymbol{r}')\cdot[\boldsymbol{T}^{+}(\partial_{\boldsymbol{r}'},\hat{\boldsymbol{n}}') - \boldsymbol{T}^{-}(\partial\boldsymbol{r}',\hat{\boldsymbol{n}}')]\,\widetilde{\boldsymbol{\Gamma}}^{+}(\boldsymbol{r},\boldsymbol{r}')\,ds(\boldsymbol{r}') \\
&\quad + \frac{1}{4\pi}\int_{S}[\boldsymbol{u}^{-}(\boldsymbol{r}')\cdot \boldsymbol{T}^{-}(\partial_{\boldsymbol{r}'},\hat{\boldsymbol{n}}')\,\widetilde{\boldsymbol{\Gamma}}^{+}(\boldsymbol{r},\boldsymbol{r}') \\
&\quad - \widetilde{\boldsymbol{\Gamma}}^{+}(\boldsymbol{r},\boldsymbol{r}')\cdot \boldsymbol{T}^{-}(\partial_{\boldsymbol{r}'},\hat{\boldsymbol{n}}')\boldsymbol{u}^{-}(\boldsymbol{r}')]\,ds(\boldsymbol{r}')
\end{aligned} \tag{4.92}$$

Then the Betti formula (2.142) is employed, identifying V with V^{-}, $\boldsymbol{u}$ with $\boldsymbol{u}^{-}$ and $\boldsymbol{v}$ with the columns of $\widetilde{\boldsymbol{\Gamma}}^{+}(\boldsymbol{r},\boldsymbol{r}')$, choosing $\boldsymbol{r} \in V^{+}$ and $\boldsymbol{r}' \in V^{-}$ and employing the relation

$$\Delta^{e-} = \frac{\rho^{+}}{\rho^{-}}\frac{\mu^{-}}{\mu^{+}}\Delta^{e+} - \frac{1}{\rho^{-}}\frac{\lambda^{+}\mu^{-} - \lambda^{-}\mu^{+}}{\mu^{+}}\nabla(\nabla\cdot) \tag{4.93}$$

to rewrite (4.92) as

$$\begin{aligned}
\boldsymbol{u}^{+}(\boldsymbol{r}) &= \boldsymbol{u}^{i}(\boldsymbol{r}) + \frac{\rho^{-}\omega^{2}}{4\pi}\left(1 - \frac{\rho^{+}}{\rho^{-}}\frac{\mu^{-}}{\mu^{+}}\right)\int_{V^{-}}\boldsymbol{u}^{-}(\boldsymbol{r}')\cdot\,\widetilde{\boldsymbol{\Gamma}}^{+}(\boldsymbol{r},\boldsymbol{r}')\,dv(\boldsymbol{r}') \\
&\quad + \frac{1}{4\pi}\frac{(\lambda^{-}\mu^{+} - \lambda^{+}\mu^{-})}{\mu^{+}}\int_{V^{-}}\boldsymbol{u}^{-}(\boldsymbol{r}')\cdot\nabla_{\boldsymbol{r}'}\nabla_{\boldsymbol{r}'}\cdot\,\widetilde{\boldsymbol{\Gamma}}^{+}(\boldsymbol{r},\boldsymbol{r}')\,dv(\boldsymbol{r}') \\
&\quad + \frac{1}{4\pi}\int_{S}\boldsymbol{u}^{-}(\boldsymbol{r}')\cdot[\boldsymbol{T}^{+}(\partial_{\boldsymbol{r}'},\hat{\boldsymbol{n}}') - \boldsymbol{T}^{-}(\partial_{\boldsymbol{r}'},\hat{\boldsymbol{n}}')]\,\widetilde{\boldsymbol{\Gamma}}^{+}(\boldsymbol{r},\boldsymbol{r}')\,ds(\boldsymbol{r}')
\end{aligned} \tag{4.94}$$

An alternative form of (4.94) may be obtained in terms of the indices of refraction (1.218) and (1.219) in which case

$$\Delta^{e-} = \frac{1}{\eta_{s}^{2}}\Delta^{e+} + (c_{p}^{+})^{2}\left(\frac{1}{\eta_{p}^{2}} - \frac{1}{\eta_{s}^{2}}\right)\nabla(\nabla\cdot) \tag{4.95}$$

and

$$\begin{aligned}
\boldsymbol{u}^{+}(\boldsymbol{r}) &= \boldsymbol{u}^{i}(\boldsymbol{r}) + \frac{(k_{s}^{-})^{2}\mu^{-}}{4\pi}\frac{(\eta_{s}^{2} - 1)}{\eta_{s}^{2}}\int_{V^{-}}\boldsymbol{u}^{-}(\boldsymbol{r}')\cdot\,\widetilde{\boldsymbol{\Gamma}}^{+}(\boldsymbol{r},\boldsymbol{r}')\,dv(\boldsymbol{r}') \\
&\quad + \frac{(\lambda^{-} + 2\mu^{-})}{4\pi}\frac{(\eta_{s}^{2} - \eta_{p}^{2})}{\eta_{s}^{2}}\int_{V^{-}}\boldsymbol{u}^{-}(\boldsymbol{r}')\cdot\nabla_{\boldsymbol{r}'}\nabla_{\boldsymbol{r}'}\cdot\,\widetilde{\boldsymbol{\Gamma}}^{+}(\boldsymbol{r},\boldsymbol{r}')\,dv(\boldsymbol{r}') \\
&\quad + \frac{1}{4\pi}\int_{S}\boldsymbol{u}^{-}(\boldsymbol{r}')\cdot[\boldsymbol{T}^{+}(\partial_{\boldsymbol{r}'},\hat{\boldsymbol{n}}') - \boldsymbol{T}^{-}(\partial_{\boldsymbol{r}'},\hat{\boldsymbol{n}}')]\,\widetilde{\boldsymbol{\Gamma}}^{+}(\boldsymbol{r},\boldsymbol{r}')\,ds(\boldsymbol{r}')
\end{aligned} \tag{4.96}$$

which holds for $\boldsymbol{r}$ in V^+. The alternative form of the integral representation of $\boldsymbol{u}^+(\boldsymbol{r})$ in V^+ (4.94) leads to alternative forms of the scattering amplitudes. When the asymptotic expressions for $\widetilde{\boldsymbol{\Gamma}}^+$, (2.149) and (2.150), together with

$$\nabla_{\boldsymbol{r}'}\nabla_{\boldsymbol{r}'} \cdot \widetilde{\boldsymbol{\Gamma}}^+(\boldsymbol{r},\boldsymbol{r}') = \frac{-ik_p^3}{\lambda^+ + 2\mu^+} e^{-ik_p\hat{\boldsymbol{r}}\cdot\boldsymbol{r}'} h(k_p r)\hat{\boldsymbol{r}}\hat{\boldsymbol{r}} + O\left(\frac{1}{r^2}\right) \quad \text{as } r \to \infty \tag{4.97}$$

and

$$\begin{aligned}
&(\boldsymbol{T}^+(\partial_{\boldsymbol{r}'},\hat{\boldsymbol{n}}') - \boldsymbol{T}^-(\partial_{\boldsymbol{r}'},\hat{\boldsymbol{n}}'))\,\widetilde{\boldsymbol{\Gamma}}^+(\boldsymbol{r},\boldsymbol{r}') \\
&= k_p^2 e^{-ik_p\hat{\boldsymbol{r}}\cdot\boldsymbol{r}'}\hat{\boldsymbol{n}}' \cdot \frac{[(\lambda^+ - \lambda^-)\tilde{\boldsymbol{I}} + 2(\mu^+ - \mu^-)\hat{\boldsymbol{r}}\hat{\boldsymbol{r}}]\hat{\boldsymbol{r}}}{\lambda^+ + 2\mu^+} h(k_p r) \\
&\quad + k_s^2 \frac{(\mu^+ - \mu^-)}{\mu^+} e^{-ik_s\hat{\boldsymbol{r}}\cdot\boldsymbol{r}'}\hat{\boldsymbol{n}}' \cdot [\tilde{\boldsymbol{I}} \times \tilde{\boldsymbol{I}} \times \hat{\boldsymbol{r}} + 2\hat{\boldsymbol{r}}\tilde{\boldsymbol{I}}] \cdot (\tilde{\boldsymbol{I}} - \hat{\boldsymbol{r}}\hat{\boldsymbol{r}}) h(k_s r) \\
&\quad + O\left(\frac{1}{r^2}\right) \quad \text{as } r \to \infty
\end{aligned} \tag{4.98}$$

are substituted into the integral representation (4.94), we again obtain the far field form (2.154). However, now the scattering amplitudes are given by

$$\begin{aligned}
g_r(\hat{\boldsymbol{r}}) &= k_p^2 \widetilde{\boldsymbol{H}}_p^+ : \frac{(\lambda^+ - \lambda^-)\tilde{\boldsymbol{I}} + 2(\mu^+ - \mu^-)\hat{\boldsymbol{r}}\hat{\boldsymbol{r}}}{\lambda^+ + 2\mu^+} \\
&\quad + ik_p^3 \frac{\rho^-}{\rho^+} \frac{(\eta_p^2 - 1)}{\eta_p^2} \boldsymbol{L}_p^- \cdot \hat{\boldsymbol{r}}
\end{aligned} \tag{4.99}$$

$$\begin{aligned}
g_\theta(\hat{\boldsymbol{r}}) &= k_s^2 \frac{(\mu^+ - \mu^-)}{\mu^+} \widetilde{\boldsymbol{H}}_s^+ : [\tilde{\boldsymbol{I}} \times \tilde{\boldsymbol{I}} \times \hat{\boldsymbol{r}} + 2\hat{\boldsymbol{r}}\tilde{\boldsymbol{I}}] \cdot \hat{\boldsymbol{\theta}} \\
&\quad + ik_s^3 \frac{\rho^-}{\rho^+} \frac{(\eta_s^2 - 1)}{\eta_s^2} \boldsymbol{L}_s^- \cdot \hat{\boldsymbol{\theta}}
\end{aligned} \tag{4.100}$$

$$\begin{aligned}
g_\varphi(\hat{\boldsymbol{r}}) &= k_s^2 \frac{(\mu^+ - \mu^-)}{\mu^+} \widetilde{\boldsymbol{H}}_s^+ : [\tilde{\boldsymbol{I}} \times \tilde{\boldsymbol{I}} \times \hat{\boldsymbol{r}} + 2\hat{\boldsymbol{r}}\tilde{\boldsymbol{I}}] \cdot \hat{\boldsymbol{\varphi}} \\
&\quad + ik_s^3 \frac{\rho^-}{\rho^+} \frac{(\eta_s^2 - 1)}{\eta_s^2} \boldsymbol{L}_s^- \cdot \hat{\boldsymbol{\varphi}}
\end{aligned} \tag{4.101}$$

where $\widetilde{\boldsymbol{H}}_p^-$ and $\widetilde{\boldsymbol{H}}_s^+$ are given by (2.158) and (2.159), while $\boldsymbol{L}_p^-$ and $\boldsymbol{L}_s^-$ are

$$\boldsymbol{L}_p^- = \frac{1}{4\pi} \int_{V^-} \boldsymbol{u}^-(\boldsymbol{r}') e^{-ik_p\hat{\boldsymbol{r}}\cdot\boldsymbol{r}'} \, dv(\boldsymbol{r}') \tag{4.102}$$

and

$$\boldsymbol{L}_s^- = \frac{1}{4\pi} \int_{V^-} \boldsymbol{u}^-(\boldsymbol{r}') e^{-ik_s\hat{\boldsymbol{r}}\cdot\boldsymbol{r}'} \, dv(\boldsymbol{r}') \tag{4.103}$$

Observe that the volume integral term in g_r is not present if $\eta_p^2 = 1$ and is absent in g_θ and g_φ if $\eta_s^2 = 1$. Expressions (4.99)–(4.101) for the scattering amplitudes are valid for all frequencies. Introducing the low frequency expansions, (3.229) and (3.231), together with the expansion of the exponential and reordering terms, we obtain alternative expressions for the quantities $g_r(\hat{\boldsymbol{r}})$, (3.264), and $\boldsymbol{g}_t(\hat{\boldsymbol{r}})$, (3.266), which are given by

$$
\begin{aligned}
g_r(\hat{\boldsymbol{r}}) = {} & \frac{1}{4\pi}\sum_{n=0}^{\infty}\frac{(ik_p)^{n+2}}{n!}\sum_{m=0}^{n}\binom{n}{m}(-1)^{m+1} \\
& \times\left[\frac{(\lambda^+-\lambda^-)\tilde{\boldsymbol{I}}+2(\mu^+-\mu^-)\hat{\boldsymbol{r}}\hat{\boldsymbol{r}}}{\lambda^++2\mu^+}\right]:\int_S \boldsymbol{u}^+_{n-m}(\boldsymbol{r}')\hat{\boldsymbol{n}}'(\hat{\boldsymbol{r}}\cdot\boldsymbol{r}')^m\,ds(\boldsymbol{r}') \\
& +\frac{(\eta_p^2-1)}{4\pi\eta_p^2}\frac{\rho^-}{\rho^+}\sum_{n=0}^{\infty}\frac{(ik_p)^{n+3}}{n!}\sum_{m=0}^{n}\binom{n}{m}(-1)^{m+1} \\
& \times\hat{\boldsymbol{r}}\cdot\int_{V^-}\boldsymbol{u}^-_{n-m}(\boldsymbol{r}')(\hat{\boldsymbol{r}}\cdot\boldsymbol{r}')^m\,dv(\boldsymbol{r}')
\end{aligned}
\tag{4.104}
$$

and

$$
\begin{aligned}
\boldsymbol{g}_t(\boldsymbol{r}) = {} & \frac{1}{4\pi\tau^+}\frac{(\mu^+-\mu^-)}{\mu^+}\sum_{n=0}^{\infty}\frac{(ik_p)^{n+2}}{n!}\sum_{m=0}^{n}\binom{n}{m}\left(-\frac{1}{\tau^+}\right)^{m+1} \\
& \times\int_S \boldsymbol{u}^+_{n-m}(\boldsymbol{r}')\hat{\boldsymbol{n}}'(\hat{\boldsymbol{r}}\cdot\boldsymbol{r}')^m\,ds(\boldsymbol{r}'):[\tilde{\boldsymbol{I}}\times\tilde{\boldsymbol{I}}\times\hat{\boldsymbol{r}}+2\hat{\boldsymbol{r}}\tilde{\boldsymbol{I}}] \\
& +\frac{(\eta_s^2-1)}{4\pi(\eta_s\tau^+)^2}\frac{\rho^-}{\rho^+}\sum_{n=0}^{\infty}\frac{(ik_p)^{n+3}}{n!}\sum_{m=0}^{n}\binom{n}{m}\left(-\frac{1}{\tau^+}\right)^{m+1} \\
& \times\int_{V^-}\boldsymbol{u}^-_{n-m}(\boldsymbol{r}')(\hat{\boldsymbol{r}}\cdot\boldsymbol{r}')^m\,dv(\boldsymbol{r}')
\end{aligned}
\tag{4.105}
$$

These expressions in turn generate alternative forms of the nonvanishing coefficients in (3.325)–(3.327) in the expansions (3.298)–(3.303), which for plane wave incidence are

$$
\begin{aligned}
{}^{\cdot}A_{r(n+2)} = {} & \frac{1}{4\pi n!(\lambda^++2\mu^+)}\sum_{m=0}^{n}\binom{n}{m}(-1)^m\int_S[(\lambda^+-\lambda^-)(\hat{\boldsymbol{n}}'\cdot\boldsymbol{u}^+_{n-m}(\boldsymbol{r}')) \\
& +2(\mu^+-\mu^-)(\hat{\boldsymbol{r}}\cdot\boldsymbol{u}^+_{n-m}(\boldsymbol{r}'))(\hat{\boldsymbol{r}}\cdot\hat{\boldsymbol{n}}')](\hat{\boldsymbol{r}}\cdot\boldsymbol{r}')^m\,ds(\boldsymbol{r}') \\
& +\frac{(\eta_p^2-1)}{4\pi\eta_p^2 n!}\frac{\rho^-}{\rho^+}\sum_{m=1}^{n}m\binom{n}{m}(-1)^{m+1}\hat{\boldsymbol{r}}\cdot\int_{V^-}\boldsymbol{u}^-_{n-m}(\boldsymbol{r}')(\hat{\boldsymbol{r}}\cdot\boldsymbol{r}')^{m-1}\,dv(\boldsymbol{r}')
\end{aligned}
\tag{4.106}
$$

$$
\begin{aligned}
A_{\theta(n+2)} &= \frac{(\mu^+ - \mu^-)}{4\pi\tau^+\mu^+ n!}\sum_{m=0}^{n}\binom{n}{m}\left(-\frac{1}{\tau^+}\right)^{m+1} \\
&\times \int_S [(\hat{\boldsymbol{\theta}}\cdot\boldsymbol{u}^+_{n-m}(\boldsymbol{r}'))(\hat{\boldsymbol{n}}'\cdot\hat{\boldsymbol{r}}) + (\hat{\boldsymbol{\theta}}\cdot\hat{\boldsymbol{n}}')(\hat{\boldsymbol{r}}\cdot\boldsymbol{u}^+_{n-m}(\boldsymbol{r}'))](\hat{\boldsymbol{r}}\cdot\boldsymbol{r}')^m\,ds(\boldsymbol{r}') \\
&+ \frac{(\eta_s^2 - 1)}{4\pi(\eta_s\tau^+)^2 n!}\frac{\rho^-}{\rho^+}\sum_{m=1}^{n} m\binom{n}{m}\left(-\frac{1}{\tau^+}\right)^m \hat{\boldsymbol{\theta}}\cdot\int_{V^-}\boldsymbol{u}^-_{n-m}(\boldsymbol{r}')(\hat{\boldsymbol{r}}\cdot\boldsymbol{r}')^{m-1}\,dv(\boldsymbol{r}')
\end{aligned}
\tag{4.107}
$$

$$
\begin{aligned}
A_{\varphi(n+2)} &= \frac{(\mu^+ - \mu^-)}{4\pi\tau^+\mu^+ n!}\sum_{m=0}^{n}\binom{n}{m}\left(-\frac{1}{\tau^+}\right)^{m+1} \\
&\times \int_S [(\hat{\boldsymbol{\varphi}}\cdot\boldsymbol{u}^+_{n-m}(\boldsymbol{r}'))(\hat{\boldsymbol{n}}'\cdot\hat{\boldsymbol{r}}) + (\hat{\boldsymbol{\varphi}}\cdot\hat{\boldsymbol{n}}')(\hat{\boldsymbol{r}}\cdot\boldsymbol{u}^+_{n-m}(\boldsymbol{r}'))](\hat{\boldsymbol{r}}\cdot\boldsymbol{r}')^m\,ds(\boldsymbol{r}') \\
&+ \frac{(\eta_s^2 - 1)}{4\pi(\eta_s\tau^+)^2 n!}\frac{\rho^-}{\rho^+}\sum_{m=1}^{n} m\binom{n}{m}\left(-\frac{1}{\tau^+}\right)^m \hat{\boldsymbol{\varphi}}\cdot\int_{V^-}\boldsymbol{u}^-_{n-m}(\boldsymbol{r}')(\hat{\boldsymbol{r}}\cdot\boldsymbol{r}')^{m-1}\,dv(\boldsymbol{r}')
\end{aligned}
\tag{4.108}
$$

for $n \geq 1$. In the expressions (4.106)–(4.108) it is assumed that every coefficient $u_n^{\pm}$ with $n < 0$ vanishes. The far field coefficients (4.106)–(4.108) can also be expressed in terms of the surface moment integrals

$$
\begin{aligned}
\widetilde{\boldsymbol{M}}_n^m(\hat{\boldsymbol{r}}) &= \int_S \hat{\boldsymbol{n}}'\boldsymbol{u}^+_{n-m}(\boldsymbol{r}')(\hat{\boldsymbol{r}}\cdot\boldsymbol{r}')^m\,ds(\boldsymbol{r}') \\
&= \int_{V^-} \nabla_{r'}[\boldsymbol{r}^-_{n-m}(\boldsymbol{r}')(\hat{\boldsymbol{r}}\cdot\boldsymbol{r}')^m]\,dv(\boldsymbol{r}')
\end{aligned}
\tag{4.109}
$$

and the volume moment integrals

$$
\boldsymbol{N}_n^m(\hat{\boldsymbol{r}}) = \int_{V^-}\boldsymbol{u}^-_{n-m}(\boldsymbol{r}')(\hat{\boldsymbol{r}}\cdot\boldsymbol{r}')^m\,dv(\boldsymbol{r}')
\tag{4.110}
$$

as follows

$$
\begin{aligned}
A_{r(n+2)} &= \frac{1}{4\pi n!}\frac{(\lambda^+ - \lambda^-)\,\tilde{\boldsymbol{I}} + 2(\mu^+ - \mu^-)\hat{\boldsymbol{r}}\hat{\boldsymbol{r}}}{\lambda^+ + 2\mu^+} : \sum_{m=0}^{n}\binom{n}{m}(-1)^m\,\widetilde{\boldsymbol{M}}_n^m(\hat{\boldsymbol{r}}) \\
&+ \frac{1}{4\pi n!}\frac{\eta_p^2 - 1}{\eta_p^2}\frac{\rho^-}{\rho^+}\hat{\boldsymbol{r}}\cdot\sum_{m=1}^{n}\binom{n}{m} m(-1)^{m+1}\boldsymbol{N}_n^{m-1}(\hat{\boldsymbol{r}})
\end{aligned}
\tag{4.111}
$$

$$A_{\theta(n+2)} = \frac{1}{4\pi n!}\frac{\mu^+ - \mu^-}{\tau^+\mu^+}(\hat{r}\hat{\theta} + \hat{\theta}\hat{r}) : \sum_{m=0}^{n}\binom{n}{m}\left(-\frac{1}{\tau^+}\right)^{m+1}\widetilde{M}_n^m(\hat{r})$$
$$+\frac{1}{4\pi n!}\frac{\eta_s^2 - 1}{(\eta_s\tau^+)^2}\frac{\rho^-}{\rho^+}\hat{\theta}\cdot\sum_{m=1}^{n}\binom{n}{m} m\left(-\frac{1}{\tau^+}\right)^{m} N_n^{m-1}(\hat{r}) \quad (4.112)$$

$$A_{\varphi(n+2)} = \frac{1}{4\pi n!}\frac{\mu^+ - \mu^-}{\tau^+\mu^+}(\hat{r}\hat{\varphi} + \hat{\varphi}\hat{r}) : \sum_{m=0}^{n}\binom{n}{m}\left(-\frac{1}{\tau^+}\right)^{m+1}\widetilde{M}_n^m(\hat{r})$$
$$+\frac{1}{4\pi n!}\frac{\eta_s^2 - 1}{(\eta_s\tau^+)^2}\frac{\rho^-}{\rho^+}\hat{\varphi}\cdot\sum_{m=1}^{n}\binom{n}{m} m\left(-\frac{1}{\tau^+}\right)^{m+1} N_n^{m-1}(\hat{r}) \quad (4.113)$$

The advantage in using this volume integral form of the scattering amplitudes lies in the fact that knowledge of the low frequency coefficients up to order n is sufficient to determine the low frequency expansion of the scattering amplitudes up to and including the term involving k_p^{n+2} as can be seen from (4.104) and (4.105). On the other hand, to determine the same term using the surface integral representation for the scattering amplitudes, (3.264)–(3.266), it is necessary to know the low frequency coefficients up to order $n+1$. Put another way, using the volume integral form allows the determination of an 'extra' term in the low frequency expansion of the scattering amplitudes.

4.E.2 *Problems in elasticity*

1. Derive the representation (4.94).
2. Prove that if the index of refractions for the longitudinal and the transverse waves coincide, then the represenation (4.96) assumes the following form

$$u^+(r) = u^i(r) + \frac{\rho^-\omega^2}{4\pi}\left(\frac{\mu^+}{\mu^-} - \frac{\rho^+}{\rho^-}\right)\int_{V^-} u^-(r')\cdot\widetilde{\Gamma}^+(r,r')\,dv(r')$$
$$+\frac{1}{4\pi}\left(\frac{\mu^+}{\mu^-} - 1\right)\int_S \widetilde{\Gamma}^+(r,r')\cdot T^+(\partial_{r'},\hat{n}')u^+(r')\,ds(r') \quad (4.114)$$

3. Derive the asymptotic forms (4.97) and (4.98).
4. Show that the scattering amplititudes given by (4.104) and (4.105) can also be written in the form (3.298)–(3.303) where the coefficients of the low frequency expansions are given by (4.106)–(4.108) and (4.111)–(4.113).

5

RAYLEIGH SCATTERING

In this chapter we particularize the general results of Chapters 3 and 4 to the first nontrivial terms in the low frequency expansions known as **Rayleigh approximations** or **Rayleigh terms** of both the near and the far fields. Also included are the leading terms in the scattering cross sections. These results are given for general shapes in terms of the potential functions to which they have been reduced in Chapters 3 and 4. For acoustics and electromagnetics, most of the material in this chapter refers to results obtained in the 1960s and the 1970s by a number of researchers such as D. S. Jones, R. E. Kleinman, T. B. A. Senior, V. Twersky, J. Van Bladel and many others. The corresponding material in elasticity was developed in the 1980s by G. Dassios, K. Kiriaki, D. Polyzos and others.

5.0 Basic potential problems

Since we have reduced our attention to the solution of potential problems it is useful to record some facts about potential functions and introduce some particularly relevant ones.

A **harmonic** or **potential function** (Kellogg 1953), u, is said to be regular in V^- if it satisfies Laplace's equation everywhere in V^- and regular in V^+ if it satisfies Laplace's equation everywhere in V^+ and it decays as r^{-1} as $r \to \infty$. Corresponding to the integral representations (2.7) and (2.5), if $u^\pm$ are regular potential functions in $V^\pm$, then for all $\boldsymbol{r} \in \mathbb{R}^3$

$$\alpha(\boldsymbol{r})u^+(\boldsymbol{r}) = \frac{1}{4\pi}\int_S \left[u^+(\boldsymbol{r}')\frac{\partial}{\partial n'}\frac{1}{|\boldsymbol{r}-\boldsymbol{r}'|} - \frac{1}{|\boldsymbol{r}-\boldsymbol{r}'|}\frac{\partial}{\partial n'}u^+(\boldsymbol{r}')\right] ds(\boldsymbol{r}') \tag{5.1}$$

and

$$(\alpha(\boldsymbol{r})-1)u^-(\boldsymbol{r}) = \frac{1}{4\pi}\int_S \left[u^-(\boldsymbol{r}')\frac{\partial}{\partial n'}\frac{1}{|\boldsymbol{r}-\boldsymbol{r}'|} - \frac{1}{|\boldsymbol{r}-\boldsymbol{r}'|}\frac{\partial}{\partial n'}u^-(\boldsymbol{r}')\right] ds(\boldsymbol{r}') \tag{5.2}$$

where $\alpha(\boldsymbol{r})$ is defined in (2.3).

When the expansion

$$\frac{1}{|\boldsymbol{r}-\boldsymbol{r}'|} = \sum_{n=0}^{\infty}\frac{r'^n}{r^{n+1}}P_n(\hat{\boldsymbol{r}}\cdot\hat{\boldsymbol{r}}'), \quad r > r' \tag{5.3}$$

is employed in the integral representation (5.1) we find that

$$u^+(\boldsymbol{r}) = \sum_{n=0}^{\infty} \frac{Y_n(\hat{\boldsymbol{r}})}{r^{n+1}}, \qquad r > \max_{\boldsymbol{r}' \in S} r' \tag{5.4}$$

where P_n are the Legendre polynomials and $Y_n(\hat{\boldsymbol{r}})$ are the spherical harmonics defined by

$$Y_n(\hat{\boldsymbol{r}}) = \frac{1}{4\pi} \int_S \left[u^+(\boldsymbol{r}') \frac{\partial}{\partial n'} (r'^n P_n(\hat{\boldsymbol{r}} \cdot \hat{\boldsymbol{r}}')) - r'^n P_n(\hat{\boldsymbol{r}} \cdot \hat{\boldsymbol{r}}') \frac{\partial}{\partial n'} u^+(\boldsymbol{r}') \right] ds(\boldsymbol{r}') \tag{5.5}$$

In particular

$$Y_0(\hat{\boldsymbol{r}}) = -\frac{1}{4\pi} \int_S \frac{\partial}{\partial n'} u^+(\boldsymbol{r}')\, ds(\boldsymbol{r}') \tag{5.6}$$

and

$$Y_1(\hat{\boldsymbol{r}}) = \frac{1}{4\pi} \int_S \left[u^+(\boldsymbol{r}')(\hat{\boldsymbol{n}}' \cdot \hat{\boldsymbol{r}}) - (\hat{\boldsymbol{r}} \cdot \boldsymbol{r}') \frac{\partial}{\partial n'} u_n^+(\boldsymbol{r}') \right] ds(\boldsymbol{r}') \tag{5.7}$$

With these general observations in mind we now define some standard potential functions. The most well-known such function is the **conductor potential** (Stratton 1941), $\phi^c(\boldsymbol{r})$, defined as the solution of the following potential problem for the surface S

$$\left. \begin{aligned} \nabla^2 \phi^c &= 0 && \text{in } \boldsymbol{r} \in V^+ \\ \phi^c &= 1 && \text{on } S \\ \phi^c &= O\left(\frac{1}{r}\right) && \text{as } r \to \infty \end{aligned} \right\} \tag{5.8}$$

In terms of ϕ^c the **capacity** (Pólya and Szegö 1951), C, of S is defined as

$$C = -\frac{1}{4\pi} \int_S \frac{\partial}{\partial n} \phi^c(\boldsymbol{r})\, ds(\boldsymbol{r}) \tag{5.9}$$

With (5.4) and (5.6) it is seen that C appears in the monopole term in the far field expansion, i.e.

$$\phi^c(\boldsymbol{r}) = \frac{C}{r} + O\left(\frac{1}{r^2}\right) \qquad \text{as } r \to \infty \tag{5.10}$$

thus we may also characterize C as

$$C = \lim_{r \to \infty} r \phi^c(\boldsymbol{r}) \tag{5.11}$$

A corresponding function for the Neumann problem, denoted by $\psi^c(\boldsymbol{r})$ is defined as the solution of

$$\left.\begin{aligned} \nabla^2\psi^c &= 0 && \text{in } V^+ \\ \frac{\partial}{\partial n}\psi^c &= 1 && \text{on } S \\ \psi^c &= O\left(\frac{1}{r}\right) && \text{as } r \to \infty \end{aligned}\right\} \tag{5.12}$$

Next, we define the vector valued potential $\boldsymbol{\Phi}(\boldsymbol{r}) = (\phi_1, \phi_2, \phi_3)$ as the solution of

$$\left.\begin{aligned} \nabla^2\boldsymbol{\Phi} &= \boldsymbol{0} && \text{in } V^+ \\ \boldsymbol{\Phi} &= \boldsymbol{r} + \boldsymbol{c} && \text{on } S \\ \boldsymbol{\Phi} &= O\left(\frac{1}{r^2}\right) && \text{as } r \to \infty \end{aligned}\right\} \tag{5.13}$$

where $\boldsymbol{c} = (c_1, c_2, c_3)$ is a constant vector. To guarantee the specified growth at infinity, $\boldsymbol{c}$ must be chosen so that

$$\int_S \frac{\partial \boldsymbol{\Phi}}{\partial n}(\boldsymbol{r})\, ds(\boldsymbol{r}) = \boldsymbol{0} \tag{5.14}$$

In terms of the conductor potential this implies that

$$\begin{aligned} \boldsymbol{0} &= \int_S \phi^c(\boldsymbol{r}) \frac{\partial}{\partial n}\boldsymbol{\Phi}(\boldsymbol{r})\, ds(\boldsymbol{r}) = \int_S (\boldsymbol{r} + \boldsymbol{c}) \frac{\partial}{\partial n}\phi^c(\boldsymbol{r})\, ds(\boldsymbol{r}) \\ &= -4\pi C\boldsymbol{c} + \int_S \boldsymbol{r} \frac{\partial}{\partial n}\phi^c(\boldsymbol{r})\, ds(\boldsymbol{r}) \end{aligned} \tag{5.15}$$

where C is the capacity, hence

$$\boldsymbol{c} = \frac{1}{4\pi C} \int_S \boldsymbol{r} \frac{\partial}{\partial n}\phi^c(\boldsymbol{r})\, ds(\boldsymbol{r}) \tag{5.16}$$

The **polarization tensor** for an isolated conductor has been defined in terms of $\boldsymbol{\Phi}$ (Schiffer and Szegö 1949), as

$$\widetilde{\boldsymbol{Q}} = -\int_S \boldsymbol{r} \frac{\partial}{\partial n}\boldsymbol{\Phi}(\boldsymbol{r})\, ds(\boldsymbol{r}) \tag{5.17}$$

Closely related is the **electric polarizability tensor** (Keller *et al.* 1972)

$$\widetilde{\boldsymbol{P}} = \widetilde{\boldsymbol{Q}} + |V^-|\widetilde{\boldsymbol{I}} \tag{5.18}$$

where $|V^-|$ is the volume of the domain V^-. This polarizability tensor appears in the leading (dipole) term in the far field expansion as

$$\boldsymbol{\Phi}(\boldsymbol{r}) = \frac{1}{4\pi}\frac{\widetilde{\boldsymbol{P}}\cdot\hat{\boldsymbol{r}}}{r^2} + O\left(\frac{1}{r^3}\right) \quad \text{as } r \to \infty \tag{5.19}$$

A corresponding vector field satisfying a basic vector Neumann condition is $\boldsymbol{\Psi}(\boldsymbol{r}) = (\psi_1(\boldsymbol{r}), \psi_2(\boldsymbol{r}), \psi_3(\boldsymbol{r}))$ defined to be the solution of

$$\left.\begin{aligned} \nabla^2\boldsymbol{\Psi} &= \boldsymbol{0} && \text{in } V^+ \\ \frac{\partial}{\partial n}\boldsymbol{\Psi} &= \hat{\boldsymbol{n}} && \text{on } S \\ \boldsymbol{\Psi} &= O\left(\frac{1}{r^2}\right) && \text{as } r \to \infty \end{aligned}\right\} \tag{5.20}$$

This function is used to define the **virtual mass tensor** $\widetilde{\boldsymbol{W}}$ as

$$\widetilde{\boldsymbol{W}} = -\int_S \hat{\boldsymbol{n}}\boldsymbol{\Psi}(\boldsymbol{r})\, ds(\boldsymbol{r}) \tag{5.21}$$

This tensor was used to describe the irrotational flow of an incompressible inviscid fluid past a rigid surface S (Taylor 1928; Schiffer and Szegö 1949). Closely related is the **magnetic polarizability tensor** (Keller *et al.* 1972) $\widetilde{\boldsymbol{M}}$ defined as

$$\widetilde{\boldsymbol{M}} = \widetilde{\boldsymbol{W}} + |V^-|\widetilde{\boldsymbol{I}} \tag{5.22}$$

Using (5.1) and (5.4)–(5.6) and the definition of $\widetilde{\boldsymbol{W}}$ and $\widetilde{\boldsymbol{M}}$ it follows that

$$\boldsymbol{\Psi}(\boldsymbol{r}) = -\frac{1}{4\pi}\frac{\widetilde{\boldsymbol{M}}\cdot\hat{\boldsymbol{r}}}{r^2} + O\left(\frac{1}{r^3}\right) \quad \text{as } r \to \infty \tag{5.23}$$

The tensors $\widetilde{\boldsymbol{Q}}$, $\widetilde{\boldsymbol{P}}$, $\widetilde{\boldsymbol{W}}$ and $\widetilde{\boldsymbol{M}}$ are all symmetric as can be seen from the alternative definitions of the elements

$$Q_{ij} = -\int_S x_i \frac{\partial}{\partial n}\phi_j(\boldsymbol{r})\, ds(\boldsymbol{r}) = \int_{V^+} \nabla\phi_i \cdot \nabla\phi_j \, dv \tag{5.24}$$

$$W_{ij} = -\int_S n_i \frac{\partial}{\partial n}\psi_j(\boldsymbol{r})\, ds(\boldsymbol{r}) = \int_{V^+} \nabla\psi_i \cdot \nabla\psi_j \, dv \tag{5.25}$$

Moreover, the elements of these tensors have been shown to satisfy a number of interesting inequalities (Keller *et al.* 1972; Kleinman and Senior 1972, 1986 and the references cited therein). When the surface is axially symmetric and the axis of symmetry coincides with the x_3 axis, the tensors are not only symmetric but diagonal. Moreover, of the three nonzero elements in each tensor, two are equal in each case, i.e. $Q_{11} = Q_{22}$ and $M_{11} = M_{22}$ and, remarkably, if the surface is simply connected, then $2M_{33} = P_{11}$. Thus, only three independent tensor elements need be found to completely determine all four of the tensors for axially symmetric scatterers. More detail and a discussion of isoperimetric inequalities involving the tensor elements are found in the references cited above.

For the transmission problem, the vector field associated with the general polarizability tensor is defined to be the solution of the following transmission problem

$$\left.\begin{aligned}
\nabla^2 \boldsymbol{v}^+ &= \mathbf{0} && \text{in } V^+ \\
\nabla^2 \boldsymbol{v}^- &= \mathbf{0} && \text{in } V^- \\
\boldsymbol{v}^+ &= \boldsymbol{v}^- + \boldsymbol{r} && \text{on } S \\
\frac{\partial}{\partial n}\boldsymbol{v}^- &= \beta \frac{\partial}{\partial n}\boldsymbol{v}^- + \hat{\boldsymbol{n}} && \text{on } S \\
\boldsymbol{v}^+ &= O\left(\frac{1}{r^2}\right) && \text{as } r \to \infty
\end{aligned}\right\} \tag{5.26}$$

Since the functions $\boldsymbol{v}^\pm$ depend on β we write $\boldsymbol{v}^\pm = \boldsymbol{v}^\pm(\boldsymbol{r}; \beta)$. The **general polarizability tensor** is defined as

$$\widetilde{X}(\beta) = (1 - \beta) \int_S \hat{\boldsymbol{n}} \boldsymbol{v}^-(\boldsymbol{r})\, ds(\boldsymbol{r}) \tag{5.27}$$

and the transmission conditions may be used to derive alternative forms (Kleinman and Senior 1986; Senior 1973). This tensor appears in the leading (dipole) term in the far field

$$\boldsymbol{v}^+(\boldsymbol{r}; \beta) = \frac{1}{4\pi} \frac{\widetilde{X}(\beta) \cdot \hat{\boldsymbol{r}}}{r^2} + O\left(\frac{1}{r^3}\right) \quad \text{as } r \to \infty \tag{5.28}$$

The vector field $\boldsymbol{v}^\pm$ may be decomposed into two fields, each embodying one of the inhomogeneous transmission conditions as follows

$$\boldsymbol{v}^+(\boldsymbol{r}; \beta) = \boldsymbol{\Phi}^+(\boldsymbol{r}, \beta) + \boldsymbol{\Psi}^+(\boldsymbol{r}, \beta), \quad \boldsymbol{r} \in V^+ \tag{5.29}$$

$$\boldsymbol{v}^-(\boldsymbol{r}; \beta) = \boldsymbol{\Phi}^-(\boldsymbol{r}, \beta) + \boldsymbol{\Psi}^-(\boldsymbol{r}, \beta), \quad \boldsymbol{r} \in V^- \tag{5.30}$$

where

$$\left.\begin{aligned}
\nabla^2 \boldsymbol{\Phi}^+ &= \mathbf{0} && \text{in } V^+ \\
\nabla^2 \boldsymbol{\Phi}^- &= \mathbf{0} && \text{in } V^- \\
\boldsymbol{\Phi}^+ &= \boldsymbol{\Phi}^- + \boldsymbol{r} && \text{on } S \\
\frac{\partial}{\partial n}\boldsymbol{\Phi}^+ &= \beta \frac{\partial}{\partial n}\boldsymbol{\Phi}^- && \text{on } S \\
\boldsymbol{\Phi}^+ &= O\left(\frac{1}{r^2}\right) && \text{as } r \to \infty
\end{aligned}\right\} \tag{5.31}$$

and

$$\left.\begin{aligned} \nabla^2 \boldsymbol{\Psi}^+ &= \mathbf{0} && \text{in } V^+ \\ \nabla^2 \boldsymbol{\Psi}^- &= \mathbf{0} && \text{in } V^- \\ \boldsymbol{\Psi}^+ &= \boldsymbol{\Psi}^- && \text{on } S \\ \frac{\partial}{\partial n}\boldsymbol{\Psi}^+ &= \beta\frac{\partial}{\partial n}\boldsymbol{\Psi}^- + \hat{\boldsymbol{n}} && \text{on } S \\ \boldsymbol{\Psi}^+ &= O\left(\frac{1}{r^2}\right) && \text{as } r \to \infty \end{aligned}\right\} \tag{5.32}$$

Remarkably, the general polarizability tensor may be expressed entirely in terms of $\boldsymbol{\Phi}^+$ or $\boldsymbol{\Psi}^+$ as

$$\begin{aligned} \widetilde{\boldsymbol{X}}(\beta) &= (1-\beta)^2 \int_S \hat{\boldsymbol{n}}\boldsymbol{\Psi}^+(\boldsymbol{r};\beta)\,ds(\boldsymbol{r}) - (1-\beta)|V^-|\widetilde{\boldsymbol{I}} \\ &= -\left(\frac{1-\beta}{\beta}\right)^2 \int_S \boldsymbol{r}\frac{\partial}{\partial n}\boldsymbol{\Phi}^+(\boldsymbol{r};\beta)\,ds(\boldsymbol{r}) - \frac{1-\beta}{\beta}|V^-|\widetilde{\boldsymbol{I}} \end{aligned} \tag{5.33}$$

To derive these expressions use the identities

$$\int_S \hat{\boldsymbol{n}}\boldsymbol{r}\,ds(\boldsymbol{r}) = |V^-|\widetilde{\boldsymbol{I}} \tag{5.34}$$

$$\int_S \hat{\boldsymbol{n}}\boldsymbol{\Phi}^+(\boldsymbol{r};\beta)ds = -\beta\int_S \hat{\boldsymbol{n}}\boldsymbol{\Psi}^+(\boldsymbol{r};\beta)\,ds(\boldsymbol{r}) \tag{5.35}$$

and

$$\int_S \boldsymbol{r}\frac{\partial}{\partial n}\boldsymbol{\Psi}^+(\boldsymbol{r};\beta)\,ds(\boldsymbol{r}) = -\frac{1}{\beta}\int_S \boldsymbol{r}\frac{\partial}{\partial n}\boldsymbol{\Phi}^+(\boldsymbol{r};\beta)\,ds(\boldsymbol{r}) \tag{5.36}$$

The first identity is a direct result of Gauss' theorem while the second and third follow from repeated use of the transmission conditions, the fact that $\hat{\boldsymbol{n}}$ is the normal derivative of $\boldsymbol{r}$, $\boldsymbol{r}$ is a regular potential function in V^- and Green's second identity.

These alternative forms of the general polarizability tensor have the advantage of involving the solution of a simpler transmission problem (only one inhomogeneous transmission condition) than the original expressions (5.27) and (5.30). An additional advantage comes from the facts that

$$\lim_{\beta\to 0} \boldsymbol{\Psi}^+(\boldsymbol{r};\beta) = \boldsymbol{\Psi}(r) \tag{5.37}$$

and

$$\lim_{\beta\to\infty} \boldsymbol{\Phi}^+(\boldsymbol{r};\beta) = \boldsymbol{\Phi}(\boldsymbol{r}) \tag{5.38}$$

where $\boldsymbol{\Psi}$ and $\boldsymbol{\Phi}$ are the vector fields associated with the virtual mass and polarization tensors, (5.21) and (5.17), respectively. These limits are relatively easy to establish heuristically but a rigorous proof is not so simple. However, if we accept them as valid it follows by letting $\beta \to 0$ in (5.33) that

$$\lim_{\beta\to 0} \widetilde{\boldsymbol{X}}(\beta) = \int_S \hat{\boldsymbol{n}}\boldsymbol{\Psi}(\boldsymbol{r})\, ds(\boldsymbol{r}) - |V^-|\widetilde{\boldsymbol{I}} = -\widetilde{\boldsymbol{W}} - |V^-|\widetilde{\boldsymbol{I}} = -\widetilde{\boldsymbol{M}} \tag{5.39}$$

and letting $\beta \to \infty$ in (5.33) that

$$\begin{aligned}\lim_{\beta\to\infty} \widetilde{\boldsymbol{X}}(\beta) &= -\int_S \boldsymbol{r}\frac{\partial}{\partial n}\boldsymbol{\Phi}(\boldsymbol{r})\, ds(\boldsymbol{r}) + |V^-|\widetilde{\boldsymbol{I}} \\ &= \widetilde{\boldsymbol{Q}} + |V^-|\widetilde{\boldsymbol{I}} = \widetilde{\boldsymbol{P}}\end{aligned} \tag{5.40}$$

where $\widetilde{\boldsymbol{P}}$ and $\widetilde{\boldsymbol{M}}$ are the electric and magnetic polarizability tensors described in (5.18) and (5.22), respectively.

5.A Acoustics

5.A.1 *The Dirichlet problem*

From the general results (3.18) of Chapter 3, we know that each term in the low frequency expansion of the total excess pressure field in V^+ may be written as

$$u_n^+(\boldsymbol{r}) = f_n^+(\boldsymbol{r}) + v_n^+(\boldsymbol{r}) \tag{5.41}$$

where v_n^+ is a solution of the exterior potential problem posed in (3.37). Here we discuss the first few terms in this expansion. From (3.35) we find that

$$f_0^+(\boldsymbol{r}) = u_0^i(\boldsymbol{r}) \tag{5.42}$$

$$f_1^+(\boldsymbol{r}) = u_1^i(\boldsymbol{r}) - \frac{1}{4\pi}\int_S \frac{\partial}{\partial n'}u_0^+(\boldsymbol{r}')\, ds(\boldsymbol{r}') \tag{5.43}$$

$$\begin{aligned}f_2^+(\boldsymbol{r}) = u_2^i(\boldsymbol{r}) &- \frac{1}{4\pi}\int_S |\boldsymbol{r}-\boldsymbol{r}'|\frac{\partial}{\partial n'}u_0^+(\boldsymbol{r}')\, ds(\boldsymbol{r}') \\ &- \frac{1}{2\pi}\int_S \frac{\partial}{\partial n'}u_1^+(\boldsymbol{r}')\, ds(\boldsymbol{r}')\end{aligned} \tag{5.44}$$

The functions defined in the previous section may be used to simplify the expressions for the f_n^+ for $n = 0, 1, 2$. For both plane waves and point sources in V^+ it is easy to prove that

$$\int_S \frac{\partial}{\partial n'}u_0^i(\boldsymbol{r}')\, ds(\boldsymbol{r}') = \int_S \frac{\partial}{\partial n'}u_1^i(\boldsymbol{r}')\, ds(\boldsymbol{r}') = 0 \tag{5.45}$$

Using these facts together with the conductor potential and Green's theorem we find that

$$\begin{aligned}\int_S \frac{\partial}{\partial n'} u_0^+(\boldsymbol{r}')\, ds(\boldsymbol{r}') &= \int_S \frac{\partial}{\partial n'} v_0^+(\boldsymbol{r}')\, ds(\boldsymbol{r}') = \int_S \frac{\partial}{\partial n'} v_0^+(\boldsymbol{r}')\phi^c(\boldsymbol{r}')\, ds(\boldsymbol{r}') \\ &= \int_S v_0^+(\boldsymbol{r}') \frac{\partial}{\partial n'} \phi^c(\boldsymbol{r}') ds(\boldsymbol{r}') \\ &= -\int_S u_0^i(\boldsymbol{r}') \frac{\partial}{\partial n'} \phi^c(\boldsymbol{r}')\, ds(\boldsymbol{r}') \end{aligned} \tag{5.46}$$

For plane wave incidence, $u_0^i = 1$, in which case (5.9) implies that

$$\int_S \frac{\partial}{\partial n'} u_0^+(\boldsymbol{r}')\, ds(\boldsymbol{r}') = -\int_S \frac{\partial}{\partial n'} \phi^c(\boldsymbol{r}')\, ds(\boldsymbol{r}') = 4\pi C \tag{5.47}$$

whereas for a point source at $\boldsymbol{r}_p \in V^+$, $u_0^i = 1/|\boldsymbol{r} - \boldsymbol{r}_p|$ and the integral representation of ϕ^c, (5.1), shows that for $\boldsymbol{r}_p \in V^+$

$$\begin{aligned}\int_S \frac{\partial}{\partial n'} u_0^+(\boldsymbol{r}')\, ds(\boldsymbol{r}') &= -\int_S \frac{1}{|\boldsymbol{r}' - \boldsymbol{r}_p|} \frac{\partial}{\partial n'} \phi^c(\boldsymbol{r}')\, ds(\boldsymbol{r}') \\ &= -\int_S \phi^c(\boldsymbol{r}') \frac{\partial}{\partial n'} \frac{1}{|\boldsymbol{r}' - \boldsymbol{r}_p|} ds(\boldsymbol{r}') + 4\pi\phi^c(\boldsymbol{r}_p) \\ &= 4\pi\phi^c(\boldsymbol{r}_p) \end{aligned} \tag{5.48}$$

where the integral term vanished because $\phi^c = 1$ on S and $1/|\boldsymbol{r}' - \boldsymbol{r}_p|$ is a regular potential function in V^- for $\boldsymbol{r}_p$ in V^+.

For plane wave incidence in the direction $\hat{\boldsymbol{k}}$ it follows first of all that $v_0^+ = -\phi^c$ so that

$$u_0^+(\boldsymbol{r}) = 1 - \phi^c(\boldsymbol{r}) \tag{5.49}$$

and

$$f_0^+(\boldsymbol{r}) = 1 \tag{5.50}$$

$$f_1^+(\boldsymbol{r}) = \hat{\boldsymbol{k}} \cdot \boldsymbol{r} - C \tag{5.51}$$

Since v_1^+ is the potential function which takes on the value $C - \hat{\boldsymbol{k}} \cdot \boldsymbol{r}$ on S it is straightforward to verify that

$$u_1^+(\boldsymbol{r}) = C(\phi^c(\boldsymbol{r}) - 1) + \hat{\boldsymbol{k}} \cdot (\boldsymbol{r} + \boldsymbol{c}\phi^c(\boldsymbol{r}) - \boldsymbol{\Phi}(\boldsymbol{r})) \tag{5.52}$$

where the vector potential $\boldsymbol{\Phi}(\boldsymbol{r})$ and the constant vector $\boldsymbol{c}$ were defined in (5.13) and (5.16), respectively. With this result we see that

$$\int_S \frac{\partial}{\partial n'} u_1^+(\boldsymbol{r}')\, ds(\boldsymbol{r}')$$

$$= \int_S \left[C\frac{\partial}{\partial n'}\phi^c(\boldsymbol{r}') + \hat{\boldsymbol{n}}'\cdot\hat{\boldsymbol{k}} + \hat{\boldsymbol{k}}\cdot\boldsymbol{c}\frac{\partial}{\partial n'}\phi^c(\boldsymbol{r}') - \hat{\boldsymbol{k}}\cdot\frac{\partial}{\partial n'}\boldsymbol{\Phi}(\boldsymbol{r}')\right] ds(\boldsymbol{r}')$$

$$= -4\pi C^2 - 4\pi C\hat{\boldsymbol{k}}\cdot\boldsymbol{c} \tag{5.53}$$

where we have used the definition of capacity and equation (5.14). Thus, for plane wave incidence we find that

$$f_2^+(\boldsymbol{r}) = (\hat{\boldsymbol{k}}\cdot\boldsymbol{r})^2 + \frac{1}{4\pi}\int_S |\boldsymbol{r}-\boldsymbol{r}'|\frac{\partial}{\partial n'}\phi^c(\boldsymbol{r}')\, ds(\boldsymbol{r}') + 2C^2 + 2C\hat{\boldsymbol{k}}\cdot\boldsymbol{c} \tag{5.54}$$

The scattering amplitude as given in (3.39) takes the form

$$g(\hat{\boldsymbol{r}}) = ikA_1 + k^2A_2 + ik^3A_3 + O(k^4) \tag{5.55}$$

where A_1, A_2 and A_3 are defined in (3.40) and (3.41) as

$$A_1 = -\frac{1}{4\pi}\int_S \frac{\partial}{\partial n'}u_0^+(\boldsymbol{r}')\, ds(\boldsymbol{r}') \tag{5.56}$$

$$A_2 = \frac{1}{4\pi}\int_S \frac{\partial}{\partial n'}u_1^+(\boldsymbol{r}')\, ds(\boldsymbol{r}') - \frac{1}{4\pi}\int_S (\hat{\boldsymbol{r}}\cdot\boldsymbol{r}')\frac{\partial}{\partial n'}u_0^+(\boldsymbol{r}')\, ds(\boldsymbol{r}') \tag{5.57}$$

and

$$A_3 = \frac{1}{8\pi}\int_S \left[\frac{\partial}{\partial n'}u_2^+(\boldsymbol{r}') - 2(\hat{\boldsymbol{r}}\cdot\boldsymbol{r}')\frac{\partial}{\partial n'}u_1^+(\boldsymbol{r}') + (\hat{\boldsymbol{r}}\cdot\boldsymbol{r}')^2\frac{\partial}{\partial n'}u_0^+(\boldsymbol{r}')\right] ds(\boldsymbol{r}') \tag{5.58}$$

From the general expression for the differential scattering cross section (3.46) we see that for plane wave incidence in the direction $\hat{\boldsymbol{k}}$

$$\sigma(\hat{\boldsymbol{r}};\hat{\boldsymbol{k}}) = 4\pi A_1^2 + 4\pi k^2\left[A_2^2 + 2A_3A_1\right] + O(k^4) \tag{5.59}$$

while the scattering cross section is

$$\sigma_s = \frac{1}{4\pi}\int_{S^2}\sigma(\hat{\boldsymbol{r}},\hat{\boldsymbol{k}})\, ds(\hat{\boldsymbol{r}}) = \int_{S^2} A_1^2(\hat{\boldsymbol{r}},\hat{\boldsymbol{k}})\, ds(\hat{\boldsymbol{r}})$$

$$+ k^2\int_{S^2}[A_2^2(\hat{\boldsymbol{r}},\hat{\boldsymbol{k}}) + 2A_1(\hat{\boldsymbol{r}},\hat{\boldsymbol{k}})A_3(\hat{\boldsymbol{r}},\hat{\boldsymbol{k}})ds(\hat{\boldsymbol{r}}) + O(k^4) \tag{5.60}$$

In this case, the absorption cross section vanishes as was noted in the comment preceding equation (3.47); that is, $\sigma_a = 0$ hence the extinction cross section σ_e (defined in (2.43)) is equal to σ_s.

The coefficients A_i may be expressed in terms of the potentials ϕ^c and ϕ_i as follows. Consider first A_1 given in (5.56). For plane wave incidence we find, using (5.47), that

$$A_1 = -C \tag{5.61}$$

whereas for a point source at $\boldsymbol{r}_p \in V^+$, using (5.48)

$$A_1 = -\phi^c(\boldsymbol{r}_p) \tag{5.62}$$

Restricting attention to plane wave incidence we present expressions for the next two coefficients. Observe that the first integral in the expression for A_2, (5.57), has already been evaluated in (5.53). Using this result together with the expression (5.49) for u_0^+ we find that

$$A_2 = -C^2 - C\hat{\boldsymbol{k}} \cdot \boldsymbol{c} + \frac{1}{4\pi} \int_S (\hat{\boldsymbol{r}} \cdot \boldsymbol{r}') \frac{\partial}{\partial n'} \phi^c(\boldsymbol{r}')\, ds(\boldsymbol{r}') \tag{5.63}$$

Alternative forms are found, using the definition of $\boldsymbol{c}$, (5.16), to be

$$A_2 = -C^2 + \frac{1}{4\pi}(\hat{\boldsymbol{r}} - \hat{\boldsymbol{k}}) \cdot \int_S \boldsymbol{r}' \frac{\partial}{\partial n'} \phi^c(\boldsymbol{r}')\, ds(\boldsymbol{r}') \tag{5.64}$$

or

$$A_2 = -C^2 + C(\hat{\boldsymbol{r}} - \hat{\boldsymbol{k}}) \cdot \boldsymbol{c} \tag{5.65}$$

Note that the first two terms in the scattering amplitude, A_1 and A_2, have been expressed in terms of the solution of one scalar potential problem, the conductor potential. From an examination of the definition of A_3 in (5.58), it would appear that in addition to u_0^+ and u_1^+, which we have already found, it is necessary to also find u_2^+ in order to completely specify A_3. However, this is not necessary as was shown by Van Bladel (1968a), see also Jones (1979a,b). The simplification, which in fact applies if u_2^+ is replaced by u_n^+ thus permitting an 'extra' term in the scattering amplitude to be determined (that is, if u_n are known for $n = 0, 1, \ldots, N$, then A_n may be determined for $n = 1, 2, \ldots, N+2$), requires some work which we outline here. Denoting the three integrals in the definition of A_3 (5.58) by $I_i, i = 1, 2, 3$, we use the decomposition (5.41) to write

$$\begin{aligned} I_1 &= \frac{1}{8\pi} \int_S \frac{\partial}{\partial n'} u_2^+(\boldsymbol{r}')\, ds(\boldsymbol{r}') \\ &= \frac{1}{8\pi} \int_S \frac{\partial}{\partial n'} f_2^+(\boldsymbol{r}')\, ds(\boldsymbol{r}') + \frac{1}{8\pi} \int_S \frac{\partial}{\partial n'} v_2^+(\boldsymbol{r}')\, ds(\boldsymbol{r}') \end{aligned} \tag{5.66}$$

First, we use the expression for f_2^+, (5.54) and Gauss' theorem to show that

$$\int_S \frac{\partial}{\partial n'} f_2^+(\boldsymbol{r}')\, ds(\boldsymbol{r}')$$

$$= \int_S \frac{\partial}{\partial n'}\left[(\hat{\boldsymbol{k}}\cdot\boldsymbol{r}')^2 + \frac{1}{4\pi}\int_S |\boldsymbol{r}'-\boldsymbol{r}''|\frac{\partial}{\partial n''}\phi^c(\boldsymbol{r}'')\, ds(\boldsymbol{r}'')\right] ds(\boldsymbol{r}')$$

$$= \int_{V^-} \nabla^2_{\boldsymbol{r}'}\left[(\hat{\boldsymbol{k}}\cdot\boldsymbol{r}')^2 + \frac{1}{4\pi}\int_S |\boldsymbol{r}'-\boldsymbol{r}''|\frac{\partial}{\partial n''}\phi^c(\boldsymbol{r}'')\, ds(\boldsymbol{r}'')\right] dv(\boldsymbol{r}') \tag{5.67}$$

But

$$\nabla^2_{\boldsymbol{r}'}(\hat{\boldsymbol{k}}\cdot\boldsymbol{r}')^2 = 2 \tag{5.68}$$

and

$$\nabla^2_{\boldsymbol{r}'}|\boldsymbol{r}'-\boldsymbol{r}''| = \frac{2}{|\boldsymbol{r}'-\boldsymbol{r}''|} \quad \text{for } \boldsymbol{r}' \in V^-,\ \boldsymbol{r}'' \in S \tag{5.69}$$

thus

$$\int_S \frac{\partial}{\partial n'} f_2^+(\boldsymbol{r}')\, ds(\boldsymbol{r}')$$

$$= 2\int_{V^-}\left[1 + \frac{1}{4\pi}\int_S \frac{1}{|\boldsymbol{r}'-\boldsymbol{r}''|}\frac{\partial}{\partial n''}\phi^c(\boldsymbol{r}'')\, ds(\boldsymbol{r}'')\right] dv(\boldsymbol{r}') \tag{5.70}$$

Now using Green's identity in V^+, the boundary values for ϕ^c and Gauss' integral in V^- we find that

$$\begin{aligned}\int_S \frac{\partial}{\partial n'} f_2^+(\boldsymbol{r}')\, ds(\boldsymbol{r}') &= 2\int_{V^-}\left[1 + \frac{1}{4\pi}\int_S \phi^c(\boldsymbol{r}'')\frac{\partial}{\partial n''}\frac{1}{|\boldsymbol{r}'-\boldsymbol{r}''|} ds(\boldsymbol{r}'')\right] dv(\boldsymbol{r}') \\ &= 2\int_{V^-}\left[1 + \frac{1}{4\pi}\int_S \frac{\partial}{\partial n''}\frac{1}{|\boldsymbol{r}'-\boldsymbol{r}''|} ds(\boldsymbol{r}'')\right] dv(\boldsymbol{r}') \\ &= 0\end{aligned} \tag{5.71}$$

Hence

$$I_1 = \frac{1}{8\pi}\int_S \frac{\partial}{\partial n'} v_2^+(\boldsymbol{r}')\, ds(\boldsymbol{r}') \tag{5.72}$$

Now we again employ the boundary conditions satisfied by ϕ^c and v_2^+ and Green's theorem in V^+ to rewrite this as

$$\begin{aligned} I_1 &= \frac{1}{8\pi}\int\limits_S \phi^c(\boldsymbol{r}')\frac{\partial}{\partial n'}v_2^+(\boldsymbol{r}')\,ds(\boldsymbol{r}') \\ &= \frac{1}{8\pi}\int\limits_S v_2^+(\boldsymbol{r}')\frac{\partial}{\partial n'}\phi^c(\boldsymbol{r}')\,ds(\boldsymbol{r}') \\ &= -\frac{1}{8\pi}\int\limits_S f_2^+(\boldsymbol{r}')\frac{\partial}{\partial n'}\phi^c(\boldsymbol{r}')\,ds(\boldsymbol{r}') \end{aligned} \tag{5.73}$$

or with the expression (5.54) for $f_2^+(\boldsymbol{r}')$

$$\begin{aligned} I_1 = -\frac{1}{8\pi}\int\limits_S \Bigg[&(\hat{\boldsymbol{k}}\cdot\hat{\boldsymbol{r}}')^2 + \frac{1}{4\pi}\int\limits_S |\boldsymbol{r}'-\boldsymbol{r}''|\frac{\partial}{\partial n''}\phi^c(\boldsymbol{r}'')\,ds(\boldsymbol{r}'') \\ &+ 2C^2 + 2C\hat{\boldsymbol{k}}\cdot\boldsymbol{c}\Bigg]\frac{\partial}{\partial n'}\phi^c(\boldsymbol{r}')\,ds(\boldsymbol{r}') \end{aligned} \tag{5.74}$$

Finally, using the definition of capacity we obtain

$$\begin{aligned} I_1 = &-\frac{1}{8\pi}\int\limits_S (\hat{\boldsymbol{k}}\cdot\hat{\boldsymbol{r}}')^2\frac{\partial}{\partial n'}\phi^c(\boldsymbol{r}')\,ds(\boldsymbol{r}') \\ &- \frac{1}{2(4\pi)^2}\int\limits_S\int\limits_S |\boldsymbol{r}'-\boldsymbol{r}''|\left(\frac{\partial}{\partial n'}\phi^c(\boldsymbol{r}')\right)\frac{\partial}{\partial n''}\phi^c(\boldsymbol{r}'')\,ds(\boldsymbol{r}')\,ds(\boldsymbol{r}'') \\ &+ C^3 + C^2\hat{\boldsymbol{k}}\cdot\boldsymbol{c} \end{aligned} \tag{5.75}$$

Thus, I_1 is expressed entirely in terms of the conductor potential. The third integral in the definition of A_3, (5.58), is easily seen, with the expression for u_0^+, (5.49), to be

$$I_3 = \frac{1}{8\pi}\int\limits_S (\hat{\boldsymbol{r}}\cdot\boldsymbol{r}')^2\frac{\partial}{\partial n'}u_0^+(\boldsymbol{r}')\,ds(\boldsymbol{r}') = -\frac{1}{8\pi}\int\limits_S (\hat{\boldsymbol{r}}\cdot\boldsymbol{r}')^2\frac{\partial}{\partial n'}\phi^c(\boldsymbol{r}')\,ds(\boldsymbol{r}') \tag{5.76}$$

Thus, both I_1 and I_3 are expressible entirely in terms of the conductor potential, ϕ^c. However, the second integral appearing in the definition of A_3, (5.58), has more structure.

Now we use the expression (5.52) for u_1^+ to rewrite the second integral in the expression (5.58) for A_3 as

$$I_2 = -\frac{1}{4\pi}\int_S (\hat{\boldsymbol{r}}\cdot\boldsymbol{r}')\frac{\partial}{\partial n'}u_1^+(\boldsymbol{r}')\,ds(\boldsymbol{r}')$$
$$= -\frac{1}{4\pi}\int_S (\hat{\boldsymbol{r}}\cdot\boldsymbol{r}')\frac{\partial}{\partial n'}[(C+\hat{\boldsymbol{k}}\cdot\boldsymbol{c})\phi^c(\boldsymbol{r}')+\hat{\boldsymbol{k}}\cdot\boldsymbol{r}'-\hat{\boldsymbol{k}}\cdot\boldsymbol{\Phi}(\boldsymbol{r}')]ds(\boldsymbol{r}') \tag{5.77}$$

Using the definitions of $\boldsymbol{c}$, $\boldsymbol{\Phi}$, $\widetilde{\boldsymbol{Q}}$ and $\widetilde{\boldsymbol{P}}$ given in section 5.0 and the fact that

$$\int_S (\hat{\boldsymbol{r}}\cdot\boldsymbol{r}')\frac{\partial}{\partial n'}(\hat{\boldsymbol{k}}\cdot\boldsymbol{r}')\,ds(\boldsymbol{r}') = |V^-|(\hat{\boldsymbol{r}}\cdot\hat{\boldsymbol{k}}) \tag{5.78}$$

we find ultimately that

$$I_2 = -C(\hat{\boldsymbol{r}}\cdot\boldsymbol{c})(C+\hat{\boldsymbol{k}}\cdot\boldsymbol{c}) - \frac{1}{4\pi}\widetilde{\boldsymbol{P}}:\hat{\boldsymbol{k}}\hat{\boldsymbol{r}} \tag{5.79}$$

Substituting the results for I_1, I_2, and I_3 in the expression for A_3, (5.58), we obtain

$$A_3 = -\frac{1}{8\pi}\int_S [(\hat{\boldsymbol{k}}\cdot\boldsymbol{r}')^2+(\hat{\boldsymbol{r}}\cdot\boldsymbol{r}')^2]\frac{\partial}{\partial n'}\phi^c(\boldsymbol{r}')\,ds(\boldsymbol{r}')$$
$$-\frac{1}{2(4\pi)^2}\int_S\int_S |\boldsymbol{r}'-\boldsymbol{r}''|\left(\frac{\partial}{\partial n'}\phi^c(\boldsymbol{r}')\right)\frac{\partial}{\partial n''}\phi^c(\boldsymbol{r}'')\,ds(\boldsymbol{r}')\,ds(\boldsymbol{r}'')$$
$$+C^3+C^2(\hat{\boldsymbol{k}}-\hat{\boldsymbol{r}})\cdot\boldsymbol{c}-C(\hat{\boldsymbol{r}}\cdot\boldsymbol{c})(\hat{\boldsymbol{k}}\cdot\boldsymbol{c})-\frac{1}{4\pi}\widetilde{\boldsymbol{P}}:\hat{\boldsymbol{k}}\hat{\boldsymbol{r}} \tag{5.80}$$

Combining this result with the expressions for g, A_1, and A_2, (5.55), (5.61) and (5.65), we find that for plane wave incidence in the direction $\hat{\boldsymbol{k}}$

$$g(\hat{\boldsymbol{r}};\hat{\boldsymbol{k}}) = -ikC + k^2[C(\hat{\boldsymbol{r}}-\hat{\boldsymbol{k}})\cdot\boldsymbol{c}-C^2] + ik^3A_3 + O(k^4) \tag{5.81}$$

The differential scattering cross section, (5.59), becomes

$$\sigma(\hat{\boldsymbol{r}};\hat{\boldsymbol{k}}) = 4\pi C^2 + 4\pi k^2C^2\Bigg[(\hat{\boldsymbol{r}}\cdot\boldsymbol{c}-\hat{\boldsymbol{k}}\cdot\boldsymbol{c}-C)^2$$
$$-\frac{2}{C}A_3(\hat{\boldsymbol{r}},\hat{\boldsymbol{k}})\Bigg] + O(k^4) \tag{5.82}$$

while the total scattering cross section (5.60) is

$$\sigma_s = 4\pi C^2 + 4\pi k^2C^2\Bigg[\frac{1}{3}|\boldsymbol{c}|^2+(C+\hat{\boldsymbol{k}}\cdot\boldsymbol{c})^2$$
$$-\frac{1}{2\pi C}\int_{S^2} A_3(\hat{\boldsymbol{r}},\hat{\boldsymbol{k}})\,ds(\hat{\boldsymbol{r}})\Bigg] + O(k^4) \tag{5.83}$$

where the capacity C and the constant vector $\boldsymbol{c}$ are defined in terms of the conductor potential in (5.9) and (5.16), respectively. As remarked earlier, the absorption cross section, σ_a, vanishes so that $\sigma_e = \sigma_s$.

For specific surfaces explicit calculations of C, $\boldsymbol{c}$, and $\widetilde{\boldsymbol{P}}$ are possible and these results may be used in the expressions given above.

5.A.2 *The Neumann problem*

As before, the general results of Chapter 3 show that the terms in the low frequency expansion are of the form (3.18), (3.51) and (3.52) where $f_n^+ = u_n^i$ for $n = 0, 1$. In this section we consider the first two terms. The exterior potential functions v_n^+ for $n = 0, 1$ are unique solutions of the boundary value problems (3.53)

$$\left.\begin{aligned} \nabla^2 v_n^+ &= 0 && \text{in } V^+ \\ \frac{\partial}{\partial n} v_n^+ &= -\frac{\partial}{\partial n} u_n^i && \text{on } S \\ v_n^+ &= O\left(\frac{1}{r^2}\right) && \text{as } r \to \infty \end{aligned}\right\} \tag{5.84}$$

For plane wave incidence $u_0^i = 1$, which implies that its normal derivative on S vanishes, hence $v_0^+ = 0$ and

$$u_0^+(\boldsymbol{r}) = 1, \quad \boldsymbol{r} \in V^+ \cup S \tag{5.85}$$

The next term exhibits more structure. Since $u_1^i = \hat{\boldsymbol{k}} \cdot \boldsymbol{r}$, the boundary condition for v_1^+ is

$$\frac{\partial}{\partial n} v_1^+ = -\frac{\partial}{\partial n}(\hat{\boldsymbol{k}} \cdot \boldsymbol{r}) = -\hat{\boldsymbol{n}} \cdot \hat{\boldsymbol{k}}, \quad \boldsymbol{r} \in S \tag{5.86}$$

In terms of the vector field $\boldsymbol{\Psi}$ defined in (5.20) it is readily seen that

$$v_1^+(\boldsymbol{r}) = -\hat{\boldsymbol{k}} \cdot \boldsymbol{\Psi}(\boldsymbol{r}) \tag{5.87}$$

and

$$u_1^+(\boldsymbol{r}) = \hat{\boldsymbol{k}} \cdot \boldsymbol{r} - \hat{\boldsymbol{k}} \cdot \boldsymbol{\Psi}(\boldsymbol{r}) \tag{5.88}$$

The expression for the scattering amplitude given in (3.57)–(3.59) is

$$g(\hat{\boldsymbol{r}}; \hat{\boldsymbol{k}}) = ik^3 A_3(\hat{\boldsymbol{r}}; \hat{\boldsymbol{k}}) + O(k^4) \tag{5.89}$$

where

$$A_3(\hat{\boldsymbol{r}}; \hat{\boldsymbol{k}}) = \frac{1}{4\pi} \int_S (\hat{\boldsymbol{n}}' \cdot \hat{\boldsymbol{r}}) u_1^+(\boldsymbol{r}')\, ds(\boldsymbol{r}') - \frac{1}{4\pi} \int_S (\hat{\boldsymbol{n}}' \cdot \hat{\boldsymbol{r}})(\hat{\boldsymbol{r}} \cdot \boldsymbol{r}') u_0^+(\boldsymbol{r}')\, ds(\boldsymbol{r}') \tag{5.90}$$

Using the expressions for u_0^+ and u_1^+, (5.85) and (5.88), we find that

$$A_3(\hat{\boldsymbol{r}};\hat{\boldsymbol{k}}) = \frac{1}{4\pi}\int_S (\hat{\boldsymbol{n}}'\cdot\hat{\boldsymbol{r}})[(\hat{\boldsymbol{k}}\cdot\boldsymbol{r}') - \hat{\boldsymbol{k}}\cdot\boldsymbol{\Psi}(\boldsymbol{r}')]ds(\boldsymbol{r}')$$
$$-\frac{1}{4\pi}\int_S (\hat{\boldsymbol{n}}'\cdot\hat{\boldsymbol{r}})(\hat{\boldsymbol{r}}\cdot\boldsymbol{r}')\,ds(\boldsymbol{r}') \tag{5.91}$$

Recalling the definition of $\widetilde{\boldsymbol{W}}$ and $\widetilde{\boldsymbol{M}}$ and employing Gauss' theorem we obtain

$$A_3(\hat{\boldsymbol{r}},\hat{\boldsymbol{k}}) = \frac{1}{4\pi}\left[(\hat{\boldsymbol{k}}\cdot\hat{\boldsymbol{r}})|V^-| + \widetilde{\boldsymbol{W}} : \hat{\boldsymbol{k}}\hat{\boldsymbol{r}} - |V^-|\right]$$
$$= \frac{1}{4\pi}\left[\widetilde{\boldsymbol{M}} : \hat{\boldsymbol{k}}\hat{\boldsymbol{r}} - |V^-|\right] \tag{5.92}$$

The differential scattering cross section is, with the general expression (3.63),

$$\sigma(\hat{\boldsymbol{r}};\hat{\boldsymbol{k}}) = \frac{k^4}{4\pi}\left[\widetilde{\boldsymbol{M}} : \hat{\boldsymbol{k}}\hat{\boldsymbol{r}} - |V^-|\right]^2 + O(k^6) \tag{5.93}$$

As in the Dirichlet case, the absorption cross section is zero and the extinction cross section is

$$\sigma_e = \sigma_s = \frac{1}{4\pi}\int_{S^2} \frac{k^4}{4\pi}\left[\widetilde{\boldsymbol{M}} : \hat{\boldsymbol{k}}\hat{\boldsymbol{r}} - |V^-|\right]^2 ds(\hat{\boldsymbol{r}}) + O(k^6)$$
$$= \frac{k^4}{12\pi}\left[|\widetilde{\boldsymbol{M}}\cdot\hat{\boldsymbol{k}}|^2 + 3|V^-|^2\right] + O(k^6) \tag{5.94}$$

From the discussion of the extinction cross section for this problem following (3.65) we know that

$$A_4(\hat{\boldsymbol{k}};\hat{\boldsymbol{k}}) = 0 \tag{5.95}$$

Moreover, for surfaces with 'inversion symmetry' the relation expressing A_{2n} in terms of previous coefficients (3.49) yields

$$A_4(\hat{\boldsymbol{r}};\hat{\boldsymbol{k}}) = -\frac{1}{4\pi}\int_{S^2} [A_1(\hat{\boldsymbol{r}}',\hat{\boldsymbol{k}})A_3(\hat{\boldsymbol{r}}',\hat{\boldsymbol{r}}) + A_2(\hat{\boldsymbol{r}}',\hat{\boldsymbol{k}})A_2(\hat{\boldsymbol{r}}',\hat{\boldsymbol{r}})$$
$$+ A_3(\hat{\boldsymbol{r}}',\hat{\boldsymbol{k}})A_1(\hat{\boldsymbol{r}}',\hat{\boldsymbol{r}})]ds(\hat{\boldsymbol{r}}')$$
$$= 0 \tag{5.96}$$

because $A_1 = A_2 = 0$.

It is not known if this relation holds for general surfaces although it is true for $\hat{\boldsymbol{r}} = \hat{\boldsymbol{k}}$.

5.A.3 *The Robin problem*

The general results of Chapter 3 may be specialized to obtain expressions for the first terms in the low frequency expansion of $u^+(\boldsymbol{r})$. Explicitly

$$u_0^+(\boldsymbol{r}) = u_0^i(\boldsymbol{r}) + v_0^+(\boldsymbol{r}) \tag{5.97}$$

where v_0^+ is a solution of the problem

$$\left.\begin{aligned} \nabla^2 v_0^+ &= 0 && \text{in } V^+ \\ \frac{\partial}{\partial n} v_0^+ &= -\frac{\partial}{\partial n} u_0^i && \text{on } S \\ v_0^+ &= O\left(\frac{1}{r^2}\right) && \text{as } r \to \infty \end{aligned}\right\} \tag{5.98}$$

which is (3.68) for $n = 0$. If u^{inc} is a plane wave of unit amplitude, then $u_0^i = 1$ and v_0^+ vanishes, hence

$$u_0^+ = 1 \tag{5.99}$$

If u^{inc} is a point source at $\boldsymbol{r}_p \in V^+$, then the problem of finding v_0^+ is essentially that of determining Green's function for the Neumann problem for the surface S, a problem which has not been solved for arbitrary surfaces.

The second term is given by

$$u_1^+(\boldsymbol{r}) = u_1^i(\boldsymbol{r}) + v_1^+(\boldsymbol{r}) \tag{5.100}$$

and $v_1^+(\boldsymbol{r})$ is a solution of the problem

$$\left.\begin{aligned} \nabla^2 v_1^+ &= 0 && \text{in } V^+ \\ \frac{\partial}{\partial n} v_1^+ &= -\nu u_0^+ - \frac{\partial}{\partial n} u_1^i && \text{on } S \\ v_1^+ &= O\left(\frac{1}{r}\right) && \text{as } r \to \infty \end{aligned}\right\} \tag{5.101}$$

Even for plane wave incidence, however, this problem is nontrivial. Since in this case

$$u_1^i(\boldsymbol{r}) = \hat{\boldsymbol{k}} \cdot \boldsymbol{r} \tag{5.102}$$

the boundary condition becomes

$$\frac{\partial}{\partial n} v_1^+ = -\nu - \hat{\boldsymbol{n}} \cdot \hat{\boldsymbol{k}} \tag{5.103}$$

then

$$v_1^+(\boldsymbol{r}) = -\nu \psi^c(\boldsymbol{r}) - \hat{\boldsymbol{k}} \cdot \boldsymbol{\Psi}(\boldsymbol{r}) \tag{5.104}$$

where $\psi^c(\boldsymbol{r})$ and $\boldsymbol{\Psi}(\boldsymbol{r})$ were introduced in (5.12) and (5.20), respectively. Furthermore

$$u_1^+(\boldsymbol{r}) = \hat{\boldsymbol{k}} \cdot \boldsymbol{r} - \nu\psi^c(\boldsymbol{r}) - \hat{\boldsymbol{k}} \cdot \boldsymbol{\Psi}(\boldsymbol{r}) \tag{5.105}$$

The first two nonvanishing terms of the scattering amplitude (3.69) are

$$g(\hat{\boldsymbol{r}}; \hat{\boldsymbol{k}}) = k^2 A_2(\hat{\boldsymbol{r}}; \hat{\boldsymbol{k}}) + ik^3 A_3(\hat{\boldsymbol{r}}; \hat{\boldsymbol{k}}) + O(k^4) \tag{5.106}$$

where

$$A_2(\hat{\boldsymbol{r}}; \hat{\boldsymbol{k}}) = \frac{1}{4\pi} \int_S (\hat{\boldsymbol{n}}' \cdot \hat{\boldsymbol{r}} - \nu) u_0^+(\boldsymbol{r}')\, ds(\boldsymbol{r}') \tag{5.107}$$

and

$$A_3(\hat{\boldsymbol{r}}; \hat{\boldsymbol{k}}) = \frac{1}{4\pi} \int_S (\hat{\boldsymbol{n}}' \cdot \hat{\boldsymbol{r}} - \nu)[u_1^+(\boldsymbol{r}') - (\hat{\boldsymbol{r}} \cdot \boldsymbol{r}') u_0^+(\boldsymbol{r}')]\, ds(\boldsymbol{r}') \tag{5.108}$$

With the fact that $u_0^+ = 1$ we find that A_2 is independent of both $\hat{\boldsymbol{r}}$ and $\hat{\boldsymbol{k}}$ and is given by

$$A_2 = -\frac{\nu|S|}{4\pi} \tag{5.109}$$

as in (3.70), where the symbol $|S|$ denotes the surface area of the surface S. The surface integral of $\hat{\boldsymbol{n}} \cdot \hat{\boldsymbol{r}}$ vanishes by virtue of Gauss' theorem. Now considering $A_3(\hat{\boldsymbol{r}}; \hat{\boldsymbol{k}})$ we use the expressions for u_0^+ and u_1^+, (5.99) and (5.105), to find that

$$A_3(\hat{\boldsymbol{r}}; \hat{\boldsymbol{k}}) = \frac{1}{4\pi} \int_S (\hat{\boldsymbol{n}}' \cdot \hat{\boldsymbol{r}} - \nu)[(\hat{\boldsymbol{k}} - \hat{\boldsymbol{r}}) \cdot \boldsymbol{r}' - \nu\psi^c(\boldsymbol{r}') - \hat{\boldsymbol{k}} \cdot \boldsymbol{\Psi}(\boldsymbol{r}')] ds(\boldsymbol{r}') \tag{5.110}$$

With the help of Gauss' theorem, the definition of ψ^c and the magnetic polarizability tensor this coefficient may be rewritten as

$$\begin{aligned} A_3(\hat{\boldsymbol{r}}; \hat{\boldsymbol{k}}) = \frac{1}{4\pi} \Bigg[& \widetilde{\boldsymbol{M}} : \hat{\boldsymbol{k}}\hat{\boldsymbol{r}} - |V^-| - \nu(\hat{\boldsymbol{k}} - \hat{\boldsymbol{r}}) \cdot \int_S \boldsymbol{r}' ds(\boldsymbol{r}') \\ & + \nu(\hat{\boldsymbol{k}} - \hat{\boldsymbol{r}}) \cdot \int_S \hat{\boldsymbol{n}}' \psi^c(\boldsymbol{r}')\, ds(\boldsymbol{r}') + \nu^2 \int_S \psi^c(\boldsymbol{r}')\, ds(\boldsymbol{r}') \Bigg] \end{aligned} \tag{5.111}$$

With the general expressions (2.33) and (2.35) we find that the differential scattering cross section is

$$\sigma(\hat{\boldsymbol{r}}; \hat{\boldsymbol{k}}) = \frac{k^2}{4\pi} \nu^2 |S^2| + O(k^4) \tag{5.112}$$

the scattering cross section is

$$\sigma_s = \frac{k^2}{4\pi} \nu^2 |S^2| + O(k^4) \tag{5.113}$$

and the absorption cross section (2.42) is

$$\sigma_a = \nu|S| + O(k^2) \tag{5.114}$$

From its definition as the sum of σ_s and σ_a we find that

$$\sigma_e = \nu|S| + O(k^2) \tag{5.115}$$

It is necessary to find the k^2 term of σ_a in order to have higher order accuracy, even though σ_s is known up to terms of order $O(k^4)$.

5.A.4 *The transmission problem*

From the general results (3.17)–(3.19) and (3.25)–(3.28) of Chapter 3, we consider the first terms in the low frequency expansions of $u^{\pm}$.

Specializing the results in (3.17) and (3.25) and using (3.26) as well as the conditions (3.80) we find that

$$\left.\begin{aligned}
f_0^+(\boldsymbol{r}) &= u_0^i(\boldsymbol{r}), \quad && \boldsymbol{r} \in V^+ \\
f_0^-(\boldsymbol{r}) &= 0, && \boldsymbol{r} \in V^- \\
f_1^+(\boldsymbol{r}) &= u_1^i(\boldsymbol{r}), && \boldsymbol{r} \in V^+ \\
f_1^-(\boldsymbol{r}) &= 0, && \boldsymbol{r} \in V^-
\end{aligned}\right\} \tag{5.116}$$

$$\left.\begin{aligned}
f_2^+(\boldsymbol{r}) &= u_2^i(\boldsymbol{r}) + \frac{1}{4\pi}\int_S \left[u_0^+(\boldsymbol{r}')\frac{\partial}{\partial n'}|\boldsymbol{r}-\boldsymbol{r}'| - |\boldsymbol{r}-\boldsymbol{r}'|\frac{\partial}{\partial n'}u_0^+(\boldsymbol{r}')\right] ds(\boldsymbol{r}'), \quad && \boldsymbol{r} \in V^+ \\
f_2^-(\boldsymbol{r}) &= \frac{\eta^2}{4\pi}\int_S \left[u_0^-(\boldsymbol{r}')\frac{\partial}{\partial n'}|\boldsymbol{r}-\boldsymbol{r}'| - |\boldsymbol{r}-\boldsymbol{r}'|\frac{\partial}{\partial n'}u_0^-(\boldsymbol{r}')\right] ds(\boldsymbol{r}'), && \boldsymbol{r} \in V^-
\end{aligned}\right\} \tag{5.117}$$

Point source incidence: If the incident wave is the point source field

$$u^i(\boldsymbol{r}) = \frac{e^{ik|\boldsymbol{r}-\boldsymbol{r}_p|}}{|\boldsymbol{r}-\boldsymbol{r}_p|}, \quad \boldsymbol{r} \neq \boldsymbol{r}_p, \ \ \boldsymbol{r}_p \in V^+ \tag{5.118}$$

then

$$\left.\begin{aligned}
\nabla^2 v_0^+ &= 0 && \text{in } V^+ \\
\nabla^2 v_0^- &= 0 && \text{in } V^- \\
v_0^+ &= v_0^- - \frac{1}{|\boldsymbol{r}-\boldsymbol{r}_p|} && \text{on } S \\
\frac{\partial}{\partial n}v_0^+ &= \beta\frac{\partial}{\partial n}v_0^- - \frac{\partial}{\partial n}\frac{1}{|\boldsymbol{r}-\boldsymbol{r}_p|} && \text{on } S \\
v_0^+ &= O\left(\frac{1}{r^2}\right) && \text{as } r \to \infty
\end{aligned}\right\} \tag{5.119}$$

which is exactly the problem of finding Green's function for the transmission problem for Laplace's equation and has been solved for only special geometries. In the far field we find, from (3.92) and the transmission condition (3.80), that

$$
\begin{aligned}
A_2(\hat{\boldsymbol{r}}) &= \frac{1}{4\pi}\int_S \left[(\hat{\boldsymbol{r}}\cdot\hat{\boldsymbol{n}}')u_0^+(\boldsymbol{r}') - (\hat{\boldsymbol{r}}\cdot\boldsymbol{r}')\frac{\partial}{\partial n'}u_0^+(\boldsymbol{r}')\right] ds(\boldsymbol{r}') \\
&= \frac{1}{4\pi}\int_S \left[(\hat{\boldsymbol{r}}\cdot\hat{\boldsymbol{n}}')u_0^-(\boldsymbol{r}') - (\hat{\boldsymbol{r}}\cdot\boldsymbol{r}')\beta\frac{\partial}{\partial n'}u_0^-(\boldsymbol{r}')\right] ds(\boldsymbol{r}') \\
&= \frac{1-\beta}{4\pi}\int_S (\hat{\boldsymbol{r}}\cdot\boldsymbol{r}')\frac{\partial}{\partial n'}v_0^-(\boldsymbol{r}')\, ds(\boldsymbol{r})
\end{aligned}
\tag{5.120}
$$

where we have used the fact that u_0^- is harmonic in V^- and $f_0^- = 0$. The coefficient A_2 is seen to vanish when $\beta = 1$ but not otherwise. Therefore, the most prominent low frequency scattering effects are due to differences in mass densities between V^+ and V^-, a result that was first observed by Rayleigh. In fact, we have

$$
g(\hat{\boldsymbol{r}}) = k^2 A_2(\hat{\boldsymbol{r}}) + O(k^3) \tag{5.121}
$$

Note that this result is valid for both lossless and lossy scatterers with $\beta = (\rho^+/\rho^-)$. If the lossy version of β, (3.96) and (3.97), is employed in (5.116)–(5.120), the resulting expression would still be accurate only to $O(k^2)$. The situation is different for plane wave incidence.

Plane wave incidence: With plane wave incidence we find that

$$
\left.\begin{aligned}
f_0^+(\boldsymbol{r}) &= 1, \quad \boldsymbol{r}\in V^+ \\
f_0^-(\boldsymbol{r}) &= 0, \quad \boldsymbol{r}\in V^-
\end{aligned}\right\} \tag{5.122}
$$

which implies that

$$
u_0^+(\boldsymbol{r}) = 1, \quad \boldsymbol{r}\in V^+ \tag{5.123}
$$

and

$$
u_0^-(\boldsymbol{r}) = 1, \quad \boldsymbol{r}\in V^- \tag{5.124}
$$

Then (5.116) and (5.117) read as

$$
\left.\begin{aligned}
f_1^+(\boldsymbol{r}) &= \hat{\boldsymbol{k}}\cdot\boldsymbol{r}, \quad \boldsymbol{r}\in V^+ \\
f_1^-(\boldsymbol{r}) &= 0, \qquad\quad \boldsymbol{r}\in V^-
\end{aligned}\right\} \tag{5.125}
$$

$$
\left.\begin{aligned}
f_2^+(\boldsymbol{r}) &= (\hat{\boldsymbol{k}}\cdot\boldsymbol{r})^2 + \frac{1}{4\pi}\int_S \frac{\partial}{\partial n'}|\boldsymbol{r}-\boldsymbol{r}'|ds(\boldsymbol{r}'), \quad \boldsymbol{r}\in V^+ \\
f_2^-(\boldsymbol{r}) &= \frac{\eta^2}{4\pi}\int_S \frac{\partial}{\partial n'}|\boldsymbol{r}-\boldsymbol{r}'|ds(\boldsymbol{r}'), \qquad\qquad\quad \boldsymbol{r}\in V^-
\end{aligned}\right\} \tag{5.126}
$$

Because of (3.84) and (3.85), the first-order terms for plane wave incidence become

$$\left.\begin{aligned} u_1^+ &= \hat{\boldsymbol{k}}\cdot\boldsymbol{r} + v_1^+ = \hat{\boldsymbol{k}}\cdot\boldsymbol{r} + (1-\beta)w_1^+ && \text{in } V^+ \\ u_1^- &= v_1^- = \hat{\boldsymbol{k}}\cdot\boldsymbol{r} + (1-\beta)w_1^- && \text{in } V^- \end{aligned}\right\} \tag{5.127}$$

where $v_1^\pm$ and $w_1^\pm$ satisfy the transmission problems

$$\left.\begin{aligned} \nabla^2 v_1^+ &= 0 && \text{in } V^+ \\ \nabla^2 v_1^- &= 0 && \text{in } V^- \\ v_1^+ &= v_1^- - \hat{\boldsymbol{k}}\cdot\boldsymbol{r} && \text{on } S \\ \frac{\partial}{\partial n}v_1^+ &= \beta\frac{\partial}{\partial n}v_1^- - \hat{\boldsymbol{k}}\cdot\hat{\boldsymbol{n}} && \text{on } S \\ v_1^+ &= O\left(\frac{1}{r^2}\right) && \text{as } r\to\infty \end{aligned}\right\} \tag{5.128}$$

and

$$\left.\begin{aligned} \nabla^2 w_1^+ &= 0 && \text{in } V^+ \\ \nabla^2 w_1^- &= 0 && \text{in } V^- \\ w_1^+ &= w_1^- && \text{on } S \\ \frac{\partial}{\partial n}w_1^+ &= \beta\frac{\partial}{\partial n}w_1^- - \hat{\boldsymbol{n}}\cdot\hat{\boldsymbol{k}} && \text{on } S \\ w_1^+ &= O\left(\frac{1}{r^2}\right) && \text{as } r\to\infty \end{aligned}\right\} \tag{5.129}$$

Employing vector fields for the general polarizability tensor (5.26) and (5.32) it is easily seen that

$$\left.\begin{aligned} v_1^\pm(\boldsymbol{r}) &= -\hat{\boldsymbol{k}}\cdot\boldsymbol{v}^\pm(\boldsymbol{r};\beta) \\ w_1^\pm(\boldsymbol{r}) &= -\hat{\boldsymbol{k}}\cdot\boldsymbol{\Psi}^\pm(\boldsymbol{r};\beta) \end{aligned}\right\} \tag{5.130}$$

hence

$$u_1^\pm(\boldsymbol{r}) = \hat{\boldsymbol{k}}\cdot\boldsymbol{r} - (1-\beta)\hat{\boldsymbol{k}}\cdot\boldsymbol{\Psi}^\pm(\boldsymbol{r},\beta) \tag{5.131}$$

or

$$\left.\begin{aligned} u_1^+ &= \hat{\boldsymbol{k}}\cdot\boldsymbol{r} - \hat{\boldsymbol{k}}\cdot\boldsymbol{v}^+ && \text{in } V^+ \\ u_1^- &= -\hat{\boldsymbol{k}}\cdot\boldsymbol{v}^- && \text{in } V^- \end{aligned}\right\} \tag{5.132}$$

In the far field, equation (5.120) is still valid for plane wave incidence, in which case, the normal derivatives of u_0^- and v_0^- vanish. Hence, $A_2 = 0$ and $g = O(k^3)$.

Comparison with the result for the Robin problem, equation (5.109) shows that for plane wave incidence, the scattered far fields for the Robin and transmission problems are fundamentally different, having different orders of k. Hence, any attempt to model a transmission problem with a Robin or impedance boundary condition will necesarily fail at low frequencies.

To find the leading term in the far field explicitly we use the general expression (3.93) to find that

$$A_3(\hat{\boldsymbol{r}};\hat{\boldsymbol{k}}) = \frac{1}{4\pi}\beta\eta^2 \int\limits_{V^-} u_0^-(\boldsymbol{r}')\,dv(\boldsymbol{r}') + \frac{1}{4\pi}\int\limits_S \Bigg[(\hat{\boldsymbol{r}}\cdot\hat{\boldsymbol{n}}')u_1^+(\boldsymbol{r}') - (\hat{\boldsymbol{r}}\cdot\boldsymbol{r}')\frac{\partial}{\partial n'}u_1^+(\boldsymbol{r}')$$

$$+ \frac{(\hat{\boldsymbol{r}}\cdot\boldsymbol{r}')^2}{2}\frac{\partial}{\partial n'}u_0^+(\boldsymbol{r}') - (\hat{\boldsymbol{r}}\cdot\hat{\boldsymbol{n}}')(\hat{\boldsymbol{r}}\cdot\boldsymbol{r}')u_0^+(\boldsymbol{r}')\Bigg]ds(\boldsymbol{r}')$$

$$= \frac{1}{4\pi}\Bigg[\beta\eta^2|V^-| - (1-\beta)\int\limits_S (\hat{\boldsymbol{r}}\cdot\hat{\boldsymbol{n}}')(\hat{\boldsymbol{k}}\cdot\boldsymbol{v}^-(\boldsymbol{r}';\beta))\,ds(\boldsymbol{r}') - |V^-|\Bigg] \tag{5.133}$$

where we have employed the transmission conditions (3.80), the fact that $u_0^+ = 1$, and relations (5.34) and (5.132). Recalling the definition of the general polarizability tensor (5.27) we find that

$$A_3(\hat{\boldsymbol{r}};\hat{\boldsymbol{k}}) = -\frac{1}{4\pi}\left[(1-\beta\eta^2)|V^-| + \hat{\boldsymbol{r}}\cdot\widetilde{\boldsymbol{X}}(\beta)\cdot\hat{\boldsymbol{k}}\right] \tag{5.134}$$

and

$$g(\hat{\boldsymbol{r}};\hat{\boldsymbol{k}}) = ik^3 A_3(\hat{\boldsymbol{r}};\hat{\boldsymbol{k}}) + O(k^4) \tag{5.135}$$

Hence, the differential scattering cross section (2.34) is

$$\sigma(\hat{\boldsymbol{r}};\hat{\boldsymbol{k}}) = \frac{k^4}{4\pi}|(1-\beta\eta^2)|V^-| + \hat{\boldsymbol{r}}\cdot\widetilde{\boldsymbol{X}}(\beta)\cdot\hat{\boldsymbol{k}}|^2 + O(k^5) \tag{5.136}$$

The scattering cross section (2.35) is

$$\sigma_s = \frac{k^4}{4\pi}\left[|1-\beta\eta^2|^2|V^-|^2 + \frac{1}{3}|\widetilde{\boldsymbol{X}}(\beta)\cdot\hat{\boldsymbol{k}}|^2\right] + O(k^6) \tag{5.137}$$

For lossy scatterers, one might expect that since $\text{Im}A_3 \neq 0$, the optical or forward scattering theorem (2.59) would imply that

$$\begin{aligned}\sigma_e &= 4\pi k \text{Im} A_3(\hat{\boldsymbol{k}},\hat{\boldsymbol{k}}) + O(k^2)\\ &= k\text{Im}\{\beta\eta^2|V^-| - \hat{\boldsymbol{k}}\cdot\widetilde{\boldsymbol{X}}(\beta)\cdot\hat{\boldsymbol{k}}\} + O(k^2)\\ &= O(k)\end{aligned} \tag{5.138}$$

But using the definitions of the dimensionless constants η and β it follows that

$$\beta\eta^2 = \frac{\gamma^-}{\gamma^+} \tag{5.139}$$

hence

$$\text{Im}\beta\eta^2 = 0 \tag{5.140}$$

Moreover, since

$$\beta = \frac{\rho^+}{\rho^-}\left(1 - ik\frac{\delta^-\gamma^-}{\sqrt{\rho^+\gamma^+}}\right) \tag{5.141}$$

it may be shown that

$$\text{Im}\widetilde{X}(\beta) = O(k) \tag{5.142}$$

Hence, even in the lossy case

$$\sigma_e = O(k^2) \tag{5.143}$$

from which, since $\sigma_s = O(k^4)$, it follows that the absorption cross section is also $O(k^2)$.

Of course, for lossless scatterers $\sigma_a = 0$ and $\sigma_e = \sigma_s$, and all of the A_n are real valued. If in addition the surface has inversion symmetry, then (3.49) remains valid which implies, since $A_1 = A_2 = 0$, that $A_4 = 0$ as well.

In the case when $\beta = 1$, it follows that $\widetilde{X}(1) = 0$ and therefore one may calculate as many terms in the far field as one wishes without solving any potential problems because of the remarkable relations (3.84) and (3.85). In particular, it can be shown that A_4, as it is given in (3.92), can be written as

$$A_4(\hat{\boldsymbol{r}}, \hat{\boldsymbol{k}}) = \frac{1}{4\pi}(1 - \eta^2)(\hat{\boldsymbol{k}} - \hat{\boldsymbol{r}}) \cdot \int_{V^-} \boldsymbol{r}'\, dv(\boldsymbol{r}') \tag{5.144}$$

and this term vanishes if the origin is chosen to coincide with the center of mass of V^-, consistent with the corresponding result for scatterers with inversion symmetry.

5.A.5 *Problems in potentials and acoustics*

1. Suppose that the scatterer S has inversion symmetry, in the sense that $\boldsymbol{r} \in S$ implies that $-\boldsymbol{r} \in S$, and prove that

$$\int_{V^-} \boldsymbol{r}\, dv(\boldsymbol{r}) = \boldsymbol{0}$$

That is, inversion symmetry has as a necessary condition the fact that the origin is located at the center of mass.

2. Prove that if $\boldsymbol{v}^+$ is the solution of (5.26) in V^+, then

$$\int_S \frac{\partial}{\partial n}\boldsymbol{v}^+ ds = \boldsymbol{0}$$

3. Explain the details that lead to the derivation of (5.48).
4. Prove that the general polarizability tensor $\widetilde{X}(\beta)$ is symmetric.
5. Prove the identities (5.34)–(5.36) and use them to derive (5.33).

6. Prove the relations (5.45) when the incident field is a plane wave, or a point source field.
7. Provide a detailed proof of (5.71).
8. Derive the relations (5.78) and (5.79).
9. Explain the asymptotic form $v_0^+ = O(1/r^2)$ as $r \to \infty$ in (5.98).
10. Prove that A_3, as it is given by (5.110), may be rewritten as (5.111).
11. Derive the forms of $f_n^{\pm}, n = 0, 1, 2$ given in (5.116)–(5.117).
12. Prove the last equality in (5.120).
13. Explain why the functions $v_1^{\pm}$ and $w_1^{\pm}$, given in (5.130), solve the transmission problems (5.128) and (5.129), respectively.
14. Consider a plane wave incidence for the acoustic transmission problem with $\rho^+ = \rho^-$ and prove formula (5.144), where the coefficient A_4 is given by (3.92) for $n = 1$.

5.EM Electromagnetics

5.EM.1 *The perfect conductor problem*

Recall that in Chapter 3 we discussed the reduction of the problem of determining the coefficients $\boldsymbol{E}_n^+$ and $\boldsymbol{H}_n^+$ in the expansions of the exterior field to a sequence of potential problems by writing

$$\left.\begin{aligned} \boldsymbol{E}_n^+ &= \boldsymbol{F}_{en}^+ - \nabla\phi_{en}^+ \\ \boldsymbol{H}_n^+ &= \boldsymbol{F}_{mn}^+ + \boldsymbol{h}_n^+ - \nabla\phi_{mn}^+ \end{aligned}\right\} \quad \text{in } V^+ \tag{5.145}$$

where the functions $\boldsymbol{F}_{en}^+$ and $\boldsymbol{F}_{mn}^+$ are given in terms of field coefficients up to order $n-1$ in (3.146) and (3.147), respectively, and

$$\boldsymbol{h}_n^+(\boldsymbol{r}) = \frac{1}{4\pi}\nabla \times \int_S \frac{\phi_{mn}^-(\boldsymbol{r}')\hat{\boldsymbol{n}}'}{|\boldsymbol{r}-\boldsymbol{r}'|} ds(\boldsymbol{r}') \quad \text{in } V^+ \tag{5.146}$$

as given in (3.159). The functions $\phi_{en}^{\pm}$ and $\phi_{mn}^{\pm}$ are all potential functions and satisfy the boundary value problems (3.154), (3.155) and (3.172). Here, we examine the lowest order terms. From (3.146) and (3.147) we see that

$$\boldsymbol{F}_{eo}^+(\boldsymbol{r}) = \boldsymbol{E}_0^i(\boldsymbol{r}) \qquad \boldsymbol{F}_{mo}^+(\boldsymbol{r}) = \boldsymbol{H}_0^i(\boldsymbol{r}) \tag{5.147}$$

For plane wave incidence (1.152) and (1.153) yield

$$\left.\begin{aligned} \boldsymbol{E}_0^i(\boldsymbol{r}) &= \hat{\boldsymbol{a}} = \nabla(\hat{\boldsymbol{a}}\cdot\boldsymbol{r}) = \nabla[\hat{\boldsymbol{a}}\cdot(\boldsymbol{r}+\boldsymbol{c})] \\ \boldsymbol{H}_0^i(\boldsymbol{r}) &= Y^+\hat{\boldsymbol{b}} \end{aligned}\right\} \tag{5.148}$$

for any constant vector $\boldsymbol{c}$. The potential function ϕ_{eo}^+ is a solution of

$$\left.\begin{aligned} &\nabla^2\phi_{eo}^+ = 0 && \text{in } V^+ \\ &\phi_{eo}^+ = \hat{\boldsymbol{a}}\cdot(\boldsymbol{r}+\boldsymbol{c}) && \text{on } S \\ &\int_S \frac{\partial}{\partial n}\phi_{eo}^+ ds = \int_S \hat{\boldsymbol{n}}\cdot\boldsymbol{F}_{eo}^+ ds = \int_S \hat{\boldsymbol{n}}\cdot\hat{\boldsymbol{a}} ds = 0 && \text{on } S \\ &\phi_{eo}^+ = O\left(\frac{1}{r^2}\right) && \text{as } r\to\infty \end{aligned}\right\} \tag{5.149}$$

which comes from (3.154) for $n = 0$ and (3.129). The integral condition implies that actually ϕ_{eo}^+ is of order $1/r^2$ as $r \to \infty$. With the vector potential associated with the electric polarizability tensor introduced in (5.13) it is easily seen that

$$\phi_{eo}^+(\boldsymbol{r}) = \hat{\boldsymbol{a}}\cdot\boldsymbol{\Phi}(\boldsymbol{r}), \quad \boldsymbol{r}\in V^+ \tag{5.150}$$

The function ϕ_{mo}^- is a regular interior potential which, by virtue of (3.155), is a constant. This constant can be chosen to be zero because $\boldsymbol{h}_n^+$ is not uniquely defined. Hence, with (3.172), ϕ_{mo}^+ is seen to be a solution of

$$\left.\begin{aligned} &\nabla^2\phi_{mo}^+ = 0 && \text{in } V^+ \\ &\frac{\partial}{\partial n}\phi_{mo}^+ = \hat{\boldsymbol{n}}\cdot\boldsymbol{F}_{mo}^+ = \hat{\boldsymbol{n}}\cdot\boldsymbol{H}_0^i = Y^+\hat{\boldsymbol{n}}\cdot\hat{\boldsymbol{b}} && \text{on } S \\ &\phi_{mo}^+ = O\left(\frac{1}{r}\right) && \text{as } r\to\infty \end{aligned}\right\} \tag{5.151}$$

Since

$$\int_S \frac{\partial}{\partial n'}\phi_{mo}^+(\boldsymbol{r}')\, ds(\boldsymbol{r}') = Y^+\int_S \hat{\boldsymbol{n}}'\cdot\hat{\boldsymbol{b}} ds(\boldsymbol{r}') = 0 \tag{5.152}$$

it follows that actually

$$\phi_{mo}^+ = O\left(\frac{1}{r^2}\right), \quad r\to\infty \tag{5.153}$$

Hence, we may immediately write the solution as

$$\phi_{mo}^+(\boldsymbol{r}) = Y^+\hat{\boldsymbol{b}}\cdot\boldsymbol{\Psi}(\boldsymbol{r}), \quad \boldsymbol{r}\in V^+ \tag{5.154}$$

where $\boldsymbol{\Psi}(\boldsymbol{r})$ is the vector field associated with the magnetic polarizability tensor, (5.20). Therefore,

$$\left.\begin{aligned} \boldsymbol{E}_0^+ &= \hat{\boldsymbol{a}} - \nabla(\hat{\boldsymbol{a}}\cdot\boldsymbol{\Phi}) \\ \boldsymbol{H}_0^+ &= Y^+\hat{\boldsymbol{b}} - Y^+\nabla(\hat{\boldsymbol{b}}\cdot\boldsymbol{\Psi}) \end{aligned}\right\} \quad \text{in } V^+ \tag{5.155}$$

The far field coefficients $\boldsymbol{g}_e(\hat{\boldsymbol{r}})$ and $\boldsymbol{g}_m(\hat{\boldsymbol{r}})$, (3.143) and (3.144), become, using the boundary conditions (3.145)

$$\boldsymbol{g}_e(\hat{\boldsymbol{r}}) = \frac{1}{4\pi}\sum_{n=0}^{\infty}\frac{(ik)^{n+3}}{n!}\sum_{m=0}^{n}\binom{n}{m}\hat{\boldsymbol{r}}\times\left\{\hat{\boldsymbol{r}}\times\int_S\left[\hat{\boldsymbol{n}}'\cdot\boldsymbol{E}^+_{n-m}(\boldsymbol{r}')\right.\right.$$
$$\left.\left.-Z^+\hat{\boldsymbol{r}}\cdot(\hat{\boldsymbol{n}}'\times\boldsymbol{H}^+_{n-m}(\boldsymbol{r}'))\right]\boldsymbol{r}'(-\hat{\boldsymbol{r}}\cdot\boldsymbol{r}')^m ds(\boldsymbol{r}')\right\} \tag{5.156}$$

$$\boldsymbol{g}_m(\hat{\boldsymbol{r}}) = Y^+\hat{\boldsymbol{r}}\times\boldsymbol{g}_e(\hat{\boldsymbol{r}})$$

Since $\boldsymbol{E}^+_n$ and $\boldsymbol{H}^+_n$ are real valued $\boldsymbol{g}_e(\hat{\boldsymbol{r}})$ may be easily decomposed into real and imaginary parts. For the lowest order term we see that

$$\boldsymbol{g}_e(\hat{\boldsymbol{r}}) = \frac{(ik)^3}{4\pi}\hat{\boldsymbol{r}}\times\left\{\hat{\boldsymbol{r}}\times\int_S[\hat{\boldsymbol{n}}'\cdot\boldsymbol{E}^+_0(\boldsymbol{r}')\right.$$
$$\left.-Z^+(\hat{\boldsymbol{r}}\times\hat{\boldsymbol{n}}')\cdot\boldsymbol{H}^+_0(\boldsymbol{r}')]\boldsymbol{r}'ds(\boldsymbol{r}')\right\} + O(k^4) \tag{5.157}$$

where $\boldsymbol{E}^+_0$ and $\boldsymbol{H}^+_0$ are given in (5.155). With the definitions of the electric and magnetic polarizability tensors and using Gauss' theorem we find that

$$\boldsymbol{g}_e(\hat{\boldsymbol{r}}) = \frac{(ik)^3}{4\pi}\hat{\boldsymbol{r}}\times[\hat{\boldsymbol{r}}\times\widetilde{\boldsymbol{P}}\cdot\hat{\boldsymbol{a}} - \widetilde{\boldsymbol{M}}\cdot\hat{\boldsymbol{b}}] + O(k^4) \tag{5.158}$$

The terms $\widetilde{\boldsymbol{P}}\cdot\hat{\boldsymbol{a}}$ and $\widetilde{\boldsymbol{M}}\cdot\hat{\boldsymbol{b}}$ can be interpreted as electric and magnetic dipole moments. That is, the far field, to order k^3, is the same as that emanating from an electric and a magnetic dipole, suitably normalized (see section 1.EM.5). Thus, a plane polarized incident wave is seen to excite both electric and magnetic dipoles in a scatterer.

The differential scattering cross section is given by

$$\begin{aligned}\sigma(\hat{\boldsymbol{r}}) &= \frac{k^4}{4\pi}|\hat{\boldsymbol{r}}\times(\hat{\boldsymbol{r}}\times\widetilde{\boldsymbol{P}}\cdot\hat{\boldsymbol{a}}) - \hat{\boldsymbol{r}}\times\widetilde{\boldsymbol{M}}\cdot\hat{\boldsymbol{b}}|^2 + O(k^6)\\ &= \frac{k^4}{4\pi}\left[|\hat{\boldsymbol{r}}\times\widetilde{\boldsymbol{P}}\cdot\hat{\boldsymbol{a}}|^2 + |\hat{\boldsymbol{r}}\times\widetilde{\boldsymbol{M}}\cdot\hat{\boldsymbol{b}}|^2 - 2\hat{\boldsymbol{r}}\cdot(\widetilde{\boldsymbol{P}}\cdot\hat{\boldsymbol{a}})\times(\widetilde{\boldsymbol{M}}\cdot\hat{\boldsymbol{b}})\right] + O(k^6)\end{aligned} \tag{5.159}$$

Note that the remainder is $O(k^6)$ rather than $O(k^5)$ since the coefficients of $(ik)^n$ in the expression for $\boldsymbol{g}_e(\hat{\boldsymbol{r}})$ are real.

The scattering cross section is given by

$$\sigma_s = \frac{k^4}{6\pi}\left\{|\widetilde{\boldsymbol{P}}\cdot\hat{\boldsymbol{a}}|^2 + |\widetilde{\boldsymbol{M}}\cdot\hat{\boldsymbol{b}}|^2\right\} + O(k^6) \tag{5.160}$$

Since the absorption cross section $\sigma_a = 0$ for a perfect conductor (see 2.117) it follows that the extinction cross section σ_e is identical with σ_s. We omit a discussion of the Rayleigh terms in the impedance problem since no general results are available.

5.EM.2 *The transmission problem*

In the case of the transmission problem it is more convenient to work with the low frequency expansions given in Chapter 4 rather than the expressions given in Chapter 3, although either lead eventually to the same results.

The lossless transmission problem: $\sigma^- = 0$. Recall that the coefficients $\boldsymbol{E}_n^\pm$ and $\boldsymbol{H}_n^\pm$ are given in (4.53)–(4.56) as

$$\boldsymbol{E}_n^\pm = \boldsymbol{F}_{en}^\pm + \nabla\phi_n^\pm \quad \text{in } V^\pm \tag{5.161}$$

and

$$\boldsymbol{H}_n^\pm = \boldsymbol{F}_{mn}^\pm + \nabla\psi_n^\pm \quad \text{in } V^\pm \tag{5.162}$$

where $\boldsymbol{F}_{en}^\pm$ and $\boldsymbol{F}_{mn}^\pm$ are given in (4.49) and (4.51) and $\phi_n^\pm$ and $\psi_n^\pm$ satisfy the transmission problems (4.57) and (4.58). For plane wave incidence $\boldsymbol{F}_{eo}^\pm$ and $\boldsymbol{F}_{mo}^\pm$ become

$$\boldsymbol{F}_{eo}^\pm = \hat{\boldsymbol{a}}, \qquad \boldsymbol{F}_{mo}^\pm = Y^+\hat{\boldsymbol{b}} \tag{5.163}$$

and $\phi_0^\pm$ satisfies the transmission problem

$$\left.\begin{aligned}
\nabla^2\phi_0^+ &= 0 && \text{in } V^+\\
\nabla^2\phi_0^- &= 0 && \text{in } V^-\\
\hat{\boldsymbol{n}}\times\nabla(\phi_0^+ - \phi_0^-) &= \boldsymbol{0} && \text{on } S\\
\frac{\partial}{\partial n}\phi_0^+ - \frac{\epsilon^-}{\epsilon^+}\frac{\partial}{\partial n}\phi_0^- &= -\left(1-\frac{\epsilon^-}{\epsilon^+}\right)\hat{\boldsymbol{a}}\cdot\hat{\boldsymbol{n}} && \text{on } S\\
\phi_0^+ &= O\left(\frac{1}{r^2}\right) && \text{as } r\to\infty
\end{aligned}\right\} \tag{5.164}$$

It is easy to see that the solution of this problem may be expressed in terms of the vector functions $\boldsymbol{\Psi}^\pm$ introduced in (5.32) as

$$\phi_0^\pm(\boldsymbol{r}) = -\left(1-\frac{\epsilon^-}{\epsilon^+}\right)\hat{\boldsymbol{a}}\cdot\boldsymbol{\Psi}^\pm\left(\boldsymbol{r};\frac{\epsilon^-}{\epsilon^+}\right) \tag{5.165}$$

where we have explicitly indicated the dependence of $\boldsymbol{\Psi}^\pm$ on $\beta = \epsilon^-/\epsilon^+$. Similarly, the solution of (4.58) for $n = 0$ is

$$\psi_0^\pm(\boldsymbol{r}) = -Y^+\left(1-\frac{\mu^-}{\mu^+}\right)\hat{\boldsymbol{b}}\cdot\boldsymbol{\Psi}^\pm\left(\boldsymbol{r};\frac{\mu^-}{\mu^+}\right) \tag{5.166}$$

Hence

$$\boldsymbol{E}_0^\pm(\boldsymbol{r}) = \hat{\boldsymbol{a}} - \left(1-\frac{\epsilon^-}{\epsilon^+}\right)\nabla\left[\hat{\boldsymbol{a}}\cdot\boldsymbol{\Psi}^\pm\left(\boldsymbol{r},\frac{\epsilon^-}{\epsilon^+}\right)\right] \tag{5.167}$$

and

$$\boldsymbol{H}_0^{\pm}(\boldsymbol{r}) = Y^+ \left\{ \hat{\boldsymbol{b}} - \left(1 - \frac{\mu^-}{\mu^+}\right) \nabla \left[\hat{\boldsymbol{b}} \cdot \boldsymbol{\Psi}^{\pm}\left(\boldsymbol{r}, \frac{\mu^-}{\mu^+}\right)\right]\right\} \tag{5.168}$$

Introducing the transmission conditions (1.142)–(1.147) with $\sigma^- = 0$ into the expression for the scattering amplititudes, (3.143), we obtain

$$\begin{aligned} \boldsymbol{g}_e(\hat{\boldsymbol{r}}) = \frac{(ik)^3}{4\pi} \hat{\boldsymbol{r}} \times \Bigg[& \frac{\epsilon^-}{\epsilon^+} \hat{\boldsymbol{r}} \times \int_S (\hat{\boldsymbol{n}}' \cdot \boldsymbol{E}_0^-(\boldsymbol{r}'))\boldsymbol{r}' \, ds(\boldsymbol{r}') \\ & - Z^+ \hat{\boldsymbol{r}} \times \int_S \hat{\boldsymbol{r}} \cdot (\hat{\boldsymbol{n}}' \times \boldsymbol{H}_0^-(\boldsymbol{r}'))\boldsymbol{r}' \, ds(\boldsymbol{r}') \\ & + Z^+ \frac{\mu^-}{\mu^+} \int_S (\hat{\boldsymbol{n}}' \cdot \boldsymbol{H}_0^-(\boldsymbol{r}'))\boldsymbol{r}' \, ds(\boldsymbol{r}') \\ & + \int_S \hat{\boldsymbol{r}} \cdot (\hat{\boldsymbol{n}}' \times \boldsymbol{E}_0^-(\boldsymbol{r}'))\boldsymbol{r}' ds(\boldsymbol{r}') \Bigg] + O(k^4) \end{aligned} \tag{5.169}$$

which, with the expressions (5.167) and (5.168) for $\boldsymbol{E}_0^-$ and $\boldsymbol{H}_0^-$, becomes

$$\begin{aligned} \boldsymbol{g}_e(\hat{\boldsymbol{r}}) = \frac{(ik)^3}{4\pi} \hat{\boldsymbol{r}} \times \Bigg\{ & \hat{\boldsymbol{r}} \times \int_S \left[\frac{\epsilon^-}{\epsilon^+} \hat{\boldsymbol{n}}' \cdot \hat{\boldsymbol{a}} - \hat{\boldsymbol{r}} \cdot (\hat{\boldsymbol{n}}' \times \hat{\boldsymbol{b}}) \right] \boldsymbol{r}' ds(\boldsymbol{r}') \\ & + \int_S \left[\frac{\mu^-}{\mu^+} \hat{\boldsymbol{n}}' \cdot \hat{\boldsymbol{b}} + \hat{\boldsymbol{r}} \cdot (\hat{\boldsymbol{n}}' \times \hat{\boldsymbol{a}}) \right] \boldsymbol{r}' ds(\boldsymbol{r}') \\ & - \frac{\epsilon^-}{\epsilon^+} \left(1 - \frac{\epsilon^-}{\epsilon^+}\right) \hat{\boldsymbol{r}} \times \int_S \hat{\boldsymbol{n}}' \cdot \nabla_{\boldsymbol{r}'} \left(\hat{\boldsymbol{a}} \cdot \boldsymbol{\Psi}^- \left(\boldsymbol{r}', \frac{\epsilon^-}{\epsilon^+}\right)\right) \boldsymbol{r}' ds(\boldsymbol{r}') \\ & + \left(1 - \frac{\mu^-}{\mu^+}\right) \hat{\boldsymbol{r}} \times \int_S \hat{\boldsymbol{r}} \cdot \left[\hat{\boldsymbol{n}}' \times \nabla_{\boldsymbol{r}'} \left(\hat{\boldsymbol{b}} \cdot \boldsymbol{\Psi}^- \left(\boldsymbol{r}', \frac{\mu^-}{\mu^+}\right)\right)\right] \boldsymbol{r}' ds(\boldsymbol{r}') \\ & - \frac{\mu^-}{\mu^+} \left(1 - \frac{\mu^-}{\mu^+}\right) \int_S \hat{\boldsymbol{n}}' \cdot \nabla_{\boldsymbol{r}'} \left(\hat{\boldsymbol{b}} \cdot \boldsymbol{\Psi}^- \left(\boldsymbol{r}', \frac{\mu^-}{\mu^+}\right)\right) \boldsymbol{r}' ds(\boldsymbol{r}') \\ & - \left(1 - \frac{\epsilon^-}{\epsilon^+}\right) \int_S \hat{\boldsymbol{r}} \cdot \left[\hat{\boldsymbol{n}}' \times \nabla_{\boldsymbol{r}'} \left(\hat{\boldsymbol{a}} \cdot \boldsymbol{\Psi}^- \left(\boldsymbol{r}', \frac{\epsilon^-}{\epsilon^+}\right)\right)\right] \boldsymbol{r}' ds(\boldsymbol{r}') \Bigg\} \\ & + O(k^4) \end{aligned} \tag{5.170}$$

Now we employ the following identities which hold for any constant vector $\boldsymbol{c}$

$$\int_S (\hat{\boldsymbol{n}}' \cdot \boldsymbol{c})\boldsymbol{r}' ds(\boldsymbol{r}') = \boldsymbol{c}|V^-| \tag{5.171}$$

$$\int_S \hat{\boldsymbol{r}} \cdot (\hat{\boldsymbol{n}}' \times \boldsymbol{c})\boldsymbol{r}' ds(\boldsymbol{r}') = -\hat{\boldsymbol{r}} \times \boldsymbol{c}|V^-| \tag{5.172}$$

and

$$\int_S \hat{\boldsymbol{r}} \cdot [\hat{\boldsymbol{n}}' \times \nabla_{\boldsymbol{r}'}(\boldsymbol{c} \cdot \boldsymbol{\Psi}^-(\boldsymbol{r}'))]\boldsymbol{r}' ds(\boldsymbol{r}') = -\hat{\boldsymbol{r}} \times \int_S (\boldsymbol{c} \cdot \boldsymbol{\Psi}^-(\boldsymbol{r}'))\hat{\boldsymbol{n}}'\, ds(\boldsymbol{r}') \tag{5.173}$$

which result from the facts that $\boldsymbol{r}'$ and $\boldsymbol{c} \cdot \boldsymbol{\Psi}^-$ are harmonic functions in V^-, Gauss' theorem and Stokes' theorem. When these identities are introduced in (5.168) we obtain, after some simplification

$$\begin{aligned} \boldsymbol{g}_e(\hat{\boldsymbol{r}}) = \frac{1}{4\pi}(ik)^3 \hat{\boldsymbol{r}} \times \Bigg\{ \hat{\boldsymbol{r}} \times \Bigg[&-\left(1 - \frac{\epsilon^-}{\epsilon^+}\right)|V^-|\widetilde{\boldsymbol{I}} \\ &+\left(1 - \frac{\epsilon^-}{\epsilon^+}\right)^2 \int_S \hat{\boldsymbol{n}}' \boldsymbol{\Psi}^-\left(\boldsymbol{r}', \frac{\epsilon^-}{\epsilon^+}\right) ds(\boldsymbol{r}')\Bigg] \cdot \hat{\boldsymbol{a}} \\ &+\Bigg[-\left(1 - \frac{\mu^-}{\mu^+}\right)|V^-|\widetilde{\boldsymbol{I}} \\ &+\left(1 - \frac{\mu^-}{\mu^+}\right)^2 \int_S \hat{\boldsymbol{n}}' \boldsymbol{\Psi}^-\left(\boldsymbol{r}', \frac{\mu^-}{\mu^+}\right) ds(\boldsymbol{r}')\Bigg] \cdot \hat{\boldsymbol{b}} \Bigg\} + O(k^4) \end{aligned} \tag{5.174}$$

With the definition of the general polarizability tensor (5.32) we write the scattering amplitudes as

$$\boldsymbol{g}_e(\hat{\boldsymbol{r}}) = \frac{1}{4\pi}(ik)^3 \left[\hat{\boldsymbol{r}} \times \left(\hat{\boldsymbol{r}} \times \widetilde{\boldsymbol{X}}\left(\frac{\epsilon^-}{\epsilon^+}\right)\right) \cdot \hat{\boldsymbol{a}} + \left(\hat{\boldsymbol{r}} \times \widetilde{\boldsymbol{X}}\left(\frac{\mu^-}{\mu^+}\right)\right) \cdot \hat{\boldsymbol{b}}\right] + O(k^4) \tag{5.175}$$

The two polarizability tensor terms may be interpreted as electric and magnetic dipole moments. The differential scattering cross section is given by

$$\begin{aligned} \sigma(\hat{\boldsymbol{r}}) = \frac{k^4}{4\pi}\Bigg[&\left|\hat{\boldsymbol{r}} \times \widetilde{\boldsymbol{X}}\left(\frac{\epsilon^-}{\epsilon^+}\right) \cdot \hat{\boldsymbol{a}}\right|^2 + \left|\hat{\boldsymbol{r}} \times \widetilde{\boldsymbol{X}}\left(\frac{\mu^-}{\mu^+}\right) \cdot \hat{\boldsymbol{b}}\right|^2 \\ &+ 2\hat{\boldsymbol{r}} \cdot \left(\widetilde{\boldsymbol{X}}\left(\frac{\epsilon^-}{\epsilon^+}\right) \cdot \hat{\boldsymbol{a}}\right) \times \left(\widetilde{\boldsymbol{X}}\left(\frac{\mu^-}{\mu^+}\right) \cdot \hat{\boldsymbol{b}}\right)\Bigg] + O(k^6) \end{aligned} \tag{5.176}$$

and the scattering cross section is

$$\sigma_s = \frac{k^4}{6\pi}\left[\left|\widetilde{\boldsymbol{X}}\left(\frac{\epsilon^-}{\epsilon^+}\right) \cdot \hat{\boldsymbol{a}}\right|^2 + \left|\widetilde{\boldsymbol{X}}\left(\frac{\mu^-}{\mu^+}\right) \cdot \hat{\boldsymbol{b}}\right|^2\right] + O(k^6) \tag{5.177}$$

In this lossless case the absorption cross section, σ_a, vanishes which means that $\sigma_e = \sigma_s$. As before the remainder term is $O(k^6)$ rather than $O(k^5)$ since the coefficients in the low frequency expansions are real.

The lossy transmission problem: $\sigma^- > 0$. In this case, the $n = 0$ terms in (4.76) and (4.77) are

$$\boldsymbol{E}_0^+ = \boldsymbol{F}_{eo}^+ + \sigma^- \boldsymbol{G}_{eo}^+ + \nabla\phi_0^+ \quad \text{in } V^+ \tag{5.178}$$

and

$$\boldsymbol{E}_0^- = \boldsymbol{F}_{eo}^- + \sigma^- \boldsymbol{G}_{eo}^- + \nabla\phi_0^- \quad \text{in } V^- \tag{5.179}$$

The terms $\boldsymbol{F}_{eo}^+$ and $\boldsymbol{F}_{eo}^-$ are the same as the lossless case which, from (5.163), are

$$\boldsymbol{F}_{eo}^+ = \boldsymbol{F}_{eo}^- = \hat{\boldsymbol{a}} \tag{5.180}$$

Moreover, the terms involving σ^- are found to vanish for $n = 0$ using (4.74); that is

$$\boldsymbol{G}_{eo}^+ = \boldsymbol{G}_{eo}^- = \boldsymbol{0} \tag{5.181}$$

Although the coefficients $\boldsymbol{E}_0^+$ and $\boldsymbol{E}_0^-$ now appear to have the same form as in the lossless case the determination of ϕ_0^+ and ϕ_0^- is different. With (4.78) we see that ϕ_0^- is the solution of the interior Neumann problem

$$\left.\begin{aligned} \nabla^2\phi_0^- &= 0 && \text{in } V^- \\ \frac{\partial}{\partial n}\phi_0^- &= -\hat{\boldsymbol{n}}\cdot\hat{\boldsymbol{a}} && \text{on } S \end{aligned}\right\} \tag{5.182}$$

The solution of this problem is easily seen to be

$$\phi_0^-(\boldsymbol{r}) = -\hat{\boldsymbol{a}}\cdot\boldsymbol{r}, \quad \boldsymbol{r}\in V^- \tag{5.183}$$

hence

$$\boldsymbol{E}_0^-(\boldsymbol{r}) = \hat{\boldsymbol{a}} - \nabla(\widetilde{\boldsymbol{a}}\cdot\boldsymbol{r}) = \boldsymbol{0}, \quad \boldsymbol{r}\in V^- \tag{5.184}$$

in agreement with the discussion following (4.69). Once ϕ_0^- has been found we turn to the problem of finding ϕ_0^+, which with (4.83) is seen to be the solution of the problem

$$\left.\begin{aligned} \nabla^2\phi_0^+ &= 0 && \text{in } V^+ \\ \phi_0^+ &= -\hat{a}\cdot\boldsymbol{r} + c && \text{on } S \\ \phi_0^+ &= O\left(\frac{1}{r^2}\right) && \text{as } r\to\infty \end{aligned}\right\} \tag{5.185}$$

In terms of the vector valued potential function $\boldsymbol{\Phi}$ introduced in (5.13) the solution may be expressed as

$$\phi_0^+ = -\hat{\boldsymbol{a}}\cdot\boldsymbol{\Phi} \tag{5.186}$$

with

$$c = -\hat{\boldsymbol{a}}\cdot\boldsymbol{c} \tag{5.187}$$

Hence

$$E_0^+ = \hat{a} - \nabla(\hat{a} \cdot \boldsymbol{\Phi}) \tag{5.188}$$

which is exactly the same as the field in the case of perfect conductivity. The problem of determining the first magnetic field coefficient in the lossy case turns out to be exactly the same as in the lossless case. Since E_0^- vanishes we see from (4.87) that

$$G_{mo}^+ = G_{mo}^- = \mathbf{0} \tag{5.189}$$

Hence (4.88) and (4.89) become, for $n = 0$

$$H_0^+ = F_{mo}^+ + \nabla\psi_0^+ \quad \text{in } V^+ \tag{5.190}$$

$$H_0^- = F_{mo}^- + \nabla\psi_0^- \quad \text{in } V^- \tag{5.191}$$

where F_{mo}^+ and F_{mo}^- are exactly the same as in the lossless case (5.163), by (4.51) and the potential functions ψ_0^+ and ψ_0^- satisfy the same transmission potential problem, (4.90), as in the lossless case. Hence, the coefficients H_0^+ and H_0^- are identical with those in the lossless case given by (5.168). With these expressions for $F_0^\pm$ and $H_0^\pm$ together with the expression for the scattering amplitude, (3.143), we find, by exactly the same analysis used for the perfectly conducting case for the E field and the lossless transmission case for the H field, that

$$g_e(\hat{r}) = \frac{(ik)^3}{4\pi}\left[\hat{r} \times (\hat{r} \times \widetilde{P}) \cdot \hat{a} + \hat{r} \times \widetilde{X}\left(\frac{\mu^-}{\mu^+}\right) \cdot \hat{b}\right] + O(k^4) \tag{5.192}$$

The cross sections become

$$\sigma(\hat{r}) = \frac{k^4}{4\pi}\left[|\hat{r} \times \widetilde{P} \cdot \hat{a}|^2 + \left|\hat{r} \times \widetilde{X}\left(\frac{\mu^-}{\mu^+}\right) \cdot \hat{b}\right|^2 \right.$$

$$\left. + 2\hat{r} \cdot (P \cdot \hat{a}) \times \left(\widetilde{X}\left(\frac{\mu^-}{\mu^+}\right) \cdot \hat{b}\right)\right] + O(k^6) \tag{5.193}$$

and

$$\sigma_s = \frac{k^4}{6\pi}\left\{|\widetilde{P} \cdot \hat{a}|^2 + \left|\widetilde{X}\left(\frac{\mu^-}{\mu^+}\right) \cdot \hat{b}\right|^2\right\} + O(k^6) \tag{5.194}$$

In the lossy case the absorption cross section σ_a does not vanish. Since

$$\operatorname{Re} g_e = O(k^4) \tag{5.195}$$

the optical theorem implies that

$$\sigma_e = O(k^2) \tag{5.196}$$

from which we infer that

$$\sigma_a = O(k^2) \tag{5.197}$$

In order to obtain the leading order term in σ_e it is necessary to determine $E_1^\pm$ and $H_1^\pm$.

5.EM.3 *Problems in electromagnetics*

1. Use $\boldsymbol{E}_0^+$ and $\boldsymbol{H}_0^+$, given by (5.155), in the expression (5.157) and prove (5.158).
2. Prove that the solution of the transmission problem (5.164) can be expressed as (5.165).
3. Prove the identities (5.172) and (5.173) and use them to derive the expression (5.174) for the electric scattering amplitude.
4. Derive expression (5.177).

5.E Elasticity

5.E.1 *The rigid problem*

The problem of finding the Rayleigh term for the rigid scatterer involves solving the elastostatic problem

$$\left.\begin{aligned} \Delta^{e+}\boldsymbol{v}_0^+ &= 0, && \text{in } V^+ \\ \boldsymbol{v}_0^+ &= -\boldsymbol{f}_0^+ && \text{on } S \\ \boldsymbol{v}_0^+ &= O\left(\frac{1}{r}\right) && \text{as } r \to \infty \end{aligned}\right\} \tag{5.198}$$

From the general expression for $\boldsymbol{f}_n^+$ (3.249) it follows that

$$\boldsymbol{f}_0^+(\boldsymbol{r}) = \boldsymbol{u}_0^i(\boldsymbol{r}) \tag{5.199}$$

and for arbitrary plane wave incidence

$$\boldsymbol{u}_0^i(\boldsymbol{r}) = \alpha_p\hat{\boldsymbol{k}} + \alpha_s\hat{\boldsymbol{b}} \tag{5.200}$$

Then the zeroth order coefficient has the form

$$\boldsymbol{u}_0^+(\boldsymbol{r}) = \boldsymbol{u}_0^i(\boldsymbol{r}) + \boldsymbol{v}_0^+(\boldsymbol{r}), \quad \boldsymbol{r} \in V^+ \tag{5.201}$$

For plane wave incidence the general expressions for the scattering amplitudes given in (3.264)–(3.266) become

$$\left.\begin{aligned} g_r(\hat{\boldsymbol{r}}) &= ik_pA_{r1}(\hat{\boldsymbol{r}}) + O(k_p^2) \\ g_\theta(\hat{\boldsymbol{r}}) &= ik_pA_{\theta1}(\hat{\boldsymbol{r}}) + O(k_p^2) \\ g_\varphi(\hat{\boldsymbol{r}}) &= ik_pA_{\varphi1}(\hat{\boldsymbol{r}}) + O(k_p^2) \end{aligned}\right\} \tag{5.202}$$

where, since $\boldsymbol{f}_0^+$ is constant

$$\left.\begin{aligned} A_{r1}(\hat{\boldsymbol{r}}) &= -\frac{1}{4\pi(\lambda^+ + 2\mu^+)}\hat{\boldsymbol{r}} \cdot \int_S \boldsymbol{T}^+(\partial_{\boldsymbol{r}'}, \hat{\boldsymbol{n}}')\boldsymbol{v}_0^+(\boldsymbol{r}')\, ds(\boldsymbol{r}') \\ A_{\theta1}(\hat{\boldsymbol{r}}) &= -\frac{1}{4\pi\mu^+\tau^+}\hat{\boldsymbol{\theta}} \cdot \int_S \boldsymbol{T}^+(\partial_{\boldsymbol{r}'}, \hat{\boldsymbol{n}}')\boldsymbol{v}_0^+(\boldsymbol{r}')\, ds(\boldsymbol{r}') \\ A_{\varphi1}(\hat{\boldsymbol{r}}) &= -\frac{1}{4\pi\mu^+\tau^+}\hat{\boldsymbol{\varphi}} \cdot \int_S \boldsymbol{T}^+(\partial_{\boldsymbol{r}'}, \hat{\boldsymbol{n}}')\boldsymbol{v}_0^+(\boldsymbol{r}')\, ds(\boldsymbol{r}') \end{aligned}\right\} \tag{5.203}$$

The differential scattering cross section is given by (3.286) as

$$\sigma(\hat{\boldsymbol{r}}) = \frac{4\pi}{\alpha_p^2 + \tau^+\alpha_s^2}[A_{r1}^2(\hat{\boldsymbol{r}}) + \tau^{+3}(A_{\theta 1}^2(\hat{\boldsymbol{r}}) + A_{\varphi 1}^2(\hat{\boldsymbol{r}}))] + O(k_p^2) \tag{5.204}$$

while the scattering cross section (3.287) is

$$\sigma_s = \frac{1}{12\pi(\alpha_p^2 + \tau^+\alpha_s^2)}\left[\frac{1}{(\lambda^+ + 2\mu^+)^2} + \frac{2\tau^+}{\mu^{+2}}\right]\left|\int_S \boldsymbol{T}^+(\partial \boldsymbol{r}', \hat{\boldsymbol{n}}')\boldsymbol{v}_0^+(\boldsymbol{r}')\,ds(\boldsymbol{r}')\right|^2 + O(k_p^2) \tag{5.205}$$

The absorption cross section for a rigid scatterer vanishes.

5.E.2 *The cavity problem*

The zeroth order term in the low frequency expansion for the cavity problem is of the form

$$\boldsymbol{u}_0^+(\boldsymbol{r}) = \boldsymbol{f}_0^+(\boldsymbol{r}) + \boldsymbol{v}_0^+(\boldsymbol{r}) \tag{5.206}$$

where $\boldsymbol{v}_0^+$ is the solution of the problem

$$\left.\begin{aligned} \Delta^{e+}\boldsymbol{v}_0^+(\boldsymbol{r}) &= \boldsymbol{0}, & \boldsymbol{r} &\in V^+ \\ \boldsymbol{T}^+(\partial_{\boldsymbol{r}}, \hat{\boldsymbol{n}})\boldsymbol{v}_0^+(\boldsymbol{r}) &= -\boldsymbol{T}^+(\partial_{\boldsymbol{r}}, \hat{\boldsymbol{n}})\boldsymbol{f}_0^+(\boldsymbol{r}), & \boldsymbol{r} &\in S \\ \boldsymbol{v}_0^+(\boldsymbol{r}) &= O\left(\frac{1}{r^2}\right) & &\text{as } r \to \infty \end{aligned}\right\} \tag{5.207}$$

where

$$\boldsymbol{f}_0^+(\boldsymbol{r}) = \boldsymbol{u}_0^i(\boldsymbol{r}) \tag{5.208}$$

For general plane wave incidence

$$\boldsymbol{u}_n^i(\boldsymbol{r}) = \left[\alpha_p\hat{\boldsymbol{k}} + \frac{1}{(\tau^+)^n}\alpha_s\hat{\boldsymbol{b}}\right](\hat{\boldsymbol{k}} \cdot \boldsymbol{r})^n \tag{5.209}$$

and

$$\boldsymbol{f}_0^+(\boldsymbol{r}) = \boldsymbol{u}_0^i(\boldsymbol{r}) = \alpha_p\hat{\boldsymbol{k}} + \alpha_s\hat{\boldsymbol{b}} \tag{5.210}$$

Since $\boldsymbol{f}_0^+$ is constant $\boldsymbol{T}^+(\partial_{\boldsymbol{r}}, \hat{\boldsymbol{n}})\boldsymbol{f}_0^+$ vanishes. Hence, $\boldsymbol{v}_0^+(\boldsymbol{r})$ is a solution of a homogeneous elastostatic cavity problem and therefore vanishes. Hence

$$\boldsymbol{u}_0^+(\boldsymbol{r}) \equiv \boldsymbol{u}_0^i(\boldsymbol{r}), \quad \boldsymbol{r} \in V^+ \tag{5.211}$$

Considering the next low frequency term

$$\boldsymbol{f}_1^+(\boldsymbol{r}) = \boldsymbol{u}_1^i(\boldsymbol{r}) = \left[\alpha_p\hat{\boldsymbol{k}} + \frac{1}{\tau^+}\alpha_s\hat{\boldsymbol{b}}\right](\hat{\boldsymbol{k}} \cdot \boldsymbol{r}) \tag{5.212}$$

and $\boldsymbol{v}_1^+(\boldsymbol{r})$ solves the elastostatic problem

$$\left.\begin{aligned}
\Delta^{e+}\boldsymbol{v}_1^+ &= \boldsymbol{0} && \text{in } V^+ \\
\boldsymbol{T}^+\boldsymbol{v}_1^+ &= -a_p\hat{\boldsymbol{n}}\cdot(\lambda^+\widetilde{\boldsymbol{I}}+2\mu^+\hat{\boldsymbol{k}}\hat{\boldsymbol{k}}) && \\
&\qquad -\alpha_s\frac{\mu^+}{\tau^+}\hat{\boldsymbol{n}}\cdot(\hat{\boldsymbol{b}}\hat{\boldsymbol{k}}+\hat{\boldsymbol{k}}\hat{\boldsymbol{b}}) && \text{on } S \\
\boldsymbol{v}_1^+ &= O\left(\frac{1}{r^2}\right) && \text{as } r\to\infty
\end{aligned}\right\} \tag{5.213}$$

This is a well-posed elastostatic problem and may also be reformulated in terms of Papkovich potentials as in (3.274). Papkovich potentials were used to obtain the specific results presented in Chapters 7 and 8.

In the far field, examination of the expressions for the scattering amplitudes given in (3.298)–(3.306) shows that

$$A_{r1}(\hat{\boldsymbol{r}}) = A_{\theta 1}(\hat{\boldsymbol{r}}) = A_{\varphi 1}(\hat{\boldsymbol{r}}) = 0 \tag{5.214}$$

and

$$\begin{aligned}
A_{r2}(\hat{\boldsymbol{r}}) = \frac{1}{4\pi(\lambda^+ + 2\mu^+)}\int_S [\lambda^+(\hat{\boldsymbol{n}}'\cdot\boldsymbol{u}_0^+(\boldsymbol{r}')) \\
+2\mu^+(\hat{\boldsymbol{r}}\cdot\boldsymbol{u}_0^+(\boldsymbol{r}'))(\hat{\boldsymbol{r}}\cdot\hat{\boldsymbol{n}}')]ds(\boldsymbol{r}')
\end{aligned} \tag{5.215}$$

$$A_{\theta 2}(\hat{\boldsymbol{r}}) = -\frac{1}{4\pi(\tau^+)^2}\hat{\boldsymbol{\theta}}\cdot\int_S [\hat{\boldsymbol{n}}'(\hat{\boldsymbol{r}}\cdot\boldsymbol{u}_0^+(\boldsymbol{r}')) + (\hat{\boldsymbol{n}}'\cdot\hat{\boldsymbol{r}})\boldsymbol{u}_0^+(\boldsymbol{r}')]ds(\boldsymbol{r}') \tag{5.216}$$

$$A_{\varphi 2}(\hat{\boldsymbol{r}}) = -\frac{1}{4\pi(\tau^+)^2}\hat{\boldsymbol{\varphi}}\cdot\int_S [\hat{\boldsymbol{n}}'(\hat{\boldsymbol{r}}\cdot\boldsymbol{u}_0^+(\boldsymbol{r}')) + (\hat{\boldsymbol{n}}'\cdot\hat{\boldsymbol{r}})\boldsymbol{u}_0^+(\boldsymbol{r}')]ds(\boldsymbol{r}') \tag{5.217}$$

However, since $\boldsymbol{u}_0^+(\boldsymbol{r}')$ is constant

$$\int_S \hat{\boldsymbol{n}}'ds(\boldsymbol{r}') = \boldsymbol{0} \tag{5.218}$$

and it follows that

$$A_{r2}(\hat{\boldsymbol{r}}) = A_{\theta 2}(\hat{\boldsymbol{r}}) = A_{\varphi 2}(\hat{\boldsymbol{r}}) = 0 \tag{5.219}$$

Hence, the scattering amplitudes become

$$\left.\begin{aligned}
g_r(\hat{\boldsymbol{r}}) &= ik_p^3A_{r3}(\hat{\boldsymbol{r}}) + O(k_p^4) \\
g_\theta(\hat{\boldsymbol{r}}) &= -ik_p^3A_{\theta 3}(\hat{\boldsymbol{r}}) + O(k_p^4) \\
g_\varphi(\hat{\boldsymbol{r}}) &= -ik_p^3A_{\varphi 3}(\hat{\boldsymbol{r}}) + O(k_p^4)
\end{aligned}\right\} \tag{5.220}$$

where

$$A_{r3}(\hat{\boldsymbol{r}}) = \frac{1}{4\pi(\lambda+2\mu)^2}\int_S [\lambda^+(\hat{\boldsymbol{n}}'\cdot \boldsymbol{u}_1^+(\boldsymbol{r}')) + 2\mu^+(\hat{\boldsymbol{r}}\cdot \boldsymbol{u}_1^+(\boldsymbol{r}'))(\hat{\boldsymbol{r}}\cdot\hat{\boldsymbol{n}}')]ds(\boldsymbol{r}')$$
$$- \frac{|V^-|}{4\pi}\hat{\boldsymbol{r}}\cdot(\alpha_p\hat{\boldsymbol{k}}+\alpha_s\hat{\boldsymbol{b}}) \tag{5.221}$$

$$A_{\theta 3}(\hat{\boldsymbol{r}}) = -\frac{1}{4\pi(\tau^+)^2}\hat{\boldsymbol{\theta}}\cdot\int_S [\hat{\boldsymbol{n}}'(\hat{\boldsymbol{r}}\cdot \boldsymbol{u}_1^+(\boldsymbol{r}')) + (\hat{\boldsymbol{n}}'\cdot\hat{\boldsymbol{r}})\boldsymbol{u}_1^+(\boldsymbol{r}')]ds(\boldsymbol{r}')$$
$$+ \frac{|V^-|}{4\pi(\tau^+)^3}\hat{\boldsymbol{\theta}}\cdot(\alpha_p\hat{\boldsymbol{k}}+\alpha_s\hat{\boldsymbol{b}}) \tag{5.222}$$

$$A_{\varphi 3}(\hat{\boldsymbol{r}}) = -\frac{1}{4\pi(\tau^+)^2}\hat{\boldsymbol{\varphi}}\cdot\int_S [\hat{\boldsymbol{n}}'(\hat{\boldsymbol{r}}\cdot \boldsymbol{u}_1^+(\boldsymbol{r}')) + (\hat{\boldsymbol{n}}'\cdot\hat{\boldsymbol{r}})\boldsymbol{u}_1^+(\boldsymbol{r}')]ds(\boldsymbol{r}')$$
$$+ \frac{|V^-|}{4\pi(\tau^+)^3}\hat{\boldsymbol{\varphi}}\cdot(\alpha_p\hat{\boldsymbol{k}}+\alpha_s\hat{\boldsymbol{b}}) \tag{5.223}$$

The differential scattering cross section is

$$\sigma(\hat{\boldsymbol{r}}) = \frac{4\pi}{\alpha_p^2+\tau^+\alpha_s^2}k_p^4[A_{r3}^2(\hat{\boldsymbol{r}}) + (\tau^+)^3(A_{\theta 3}^2(\hat{\boldsymbol{r}}) + A_{\varphi 3}^2(\hat{\boldsymbol{r}}))] + O(k_p^6) \tag{5.224}$$

and the scattering cross section is

$$\sigma_s = \frac{k_p^4}{\alpha_p^2+\tau^+\alpha_s^2}\int_{S^2} [A_{r3}^2(\hat{\boldsymbol{r}}) + (\tau^+)^3(A_{\theta 3}^2(\hat{\boldsymbol{r}}) + A_{\varphi 3}^2(\hat{\boldsymbol{r}}))]\, ds(\hat{\boldsymbol{r}}) + O(k_p^6) \tag{5.225}$$

As with the case of a rigid scatterer the absorption cross section for the cavity problem vanishes.

5.E.3 *The transmission problem*

The low frequency representation of the displacement field, in the case of the elastic scatterer, involves fields both interior and exterior to the scatterer and hence we need the expansions of both $\boldsymbol{u}^+$ and $\boldsymbol{u}^-$, which are given in (3.229) and (3.231), respectively. The coefficients in the expansions satisfy the transmission conditions on S

$$\left.\begin{aligned} \boldsymbol{u}_n^+(\boldsymbol{r}) &= \boldsymbol{u}_n^-(\boldsymbol{r}) \\ \boldsymbol{T}^+(\partial_r,\hat{\boldsymbol{n}})\boldsymbol{u}_n^+(\boldsymbol{r}) &= \boldsymbol{T}^-(\partial_r,\hat{\boldsymbol{n}})\boldsymbol{u}_n^-(\boldsymbol{r}) \end{aligned}\right\} \tag{5.226}$$

The coefficients have the representations (3.248) and (3.251), which are

$$\left.\begin{aligned} \boldsymbol{u}_n^+ &= \boldsymbol{f}_n^+ + \boldsymbol{v}_n^+ \quad && \text{in } V^+ \\ \boldsymbol{u}_n^- &= -\boldsymbol{f}_n^- - \boldsymbol{v}_n^- \quad && \text{in } V^- \end{aligned}\right\} \tag{5.227}$$

where $\boldsymbol{f}_n^{\pm}$ are given explicitly in (3.249) and (3.252) and $\boldsymbol{v}_n^{\pm}(\boldsymbol{r})$ is the solution of the following elastostatic transmission problem

$$\left.\begin{aligned}\Delta^{e+}\boldsymbol{v}_n^{+} &= \mathbf{0} && \text{in } V^{+}\\ \Delta^{e-}\boldsymbol{v}_n^{-} &= \mathbf{0} && \text{in } V^{-}\\ \boldsymbol{v}_n^{+} &= -\boldsymbol{v}_n^{-} - \boldsymbol{f}_n^{-} - \boldsymbol{f}_n^{+} && \text{on } S\\ \boldsymbol{T}^{+}\boldsymbol{v}_n^{+} &= -\boldsymbol{T}^{-}\boldsymbol{v}_n^{-} - \boldsymbol{T}^{-}\boldsymbol{f}_n^{-} - \boldsymbol{T}^{+}\boldsymbol{f}_n^{+} && \text{on } S\\ \boldsymbol{v}_n^{+} &= O\left(\frac{1}{r}\right) && \text{as } r\to\infty\end{aligned}\right\} \tag{5.228}$$

For general plane wave incidence

$$\boldsymbol{u}_0^{i} = \alpha_p\hat{\boldsymbol{k}} + \alpha_s\hat{\boldsymbol{b}} \tag{5.229}$$

$$\left.\begin{aligned}\boldsymbol{f}_0^{+} &= \alpha_p\hat{\boldsymbol{k}} + \alpha_s\hat{\boldsymbol{b}}\\ \boldsymbol{f}_0^{-} &= \mathbf{0}\end{aligned}\right\} \tag{5.230}$$

and the problem of finding $\boldsymbol{v}_0^{\pm}$ reduces to solving

$$\left.\begin{aligned}\Delta^{e+}\boldsymbol{v}_0^{+} &= \mathbf{0} && \text{in } V^{+}\\ \Delta^{e-}\boldsymbol{v}_0^{-} &= \mathbf{0} && \text{in } V^{-}\\ \boldsymbol{v}_0^{+} &= -\boldsymbol{v}_0^{-} - (\alpha_p\hat{\boldsymbol{k}} + \alpha_s\hat{\boldsymbol{b}}) && \text{on } S\\ \boldsymbol{T}^{+}\boldsymbol{v}_0^{+} &= -\boldsymbol{T}^{-}\boldsymbol{v}_0^{-} && \text{on } S\\ \boldsymbol{v}_0^{+}(\boldsymbol{r}) &= O\left(\frac{1}{r}\right) && \text{as } r\to\infty\end{aligned}\right\} \tag{5.231}$$

since $\boldsymbol{T}^{+}\boldsymbol{u}_0^{i} = \mathbf{0}$.

This problem has the obvious solution

$$\boldsymbol{v}_0^{+} = 0, \qquad \boldsymbol{v}_0^{-} = -(\alpha_p\hat{\boldsymbol{k}} + \alpha_s\hat{\boldsymbol{b}}) \tag{5.232}$$

in which case

$$\boldsymbol{u}_0^{\pm} = \alpha_p\hat{\boldsymbol{k}} + \alpha_s\hat{\boldsymbol{b}} \quad \text{in } V^{\pm} \tag{5.233}$$

To determine the next term much more work is required since there is no simple solution, even for very simple geometries. Since $\boldsymbol{T}^{\pm}\boldsymbol{u}_0^{\pm} = \mathbf{0}$ and $\boldsymbol{T}^{\pm}\tilde{\boldsymbol{\gamma}}_1^{\pm} = \tilde{\mathbf{0}}$ by (3.240), the expressions for $\boldsymbol{f}_1^{\pm}$ become

$$\left.\begin{aligned}\boldsymbol{f}_1^{+} &= \left(\alpha_p\hat{\boldsymbol{k}} + \frac{1}{\tau^{+}}\alpha_s\hat{\boldsymbol{b}}\right)(\hat{\boldsymbol{k}}\cdot\boldsymbol{r})\\ \boldsymbol{f}_1^{-} &= \mathbf{0}\end{aligned}\right\} \tag{5.234}$$

and the elastostatic transmission problem for $\boldsymbol{v}_1^{\pm}$ is that of determining $\boldsymbol{v}_1^{\pm}$ to satisfy

$$\left.\begin{aligned} \Delta^{e+}\boldsymbol{v}_1^{+} &= \mathbf{0} && \text{in } V^+ \\ \Delta^{e-}\boldsymbol{v}_1^{-} &= \mathbf{0} && \text{in } V^- \\ \boldsymbol{v}_1^{+} &= \boldsymbol{v}_1^{-} - \left(\alpha_p\hat{\boldsymbol{k}} + \frac{1}{\tau^+}\alpha_s\hat{\boldsymbol{b}}\right)(\hat{\boldsymbol{k}}\cdot\boldsymbol{r}) && \text{on } S \\ \boldsymbol{T}^{+}\boldsymbol{v}_1^{+} &= \boldsymbol{T}^{-}\boldsymbol{v}_1^{-} - \alpha_p\hat{\boldsymbol{n}}\cdot(\lambda^+\widetilde{\boldsymbol{I}} + 2\mu^+\hat{\boldsymbol{k}}\hat{\boldsymbol{k}}) && \\ &\quad - \frac{\alpha_s\mu^+}{\tau^+}\hat{\boldsymbol{n}}\cdot(\hat{\boldsymbol{b}}\hat{\boldsymbol{k}} + \hat{\boldsymbol{k}}\hat{\boldsymbol{b}}) && \text{on } S \\ \boldsymbol{v}_1^{+} &= O\left(\frac{1}{r}\right) && \text{as } r \to \infty \end{aligned}\right\} \tag{5.235}$$

In the far field the expressions for the scattering amplitudes are given by (3.264)–(3.266) with coefficients given in (3.325)–(3.327) or, alternatively, using the approach in Chapter 4, by (4.104)–(4.108). Since $\boldsymbol{u}_0^{+}$ is constant, the terms $A_{r2}(\hat{\boldsymbol{r}})$, $A_{\theta 2}(\hat{\boldsymbol{r}})$ and $A_{\varphi 2}(\hat{\boldsymbol{r}})$ vanish in the same way as in the cavity case, and as in that case, the scattering amplitudes are given by

$$\left.\begin{aligned} g_r(\hat{\boldsymbol{r}}) &= ik_p^3 A_{r3}(\hat{\boldsymbol{r}}) + O(k_p^4) \\ g_\theta(\hat{\boldsymbol{r}}) &= -ik_p^3 A_{\theta 3}(\hat{\boldsymbol{r}}) + O(k_p^4) \\ g_\varphi(\hat{\boldsymbol{r}}) &= -ik_p^3 A_{\varphi 3}(\hat{\boldsymbol{r}}) + O(k_p^4) \end{aligned}\right\} \tag{5.236}$$

where now

$$\begin{aligned} A_{r3}(\hat{\boldsymbol{r}}) = \frac{1}{4\pi(\lambda^+ + 2\mu^+)} &\left[2(\mu^+ - \mu^-)\int_S (\hat{\boldsymbol{r}}\cdot\boldsymbol{u}_1^{+}(\boldsymbol{r}'))(\hat{\boldsymbol{r}}\cdot\hat{\boldsymbol{n}}')\,ds(\boldsymbol{r}') \right. \\ &\left. + (\lambda^+ - \lambda^-)\int_S \hat{\boldsymbol{n}}'\cdot\boldsymbol{u}_1^{+}(\boldsymbol{r}')\,ds(\boldsymbol{r}') \right] \\ &- \frac{1}{4\pi}\left(1 - \frac{\rho^-}{\rho^+}\right)\hat{\boldsymbol{r}}\cdot(\alpha_p\hat{\boldsymbol{k}} + \alpha_s\hat{\boldsymbol{b}})|V^-| \end{aligned} \tag{5.237}$$

$$\begin{aligned} A_{\theta 3}(\hat{\boldsymbol{r}}) = -\frac{1}{4\pi(\tau^+)^2}\left(1 - \frac{\mu^-}{\mu^+}\right) &\int_S \Big[(\hat{\boldsymbol{\theta}}\cdot\boldsymbol{u}_1^{+}(\boldsymbol{r}'))(\hat{\boldsymbol{n}}'\cdot\hat{\boldsymbol{r}}) \\ &+ (\hat{\boldsymbol{r}}\cdot\boldsymbol{u}_1^{+}(\boldsymbol{r}'))(\hat{\boldsymbol{n}}'\cdot\hat{\boldsymbol{\theta}})\Big]\,ds(\boldsymbol{r}') \\ &+ \frac{1}{4\pi(\tau^+)^3}\left(1 - \frac{\rho^-}{\rho^+}\right)\hat{\boldsymbol{\theta}}\cdot(\alpha_p\hat{\boldsymbol{k}} + \alpha_s\hat{\boldsymbol{b}})|V^-| \end{aligned} \tag{5.238}$$

$$A_{\varphi 3}(\hat{\boldsymbol{r}}) = -\frac{1}{4\pi(\tau^+)^2}\left(1-\frac{\mu^-}{\mu^+}\right)\int_S \left[(\hat{\boldsymbol{\varphi}}\cdot\boldsymbol{u}_1^+(\boldsymbol{r}'))(\hat{\boldsymbol{n}}'\cdot\hat{\boldsymbol{r}}) + (\hat{\boldsymbol{r}}\cdot\hat{\boldsymbol{u}}_1^+(\boldsymbol{r}'))(\hat{\boldsymbol{n}}'\cdot\hat{\boldsymbol{\varphi}})\right] ds(\boldsymbol{r}')$$
$$+\frac{1}{4\pi(\tau^+)^3}\left(1-\frac{\rho^-}{\rho^+}\right)\hat{\boldsymbol{\varphi}}\cdot(\alpha_p\hat{\boldsymbol{k}}+\alpha_s\hat{\boldsymbol{b}})|V^-| \tag{5.239}$$

Using the expressions (4.111)–(4.113) we can also write (5.237)–(5.239) as follows

$$A_{r3}(\hat{\boldsymbol{r}}) = \frac{1}{4\pi}\frac{(\lambda^+-\lambda^-)\widetilde{\boldsymbol{I}} + 2(\mu^+-\mu^-)\hat{\boldsymbol{r}}\hat{\boldsymbol{r}}}{\lambda^+ + 2\mu^+} : \widetilde{\boldsymbol{M}}_1^0(\hat{\boldsymbol{r}})$$
$$+\frac{1}{4\pi}\left(1-\frac{\rho^-}{\rho^+}\right)(\alpha_p\hat{\boldsymbol{k}}+\alpha_s\hat{\boldsymbol{b}})\cdot\hat{\boldsymbol{r}}|V^-| \tag{5.240}$$

$$A_{\theta 3}(\hat{\boldsymbol{r}}) = -\frac{1}{4\pi(\tau^+)^2}\left(1-\frac{\mu^-}{\mu^+}\right)(\hat{\boldsymbol{r}}\hat{\boldsymbol{\theta}}+\hat{\boldsymbol{\theta}}\hat{\boldsymbol{r}}) : \widetilde{\boldsymbol{M}}_1^0(\hat{\boldsymbol{r}})$$
$$+\frac{1}{4\pi(\tau^+)^2}\left(1-\frac{\rho^-}{\rho^+}\right)(\alpha_p\hat{\boldsymbol{k}}+\alpha_s\hat{\boldsymbol{b}})\cdot\widetilde{\boldsymbol{\theta}}|V^-| \tag{5.241}$$

$$A_{\varphi 3}(\hat{\boldsymbol{r}}) = -\frac{1}{4\pi(\tau^+)^2}\left(1-\frac{\mu^-}{\mu^+}\right)(\hat{\boldsymbol{r}}\hat{\boldsymbol{\varphi}}+\hat{\boldsymbol{\varphi}}\hat{\boldsymbol{r}}) : \widetilde{\boldsymbol{M}}_1^0(\hat{\boldsymbol{r}})$$
$$+\frac{1}{4\pi(\tau^+)^3}\left(1-\frac{\rho^-}{\rho^+}\right)(\alpha_p\hat{\boldsymbol{k}}+\alpha_s\hat{\boldsymbol{b}})\cdot\hat{\boldsymbol{\varphi}}|V^-| \tag{5.242}$$

where

$$\widetilde{\boldsymbol{M}}_1^0(\hat{\boldsymbol{r}}) = \int_S \hat{\boldsymbol{n}}'\boldsymbol{u}_1^+(\boldsymbol{r}')\,ds(\boldsymbol{r}') = \int_{V^-}\nabla_{\boldsymbol{r}'}\boldsymbol{u}_1^-(\boldsymbol{r}')\,dv(\boldsymbol{r}') \tag{5.243}$$

The differential scattering cross section has the same form as in the cavity case

$$\sigma(\hat{\boldsymbol{r}}) = 4\pi\frac{k_p^4}{\alpha_p^2+\tau^+\alpha_s^2}[A_{r3}^2(\hat{\boldsymbol{r}}) + (\tau^+)^3(A_{\theta 3}^2(\hat{\boldsymbol{r}}) + A_{\varphi 3}^2(\hat{\boldsymbol{r}}))] + O(k_p^6) \tag{5.244}$$

where the coefficients A_{r3}, $A_{\theta 3}$ and $A_{\varphi 3}$ are now given by (5.237)–(5.239) or by (5.240)–(5.242). The scattering cross section is obtained from the definition

$$\sigma_s = \frac{1}{4\pi}\int_{S^2}\sigma(\hat{\boldsymbol{r}})\,ds(\hat{\boldsymbol{r}}) \tag{5.245}$$

and since λ^- and μ^- are real, the absorption cross section vanishes. Lossy media, with complex Lamé constants accounting for viscous effects, have not been analyzed in the low frequency regime.

5.E.4 *Problems in elasticity*

1. Derive formula (5.205).
2. Show that the incident field (5.212) generates the following surface traction

$$T^{+}(\partial_{\boldsymbol{r}}, \hat{\boldsymbol{n}})\boldsymbol{u}_1^i(\boldsymbol{r}) = \alpha_p \hat{\boldsymbol{n}} \cdot (\lambda^{+}\widetilde{\boldsymbol{I}} + 2\mu^{+}\hat{\boldsymbol{r}}\hat{\boldsymbol{r}})$$
$$+ \alpha_s \frac{\mu^{+}}{\tau^{+}} \hat{\boldsymbol{n}} \cdot (\hat{\boldsymbol{k}}\hat{\boldsymbol{b}} + \hat{\boldsymbol{b}}\hat{\boldsymbol{k}})$$

3. Use (3.304)–(3.306) to derive the expressions (5.221)–(5.223).
4. Derive formulae (5.237)–(5.239) and show that they can be written in the form (5.240)–(5.243).

6

INTEGRAL EQUATIONS

In Chapters 3–5 it was shown how the problems of determining the coefficients in the low frequency expansion of the field quantities in acoustics and electromagnetics could be formulated as standard problems for Laplace's equation while in elastodynamics the terms could be found as solutions of standard problems for the static Navier equation. Techniques for solving these boundary value and transmission problems have been central to the study of partial differential equations in mathematical physics for more than a hundred years. In recent years they have served as the benchmark problems for numerical approaches such as finite element methods. It is not intended to discuss these methods here nor do we presume to suggest references. There are hundreds of texts and thousands of papers on the subject and it is inconceivable that any reader who has progressed to this stage of the present work would not be familiar with many of them.

An alternative to differential equation based methods are **integral equation techniques**. This chapter is devoted to a presentation of the integral equation formulation of each of the problems discussed in Chapters 3, 4 and 5. There are many excellent books on integral equations. Among them we refer to Kress (1989), which is tightly connected to scattering theory and Zabreyko *et al.* (1975) which is a reference book.

6.0 Fundamental relations

It is not intended to present a complete theory of integral equations. In keeping with the spirit of this text we avoid the heavy technical details regarding the appropriate function spaces in which solutions lie and the minimal smoothness requirements on S for which the integral equations and representations remain valid. For smooth boundaries one may consult the classic works of Günter (1967) and Mikhlin (1970) in potential theory, Colton and Kress (1983) in acoustics and electromagnetics, Kupradze (1963, 1979) in elasticity, and for nonsmooth boundaries, Van Bladel (1995).

Here we record some useful relations and properties. If $w(\boldsymbol{r})$ is a continuous function on S the **single layer potential** with **density** $w(\boldsymbol{r})$

$$u_s(\boldsymbol{r}) = \frac{1}{4\pi} \int_S \frac{w(\boldsymbol{r}')}{|\boldsymbol{r}-\boldsymbol{r}'|}\, ds(\boldsymbol{r}') \tag{6.1}$$

is continuous as a function of $\boldsymbol{r}$ in all of $\mathbb{R}^3$ and obeys the well-known **jump condition**

$$\frac{\partial}{\partial n^{\pm}} \frac{1}{4\pi} \int_S \frac{w(\boldsymbol{r}')}{|\boldsymbol{r}-\boldsymbol{r}'|}\, ds(\boldsymbol{r}') = \mp \frac{w(\boldsymbol{r})}{2} + \frac{1}{4\pi} \int_S w(\boldsymbol{r}') \frac{\partial}{\partial n} \frac{1}{|\boldsymbol{r}-\boldsymbol{r}'|}\, ds(\boldsymbol{r}') \tag{6.2}$$

where $\partial/\partial n^{\pm}$ notes the one-sided directional derivative in the direction of the normal into V^+ as $\boldsymbol{r}$ approaches S from $V^{\pm}$. The term involving the normal derivative of $1/|\boldsymbol{r}-\boldsymbol{r}'|$ occurring in the integral is defined as

$$\frac{\partial}{\partial n}\frac{1}{|\boldsymbol{r}-\boldsymbol{r}'|} = -\hat{\boldsymbol{n}}\cdot\frac{(\boldsymbol{r}-\boldsymbol{r}')}{|\boldsymbol{r}-\boldsymbol{r}'|^3}, \quad \boldsymbol{r},\boldsymbol{r}' \in S \tag{6.3}$$

and constitutes a so-called **weak singularity** if S is smooth, in which case the integral exists as an improper integral. By smooth is meant that S admits a twice differentiable parameterization. Actually weaker smoothness such as Lyapunov conditions (Günter 1967) suffices and the relation remains valid in different senses under a variety of weaker conditions on w and S.

Similarily, the **double layer potential** with continuous **density** $w(\boldsymbol{r})$ is defined for $\boldsymbol{r} \notin S$ as

$$u_d(\boldsymbol{r}) = \frac{1}{4\pi}\int_S w(\boldsymbol{r}')\frac{\partial}{\partial n'}\frac{1}{|\boldsymbol{r}-\boldsymbol{r}'|}\,ds(\boldsymbol{r}') \tag{6.4}$$

and satisfies the jump condition as $\boldsymbol{r}$ approaches S from V^+ or V^-

$$\begin{aligned}\lim_{\substack{\boldsymbol{r}\to S\\ \boldsymbol{r}\in V^{\pm}}} \frac{1}{4\pi}\int_S w(\boldsymbol{r}')\frac{\partial}{\partial n'}\frac{1}{|\boldsymbol{r}-\boldsymbol{r}'|}\,ds(\boldsymbol{r}') \\ = \pm\frac{w(\boldsymbol{r})}{2} + \frac{1}{4\pi}\int_S w(\boldsymbol{r}')\frac{\partial}{\partial n'}\frac{1}{|\boldsymbol{r}-\boldsymbol{r}'|}\,ds(\boldsymbol{r}') \\ = \pm\frac{w(\boldsymbol{r})}{2} + \frac{1}{4\pi}\int_S w(\boldsymbol{r}')\hat{\boldsymbol{n}}'\cdot\frac{(\boldsymbol{r}-\boldsymbol{r}')}{|\boldsymbol{r}-\boldsymbol{r}'|^3}\,ds(\boldsymbol{r}')\end{aligned} \tag{6.5}$$

To complete the layer lexicon it is noted that if w is differentiable, then the normal derivative of the double layer exists and

$$\frac{\partial}{\partial n^+}\int_S w(\boldsymbol{r}')\frac{\partial}{\partial n'}\frac{1}{|\boldsymbol{r}-\boldsymbol{r}'|}\,ds(\boldsymbol{r}') = \frac{\partial}{\partial n^-}\int_S w(\boldsymbol{r}')\frac{\partial}{\partial n'}\frac{1}{|\boldsymbol{r}-\boldsymbol{r}'|}\,ds(\boldsymbol{r}') \tag{6.6}$$

This relation also remains valid under weaker conditions in various senses, see, for example, Colton and Kress (1983) and Kirsch (1989). It should be noted that formally interchanging integration and differentiation in (6.6) leads to an integral with a **singularity** so high that the integral does not exist as an improper integral nor in the **Cauchy principal value** sense. For this reason the normal derivative of a double layer is a so-called **hypersingular integral operator** on the density $w(\boldsymbol{r})$.

Corresponding to the jump condition for the single layer there is the following vector relation (Colton and Kress 1983). If $\hat{\boldsymbol{n}} \times \boldsymbol{u}(\boldsymbol{r})$ is continuous on S, then

$$\lim_{\substack{\boldsymbol{r}\to S\\ \boldsymbol{r}\in V^{\pm}}} \hat{\boldsymbol{n}} \times \left[\nabla \times \frac{1}{4\pi}\int_S \frac{\hat{\boldsymbol{n}}' \times \boldsymbol{u}(\boldsymbol{r}')}{|\boldsymbol{r}-\boldsymbol{r}'|}\, ds(\boldsymbol{r}')\right]$$

$$= \pm\frac{\hat{\boldsymbol{n}} \times \boldsymbol{u}(\boldsymbol{r}')}{2} + \frac{1}{4\pi}\int_S \hat{\boldsymbol{n}} \times \left[\left(\nabla\frac{1}{|\boldsymbol{r}-\boldsymbol{r}'|}\right) \times (\hat{\boldsymbol{n}}' \times \boldsymbol{u}(\boldsymbol{r}'))\right] ds(\boldsymbol{r}') \tag{6.7}$$

The single and double layers have exact analog in elastostatics where the fundamental dyadics $\widetilde{\gamma}_0^{\pm}(\boldsymbol{r}, \boldsymbol{r}')$ are defined in (3.233), the traction operators $\boldsymbol{T}^{\pm}(\partial_{\boldsymbol{r}}, \hat{\boldsymbol{n}})$ are defined in (1.213) and $\boldsymbol{T}^{\pm}(\partial_{\boldsymbol{r}'}, \hat{\boldsymbol{n}}')\widetilde{\gamma}_0^{\pm}(\boldsymbol{r}, \boldsymbol{r}')$ is given in (3.239). Explicitly

$$\widetilde{\gamma}_0^{\pm}(\boldsymbol{r}, \boldsymbol{r}') = \frac{1}{2|\boldsymbol{r}-\boldsymbol{r}'|}\left\{(1+(\tau^{\pm})^2)\widetilde{\boldsymbol{I}} + (1-(\tau^{\pm})^2)\frac{(\boldsymbol{r}-\boldsymbol{r}')(\boldsymbol{r}-\boldsymbol{r}')}{|\boldsymbol{r}-\boldsymbol{r}'|^2}\right\} \tag{6.8}$$

$$\boldsymbol{T}^{\pm}(\partial_{\boldsymbol{r}}, \hat{\boldsymbol{n}}) = 2\mu^{\pm}\hat{\boldsymbol{n}}\cdot\nabla_{\boldsymbol{r}} + \lambda^{\pm}\hat{\boldsymbol{n}}\nabla_{\boldsymbol{r}}\cdot + \mu^{\pm}\hat{\boldsymbol{n}} \times (\nabla_{\boldsymbol{r}}\times) \tag{6.9}$$

and

$$\boldsymbol{T}^{\pm}(\partial_{\boldsymbol{r}'}, \hat{\boldsymbol{n}}')\widetilde{\gamma}_0^{\pm}(\boldsymbol{r}, \boldsymbol{r}')$$

$$= \frac{1}{|\boldsymbol{r}-\boldsymbol{r}'|^2}\left\{\mu^{\pm}(\tau^{\pm})^2\left[\frac{\hat{\boldsymbol{n}}'\cdot(\boldsymbol{r}-\boldsymbol{r}')}{|\boldsymbol{r}-\boldsymbol{r}'|}\widetilde{\boldsymbol{I}} + \frac{(\boldsymbol{r}-\boldsymbol{r}')\hat{\boldsymbol{n}}'}{|\boldsymbol{r}-\boldsymbol{r}'|}\right]\right.$$

$$+ 3\mu^{\pm}(1-(\tau^{\pm})^2)\frac{\hat{\boldsymbol{n}}'\cdot(\boldsymbol{r}-\boldsymbol{r}')(\boldsymbol{r}-\boldsymbol{r}')(\boldsymbol{r}-\boldsymbol{r}')}{|\boldsymbol{r}-\boldsymbol{r}'|^3}$$

$$\left. + [\lambda^{\pm}(\tau^{\pm})^2 + \mu^{\pm}((\tau^{\pm})^2 - 1)]\frac{\hat{\boldsymbol{n}}'(\boldsymbol{r}-\boldsymbol{r}')}{|\boldsymbol{r}-\boldsymbol{r}'|}\right\} \tag{6.10}$$

The **elastic single layer** with continuous vector valued **density** $\boldsymbol{w}(\boldsymbol{r})$ is

$$\boldsymbol{u}_s^{\pm}(\boldsymbol{r}) = \frac{1}{4\pi\mu^{\pm}}\int_S \widetilde{\gamma}_0^{\pm}(\boldsymbol{r}, \boldsymbol{r}')\cdot\boldsymbol{w}(\boldsymbol{r}')\, ds(\boldsymbol{r}') \tag{6.11}$$

and is continuous in $\mathbb{R}^3$. If the density is somewhat smoother (Hölder continuous) then the corresponding jump condition is

$$\lim_{\substack{r\to S\\ r\in V^\pm}} \frac{1}{4\pi\mu^\pm}\boldsymbol{T}^\pm(\partial_{\boldsymbol{r}},\hat{\boldsymbol{n}})\int_S \widetilde{\gamma}_0^\pm(\boldsymbol{r},\boldsymbol{r}')\cdot\boldsymbol{w}(\boldsymbol{r}')\,ds(\boldsymbol{r}')$$

$$=\pm\frac{\boldsymbol{w}(\boldsymbol{r})}{2}+\frac{1}{4\pi\mu^\pm}\int_S(\boldsymbol{T}^\pm(\partial_{\boldsymbol{r}},\hat{\boldsymbol{n}})\widetilde{\gamma}_0^\pm(\boldsymbol{r},\boldsymbol{r}'))\cdot\boldsymbol{w}(\boldsymbol{r}')\,ds(\boldsymbol{r}') \tag{6.12}$$

Whereas the **elastic double layer**

$$\boldsymbol{u}_d^\pm(\boldsymbol{r})=\frac{1}{4\pi\mu^\pm}\int_S \boldsymbol{w}(\boldsymbol{r}')\cdot\boldsymbol{T}^\pm(\partial_{\boldsymbol{r}'},\hat{\boldsymbol{n}}')\widetilde{\gamma}_0^\pm(\boldsymbol{r},\boldsymbol{r}')\,ds(\boldsymbol{r}') \tag{6.13}$$

has the following property as $\boldsymbol{r}\to S$

$$\lim_{\substack{r\to S\\ r\in V^\pm}} \frac{1}{4\pi\mu^\pm}\int_S \boldsymbol{w}(\boldsymbol{r}')\cdot\boldsymbol{T}^\pm(\partial_{\boldsymbol{r}'},\hat{\boldsymbol{n}}')\widetilde{\gamma}_0^\pm(\boldsymbol{r},\boldsymbol{r}')\,ds(\boldsymbol{r}')$$

$$=\pm\frac{\boldsymbol{w}(\boldsymbol{r})}{2}+\frac{1}{4\pi\mu^\pm}\int_S \boldsymbol{w}(\boldsymbol{r}')\cdot\boldsymbol{T}^\pm(\partial_{\boldsymbol{r}'},\hat{\boldsymbol{n}}')\widetilde{\gamma}_0^\pm(\boldsymbol{r},\boldsymbol{r}')\,ds(\boldsymbol{r}') \tag{6.14}$$

The fact that the traction of the fundamental dyadic $\widetilde{\gamma}_0^\pm$ has a Cauchy-type singularity means that the integrals on the right-hand sides of (6.12) and (6.14) must be interpreted in the principal value sense and require additional smoothness on the density $\boldsymbol{w}$ (Hölder continuity). Finally, the traction of the double layer, which is the hypersingular operator

$$\boldsymbol{t}_d^\pm(\boldsymbol{r})=\boldsymbol{T}^\pm(\partial_{\boldsymbol{r}},\hat{\boldsymbol{n}})\int_S \boldsymbol{w}(\boldsymbol{r}')\cdot\boldsymbol{T}^\pm(\partial_{\boldsymbol{r}'},\hat{\boldsymbol{n}}')\widetilde{\gamma}_0^\pm(\boldsymbol{r},\boldsymbol{r}')\,ds(\boldsymbol{r}') \tag{6.15}$$

has meaning in both classical and Sobolev spaces (Kupradze 1979).

6.A Acoustics

6.A.1 *Basic equations in acoustics*

With the jump conditions, the expressions (3.16), (3.17), (3.24) and (3.25) for the coefficients $u_n^\pm$ and their normal derivatives in the low frequency expansion of the field lead to the following sets of integral equations for $\boldsymbol{r}$ on S

$$u_n^+(\boldsymbol{r})=2f_n^+(\boldsymbol{r})+\frac{1}{2\pi}\int_S\left[u_n^+(\boldsymbol{r}')\frac{\partial}{\partial n'}\frac{1}{|\boldsymbol{r}-\boldsymbol{r}'|}-\frac{1}{|\boldsymbol{r}-\boldsymbol{r}'|}\frac{\partial}{\partial n'}u_n^+(\boldsymbol{r}')\right]ds(\boldsymbol{r}') \tag{6.16}$$

$$\frac{\partial}{\partial n}u_n^+(\boldsymbol{r})=2\frac{\partial}{\partial n}f_n^+(\boldsymbol{r})+\frac{1}{2\pi}\frac{\partial}{\partial n}\int_S u_n^+(\boldsymbol{r}')\frac{\partial}{\partial n'}\frac{1}{|\boldsymbol{r}-\boldsymbol{r}'|}\,ds(\boldsymbol{r}')$$

$$-\frac{1}{2\pi}\int_S\left(\frac{\partial}{\partial n}\frac{1}{|\boldsymbol{r}-\boldsymbol{r}'|}\right)\frac{\partial}{\partial n'}u_n^+(\boldsymbol{r}')\,ds(\boldsymbol{r}') \tag{6.17}$$

$$-u_n^-(\boldsymbol{r}) = 2f_n^-(\boldsymbol{r}) + \frac{1}{2\pi}\int_S \left[u_n^-(\boldsymbol{r}')\frac{\partial}{\partial n'}\frac{1}{|\boldsymbol{r}-\boldsymbol{r}'|} - \frac{1}{|\boldsymbol{r}-\boldsymbol{r}'|}\frac{\partial}{\partial n'}u_n^-(\boldsymbol{r}')\right] ds(\boldsymbol{r}') \tag{6.18}$$

$$-\frac{\partial}{\partial n}u_n^-(\boldsymbol{r}) = 2\frac{\partial}{\partial n}f_n^-(\boldsymbol{r}) + \frac{1}{2\pi}\frac{\partial}{\partial n}\int_S u_n^-(\boldsymbol{r}')\frac{\partial}{\partial n'}\frac{1}{|r-r'|}\, ds(r')$$
$$-\frac{1}{2\pi}\int_S \left(\frac{\partial}{\partial n}\frac{1}{|\boldsymbol{r}-\boldsymbol{r}'|}\right)\frac{\partial}{\partial n'}u_n^-(\boldsymbol{r}')\, ds(\boldsymbol{r}') \tag{6.19}$$

These then may be used together with the boundary and transmission conditions to obtain the following integral equations for the unknown surface field quantity in each problem.

6.A.2 *The Dirichlet problem*

If $u_n^+(\boldsymbol{r}) = 0$ on S, then (6.16) and (6.17) become

$$\frac{1}{2\pi}\int_S \frac{1}{|\boldsymbol{r}-\boldsymbol{r}'|}\frac{\partial}{\partial n'}u_n^+(\boldsymbol{r}')\, ds(\boldsymbol{r}') = 2f_n^+(\boldsymbol{r}) \tag{6.20}$$

$$\frac{\partial}{\partial n}u_n^+(\boldsymbol{r}) + \frac{1}{2\pi}\int_S \left(\frac{\partial}{\partial n}\frac{1}{|\boldsymbol{r}-\boldsymbol{r}'|}\right)\frac{\partial}{\partial n'}u_n^+(\boldsymbol{r}')\, ds(\boldsymbol{r}') = 2\frac{\partial}{\partial n}f_n^+(\boldsymbol{r}) \tag{6.21}$$

where $f_n^+(\boldsymbol{r})$ is defined in (3.35) and (6.20) and (6.21) hold for all $\boldsymbol{r}$ on S. Equations (6.20) and (6.21) constitute first and second kind integral equations for the normal derivative of u_n^+ which may be used to represent u_n^+ in V^+ as

$$u_n^+(\boldsymbol{r}) = f_n^+(\boldsymbol{r}) - \frac{1}{4\pi}\int_S \frac{1}{|\boldsymbol{r}-\boldsymbol{r}'|}\frac{\partial}{\partial n'}u_n^+(\boldsymbol{r}')\, ds(\boldsymbol{r}') \tag{6.22}$$

Alternatively, one might seek a representation of $u_n^+(\boldsymbol{r})$ in terms of a double layer with an unknown density w as

$$u_n^+(\boldsymbol{r}) = f_n^+(\boldsymbol{r}) - \frac{1}{4\pi}\int_S w(\boldsymbol{r}')\frac{\partial}{\partial n'}\frac{1}{|\boldsymbol{r}-\boldsymbol{r}'|}\, ds(\boldsymbol{r}'), \quad \boldsymbol{r} \in V^+ \tag{6.23}$$

which with the boundary condition and the jump condition (6.5), leads to the boundary integral equation for $w(\boldsymbol{r})$

$$w(\boldsymbol{r}) + \frac{1}{2\pi}\int_S w(\boldsymbol{r}')\frac{\partial}{\partial n'}\frac{1}{|\boldsymbol{r}-\boldsymbol{r}'|}\,ds(r') = 2f_n^+(\boldsymbol{r}) \tag{6.24}$$

Traditionally, second kind integral equations have been preferred over first kind equations. One reason for this lies in the fact that when the integral operators are compact on the space $L_2(S)$, which is the present case when S is smooth, then they have unbounded inverses. Therefore, the first kind integral equation cannot be solved by inverting the integral operator, whereas the operators in second kind equations, which are of the form identity plus a compact operator, have bounded inverses and hence the Fredholm theory is readily applicable. However, the recent work of Hsiao (1991) has demonstrated that when viewed in the appropriate Sobolev spaces the preference for second kind equations loses its significance. In fact first kind equations may have some advantages. For example, in the present case the well-known Gauss integral

$$\frac{1}{4\pi}\int_S \frac{\partial}{\partial n'}\frac{1}{|\boldsymbol{r}-\boldsymbol{r}'|}\,ds(\boldsymbol{r}') = \alpha(\boldsymbol{r}) - 1, \qquad \boldsymbol{r}\in S \tag{6.25}$$

shows that the homogeneous equation

$$w(\boldsymbol{r}) + \frac{1}{2\pi}\int_S w(\boldsymbol{r}')\frac{\partial}{\partial n'}\frac{1}{|\boldsymbol{r}-\boldsymbol{r}'|}\,ds(\boldsymbol{r}') = 0 \tag{6.26}$$

has the nontrivial solution $w = 1$, which means that the adjoint also has nontrivial solutions. This means that neither (6.21) nor (6.24) are uniquely solvable. However, it may be shown that the first kind equation (6.20) is uniquely solvable in the present case. Alternatively, if a second kind equation is desired, it may be obtained by a variation in the assumption (6.23). That is, by assuming that u_n^+ has the form

$$u_n^+(\boldsymbol{r}) = f_n^+(\boldsymbol{r}) - \frac{1}{4\pi}\int_S \left(\frac{\partial}{\partial n'}\frac{1}{|\boldsymbol{r}-\boldsymbol{r}'|} + \frac{1}{r}\right) w(\boldsymbol{r}')\,ds(\boldsymbol{r}') \quad \text{in } V^+ \tag{6.27}$$

the jump condition (6.5) leads to the boundary integral equation for $w(\boldsymbol{r}')$

$$w(\boldsymbol{r}) + \frac{1}{2\pi}\int_S \left(\frac{\partial}{\partial n'}\frac{1}{|\boldsymbol{r}-\boldsymbol{r}'|} + \frac{1}{r}\right) w(\boldsymbol{r}')\,ds(\boldsymbol{r}') = 2f_n^+(\boldsymbol{r}) \tag{6.28}$$

which is uniquely solvable in $L_2(S)$.

6.A.3 *The Neumann problem*

Since the normal derivative of u_n^+ vanishes on S, equations (6.16) and (6.17) assume the form

$$u_n^+(\boldsymbol{r}) - \frac{1}{2\pi}\int_S u_n^+(\boldsymbol{r}')\frac{\partial}{\partial n'}\frac{1}{|\boldsymbol{r}-\boldsymbol{r}'|}\,ds(\boldsymbol{r}') = 2f_n^+(\boldsymbol{r}) \tag{6.29}$$

$$-\frac{1}{2\pi}\frac{\partial}{\partial n}\int_S u_n^+(\boldsymbol{r}')\frac{\partial}{\partial n'}\frac{1}{|\boldsymbol{r}-\boldsymbol{r}'|}\,ds(\boldsymbol{r}') = 2\frac{\partial}{\partial n}f_n^+(\boldsymbol{r}) \tag{6.30}$$

where f_n^+ is now defined in (3.51) and (6.29) and (6.30) hold for all $\boldsymbol{r}$ on S. Here again there are two boundary integral equations, one of the first kind and one of the second kind. In this case, the second kind equation is uniquely solvable whereas the first kind equation is not. Alternative equations may be obtained using a layer ansatz but they involve the adjoints of the integral operators in (6.29) and (6.30) with no obvious advantage. If (6.29) is solved for $u_n^+(\boldsymbol{r})$ on S, then with (3.16) it follows that

$$u_n^+(\boldsymbol{r}) = f_n^+(\boldsymbol{r}) + \frac{1}{4\pi}\int_S u_n^+(\boldsymbol{r})\frac{\partial}{\partial n'}\frac{1}{|\boldsymbol{r}-\boldsymbol{r}'|}\,ds(\boldsymbol{r}'), \quad \boldsymbol{r}\in V^+ \tag{6.31}$$

6.A.4 *The Robin problem*

If the boundary condition (3.66) is employed, then (6.16) and (6.17) lead to the following boundary integral equations

$$u_n^+(\boldsymbol{r}) - \frac{1}{2\pi}\int_S u_n^+(\boldsymbol{r}')\frac{\partial}{\partial n'}\frac{1}{|\boldsymbol{r}-\boldsymbol{r}'|}\,ds(\boldsymbol{r}') = 2f_n^+(\boldsymbol{r}) + \frac{n\nu}{2\pi}\int_S \frac{u_{n-1}^+(\boldsymbol{r}')}{|\boldsymbol{r}-\boldsymbol{r}'|}\,ds(r') \tag{6.32}$$

and

$$-\frac{1}{2\pi}\frac{\partial}{\partial n}\int_S u_n^+(\boldsymbol{r}')\frac{\partial}{\partial n'}\frac{1}{|\boldsymbol{r}-\boldsymbol{r}'|}\,ds(\boldsymbol{r}')$$
$$= 2\frac{\partial}{\partial n}f_n^+(\boldsymbol{r}) + n\nu u_{n-1}^+(\boldsymbol{r}) + \frac{n\nu}{2\pi}\int_S u_{n-1}^+(\boldsymbol{r}')\frac{\partial}{\partial n}\frac{1}{|\boldsymbol{r}-\boldsymbol{r}'|}\,ds(\boldsymbol{r}') \tag{6.33}$$

where $f_n^+(\boldsymbol{r})$ is now given by (3.67) and (6.32) and (6.33) hold for all $\boldsymbol{r}$ on S. These integral equations are exactly the same as those obtained for the Neumann problem save that the known terms on the right-hand side are amended. As in that case the second kind equation (6.32) is uniquely solvable but the first kind equation (6.33) is not. Once $u_n^+(\boldsymbol{r})$ is found on S the representation in V^+ is given by

$$u_n^+(\boldsymbol{r}) = f_n^+(\boldsymbol{r}) + \frac{1}{4\pi}\int_S u_n^+(\boldsymbol{r}')\frac{\partial}{\partial n'}\frac{1}{|\boldsymbol{r}-\boldsymbol{r}'|}\,ds(\boldsymbol{r}') + \frac{n\nu}{4\pi}\int_S \frac{u_{n-1}^+(\boldsymbol{r}')}{|\boldsymbol{r}-\boldsymbol{r}'|}\,ds(\boldsymbol{r}') \tag{6.34}$$

6.A.5 *The transmission problem*

When the transmission conditions on S, (3.102) and (3.103), rewritten as

$$u_n^+(\boldsymbol{r}) = u_n^-(\boldsymbol{r}), \quad \boldsymbol{r} \in S \tag{6.35}$$

$$\frac{\partial}{\partial n} u_n^-(\boldsymbol{r}) = \frac{\rho^-}{\rho^+}\frac{\partial}{\partial n} u_n^+(\boldsymbol{r}) + n\frac{\delta^-\gamma^-}{\sqrt{\gamma^+\rho^+}}\frac{\partial}{\partial n} u_{n-1}^-(\boldsymbol{r}), \quad \boldsymbol{r} \in S \tag{6.36}$$

are used in (6.18) and (6.19), equations (6.16)–(6.19) may be grouped as

$$\begin{aligned} u_n^+(\boldsymbol{r}) = {} & \frac{1}{2\pi}\int_S u_n^+(\boldsymbol{r}')\frac{\partial}{\partial n'}\frac{1}{|\boldsymbol{r}-\boldsymbol{r}'|}\,ds(\boldsymbol{r}') \\ & -\frac{1}{2\pi}\int_S \frac{1}{|\boldsymbol{r}-\boldsymbol{r}'|}\frac{\partial}{\partial n'} u_n^+(\boldsymbol{r}')\,ds(\boldsymbol{r}') + 2f_n^+(\boldsymbol{r}) \end{aligned} \tag{6.37}$$

$$\begin{aligned} -u_n^+(\boldsymbol{r}) = {} & \frac{1}{2\pi}\int_S u_n^+(\boldsymbol{r}')\frac{\partial}{\partial n'}\frac{1}{|\boldsymbol{r}-\boldsymbol{r}'|}\,ds(\boldsymbol{r}') \\ & -\frac{1}{2\pi}\frac{\rho^-}{\rho^+}\int_S \frac{1}{|\boldsymbol{r}-\boldsymbol{r}'|}\frac{\partial}{\partial n'} u_n^+(\boldsymbol{r}')\,ds(\boldsymbol{r}') + 2f_n^-(\boldsymbol{r}) \\ & -\frac{n}{2\pi}\frac{\delta^-\gamma^-}{\sqrt{\gamma^+\rho^+}}\int_S \frac{1}{|\boldsymbol{r}-\boldsymbol{r}'|}\frac{\partial}{\partial n'} u_{n-1}^-(\boldsymbol{r}')\,ds(\boldsymbol{r}') \end{aligned} \tag{6.38}$$

$$\begin{aligned} \frac{\partial}{\partial n} u_n^+(\boldsymbol{r}) = {} & \frac{1}{2\pi}\frac{\partial}{\partial n}\int_S u_n^+(\boldsymbol{r}')\frac{\partial}{\partial n'}\frac{1}{|\boldsymbol{r}-\boldsymbol{r}'|}\,ds(\boldsymbol{r}') \\ & -\frac{1}{2\pi}\int_S \left(\frac{\partial}{\partial n}\frac{1}{|\boldsymbol{r}-\boldsymbol{r}'|}\right)\frac{\partial}{\partial n'} u_n^+(\boldsymbol{r}')\,ds(\boldsymbol{r}') + 2\frac{\partial}{\partial n} f_n^+(\boldsymbol{r}) \end{aligned} \tag{6.39}$$

$$\begin{aligned} -\frac{\rho^-}{\rho^+}\frac{\partial}{\partial n} u_n^+(\boldsymbol{r}) = {} & \frac{1}{2\pi}\frac{\partial}{\partial n}\int_S u_n^+(\boldsymbol{r}')\frac{\partial}{\partial n'}\frac{1}{|\boldsymbol{r}-\boldsymbol{r}'|}\,ds(\boldsymbol{r}') \\ & -\frac{1}{2\pi}\frac{\rho^-}{\rho^+}\int_S \left(\frac{\partial}{\partial n}\frac{1}{|\boldsymbol{r}-\boldsymbol{r}'|}\right)\frac{\partial}{\partial n'} u_n^+(\boldsymbol{r}')\,ds(\boldsymbol{r}') + 2\frac{\partial}{\partial n} f_n^-(\boldsymbol{r}) \\ & +n\frac{\delta^-\gamma^-}{\sqrt{\gamma^+\rho^+}}\left[\frac{\partial}{\partial n} u_{n-1}^-(\boldsymbol{r}) - \frac{1}{2\pi}\int_S \left(\frac{\partial}{\partial n}\frac{1}{|\boldsymbol{r}-\boldsymbol{r}'|}\right)\frac{\partial}{\partial n'} u_{n-1}^-(\boldsymbol{r}')\,ds(\boldsymbol{r}')\right] \end{aligned} \tag{6.40}$$

and all four equations hold for $\boldsymbol{r}$ on S. These equations may be combined pairwise to eliminate either one or the other of the integrals involving u_n^+, resulting in the following four integral equations which hold on S

$$2u_n^+(\boldsymbol{r}) = -\frac{1}{2\pi}\left(1-\frac{\rho^-}{\rho^+}\right)\int_S \frac{1}{|\boldsymbol{r}-\boldsymbol{r}'|}\frac{\partial}{\partial n'}u_n^+(\boldsymbol{r}')\,ds(\boldsymbol{r}')$$
$$+2f_n^+(\boldsymbol{r}) - 2f_n^-(\boldsymbol{r})$$
$$+\frac{n}{2\pi}\frac{\delta^-\gamma^-}{\sqrt{\gamma^+\rho^+}}\int_S \frac{1}{|\boldsymbol{r}-\boldsymbol{r}'|}\frac{\partial}{\partial n'}u_{n-1}^-(\boldsymbol{r}')\,ds(\boldsymbol{r}') \tag{6.41}$$

$$\left(1+\frac{\rho^+}{\rho^-}\right)u_n^+(\boldsymbol{r}) = \frac{1}{2\pi}\left(1-\frac{\rho^+}{\rho^-}\right)\int_S u_n^+(\boldsymbol{r}')\frac{\partial}{\partial n'}\frac{1}{|\boldsymbol{r}-\boldsymbol{r}'|}\,ds(\boldsymbol{r}')$$
$$+2f_n^+(\boldsymbol{r}) - 2\frac{\rho^+}{\rho^-}f_n^-(\boldsymbol{r})$$
$$+\frac{n}{2\pi}\frac{\rho^+}{\rho^-}\frac{\delta^-\gamma^-}{\sqrt{\gamma^+\rho^+}}\int_S \frac{1}{|\boldsymbol{r}-\boldsymbol{r}'|}\frac{\partial}{\partial n'}u_{n-1}^-(\boldsymbol{r}')\,ds(\boldsymbol{r}') \tag{6.42}$$

$$2\frac{\partial}{\partial n}u_n^+(\boldsymbol{r}) = \frac{1}{2\pi}\left(1-\frac{\rho^+}{\rho^-}\right)\frac{\partial}{\partial n}\int_S u_n^+(\boldsymbol{r}')\frac{\partial}{\partial n'}\frac{1}{|\boldsymbol{r}-\boldsymbol{r}'|}\,ds(\boldsymbol{r}')$$
$$+2\frac{\partial}{\partial n}f_n^+(\boldsymbol{r}) - 2\frac{\rho^+}{\rho^-}\frac{\partial}{\partial n}f_n^-(\boldsymbol{r}) - n\frac{\rho^+}{\rho^-}\frac{\delta^-\gamma^-}{\sqrt{\gamma^+\rho^+}}\frac{\partial}{\partial n}u_{n-1}^-(\boldsymbol{r})$$
$$+\frac{n}{2\pi}\frac{\rho^+}{\rho^-}\frac{\delta^-\gamma^-}{\sqrt{\gamma^+\rho^+}}\int_S\left(\frac{\partial}{\partial n}\frac{1}{|\boldsymbol{r}-\boldsymbol{r}'|}\right)\frac{\partial}{\partial n'}u_{n-1}^-(\boldsymbol{r}')\,ds(\boldsymbol{r}') \tag{6.43}$$

$$\left(1+\frac{\rho^-}{\rho^+}\right)\frac{\partial}{\partial n}u_n^+(\boldsymbol{r}) = -\frac{1}{2\pi}\left(1-\frac{\rho^-}{\rho^+}\right)\int_S\left(\frac{\partial}{\partial n}\frac{1}{|\boldsymbol{r}-\boldsymbol{r}'|}\right)\frac{\partial}{\partial n'}u_n^+(\boldsymbol{r}')\,ds(\boldsymbol{r}')$$
$$+2\frac{\partial}{\partial n}f_n^+(\boldsymbol{r}) - 2\frac{\partial}{\partial n}f_n^-(\boldsymbol{r}) - n\frac{\delta^-\gamma^-}{\sqrt{\gamma^+\rho^+}}\frac{\partial}{\partial n}u_{n-1}^-(\boldsymbol{r})$$
$$+\frac{n}{2\pi}\frac{\delta^-\gamma^-}{\sqrt{\gamma^+\rho^+}}\int_S\left(\frac{\partial}{\partial n}\frac{1}{|\boldsymbol{r}-\boldsymbol{r}'|}\right)\frac{\partial}{\partial n'}u_{n-1}^-(\boldsymbol{r}')\,ds(\boldsymbol{r}') \tag{6.44}$$

In these equations, the quantities involving u_{n-1}^- as well as f_n^+ and f_n^- are regarded as known. Note that the lossless case is easily recovered by setting $\delta^- = 0$; however, this only affects the known term in these integral equations. Hence, the lossless and lossy problems have the same degree of difficulty. While it is true that in this problem it is required to determine both u_n^+ and the normal derivative of u_n^+ on S it is only necessary to solve one integral equation rather than a coupled system. For example, one may solve (6.41) for u_n^+ on S and then employ (6.44) as a representation for the normal derivative of u_n^+. Alternatively, one could solve (6.43) for the normal derivative of u_n^+ and then employ (6.41) as a representation for u_n^+. Either way, it is necessary to solve only one integral equation to solve the transmission problem in potential theory. Without the low frequency expansion the reduction of the transmission problem to a single integral equation is considerably more complicated. Of course, if $\rho^+ = \rho^-$, then (6.41)–(6.44) clearly show that it is unnecessary to solve any integral equations at all, each coefficient in the low frequency expansion being expressed entirely in terms of the preceding terms. Observe that once both u_n^+ and the normal derivative of u_n^+ are found on S, equation (3.16) provides a representation of u_n^+ in V^+.

The question of unique solvability of the second kind equations (6.42) and (6.44) is settled due to a theorem of Plemelj (Colton and Kress 1983) which states that the values of λ for which the equation

$$\lambda w(\boldsymbol{r}) - \frac{1}{2\pi}\int_S w(\boldsymbol{r}')\frac{\partial}{\partial n'}\frac{1}{|\boldsymbol{r}-\boldsymbol{r}'|}\,ds(\boldsymbol{r}') = 0 \tag{6.45}$$

has nontrivial solutions lie in the half-open interval $[-1, 1)$ and the same is true for the adjoint equation, where $\partial/\partial n$ replaces $\partial/\partial n'$. This means that (6.42) and (6.44) are uniquely solvable for every $\rho^\pm > 0$.

It should be noted that in terms of numerical methods, one need only discretize one integral operator in order to solve all the integral equations in low frequency acoustic scattering theory. Symbolically, if we denote by K the integral operator

$$(Kw)(\boldsymbol{r}) = \frac{1}{2\pi}\int_S w(\boldsymbol{r}')\frac{\partial}{\partial n'}\frac{1}{|\boldsymbol{r}-\boldsymbol{r}'|}\,ds(\boldsymbol{r}') \tag{6.46}$$

then all the integral equations (6.28), (6.29), (6.32) and (6.42) may be written in the form

$$\lambda w - Kw + \frac{\alpha}{r}\int_S w(\boldsymbol{r}')\,ds(\boldsymbol{r}') = F \tag{6.47}$$

where for the Dirichlet problem, $\lambda = -1$ and $\alpha = -1/2\pi$; for the Neumann and Robin problems, $\lambda = 1$ and $\alpha = 0$, while for the transmission problem, $\lambda = (\rho^- + \rho^+)/(\rho^- - \rho^+)$ and $\alpha = 0$. Of course, the right-hand side, F, is different in each case. Note that for the Dirichlet problem, we may set $\lambda = -1$, despite the Plemelj theorem, because $\alpha \neq 0$.

6.A.6 *Problems in acoustics*

1. Show that equation (6.20) has at most one solution.
2. Show that equation (6.29) has at most one solution.
3. Derive Gauss' integral formula (6.25)

4. Use the Gauss integral (6.25) to determine a nontrivial solution of the integral equation

$$\frac{\partial}{\partial n}\int\limits_S u_n^+(\boldsymbol{r}')\frac{\partial}{\partial n'}\frac{1}{|\boldsymbol{r}-\boldsymbol{r}'|}\,ds(\boldsymbol{r}') = 0$$

on S.

5. Use the integral equation (6.21) to find u_0^+ and u_1^+ for the soft sphere of radius a, which is excited by a plane wave.
6. Use the integral equation (6.30) to find u_0^+ and u_1^+ for the hard sphere of radius a, which is excited by a plane wave.
7. Use the integral equation (6.32) to find u_0^+ and u_1^+ for the Robin problem for the sphere of radius a, which is excited by a plane wave.
8. Use the Plemelj theorem to show that the integral equations (6.42) and (6.44) are uniquely solvable for positive mass densities $\rho^\pm$.
9. Use integral equation methods to obtain $u_0^\pm$ and $u_1^\pm$ for the lossy penetrable sphere of radius a, when the incident field is a plane wave.

6.EM Electromagnetics

6.EM.1 *Basic equations in electromagnetics*

In electromagnetics, the fact that $\boldsymbol{E}$ and $\boldsymbol{H}$ form a coupled system leads to slightly different integral equations. These may be obtained directly from the representations (3.130)–(3.134) and (3.137)–(3.141) which we rewrite here for convenience

$$\alpha(\boldsymbol{r})\boldsymbol{E}_n^+(\boldsymbol{r}) = \boldsymbol{F}_{en}^+(\boldsymbol{r}) + \frac{1}{4\pi}\nabla\times\int\limits_S\frac{\hat{\boldsymbol{n}}'\times\boldsymbol{E}_n^+(\boldsymbol{r}')}{|\boldsymbol{r}-\boldsymbol{r}'|}\,ds(\boldsymbol{r}')$$
$$-\frac{1}{4\pi}\nabla\int\limits_S\frac{\hat{\boldsymbol{n}}'\cdot\boldsymbol{E}_n^+(\boldsymbol{r}')}{|\boldsymbol{r}-\boldsymbol{r}'|}\,ds(\boldsymbol{r}') \tag{6.48}$$

$$\alpha(\boldsymbol{r})\boldsymbol{H}_n^+(\boldsymbol{r}) = \boldsymbol{F}_{mn}^+(\boldsymbol{r}) + \frac{1}{4\pi}\nabla\times\int\limits_S\frac{\hat{\boldsymbol{n}}'\times\boldsymbol{H}_n^+(\boldsymbol{r}')}{|\boldsymbol{r}-\boldsymbol{r}'|}\,ds(\boldsymbol{r}')$$
$$-\frac{1}{4\pi}\nabla\int\limits_S\frac{\hat{\boldsymbol{n}}'\cdot\boldsymbol{H}_n^+(\boldsymbol{r}')}{|\boldsymbol{r}-\boldsymbol{r}'|}\,ds(\boldsymbol{r}') \tag{6.49}$$

$$(\alpha(\boldsymbol{r})-1)\boldsymbol{E}_n^-(\boldsymbol{r}) = \boldsymbol{F}_{en}^-(\boldsymbol{r}) + \frac{1}{4\pi}\nabla\times\int\limits_S\frac{\hat{\boldsymbol{n}}'\times\boldsymbol{E}_n^-(\boldsymbol{r}')}{|\boldsymbol{r}-\boldsymbol{r}'|}\,ds(\boldsymbol{r}')$$
$$-\frac{1}{4\pi}\nabla\int\limits_S\frac{\hat{\boldsymbol{n}}'\cdot\boldsymbol{E}_n^-(\boldsymbol{r}')}{|\boldsymbol{r}-\boldsymbol{r}'|}\,ds(\boldsymbol{r}') \tag{6.50}$$

$$(\alpha(\boldsymbol{r}) - 1)\boldsymbol{H}_n^-(\boldsymbol{r}) = \boldsymbol{F}_{mn}^-(\boldsymbol{r}) + \frac{1}{4\pi}\nabla \times \int_S \frac{\hat{\boldsymbol{n}}' \times \boldsymbol{H}_n^-(\boldsymbol{r}')}{|\boldsymbol{r} - \boldsymbol{r}'|} ds(\boldsymbol{r}')$$

$$-\frac{1}{4\pi}\nabla \int_S \frac{\hat{\boldsymbol{n}}' \cdot \boldsymbol{H}_n^-(\boldsymbol{r}')}{|\boldsymbol{r} - \boldsymbol{r}'|} ds(\boldsymbol{r}') \tag{6.51}$$

These equations give rise to the appropriate boundary integral equations when particular boundary or transmission conditions are incorporated into them, and $\boldsymbol{r}$ is chosen to lie on S.

6.EM.2 *The perfect conductor problem*

When the conditions $\hat{\boldsymbol{n}} \times \boldsymbol{E}_n^+(\boldsymbol{r}) = \boldsymbol{0}$ and $\hat{\boldsymbol{n}} \cdot \boldsymbol{H}_n^+(\boldsymbol{r}) = 0$ are incorporated into (6.48) and (6.49) the following boundary integral equations result for $\boldsymbol{r}$ on S

$$\hat{\boldsymbol{n}} \cdot \boldsymbol{E}_n^+(\boldsymbol{r}) + \frac{1}{2\pi}\int_S \hat{\boldsymbol{n}}' \cdot \boldsymbol{E}_n^+(\boldsymbol{r}')\frac{\partial}{\partial n}\frac{1}{|\boldsymbol{r} - \boldsymbol{r}'|} ds(\boldsymbol{r}') = 2\hat{\boldsymbol{n}} \cdot \boldsymbol{F}_{en}^+(\boldsymbol{r}) \tag{6.52}$$

$$\frac{1}{4\pi}\hat{\boldsymbol{n}} \times \nabla \int_S \frac{\hat{\boldsymbol{n}}' \cdot \boldsymbol{E}_n^+(\boldsymbol{r}')}{|\boldsymbol{r} - \boldsymbol{r}'|} ds(\boldsymbol{r}') = \hat{\boldsymbol{n}} \times \boldsymbol{F}_{en}^+(\boldsymbol{r}) \tag{6.53}$$

$$\frac{1}{4\pi}\hat{\boldsymbol{n}} \cdot \nabla \times \int_S \frac{\hat{\boldsymbol{n}}' \cdot \boldsymbol{H}_n^+(\boldsymbol{r}')}{|\boldsymbol{r} - \boldsymbol{r}'|} ds(\boldsymbol{r}') = -\hat{\boldsymbol{n}} \cdot \boldsymbol{F}_{mn}^+(\boldsymbol{r}) \tag{6.54}$$

$$\hat{\boldsymbol{n}} \times \boldsymbol{H}_n^+(\boldsymbol{r}) - \frac{1}{2\pi}\hat{\boldsymbol{n}} \times \left[\nabla \times \int_S \frac{\hat{\boldsymbol{n}}' \times \boldsymbol{H}_n^+(\boldsymbol{r}')}{|\boldsymbol{r} - \boldsymbol{r}'|} ds(\boldsymbol{r}')\right]$$

$$= 2\hat{\boldsymbol{n}} \times \boldsymbol{F}_{mn}^+(\boldsymbol{r}) \tag{6.55}$$

where $\boldsymbol{F}_{en}^+$ and $\boldsymbol{F}_{mn}^+$ are given in (3.146) and (3.147). The integral equation of the second kind for $\hat{\boldsymbol{n}} \cdot \boldsymbol{E}_n^+(\boldsymbol{r})$, (6.52), is seen to be of the same form as (6.21) which as has been noted is not uniquely solvable. Since Fredholm theory is applicable, a necessary condition that any solutions exist is that the right-hand side be orthogonal to all solutions of the homogeneous adjoint equation. But the homogeneous adjoint equation has only constant solutions (Colton and Kress 1983) and the required orthogonality is assured by (3.153). Uniqueness is restored by imposing the condition (3.129), $\int_S \hat{\boldsymbol{n}} \cdot \boldsymbol{E}_n^+ \, ds = 0$, on the solution of (6.52). Alternatively, the first kind equation (6.53) is solvable only up to a multiple of the solution of the problem

$$\int_S \frac{w(\boldsymbol{r}')}{|\boldsymbol{r} - \boldsymbol{r}'|} ds(\boldsymbol{r}') = 1 \tag{6.56}$$

Again, the ambiguity is removed by imposing the condition (3.129). Once $\hat{\boldsymbol{n}} \cdot \boldsymbol{E}_n^+$ has been determined on S the electric field coefficient in V^+ is given by

$$\boldsymbol{E}_n^+(\boldsymbol{r}) = \boldsymbol{F}_{en}^+(\boldsymbol{r}) - \frac{1}{4\pi}\nabla \int_S \frac{\hat{\boldsymbol{n}}' \cdot \boldsymbol{E}_n^+(\boldsymbol{r}')}{|\boldsymbol{r} - \boldsymbol{r}'|} ds(\boldsymbol{r}'), \qquad \boldsymbol{r} \in V^+ \tag{6.57}$$

To determine $\boldsymbol{H}_n^+$ the equation of the second kind (6.55) has been shown to be uniquely solvable for $\hat{\boldsymbol{n}} \times \boldsymbol{H}_n^+$ on S if V^- is simply connected (see Section 5.4 in Colton and Kress 1983). The magnetic field coefficient in V^+ is given by

$$\boldsymbol{H}_n^+(\boldsymbol{r}) = \boldsymbol{F}_{mn}^+(\boldsymbol{r}) + \frac{1}{4\pi}\nabla \times \int_S \frac{\hat{\boldsymbol{n}}' \times \boldsymbol{H}_n^+(\boldsymbol{r}')}{|\boldsymbol{r}-\boldsymbol{r}'|}\, ds(\boldsymbol{r}'), \quad \boldsymbol{r} \in V^+ \tag{6.58}$$

6.EM.3 *The transmission problem*

The lossless transmission problem: $\sigma^- = 0$. Introducing the transmission conditions (3.180) and (3.183)

$$\left.\begin{array}{ll} \hat{\boldsymbol{n}} \times \boldsymbol{E}_n^+ = \hat{\boldsymbol{n}} \times \boldsymbol{E}_n^-, & \hat{\boldsymbol{n}} \times \boldsymbol{H}_n^+ = \hat{\boldsymbol{n}} \times \boldsymbol{H}_n^- \\ \epsilon^+\hat{\boldsymbol{n}} \cdot \boldsymbol{E}_n^+ = \epsilon^-\hat{\boldsymbol{n}} \cdot \boldsymbol{E}_n^-, & \mu^+\hat{\boldsymbol{n}} \cdot \boldsymbol{H}_n^+ = \mu^-\hat{\boldsymbol{n}} \cdot \boldsymbol{H}_n^- \end{array}\right\} \quad \text{on } S \tag{6.59}$$

into the representations obtained by subtracting (6.50) from (6.48) and (6.51) from (6.49) results in the alternative representation formulae

$$\begin{aligned} &\alpha(\boldsymbol{r})\boldsymbol{E}_n^+(\boldsymbol{r}) - (\alpha(\boldsymbol{r})-1)\boldsymbol{E}_n^-(\boldsymbol{r}) \\ &\quad = \boldsymbol{F}_{en}^+(\boldsymbol{r}) - \boldsymbol{F}_{en}^-(\boldsymbol{r}) - \frac{1}{4\pi}\left(1-\frac{\epsilon^+}{\epsilon^-}\right)\nabla\int_S \frac{\hat{\boldsymbol{n}}' \cdot \boldsymbol{E}_n^+(\boldsymbol{r}')}{|\boldsymbol{r}-\boldsymbol{r}'|}\, ds(\boldsymbol{r}') \end{aligned} \tag{6.60}$$

$$\begin{aligned} &\alpha(\boldsymbol{r})\boldsymbol{H}_n^+(\boldsymbol{r}) - (\alpha(\boldsymbol{r})-1)\boldsymbol{H}_n^-(\boldsymbol{r}) \\ &\quad = \boldsymbol{F}_{mn}^+(\boldsymbol{r}) - \boldsymbol{F}_{mn}^-(\boldsymbol{r}) - \frac{1}{4\pi}\left(1-\frac{\mu^+}{\mu^-}\right)\nabla\int_S \frac{\hat{\boldsymbol{n}}' \cdot \boldsymbol{H}_n^+(\boldsymbol{r}')}{|\boldsymbol{r}-\boldsymbol{r}'|}\, ds(\boldsymbol{r}') \end{aligned} \tag{6.61}$$

where both equations hold for all $\boldsymbol{r}$ in $\mathbb{R}^3$ and $\boldsymbol{F}_{en}^{\pm}(\boldsymbol{r})$ and $\boldsymbol{F}_{mn}^{\pm}(\boldsymbol{r})$ are defined in (3.131), (3.132), (3.138) and (3.139). These representations in themselves are surprising in that the full vector coefficients are expressed entirely in terms of the two (unknown) scalar quantities $\hat{\boldsymbol{n}} \cdot \boldsymbol{E}_n^+$ and $\hat{\boldsymbol{n}} \cdot \boldsymbol{H}_n^+$. Moreover, the unknown electric field term disappears if $\epsilon^+ = \epsilon^-$, while the unknown magnetic field term vanishes whenever $\mu^+ = \mu^-$. Using the transmission conditions once more in the representations (6.60) and (6.61) when $\boldsymbol{r} \in S$ leads to two uncoupled second kind boundary integral equations for the two unknown scalar quantities

$$\begin{aligned} \left(1+\frac{\epsilon^+}{\epsilon^-}\right)\hat{\boldsymbol{n}} \cdot \boldsymbol{E}_n^+(\boldsymbol{r}) &= 2\hat{\boldsymbol{n}} \cdot (\boldsymbol{F}_{en}^+(\boldsymbol{r}) - \boldsymbol{F}_{en}^-(\boldsymbol{r})) \\ &\quad - \frac{1}{2\pi}\left(1-\frac{\epsilon^+}{\epsilon^-}\right)\int_S \hat{\boldsymbol{n}}' \cdot \boldsymbol{E}_n^+(\boldsymbol{r}')\frac{\partial}{\partial n}\frac{1}{|\boldsymbol{r}-\boldsymbol{r}'|}\, ds(\boldsymbol{r}') \end{aligned} \tag{6.62}$$

and

$$\left(1+\frac{\mu^+}{\mu^-}\right)\hat{\boldsymbol{n}}\cdot\boldsymbol{H}_n^+(\boldsymbol{r}) = 2\hat{\boldsymbol{n}}\cdot(\boldsymbol{F}_{mn}^+(\boldsymbol{r})-\boldsymbol{F}_{mn}^-(\boldsymbol{r})) - \frac{1}{2\pi}\left(1-\frac{\mu^+}{\mu^-}\right)\int_S \hat{\boldsymbol{n}}'\cdot\boldsymbol{H}_n^+(\boldsymbol{r}')\frac{\partial}{\partial n}\frac{1}{|\boldsymbol{r}-\boldsymbol{r}'|}\,ds(\boldsymbol{r}') \tag{6.63}$$

The integral equation (6.62) is uniquely solvable for nonnegative $\epsilon^{\pm}$. Similarly, equation (6.63) is uniquely solvable for nonnegative $\mu^{\pm}$. This may be seen by following the same argument used to establish the solvability of the second kind equation (6.44) in the acoustic transmission problem. Once the equations are solved for $\hat{\boldsymbol{n}}\cdot\boldsymbol{E}_n^+$ and $\hat{\boldsymbol{n}}\cdot\boldsymbol{H}_n^+$ the representations (6.60) and (6.61) may be used to obtain $\boldsymbol{E}_n^{\pm}$ and $\boldsymbol{H}_n^{\pm}$ in $V^{\pm}$.

The lossy transmission problem: $\sigma^- > 0$. When σ^- is positive, all the transmission conditions in (6.59) remain the same except for that involving the normal components of the electric field which for convenience is repeated here

$$\hat{\boldsymbol{n}}\cdot\boldsymbol{E}_n^- = \frac{n}{\sigma^-\sqrt{\epsilon^+\mu^+}}(\epsilon^-\hat{\boldsymbol{n}}\cdot\boldsymbol{E}_{n-1}^- - \epsilon^+\hat{\boldsymbol{n}}\cdot\boldsymbol{E}_{n-1}^+) \quad \text{on } S \tag{6.64}$$

The treatment of the magnetic field is exactly the same as in the lossless case. The integral equation (6.63) remains valid and once it is solved for $\hat{\boldsymbol{n}}\cdot\boldsymbol{H}_n^+$, this may be used in (6.61) to represent the magnetic field in $V^{\pm}$. The electric field equations are different however. In place of the representation (6.60) we obtain

$$\alpha(\boldsymbol{r})\boldsymbol{E}_n^+(\boldsymbol{r}) - (\alpha(\boldsymbol{r})-1)\boldsymbol{E}_n^-(\boldsymbol{r}) = -\frac{1}{4\pi}\nabla\int_S \frac{\hat{\boldsymbol{n}}'\cdot\boldsymbol{E}_n^+(\boldsymbol{r}')}{|\boldsymbol{r}-\boldsymbol{r}'|}\,ds(\boldsymbol{r}') + \boldsymbol{F}_{en}^+(\boldsymbol{r}) - \boldsymbol{F}_{en}^-(\boldsymbol{r}) + \frac{n}{4\pi\sigma^-\sqrt{\epsilon^+\mu^+}}\nabla\int_S \frac{1}{|\boldsymbol{r}-\boldsymbol{r}'|}[\epsilon^-\hat{\boldsymbol{n}}'\cdot\boldsymbol{E}_{n-1}^-(\boldsymbol{r}') - \epsilon^+\hat{\boldsymbol{n}}'\cdot\boldsymbol{E}_{n-1}^+(\boldsymbol{r}')]\,ds(\boldsymbol{r}') \tag{6.65}$$

which with (6.64) leads to the boundary integral equation

$$\hat{\boldsymbol{n}}\cdot\boldsymbol{E}_n^+(\boldsymbol{r}) + \frac{1}{2\pi}\int_S \hat{\boldsymbol{n}}'\cdot\boldsymbol{E}_n^+(\boldsymbol{r}')\frac{\partial}{\partial n}\frac{1}{|\boldsymbol{r}-\boldsymbol{r}'|}\,ds(\boldsymbol{r}') = 2\hat{\boldsymbol{n}}\cdot(\boldsymbol{F}_{en}^+(\boldsymbol{r})-\boldsymbol{F}_{en}^-(\boldsymbol{r})) - \frac{n}{\sigma^-\sqrt{\epsilon^+\mu^+}}\Bigg[\hat{\boldsymbol{n}}\cdot(\epsilon^-\boldsymbol{E}_{n-1}^-(\boldsymbol{r}) - \epsilon^+\boldsymbol{E}_{n-1}^+(\boldsymbol{r})) - \frac{1}{2\pi}\int_S \hat{\boldsymbol{n}}'\cdot(\epsilon^-\boldsymbol{E}_{n-1}^-(\boldsymbol{r}') - \epsilon^+\boldsymbol{E}_{n-1}^+(\boldsymbol{r}'))\frac{\partial}{\partial n}\frac{1}{|\boldsymbol{r}-\boldsymbol{r}'|}\,ds(\boldsymbol{r}')\Bigg] \tag{6.66}$$

This equation is, except for the known term on the right-hand side, the same as that obtained for the perfect conductor, (6.52). As discussed in that case this equation is not uniquely solvable but will have solutions provided the integral of the right-hand side over S vanishes. That this is indeed the case is left for the problems.

6.EM.4 *Problems in electromagnetics*

1. Show that the integral over S of the right-hand side of equation (6.66) vanishes.
2. Use integral equation methods to obtain the Rayleigh approximation for a perfectly conducting sphere of radius a, which is excited by a plane wave.
3. Use integral equation methods to obtain the Rayleigh approximation for the lossy transmission problem when the scatterer is a sphere of radius a, which is excited by a plane wave.

6.E Elasticity

6.E.1 *Basic equations in elasticity*

In the case of an elastic scatterer, for $\boldsymbol{r} \in S$ the representations (3.248)–(3.253) together with jump conditions (6.12) and (6.14) lead to the following set of boundary integral equations which are completely analogous to the equations of acoustics (6.16)–(6.19).

$$\boldsymbol{u}_n^+(\boldsymbol{r}) = 2\boldsymbol{f}_n^+(\boldsymbol{r}) + \frac{1}{2\pi\mu^+}\int_S [\boldsymbol{u}_n^+(\boldsymbol{r}') \cdot \boldsymbol{T}^+(\partial_{\boldsymbol{r}'}, \hat{\boldsymbol{n}}')\widetilde{\boldsymbol{\gamma}}_0^+(\boldsymbol{r}, \boldsymbol{r}') - \widetilde{\boldsymbol{\gamma}}_0^+(\boldsymbol{r}, \boldsymbol{r}') \cdot \boldsymbol{T}^+(\partial_{\boldsymbol{r}'}, \hat{\boldsymbol{n}}')\boldsymbol{u}_n^+(\boldsymbol{r}')]\, ds(\boldsymbol{r}') \tag{6.67}$$

$$\begin{aligned}\boldsymbol{T}^+(\partial_{\boldsymbol{r}}, \hat{\boldsymbol{n}})\boldsymbol{u}_n^+(\boldsymbol{r}) = {} & 2\boldsymbol{T}^+(\partial_{\boldsymbol{r}}, \hat{\boldsymbol{n}})\boldsymbol{f}_n^+(\boldsymbol{r}) \\ & + \frac{1}{2\pi\mu^+}\boldsymbol{T}^+(\partial_{\boldsymbol{r}}, \hat{\boldsymbol{n}})\int_S \boldsymbol{u}_n^+(\boldsymbol{r}') \cdot \boldsymbol{T}^+(\partial_{\boldsymbol{r}'}, \hat{\boldsymbol{n}}')\widetilde{\boldsymbol{\gamma}}_0^+(\boldsymbol{r}, \boldsymbol{r}')\, ds(\boldsymbol{r}') \\ & - \frac{1}{2\pi\mu^+}\int_S \left(\boldsymbol{T}^+(\partial_{\boldsymbol{r}}, \hat{\boldsymbol{n}})\widetilde{\boldsymbol{\gamma}}_0^+(\boldsymbol{r}, \boldsymbol{r}')\right) \cdot \boldsymbol{T}^+(\partial_{\boldsymbol{r}'}, \hat{\boldsymbol{n}}')\boldsymbol{u}_n^+(\boldsymbol{r}')\, ds(\boldsymbol{r}')\end{aligned} \tag{6.68}$$

$$-\boldsymbol{u}_n^-(\boldsymbol{r}) = 2\boldsymbol{f}_n^-(\boldsymbol{r}) + \frac{1}{2\pi\mu^-}\int_S [\boldsymbol{u}_n^-(\boldsymbol{r}') \cdot \boldsymbol{T}^-(\partial_{\boldsymbol{r}'}, \hat{\boldsymbol{n}}')\widetilde{\boldsymbol{\gamma}}_0^-(\boldsymbol{r}, \boldsymbol{r}') - \widetilde{\boldsymbol{\gamma}}_0^-(\boldsymbol{r}, \boldsymbol{r}') \cdot \boldsymbol{T}^-(\partial_{\boldsymbol{r}'}, \hat{\boldsymbol{n}}')\boldsymbol{u}_n^-(\boldsymbol{r}')]\, ds(\boldsymbol{r}') \tag{6.69}$$

$$\begin{aligned}-\boldsymbol{T}^-(\partial_{\boldsymbol{r}}, \hat{\boldsymbol{n}})\boldsymbol{u}_n^-(\boldsymbol{r}) = {} & 2\boldsymbol{T}^-(\partial_{\boldsymbol{r}}, \hat{\boldsymbol{n}})\boldsymbol{f}_n^-(\boldsymbol{r}) \\ & + \frac{1}{2\pi\mu^-}\boldsymbol{T}^-(\partial_{\boldsymbol{r}}, \hat{\boldsymbol{n}})\int_S \boldsymbol{u}_n^-(\boldsymbol{r}') \cdot \boldsymbol{T}^-(\partial_{\boldsymbol{r}'}, \hat{\boldsymbol{n}}')\widetilde{\boldsymbol{\gamma}}_0^-(\boldsymbol{r}, \boldsymbol{r}')\, ds(\boldsymbol{r}') \\ & - \frac{1}{2\pi\mu^-}\int_S \left(\boldsymbol{T}^-(\partial_{\boldsymbol{r}}, \hat{\boldsymbol{n}})\widetilde{\boldsymbol{\gamma}}_0^-(\boldsymbol{r}, \boldsymbol{r}')\right) \cdot \boldsymbol{T}^-(\partial_{\boldsymbol{r}'}, \hat{\boldsymbol{n}}')\boldsymbol{u}_n^-(\boldsymbol{r}')\, ds(\boldsymbol{r}')\end{aligned} \tag{6.70}$$

These may be used together with the boundary and transmission conditions to obtain integral equations in each case.

6.E.2 *The rigid problem*

If $\boldsymbol{u}_n^+(\boldsymbol{r}) = 0$ on S, then (6.67) and (6.68) become

$$\frac{1}{2\pi\mu^+}\int_S \widetilde{\boldsymbol{\gamma}}_0^+(\boldsymbol{r},\boldsymbol{r}')\cdot \boldsymbol{T}^+(\partial_{\boldsymbol{r}'},\hat{\boldsymbol{n}}')\boldsymbol{u}_n^+(\boldsymbol{r}')\,ds(\boldsymbol{r}') = 2\boldsymbol{f}_n^+(\boldsymbol{r}) \tag{6.71}$$

and

$$\boldsymbol{T}^+(\partial_{\boldsymbol{r}},\hat{\boldsymbol{n}})\boldsymbol{u}_n^+(\boldsymbol{r}) + \frac{1}{2\pi\mu^+}\int_S \left(\boldsymbol{T}^+(\partial_{\boldsymbol{r}},\hat{\boldsymbol{n}})\widetilde{\boldsymbol{\gamma}}_0(\boldsymbol{r},\boldsymbol{r}')\right)\cdot \boldsymbol{T}^+(\partial_{\boldsymbol{r}'},\hat{\boldsymbol{n}}')\boldsymbol{u}_n^+(\boldsymbol{r}')\,ds(\boldsymbol{r}')$$
$$= 2\boldsymbol{T}^+(\partial_{\boldsymbol{r}},\hat{\boldsymbol{n}})\boldsymbol{f}_n^+(\boldsymbol{r}) \tag{6.72}$$

These equations constitute first and second kind equations for the unknown stress $\boldsymbol{T}^+\boldsymbol{u}_n^+$. However, because $\widetilde{\boldsymbol{\gamma}}_0$ and $\boldsymbol{T}^+\widetilde{\boldsymbol{\gamma}}_0$ have Cauchy singularities the operators are not compact and Fredholm theory does not apply directly. Nevertheless, these integral equations may still be used as the basis of numerical solutions. Uniqueness for the first kind equation may be established in the same manner as in the acoustic case, using uniqueness for the rigid body problem in both V^+ and V^- and the jump conditions (6.14) to show that the zero vector is the only solution of the homogeneous form of equation (6.71). Unique solvability of the second equation is not possible as can be seen from the elastostatic analog of the Gauss integral. It is easy to verify that there exist so-called 'traction-free' solutions of the elastostatic equation corresponding to rigid body displacements of the form

$$\boldsymbol{u}(\boldsymbol{r}) = \boldsymbol{c}_1 + \boldsymbol{c}_2\times\boldsymbol{r} \tag{6.73}$$

where the constant vector $\boldsymbol{c}_1$ describes a translation and the constant vector $\boldsymbol{c}_2$ describes a rotation which comprise six linearly independent functions. That is, $\Delta^e\boldsymbol{u} = \boldsymbol{0}$ in $\mathbb{R}^3$ and $\boldsymbol{T}\boldsymbol{u} = \boldsymbol{0}$ on any surface, where the Lamé parameters in Δ^e and $\boldsymbol{T}$ are arbitrary and may be chosen as λ^+ and μ^+. Since $\boldsymbol{u}$ is a regular elastostatic displacement in V^-, the $k=0$ form of the integral representation (3.251)–(3.253) yields

$$(\alpha(\boldsymbol{r})-1)(\boldsymbol{c}_1+\boldsymbol{c}_2\times\boldsymbol{r}) = \frac{1}{4\pi\mu^+}\int_S (\boldsymbol{c}_1+\boldsymbol{c}_2\times\boldsymbol{r}')\cdot \boldsymbol{T}^+(\partial_{\boldsymbol{r}'},\hat{\boldsymbol{n}}')\widetilde{\boldsymbol{\gamma}}_0(\boldsymbol{r},\boldsymbol{r}')\,ds(\boldsymbol{r}') \tag{6.74}$$

and restricting $\boldsymbol{r}$ to lie on S

$$\boldsymbol{c}_1+\boldsymbol{c}_2\times\boldsymbol{r} + \frac{1}{2\pi\mu^+}\int_S (\boldsymbol{c}_1+\boldsymbol{c}_2\times\boldsymbol{r}')\cdot \boldsymbol{T}^+(\partial_{\boldsymbol{r}'},\hat{\boldsymbol{n}}')\widetilde{\boldsymbol{\gamma}}_0(\boldsymbol{r},\boldsymbol{r}')\,ds(\boldsymbol{r}') = \boldsymbol{0} \tag{6.75}$$

Thus, corresponding to the components of $\boldsymbol{c}_1$ and $\boldsymbol{c}_2$, there are six linearly independent solutions of this equation and hence six linearly independent solutions of the homogeneous adjoint equation obtained as the homogeneous form of (6.72). Solvability, though not unique solvability, of (6.72) has been established (Kupradze 1979) using regularizers of the Cauchy integral operators. To employ the Fredholm alternative it is necessary to demonstrate that the right side (6.72) is orthogonal to all solutions of the homogeneous adjoint equation, which are the traction-free rigid body displacements defined in (6.73).

But, with Betti's third formula

$$\begin{aligned}
&\int_S (\boldsymbol{c}_1 + \boldsymbol{c}_2 \boldsymbol{r}) \cdot \boldsymbol{T}^+(\partial_{\boldsymbol{r}}, \hat{\boldsymbol{n}}) \boldsymbol{f}_n^+(\boldsymbol{r})\, ds(\boldsymbol{r}) \\
&= \int_S \boldsymbol{f}_n^+(\boldsymbol{r}) \cdot \boldsymbol{T}^+(\partial_{\boldsymbol{r}}, \hat{\boldsymbol{n}})(\boldsymbol{c}_1 + \boldsymbol{c}_2 \times \boldsymbol{r})\, ds(\boldsymbol{r}) \\
&+ \int_{V^-} [(\boldsymbol{c}_1 + \boldsymbol{c}_2 \times \boldsymbol{r}) \cdot \Delta^{e^+} \boldsymbol{f}_n^+(\boldsymbol{r}) - \boldsymbol{f}_n^+(\boldsymbol{r}) \cdot \Delta^{e^+} (\boldsymbol{c}_1 + \boldsymbol{c}_2 \times \boldsymbol{r})]\, dv(\boldsymbol{r}) \\
&= \boldsymbol{0}
\end{aligned} \tag{6.76}$$

This last result follows since in addition to being traction-free, $\boldsymbol{c}_1 + \boldsymbol{c}_2 \times \boldsymbol{r}$ are solutions of the Navier equation in V^- as is $\boldsymbol{f}_n^+(\boldsymbol{r})$, (see (3.257)). Thus, the necessary condition for solvability is fulfilled.

In this case, as in the Dirichlet problem in section 6.A.1, it would appear that the first kind equation is preferable. If a uniquely solvable second kind equation is desired one may be obtained by forming a linear combination of (6.71) and (6.72). One may also obtain integral equations adjoint to those in (6.71) and (6.72) using a layer anzatz and the jump conditions (6.12) and (6.14).

6.E.3 *The cavity problem*

When $\boldsymbol{T}^+\boldsymbol{u}_n^+ = \boldsymbol{0}$ on S, equations (6.67) and (6.68) become

$$\boldsymbol{u}_n^+(\boldsymbol{r}) - \frac{1}{2\pi\mu^+} \int_S \boldsymbol{u}_n^+(\boldsymbol{r}') \cdot \boldsymbol{T}^+(\partial_{\boldsymbol{r}'}, \hat{\boldsymbol{n}}') \widetilde{\boldsymbol{\gamma}}_0^+(\boldsymbol{r}, \boldsymbol{r}')\, ds(\boldsymbol{r}') = 2\boldsymbol{f}_n^+(\boldsymbol{r}) \tag{6.77}$$

$$-\frac{1}{2\pi\mu^+} \boldsymbol{T}^+(\partial_{\boldsymbol{r}}, \hat{\boldsymbol{n}}) \int_S \boldsymbol{u}_n^+(\boldsymbol{r}') \cdot \boldsymbol{T}^+(\partial_{\boldsymbol{r}'}, \hat{\boldsymbol{n}}') \widetilde{\boldsymbol{\gamma}}_0^+(\boldsymbol{r}, \boldsymbol{r}')\, ds(\boldsymbol{r}') = 2\boldsymbol{T}^+(\partial_{\boldsymbol{r}}, \hat{\boldsymbol{n}}) \boldsymbol{f}_n^+(\boldsymbol{r}) \tag{6.78}$$

for all $\boldsymbol{r}$ on S. Again, there are two boundary integral equations, one of the first kind and one of the second kind. However, in this case, the first kind equation not only involves a hypersingular operator, in itself not such an inhibiting factor as previously thought, but also is not uniquely solvable. On the other hand, the second kind equation, (6.77), is always uniquely solvable (Kupradze 1979). As in the rigid case, integral equations adjoint to (6.77) and (6.78) may be obtained with a layer ansatz.

6.E.4 *The transmission problem*

Introducing the transmission conditions

$$\left.\begin{aligned} \boldsymbol{u}_n^+ &= \boldsymbol{u}_n^- \\ \boldsymbol{T}^+\boldsymbol{u}_n^+ &= \boldsymbol{T}^-\boldsymbol{u}_n^- \end{aligned}\right\} \quad \text{on } S \tag{6.79}$$

into the equations (6.67)–(6.70) results in the following

$$\boldsymbol{u}_n^+(\boldsymbol{r}) = 2\boldsymbol{f}_n^+(\boldsymbol{r}) + \frac{1}{2\pi\mu^+}\int_S [\boldsymbol{u}_n^+(\boldsymbol{r}') \cdot \boldsymbol{T}^+(\partial_{\boldsymbol{r}'}, \hat{\boldsymbol{n}}')\widetilde{\boldsymbol{\gamma}}_0^+(\boldsymbol{r}, \boldsymbol{r}')$$

$$-\widetilde{\boldsymbol{\gamma}}_0^+(\boldsymbol{r}, \boldsymbol{r}') \cdot \boldsymbol{T}^+(\partial_{\boldsymbol{r}'}, \hat{\boldsymbol{n}}')\boldsymbol{u}_n^+(\boldsymbol{r}')]\, ds(\boldsymbol{r}') \tag{6.80}$$

$$-\boldsymbol{u}_n^+(\boldsymbol{r}) = 2\boldsymbol{f}_n^-(\boldsymbol{r}) + \frac{1}{2\pi\mu^-}\int_S [\boldsymbol{u}_n^+(\boldsymbol{r}') \cdot \boldsymbol{T}^-(\partial_{\boldsymbol{r}'}, \hat{\boldsymbol{n}}')\widetilde{\boldsymbol{\gamma}}_0^-(\boldsymbol{r}, \boldsymbol{r}')$$

$$-\widetilde{\boldsymbol{\gamma}}_0^-(\boldsymbol{r}, \boldsymbol{r}') \cdot \boldsymbol{T}^+(\partial_{\boldsymbol{r}'}, \hat{\boldsymbol{n}}')\boldsymbol{u}_n^+(\boldsymbol{r}')]\, ds(\boldsymbol{r}') \tag{6.81}$$

$$\boldsymbol{T}^+(\partial_{\boldsymbol{r}}, \hat{\boldsymbol{n}})\boldsymbol{u}_n^+(\boldsymbol{r}) = 2\boldsymbol{T}^+(\partial_{\boldsymbol{r}}, \hat{\boldsymbol{n}})\boldsymbol{f}_n^+(\boldsymbol{r})$$

$$+\frac{1}{2\pi\mu^+}\boldsymbol{T}^+(\partial_{\boldsymbol{r}}, \hat{\boldsymbol{n}})\int_S \boldsymbol{u}_n^+(\boldsymbol{r}') \cdot \boldsymbol{T}^+(\partial_{\boldsymbol{r}'}, \hat{\boldsymbol{n}}')\widetilde{\boldsymbol{\gamma}}_0^+(\boldsymbol{r}, \boldsymbol{r}')\, ds(\boldsymbol{r}')$$

$$-\frac{1}{2\pi\mu^+}\int_S \left(\boldsymbol{T}^+(\partial_{\boldsymbol{r}}, \hat{\boldsymbol{n}})\widetilde{\boldsymbol{\gamma}}_0^+(\boldsymbol{r}, \boldsymbol{r}')\right) \cdot \boldsymbol{T}^+(\partial_{\boldsymbol{r}'}, \hat{\boldsymbol{n}}')\boldsymbol{u}_n^+(\boldsymbol{r}')\, ds(\boldsymbol{r}') \tag{6.82}$$

$$-\boldsymbol{T}^+(\partial_{\boldsymbol{r}}, \hat{\boldsymbol{n}})\boldsymbol{u}_n^+(\boldsymbol{r}) = 2\boldsymbol{T}^-(\partial_{\boldsymbol{r}}, \hat{\boldsymbol{n}})\boldsymbol{f}_n^-(\boldsymbol{r})$$

$$+\frac{1}{2\pi\mu^-}\boldsymbol{T}^-(\partial_{\boldsymbol{r}}, \hat{\boldsymbol{n}})\int_S \boldsymbol{u}_n^+(\boldsymbol{r}') \cdot \boldsymbol{T}^-(\partial_{\boldsymbol{r}'}, \hat{\boldsymbol{n}}')\widetilde{\boldsymbol{\gamma}}_0^-(\boldsymbol{r}, \boldsymbol{r}')\, ds(\boldsymbol{r}')$$

$$-\frac{1}{2\pi\mu^-}\int_S \left(\boldsymbol{T}^-(\partial_{\boldsymbol{r}}, \hat{\boldsymbol{n}})\widetilde{\boldsymbol{\gamma}}_0^-(\boldsymbol{r}, \boldsymbol{r}')\right) \cdot \boldsymbol{T}^+(\partial_{\boldsymbol{r}'}, \hat{\boldsymbol{n}}')\boldsymbol{u}_n^+(\boldsymbol{r}')\, ds(\boldsymbol{r}') \tag{6.83}$$

In general, it is necessary to solve the coupled pair of integral equations (6.81) and (6.83), which is the pair where the transmission conditions have been introduced, for the two unknown functions $\boldsymbol{u}_n^+$ and $\boldsymbol{T}^+\boldsymbol{u}_n^+$. However, in the special case when $\tau^+ = \tau^-$ or, equivalently, $\mu^+\lambda^- = \mu^-\lambda^+$, it may be seen from their definitions (6.8)–(6.10) that

$$\frac{1}{\mu^+}\boldsymbol{T}^+(\partial_{\boldsymbol{r}'}, \hat{\boldsymbol{n}}') = \frac{1}{\mu^-}\boldsymbol{T}^-(\partial_{\boldsymbol{r}'}, \hat{\boldsymbol{n}}') \tag{6.84}$$

$$\widetilde{\boldsymbol{\gamma}}_0^+(\boldsymbol{r}, \boldsymbol{r}') = \widetilde{\boldsymbol{\gamma}}_0^-(\boldsymbol{r}, \boldsymbol{r}') \tag{6.85}$$

and

$$\frac{1}{\mu^+}\boldsymbol{T}^+(\partial_{\boldsymbol{r}'}, \hat{\boldsymbol{n}}')\widetilde{\boldsymbol{\gamma}}_0^+(\boldsymbol{r}, \boldsymbol{r}') = \frac{1}{\mu^-}\boldsymbol{T}^-(\partial_{\boldsymbol{r}'}, \hat{\boldsymbol{n}}')\widetilde{\boldsymbol{\gamma}}_0^-(\boldsymbol{r}, \boldsymbol{r}') \tag{6.86}$$

In this special case, (6.81) and (6.83) may be rewritten as

$$-\boldsymbol{u}_n^+(\boldsymbol{r}) = 2\boldsymbol{f}_n^-(\boldsymbol{r}) + \frac{1}{2\pi\mu^+}\int_S \boldsymbol{u}_n^+(\boldsymbol{r}')\cdot \boldsymbol{T}^+(\partial_{r'},\hat{\boldsymbol{n}}')\widetilde{\boldsymbol{\gamma}}_0^+(\boldsymbol{r},\boldsymbol{r}')\,ds(\boldsymbol{r}')$$
$$-\frac{1}{2\pi\mu^-}\int_S \widetilde{\boldsymbol{\gamma}}_0^+(\boldsymbol{r},\boldsymbol{r}')\cdot \boldsymbol{T}^+(\partial_{r'},\hat{\boldsymbol{n}}')\boldsymbol{u}_n^+(\boldsymbol{r}')\,ds(\boldsymbol{r}') \qquad (6.87)$$

and

$$-\boldsymbol{T}^+(\partial_r,\hat{\boldsymbol{n}})\boldsymbol{u}_n^+(\boldsymbol{r}) = 2\frac{\mu^-}{\mu^+}\boldsymbol{T}^+(\partial_r,\hat{\boldsymbol{n}})\boldsymbol{f}_n^-(\boldsymbol{r})$$
$$+\frac{1}{2\pi\mu^+}\frac{\mu^-}{\mu_+}\boldsymbol{T}^+(\partial_r,\hat{\boldsymbol{n}})\int_S \boldsymbol{u}_n^+(\boldsymbol{r}')\cdot \boldsymbol{T}^+(\partial_{r'},\hat{\boldsymbol{n}}')\widetilde{\boldsymbol{\gamma}}_0^+(\boldsymbol{r},\boldsymbol{r}')\,ds(\boldsymbol{r}')$$
$$-\frac{1}{2\pi\mu^+}\int_S \left(\boldsymbol{T}^+(\partial_r,\hat{\boldsymbol{n}})\widetilde{\boldsymbol{\gamma}}_0^+(\boldsymbol{r},\boldsymbol{r}')\right)\cdot \boldsymbol{T}^+(\partial_{r'},\hat{\boldsymbol{n}}')\boldsymbol{u}_n^+(\boldsymbol{r}')\,ds(\boldsymbol{r}') \quad (6.88)$$

Then (6.81) and (6.87) may be combined to eliminate one or the other of the integrals resulting in

$$\boldsymbol{u}_n^+(\boldsymbol{r}) = \boldsymbol{f}_n^+(\boldsymbol{r}) - \boldsymbol{f}_n^-(\boldsymbol{r})$$
$$-\frac{1}{4\pi}\left(\frac{1}{\mu^+}-\frac{1}{\mu^-}\right)\int_S \widetilde{\boldsymbol{\gamma}}_0^+(\boldsymbol{r},\boldsymbol{r}')\cdot \boldsymbol{T}^+(\partial_{r'},\hat{\boldsymbol{n}}')\boldsymbol{u}_n^+(\boldsymbol{r}')\,ds(\boldsymbol{r}') \qquad (6.89)$$

and

$$(\mu^+ + \mu^-)\boldsymbol{u}_n^+(\boldsymbol{r}) = 2\mu^+\boldsymbol{f}_n^+(\boldsymbol{r}) - 2\mu^-\boldsymbol{f}_n^-(\boldsymbol{r})$$
$$+\frac{1}{2\pi}\left(1-\frac{\mu^-}{\mu^+}\right)\int_S \boldsymbol{u}_n^+(\boldsymbol{r}')\cdot \boldsymbol{T}^+(\partial_{r'},\hat{\boldsymbol{n}}')\widetilde{\boldsymbol{\gamma}}_0^+(\boldsymbol{r},\boldsymbol{r}')\,ds(\boldsymbol{r}') \quad (6.90)$$

Similarly (6.82) and (6.88) may be combined to yield

$$\boldsymbol{T}^+(\partial_r,\hat{\boldsymbol{n}})\boldsymbol{u}_n^+(\boldsymbol{r}) = \boldsymbol{T}^+(\partial_r,\hat{\boldsymbol{n}})\boldsymbol{f}_n^+(\boldsymbol{r}) - \boldsymbol{T}^-(\partial_r,\hat{\boldsymbol{n}})\boldsymbol{f}_n^-(\boldsymbol{r})$$
$$+\frac{1}{4\pi\mu^+}\left(1-\frac{\mu^-}{\mu^+}\right)\boldsymbol{T}^+(\partial_r,\hat{\boldsymbol{n}})\int_S \boldsymbol{u}_n^+(\boldsymbol{r}')\cdot \boldsymbol{T}^+(\partial_{r'},\hat{\boldsymbol{n}}')\widetilde{\boldsymbol{\gamma}}_0^+(\boldsymbol{r},\boldsymbol{r}')\,ds(\boldsymbol{r}') \quad (6.91)$$

and

$$\left(1+\frac{\mu^+}{\mu^-}\right)\boldsymbol{T}^+(\partial_r,\hat{\boldsymbol{n}})\boldsymbol{u}_n^+(\boldsymbol{r}) = 2\boldsymbol{T}^+(\partial_r,\hat{\boldsymbol{n}})(\boldsymbol{f}_n^+(\boldsymbol{r}) - \boldsymbol{f}_n^-(\boldsymbol{r}))$$
$$-\frac{1}{2\pi\mu^+}\left(1-\frac{\mu^+}{\mu^-}\right)\int_S \left(\boldsymbol{T}^+(\partial_r,\hat{\boldsymbol{n}})\widetilde{\boldsymbol{\gamma}}_0^+(\boldsymbol{r},\boldsymbol{r}')\right)\cdot \boldsymbol{T}^+(\partial_{r'},\hat{\boldsymbol{n}}')\boldsymbol{u}_n^+(\boldsymbol{r}')\,ds(\boldsymbol{r}') \quad (6.92)$$

While it is still true that the two functions $\boldsymbol{u}_n^+$ and $\boldsymbol{T}^+\boldsymbol{u}_n^+$ must be found, it is only necessary to solve one integral equation rather than coupled pairs. That is, (6.90) can be solved for $\boldsymbol{u}_n^+$ and then this quantity substituted into (6.91) to yield $\boldsymbol{T}^+\boldsymbol{u}_n^+$. Alternatively, (6.92) can be solved for $\boldsymbol{T}^+\boldsymbol{u}_n^+$ and this quantity substituted into (6.89) to determine $\boldsymbol{u}_n^+$. This parallels the behavior for the transmission problem in acoustics where one need solve only one equation, rather than a coupled pair.

6.E.5 *Problems in elasticity*

1. Prove that the rigid body displacement field (6.73) generates no traction on any surface.
2. Use integral equation methods to obtain the Rayleigh approximation for a rigid sphere of radius a, which is excited by a plane wave.
3. Use integral equation methods to prove that the Rayleigh approximation of a cavity, excited by a plane wave, is the constant field $\boldsymbol{u}_0^i$.
4. Use integral equation methods to prove that when a penetrable body is excited by a plane wave, then the Rayleigh approximation is the constant field $\boldsymbol{u}_0^i$ both in V^+ and in V^-.
5. Use the uniqueness of the rigid body problem in both V^+ and V^- and the jump condition (6.14) to demonstrate that the zero vector is the only solution of equation

$$\int_S \tilde{\gamma}_0^+(\boldsymbol{r}, \boldsymbol{r}') \cdot \boldsymbol{T}^+(\partial_{\boldsymbol{r}'}, \hat{\boldsymbol{n}}')\boldsymbol{u}_n^+(\boldsymbol{r}')\, ds(\boldsymbol{r}') = \boldsymbol{0}$$

7

THE SPHERE

7.0 Basics

In this chapter we present explicit results for low frequency scattering by a sphere of radius a. All the results, whether or not known before, have been rederived using techniques among those developed in Chapters 3 to 6. After all the results are given for plane wave incidence along the x_3 axis in acoustics, electromagnetics and elasticity, we present some results for point source incidence in acoustics. The propagation vector $\hat{\boldsymbol{k}}$ is always given by

$$\hat{\boldsymbol{k}} = \hat{\boldsymbol{x}}_3 \tag{7.1}$$

when the incidence is a plane wave. Therefore, the incidence fields are as follows

Acoustics:

$$u^i(\boldsymbol{r}) = e^{i\boldsymbol{k}\cdot\boldsymbol{r}} = e^{ikx_3} \tag{7.2}$$

Electromagnetics:

$$\boldsymbol{E}^i(\boldsymbol{r}) = \hat{\boldsymbol{a}}e^{i\boldsymbol{k}\cdot\boldsymbol{r}} = \hat{\boldsymbol{a}}e^{ikx_3} \tag{7.3}$$

$$\boldsymbol{H}^i(\boldsymbol{r}) = Y^+\hat{\boldsymbol{b}}e^{i\boldsymbol{k}\cdot\boldsymbol{r}} = Y^+\hat{\boldsymbol{b}}e^{ikx_3} \tag{7.4}$$

$$\hat{\boldsymbol{a}} \times \hat{\boldsymbol{b}} = \hat{\boldsymbol{k}} \tag{7.5}$$

Elasticity:

$$\begin{aligned}\boldsymbol{u}^i(\boldsymbol{r}) &= \alpha_p\hat{\boldsymbol{k}}e^{ik_p\hat{\boldsymbol{k}}\cdot\boldsymbol{r}} + \alpha_s\hat{\boldsymbol{b}}e^{ik_s\hat{\boldsymbol{k}}\cdot\boldsymbol{r}} \\ &= \alpha_p\hat{\boldsymbol{x}}_3e^{ik_px_3} + \alpha_s\hat{\boldsymbol{b}}e^{ik_sx_3}\end{aligned} \tag{7.6}$$

$$\hat{\boldsymbol{k}} \cdot \hat{\boldsymbol{b}} = 0 \tag{7.7}$$

These results will be expressed in terms of spherical coordinates (r, θ, φ) and spherical unit vectors $(\hat{\boldsymbol{r}}, \hat{\boldsymbol{\theta}}, \hat{\boldsymbol{\varphi}})$ defined by

$$\left.\begin{aligned} x_1 &= r\sin\theta\cos\varphi \\ x_2 &= r\sin\theta\sin\varphi \\ x_3 &= r\cos\theta \end{aligned}\right\} \tag{7.8}$$

for $r \geq 0, 0 \leq \theta \leq \pi, 0 \leq \varphi < 2\pi$

$$\left.\begin{aligned} \hat{\boldsymbol{r}} &= (\sin\theta\cos\varphi, \sin\theta\sin\varphi, \cos\theta) \\ \hat{\boldsymbol{\theta}} &= (\cos\theta\cos\varphi, \cos\theta\sin\varphi, -\sin\theta) \\ \hat{\boldsymbol{\varphi}} &= (-\sin\varphi, \cos\varphi, 0) \end{aligned}\right\} \tag{7.9}$$

The surface of the sphere, S, is specified by

$$r = a \tag{7.10}$$

On the surface $r = a$, the unit normal, directed into V^+, is

$$\hat{\boldsymbol{n}} = \hat{\boldsymbol{r}} \tag{7.11}$$

and the normal derivative is

$$\frac{\partial}{\partial n} = \frac{\partial}{\partial r} \tag{7.12}$$

The element of surface area on $r = a$ is

$$ds = a^2 \sin\theta \, d\theta \, d\phi \tag{7.13}$$

The volume of the sphere is

$$|V^-| = \frac{4\pi}{3} a^3 \tag{7.14}$$

and the surface area is

$$|S| = 4\pi a^2 \tag{7.15}$$

Before presenting explicit low frequency results for each of the boundary value and transmission problems for the sphere, we record the realizations of the various quantities introduced in Chapter 5 which may be used to obtain most of the leading order terms.

The conductor potential (5.8):

$$\phi^c(\boldsymbol{r}) = \frac{a}{r}, \quad r > a \tag{7.16}$$

Capacity (5.9):

$$C = a \tag{7.17}$$

Constant *c* (5.16):

$$\boldsymbol{c} = \boldsymbol{0} \tag{7.18}$$

The basic potential functions (5.13), (5.20), (5.26), (5.31) and (5.32):

$$\boldsymbol{\Phi} = \frac{a^3}{r^3} \boldsymbol{r}, \quad r > a \tag{7.19}$$

$$\boldsymbol{\Psi} = -\frac{a^3}{2r^3}\hat{\boldsymbol{r}}, \quad r > a \tag{7.20}$$

$$\left.\begin{aligned} \boldsymbol{v}^+(\boldsymbol{r}) &= \frac{\beta - 1}{\beta + 2}\frac{a^3}{r^3}\boldsymbol{r}, \quad r > a \\ \boldsymbol{v}^-(\boldsymbol{r}) &= -\frac{3}{\beta + 2}\boldsymbol{r}, \quad r < a \end{aligned}\right\} \tag{7.21}$$

$$\left.\begin{aligned} \boldsymbol{\Phi}^+(\boldsymbol{r}) &= \frac{\beta}{\beta + 2}\frac{a^3}{r^3}\boldsymbol{r}, \quad r > a \\ \boldsymbol{\Phi}^-(\boldsymbol{r}) &= -\frac{2}{\beta + 2}\boldsymbol{r}, \quad r < a \end{aligned}\right\} \tag{7.22}$$

$$\left.\begin{aligned} \boldsymbol{\Psi}^+(\boldsymbol{r}) &= -\frac{1}{\beta + 2}\frac{a^3}{r^3}\boldsymbol{r}, \quad r > a \\ \boldsymbol{\Psi}^-(\boldsymbol{r}) &= -\frac{1}{\beta + 2}\boldsymbol{r}, \quad r < a \end{aligned}\right\} \tag{7.23}$$

The polarization tensor for an isolated conductor (5.17):

$$\boldsymbol{Q} = 2|V^-|\widetilde{\boldsymbol{I}} \tag{7.24}$$

The electric polarizability tensor (5.18):

$$\widetilde{\boldsymbol{P}} = 3|V^-|\widetilde{\boldsymbol{I}} \tag{7.25}$$

The virtual mass tensor (5.21):

$$\widetilde{\boldsymbol{W}} = \frac{1}{2}|V^-|\widetilde{\boldsymbol{I}} \tag{7.26}$$

The magnetic polarizability tensor (5.22):

$$\widetilde{\boldsymbol{M}} = \frac{3}{2}|V^-|\widetilde{\boldsymbol{I}} \tag{7.27}$$

The general polarizability tensor (5.27):

$$\widetilde{\boldsymbol{X}}(\beta) = 3|V^-|\frac{(\beta - 1)}{\beta + 2}\widetilde{\boldsymbol{I}} \tag{7.28}$$

Specific results for each problem follow.

7.A Acoustics

7.A.1 *The Dirichlet problem*

When the plane acoustic wave (7.2) is incident upon an acoustically soft sphere of radius a, the first few terms in the low frequency expansion (3.2) are found to be

$$u_0^+(\boldsymbol{r}) = 1 - \frac{a}{r}, \qquad r > a \tag{7.29}$$

$$u_1^+(\boldsymbol{r}) = -a\left(1 - \frac{a}{r}\right) + r\left(1 - \frac{a^3}{r^3}\right) P_1(\cos\theta), \qquad r > a \tag{7.30}$$

$$\begin{aligned} u_2^+(\boldsymbol{r}) = &\left(1 - \frac{a}{r}\right)\left[a^2 + \frac{1}{3}(a-r)^2\right] \\ &+ \frac{2}{3}r^2\left(1 - \frac{a^5}{r^5}\right) P_2(\cos\theta), \qquad r > a \end{aligned} \tag{7.31}$$

and

$$\begin{aligned} u_3^+(\boldsymbol{r}) = &-a\left(1 - \frac{a}{r}\right)[a^2 + (a-r)^2] \\ &+ 3\left[a^3\left(1 - \frac{a^2}{r^2}\right) + \frac{r^3}{5}\left(1 - \frac{a^5}{r^5}\right)\right] P_1(\cos\theta) \\ &+ \frac{2r^3}{5}\left(1 - \frac{a^7}{r^7}\right) P_3(\cos\theta), \qquad r > a \end{aligned} \tag{7.32}$$

where P_n are Legendre polynomials.

The scattering amplitude (3.38) is

$$\begin{aligned} g(\hat{\boldsymbol{r}}, \hat{\boldsymbol{k}}) = &-ika - k^2a^2 + ik^3a^3\left(\frac{2}{3} - P_1(\cos\theta)\right) \\ &+ \frac{1}{3}k^4a^4 + O(k^5) \end{aligned} \tag{7.33}$$

The differential scattering cross section (3.46) is

$$\begin{aligned} \sigma(\hat{\boldsymbol{r}}) = &4\pi a^2\left[1 - \frac{1}{3}k^2a^2 + 2k^2a^2 P_1(\cos\theta)\right] \\ &+ O(k^4) \end{aligned} \tag{7.34}$$

while the (total) scattering cross section (3.47) is

$$\sigma_s = 4\pi a^2 - \frac{4\pi}{3}a^4k^2 + O(k^4)$$
$$= 4\pi a^2\left[1 - \frac{1}{3}k^2a^2 + O(k^4)\right] \tag{7.35}$$

In this case, the absorption cross section vanishes

$$\sigma_a = 0 \tag{7.36}$$

therefore the extinction cross section is simply

$$\sigma_e = \sigma_s \tag{7.37}$$

For higher order terms see Bowman *et al.* (1969).

7.A.2 *The Neumann problem*

When the surface of the sphere is acoustically hard the first few terms of (3.2) are

$$u_0^+(\boldsymbol{r}) = 1, \quad r > a \tag{7.38}$$

$$u_1^+(\boldsymbol{r}) = r\left(1 + \frac{a^3}{2r^3}\right)P_1(\cos\theta), \quad r > a \tag{7.39}$$

$$u_2^+(\boldsymbol{r}) = \frac{r^2}{3}\left(1 + \frac{2a^3}{r^3}\right) + \frac{2r^2}{3}\left(1 + \frac{2a^5}{3r^5}\right)P_2(\cos\theta), \quad r > a \tag{7.40}$$

$$u_3^+(\boldsymbol{r}) = 2a^3 + \frac{3r^3}{10}\left(2 - 5\frac{a^3}{r^3} + 3\frac{a^5}{r^5}\right)P_1(\cos\theta)$$
$$+ \frac{2r^3}{5}\left(1 + \frac{3a^7}{4r^7}\right)P_3(\cos\theta), \quad r > a \tag{7.41}$$

The scattering amplitude (3.54) is

$$g(\hat{\boldsymbol{r}}, \hat{\boldsymbol{k}}) = -ik^3a^3\left(\frac{1}{3} - \frac{1}{2}P_1(\cos\theta)\right)$$
$$+ik^5a^5\left(\frac{1}{5} - \frac{3}{20}P_1(\cos\theta) + \frac{2}{27}P_2(\cos\theta)\right)$$
$$-k^6a^6\left(\frac{1}{9} + \frac{1}{12}P_1(\cos\theta)\right)$$
$$+k^8a^8\left(\frac{2}{15} + \frac{1}{10}P_1(\cos\theta)\right) + O(k^9) \tag{7.42}$$

The differential scattering cross section is

$$\sigma(\hat{\boldsymbol{r}}) = 4\pi k^4 a^6 \left[\left(\frac{1}{3} - \frac{1}{2} P_1(\cos\theta) \right)^2 - 2k^2 a^2 \left(\frac{1}{3} - \frac{1}{2} P_1(\cos\theta) \right) \left(\frac{1}{5} - \frac{3}{20} P_1(\cos\theta) + \frac{2}{27} P_2(\cos\theta) \right) \right] + O(k^8) \tag{7.43}$$

and the (total) cross section is

$$\sigma_s = \frac{7\pi}{9} k^4 a^6 - \frac{11\pi}{15} k^6 a^8 + O(k^8) \tag{7.44}$$

Again

$$\sigma_a = 0 \tag{7.45}$$

and

$$\sigma_e = \sigma_s \tag{7.46}$$

For higher order terms see Bowman *et al.* (1969).

7.A.3 *The Robin problem*

Recall, (1.51), that we use the dimensionless parameter

$$\nu = \frac{c^+ \rho^+}{Z^+} \tag{7.47}$$

in the Robin boundary condition (1.49)

$$\frac{\partial u^+}{\partial n} + ik\nu u^+ = 0 \quad \text{on } S \tag{7.48}$$

The first few terms in the low frequency expansion (3.2) are found to be

$$u_0^+(\boldsymbol{r}) = 1, \quad r > a \tag{7.49}$$

$$u_1^+(\boldsymbol{r}) = \nu \frac{a^2}{r} + r \left(1 + \frac{a^3}{2r^3} \right) P_1(\cos\theta), \quad r > a \tag{7.50}$$

$$u_2^+(\boldsymbol{r}) = r^2 \left(\frac{1}{3} + \frac{2a^3}{3r^3} + 2\nu \frac{a^2}{r^2} + 2\nu^2 \frac{a^3}{r^3} \right) + \frac{3}{2} \nu \frac{a^4}{r^2} P_1(\cos\theta) + \frac{2}{3} r^2 \left(1 + \frac{2a^5}{3r^5} \right) P_2(\cos\theta), \quad r > a \tag{7.51}$$

$$u_3^+(\boldsymbol{r}) = a^3\left[2 + 6\nu^2 + 3\nu\frac{r}{a} + 6(\nu + \nu^2 + \nu^3)\frac{a}{r}\right]$$
$$+ \frac{3r^3}{10}\left[2 - 5\frac{a^3}{r^3} + 3\left(1 + \frac{5}{2}\nu^2\right)\frac{a^5}{r^5}\right]P_1(\cos\theta)$$
$$+ \frac{10}{9}\nu\frac{a^6}{r^3}P_2(\cos\theta) + \frac{2r^3}{5}\left(1 + \frac{3a^7}{4r^7}\right)P_3(\cos\theta), \quad r > a \tag{7.52}$$

The scattering amplitude (3.69) is

$$g(\hat{\boldsymbol{r}}, \hat{\boldsymbol{k}}) = -\nu k^2a^2 + ik^3a^3\left(-\frac{1}{3} - \nu^2 + \frac{1}{2}P_1(\cos\theta)\right)$$
$$+ \nu a^4k^4\left(1 + \nu + \nu^2 - \frac{3}{4}P_1(\cos\theta)\right)$$
$$+ ik^5a^5\left[\frac{1}{5} + \nu\left(\frac{2}{3} + 2\nu + 2\nu^2 + \nu^3\right)\right.$$
$$\left. - \frac{3}{4}\left(\frac{1}{5} + \frac{\nu^2}{2}\right)P_1(\cos\theta) + \frac{2}{27}P_2(\cos\theta)\right] + O(k^6) \tag{7.53}$$

The differential cross section is

$$\sigma(\hat{\boldsymbol{r}}) = 4\pi\nu^2k^2a^4 + 4\pi k^4a^6\left[\frac{1}{9} + \frac{1}{4}(P_1(\cos\theta))^2 - \frac{4}{3}\nu^2 - 2\nu^3 - \nu^4\right.$$
$$\left. + \left(-\frac{1}{3} + \frac{\nu^2}{2}\right)P_1(\cos\theta)\right] + O(k^6) \tag{7.54}$$

while the (total) scattering cross section is

$$\sigma_s = 4\pi\nu^2k^2a^4 + 4\pi\left(\frac{7}{36} - \frac{4}{3}\nu^2 - 2\nu^3 - \nu^4\right)k^4a^6 + O(k^6) \tag{7.55}$$

In this case, the absorption cross section (3.70) does not vanish and is given by

$$\sigma_a = 4\pi\nu a^2 - \pi\nu k^2a^4(1 + 8\nu + 4\nu^2) + O(k^4) \tag{7.56}$$

Observe that σ_s is $O(k^2)$ while σ_a is $O(k^0)$. The extinction cross section is

$$\sigma_e = 4\pi\nu a^2 - \pi\nu k^2a^4(1 + 2\nu)^2 + O(k^4) \tag{7.57}$$

a result which may be verified with the optical theorem (2.59).

7.A.4 *The lossless transmission problem*

When $\delta^- = 0$, the first few terms in the expansion (3.2) and (3.3) are

$$u_0^+(\boldsymbol{r}) = 1, \quad r > a \tag{7.58}$$

$$u_0^-(\boldsymbol{r}) = 1, \quad r < a \tag{7.59}$$

$$u_1^+(\boldsymbol{r}) = r\left(1 - \frac{\beta - 1}{\beta + 2}\frac{a^3}{r^3}\right) P_1(\cos\theta), \quad r > a \tag{7.60}$$

$$u_1^-(\boldsymbol{r}) = \frac{3}{\beta + 2} r P_1(\cos\theta), \quad r < a \tag{7.61}$$

$$\begin{aligned} u_2^+(\boldsymbol{r}) &= \frac{r^2}{3} - \frac{2a^3}{3r}(1 - \beta\eta^2) \\ &\quad + \frac{2}{3}\left(1 - \frac{2(\beta - 1)}{2\beta + 3}\frac{a^5}{r^5}\right) r^2 P_2(\cos\theta), \quad r > a \end{aligned} \tag{7.62}$$

$$u_2^-(\boldsymbol{r}) = \frac{3 - 2\beta\eta^2 - \eta^2}{3} a^2 + \frac{\eta^2 r^2}{3} + \frac{10r^2}{3(2\beta + 3)} P_2(\cos\theta), \quad r < a \tag{7.63}$$

$$\begin{aligned} u_3^+(\boldsymbol{r}) &= 2(1 - \beta\eta^2)a^3 \\ &\quad + 3r^3\left(\frac{1}{5} + \frac{\beta - 1}{\beta + 2}\frac{a^3}{r^3} - \frac{6}{5}\frac{\beta^2 - \beta + \beta\eta^2 - 1}{(\beta + 2)^2}\frac{a^5}{r^5}\right) P_1(\cos\theta) \\ &\quad + \frac{2}{5}r^3\left(1 - \frac{3(\beta - 1)}{3\beta + 4}\frac{a^7}{r^7}\right) P_3(\cos\theta), \quad r > a \end{aligned} \tag{7.64}$$

$$\begin{aligned} u_3^-(\boldsymbol{r}) &= 2(1 - \beta\eta^2)a^3 \\ &\quad + \frac{9}{5(\beta + 2)} r^3\left(\frac{5\beta - 3\beta\eta^2}{\beta + 2}\frac{a^2}{r^2} - \frac{2\eta^2}{(\beta + 2)}\frac{a^2}{r^2} + \eta^2\right) P_1(\cos\theta) \\ &\quad + \frac{14}{5(3\beta + 4)} r^3 P_3(\cos\theta), \quad r < a \end{aligned} \tag{7.65}$$

The scattering amplitude is

$$\begin{aligned} g(\hat{\boldsymbol{r}}, \hat{\boldsymbol{k}}) &= ik^3a^3\left[\frac{\beta\eta^2 - 1}{3} - \frac{\beta - 1}{\beta + 2}P_1(\cos\theta)\right] \\ &\quad + ik^5a^5\left[\frac{(\beta\eta^2 - 1)(5\beta\eta^2 - 9) + \beta\eta^2(\eta^2 - 1)}{45}\right. \\ &\quad \left. + \frac{3(\beta^2 - \beta + \beta\eta^2 - 1)}{5(\beta + 2)^2}P_1(\cos\theta) - \frac{2(\beta - 1)}{9(2\beta + 3)}P_2(\cos\theta)\right] \\ &\quad - k^6a^6\left[\frac{(\beta\eta^2 - 1)^2}{9} + \frac{(\beta - 1)^2}{3(\beta + 2)^2}P_1(\cos\theta)\right] \end{aligned}$$

$$- k^8a^8 \left[\frac{2(\beta\eta^2 - 1)(5\beta^2\eta^4 + \beta\eta^4 - 15\beta\eta^2 + 9)}{135} - \frac{2(\beta - 1)(\beta^2 - \beta + \beta\eta^2 - 1)}{5(\beta + 2)^2} P_1(\cos\theta) \right] + O(k^9) \tag{7.66}$$

The differential cross section is

$$\sigma(\hat{\boldsymbol{r}}) = 4\pi k^4 a^6 \left[\frac{(\beta\eta^2 - 1)^2}{9} - \frac{2(\beta - 1)(\beta\eta^2 - 1)}{3(\beta + 2)} P_1(\cos\theta) + \frac{(\beta - 1)^2}{(\beta + 2)^2} (P_1(\cos\theta))^2 \right] + O(k^6) \tag{7.67}$$

Higher order terms may be obtained by employing (7.66) in the definition of $\sigma(\hat{\boldsymbol{r}})$, (2.33).

The (total) scattering cross section is

$$\sigma_s = 4\pi k^4 a^6 \left[\frac{(\beta\eta^2 - 1)^2}{9} + \frac{(\beta - 1)^2}{3(\beta + 2)^2} \right] + 4\pi k^6 a^8 \left[\frac{2(\beta\eta^2 - 1)(5\beta^2\eta^4 + \beta\eta^4 - 15\beta\eta^2 + 9)}{135} - \frac{2(\beta - 1)(\beta^2 - \beta + \beta\eta^2 - 1)}{5(\beta + 2)^3} \right] + O(k^8) \tag{7.68}$$

For lossless scatterers

$$\sigma_a = 0 \tag{7.69}$$

Hence

$$\sigma_e = \sigma_s \tag{7.70}$$

7.A.5 *The lossy transmission problem*

In this case, where $\delta^+ > 0$, the first few terms of the low frequency expansions are

$$u_0^+(\boldsymbol{r}) = 1, \quad r > a \tag{7.71}$$

$$u_0^-(\boldsymbol{r}) = 1, \quad r < a \tag{7.72}$$

$$u_1^+(\boldsymbol{r}) = r \left(1 - \frac{\rho^+ - \rho^-}{\rho^+ + 2\rho^-} \frac{a^3}{r^3} \right) P_1(\cos\theta), \quad r > a \tag{7.73}$$

$$u_1^-(\boldsymbol{r}) = \frac{3\rho^-}{\rho^+ + 2\rho^-} r P_1(\cos\theta), \quad r < a \tag{7.74}$$

$$u_2^+(\boldsymbol{r}) = \frac{r^2}{3} - \frac{2}{3}\frac{a^3}{r}\left(\frac{\gamma^-}{\gamma^+} - 1\right)$$
$$+ 6a^3 \frac{\delta^- \gamma^- \rho^-}{c^+ \gamma^+ (\rho^+ + 2\rho^-)^2} \frac{P_1(\cos\theta)}{r^2}$$
$$+ \frac{2}{3}\left[1 - 2\frac{\rho^+ - \rho^-}{2\rho^+ + 3\rho^-}\frac{a^5}{r^5}\right] r^2 P_2(\cos\theta), \quad r > a \tag{7.75}$$

$$u_2^-(\boldsymbol{r}) = \frac{r^2}{3}\frac{\rho^-\gamma^-}{\rho^+\gamma^+} + \frac{a^2}{3}\left[3 - \frac{\gamma^-}{\gamma^+}\left(\frac{\rho^-}{\rho^+} + 2\right)\right]$$
$$+ 6\frac{\delta^-\gamma^-\rho^-}{c^+\gamma^+(\rho^+ + 2\rho^-)^2} r P_1(\cos\theta)$$
$$+ \frac{10}{3}\frac{\rho^-}{2\rho^+ + 3\rho^-} r^2 P_2(\cos\theta), \quad r < a \tag{7.76}$$

$$u_3^+(\boldsymbol{r}) = 2a^3\left(1 - \frac{\gamma^-}{\gamma^+}\right)$$
$$+ \frac{3}{5}\left[1 + 5\frac{\rho^+ - \rho^-}{\rho^+ + 2\rho^-}\frac{a^3}{r^3} - 6\frac{(\rho^+)^2 - (\rho^-)^2 - \rho^+\rho^-}{(\rho^+ + 2\rho^-)^2}\frac{a^5}{r^5}\right.$$
$$- 6\frac{\gamma^-}{\gamma^+}\left(\frac{\rho^-}{\rho^+ + 2\rho^-}\right)^2 \frac{a^5}{r^5}$$
$$\left. + 30\frac{\rho^+\rho^-}{\gamma^+(\rho^+ + 2\rho^-)}\left(\frac{\delta^-\gamma^-}{\rho^+ + 2\rho^-}\right)^2 \frac{a^3}{r^5}\right] r^3 P_1(\cos\theta)$$
$$+ 20\frac{\rho^+\rho^-}{(\rho^+ + 2\rho^-)^2} c^+\delta^-\gamma^- \frac{a^5}{r^3} P_2(\cos\theta)$$
$$+ \frac{2}{5}\left[1 - 3\frac{\rho^+ - \rho^-}{3\rho^+ + 4\rho^-}\frac{a^7}{r^7}\right] r^3 P_3(\cos\theta), \quad r > a \tag{7.77}$$

$$u_3^-(\boldsymbol{r}) = 2a^3\left(1 - \frac{\gamma^-}{\gamma^+}\right)$$
$$+ \frac{\rho^-\gamma^-}{\rho^+\gamma^+} c^+\delta^-\gamma^-(r^2 - a^2)$$
$$+ \frac{9}{5}\frac{\rho^-}{\rho^+ + 2\rho^-}\left[\frac{\rho^-\gamma^-}{\rho^+\gamma^+} + 5\frac{\rho^+}{\rho^+ + 2\rho^-}\frac{a^2}{r^2}\right.$$
$$\left. - \frac{\rho^-\gamma^-}{\rho^+\gamma^+}\frac{3\rho^+ + 2\rho^-}{\rho^+ + 2\rho^-}\frac{a^2}{r^2} + \frac{10}{r^2}\frac{\rho^+}{\gamma^+}\left(\frac{\delta^-\gamma^-}{\rho^+ + 2\rho^-}\right)^2\right] r^3 P_1(\cos\theta)$$

$$+20\frac{\rho^+\rho^-}{(\rho^+ + 2\rho^-)^2}c^+\delta^-\gamma^- r^2 P_2(\cos\theta)$$
$$+\frac{14}{5}\frac{\rho^-}{3\rho^+ + 4\rho^-}r^3 P_3(\cos\theta), \quad r < a \tag{7.78}$$

Note that if we set $\delta^- = 0$ and use the fact that this implies that $\gamma^-/\gamma^+ = \beta\eta^2$ and $\rho^+/\rho^- = \beta$, then we recover the terms given above in the lossless case. The scattering amplitude is

$$g(\hat{\boldsymbol{r}}, \hat{\boldsymbol{k}}) = ik^3a^3\left[\frac{1}{3}\left(\frac{\gamma^-}{\gamma^+} - 1\right) - \frac{\rho^+ - \rho^-}{\rho^+ + 2\rho^-}P_1(\cos\theta)\right]$$
$$- 3k^4a^3\frac{\delta^-\gamma^-\rho^-}{c^+\gamma^+(\rho^+ + 2\rho^-)^2}P_1(\cos\theta) + O(k^5) \tag{7.79}$$

Observe that $\delta/c\rho$ has the dimension of length and this is the reason for the factor a^3 (instead of a^4) in the k^4-order term of g.

The differential scattering cross section is

$$\sigma(\hat{\boldsymbol{r}}) = 4\pi k^4a^6\left[\frac{1}{9}\left(\frac{\gamma^-}{\gamma^+} - 1\right)^2 - \frac{2}{3}\frac{\rho^+ - \rho^-}{\rho^+ + 2\rho^-}\left(\frac{\gamma^-}{\gamma^+} - 1\right)P_1(\cos\theta)\right.$$
$$\left. + \left(\frac{\rho^+ - \rho^-}{\rho^+ + 2\rho^-}\right)^2(P_1(\cos\theta))^2\right] + O(k^6) \tag{7.80}$$

while the (total) scattering cross section is

$$\sigma_s = 4\pi k^4a^6\left[\frac{1}{9}\left(\frac{\gamma^-}{\gamma^+} - 1\right)^2 + \frac{1}{3}\left(\frac{\rho^+ - \rho^-}{\rho^+ + 2\rho^-}\right)^2\right] + O(k^6) \tag{7.81}$$

and the absorption cross section is

$$\sigma_a = 12\pi k^2a^3c^+\delta^-\gamma^-\frac{\rho^+\rho^-}{(\rho^+ + 2\rho^-)^2} + O(k^4) \tag{7.82}$$

The extinction cross section is

$$\sigma_e = 12\pi k^2a^3c^+\delta^-\gamma^-\frac{\rho^+\rho^-}{(\rho^+ + 2\rho^-)^2} + O(k^4) \tag{7.83}$$

This last result reflects the fact that $\sigma_a = O(k^2)$ whereas $\sigma_s = O(k^4)$. The expressions for σ_e may be verified using the optical theorem (2.59) and the fact that

$$\rho^+ = \frac{1}{(c^+)^2\gamma^+} \tag{7.84}$$

7.EM Electromagnetics

7.EM.1 *The perfect conductor problem*

The first terms in the expansions (3.117) are

$$\boldsymbol{E}_0^+(\boldsymbol{r}) = \hat{\boldsymbol{a}} - \frac{a^3}{r^3}\hat{\boldsymbol{a}} \cdot (\widetilde{\boldsymbol{I}} - 3\hat{\boldsymbol{r}}\hat{\boldsymbol{r}}), \quad r > a \tag{7.85}$$

$$\boldsymbol{H}_0^+(\boldsymbol{r}) = Y^+\hat{\boldsymbol{b}} + \frac{1}{2}Y^+\frac{a^3}{r^3}\hat{\boldsymbol{b}} \cdot (\widetilde{\boldsymbol{I}} - 3\hat{\boldsymbol{r}}\hat{\boldsymbol{r}}), \quad r > a \tag{7.86}$$

$$\begin{aligned}\boldsymbol{E}_1^+(\boldsymbol{r}) = {} & \hat{\boldsymbol{a}}(\hat{\boldsymbol{k}} \cdot \hat{\boldsymbol{r}}) - \frac{1}{2}\frac{a^3}{r^3}(\hat{\boldsymbol{b}} \times \boldsymbol{r}) + \frac{5a^5}{2r^7}\hat{\boldsymbol{a}} \cdot \boldsymbol{rrr} \cdot \hat{\boldsymbol{k}} \\ & - \frac{a^5}{2r^5}(\hat{\boldsymbol{a}}\hat{\boldsymbol{k}} + \hat{\boldsymbol{k}}\hat{\boldsymbol{a}}) \cdot \boldsymbol{r}, \quad r > a\end{aligned} \tag{7.87}$$

$$\begin{aligned}\boldsymbol{H}_1^+(\boldsymbol{r}) = {} & Y^+\hat{\boldsymbol{b}}(\hat{\boldsymbol{k}} \cdot \boldsymbol{r}) - Y^+\frac{a^3}{r^3}(\hat{\boldsymbol{a}} \times \boldsymbol{r}) - \frac{5}{3}Y^+\frac{a^5}{r^7}\hat{\boldsymbol{b}} \cdot \boldsymbol{rrr} \cdot \hat{\boldsymbol{k}} \\ & + \frac{1}{3}Y^+\frac{a^5}{r^5}(\hat{\boldsymbol{b}}\hat{\boldsymbol{k}} + \hat{\boldsymbol{k}}\hat{\boldsymbol{b}}) \cdot \boldsymbol{r}, \quad r > a\end{aligned} \tag{7.88}$$

The scattering amplitudes (3.173) and (3.174) are

$$\boldsymbol{g}_e(\hat{\boldsymbol{r}}, \hat{\boldsymbol{k}}) = -ik^3a^3\left[\hat{\boldsymbol{r}} \times (\hat{\boldsymbol{r}} \times \hat{\boldsymbol{a}}) - \frac{1}{2}\hat{\boldsymbol{r}} \times \hat{\boldsymbol{b}}\right] + O(k^5) \tag{7.89}$$

$$\boldsymbol{g}_m(\hat{\boldsymbol{r}}, \hat{\boldsymbol{k}}) = ik^3a^3Y^+\left[\frac{1}{2}\hat{\boldsymbol{r}} \times (\hat{\boldsymbol{r}} \times \hat{\boldsymbol{b}}) + \hat{\boldsymbol{r}} \times \hat{\boldsymbol{a}}\right] + O(k^5) \tag{7.90}$$

The differential scattering cross section (2.114) is

$$\sigma(\hat{\boldsymbol{r}}) = 4\pi k^4a^6\left[\frac{5}{4} - \hat{\boldsymbol{r}} \cdot \hat{\boldsymbol{k}} - (\hat{\boldsymbol{r}} \cdot \hat{\boldsymbol{a}})^2 - \frac{1}{4}(\hat{\boldsymbol{r}} \cdot \hat{\boldsymbol{b}})^2\right] + O(k^6) \tag{7.91}$$

and the (total) scattering cross section is

$$\sigma_s = \frac{10\pi}{3}k^4a^6 + O(k^6) \tag{7.92}$$

while in this case

$$\sigma_a = 0 \tag{7.93}$$

and

$$\sigma_e = \sigma_s \tag{7.94}$$

Note that knowledge of the first-order near fields $\boldsymbol{E}_1^+$ and $\boldsymbol{H}_1^+$ does not help to improve the far field approximation since the k^4 term in $\boldsymbol{g}_e$ and $\boldsymbol{g}_m$ for the sphere vanishes. For higher order terms see Bowman *et al.* (1969).

7.EM.2 *The impedance problem*

While no general results for the impedance boundary condition are known it is possible to obtain some results in the special case of a spherical scatterer. From (3.117), the Rayleigh approximation is given by

$$\boldsymbol{E}_0^+(\boldsymbol{r}) = \hat{\boldsymbol{a}} - \frac{a^3}{r^3}\hat{\boldsymbol{a}} \cdot (\widetilde{\boldsymbol{I}} - 3\hat{\boldsymbol{r}}\hat{\boldsymbol{r}}), \quad r > a \tag{7.95}$$

$$\boldsymbol{H}_0^+(\boldsymbol{r}) = Y^+\hat{\boldsymbol{b}} - Y^+\frac{a^3}{r^3}\hat{\boldsymbol{b}} \cdot (\widetilde{\boldsymbol{I}} - 3\hat{\boldsymbol{r}}\hat{\boldsymbol{r}}), \quad r > a \tag{7.96}$$

The scattering amplitudes are

$$g_e(\hat{\boldsymbol{r}}, \hat{\boldsymbol{k}}) = -ik^3a^3[\hat{\boldsymbol{r}} \times (\hat{\boldsymbol{r}} \times \hat{\boldsymbol{a}}) + \hat{\boldsymbol{r}} \times \hat{\boldsymbol{b}}] + O(k^4) \tag{7.97}$$

$$g_m(\hat{\boldsymbol{r}}, \hat{\boldsymbol{k}}) = ik^3a^3Y^+[\hat{\boldsymbol{r}} \times \hat{\boldsymbol{a}} - \hat{\boldsymbol{r}} \times (\hat{\boldsymbol{r}} \times \hat{\boldsymbol{b}})] + O(k^4) \tag{7.98}$$

while the differential scattering cross section is

$$\sigma(\hat{\boldsymbol{r}}) = 4\pi k^4a^6[2 + 2\hat{\boldsymbol{r}} \cdot \hat{\boldsymbol{k}} - (\hat{\boldsymbol{r}} \cdot \hat{\boldsymbol{a}})^2 - (\hat{\boldsymbol{r}} \cdot \hat{\boldsymbol{b}})^2] + O(k^6) \tag{7.99}$$

and the (total) scattering cross section is

$$\sigma_s = \frac{16\pi}{3}k^4a^6 + O(k^6) \tag{7.100}$$

There are insufficient low frequency terms available to calculate the first term in the absorption cross section although with the optical (2.59) theorem it is assured that

$$\sigma_a = O(k^2) \tag{7.101}$$

7.EM.3 *The lossless transmission problem*

The first terms in the low frequency expansions (3.117) and (3.118) of the total fields in both V^+ and V^- are

$$\boldsymbol{E}_0^+(\boldsymbol{r}) = \hat{\boldsymbol{a}} - \frac{\mu^+\eta^2 - \mu^-}{\mu^+\eta^2 + 2\mu^-}\frac{a^3}{r^3}\hat{\boldsymbol{a}} \cdot (\widetilde{\boldsymbol{I}} - 3\hat{\boldsymbol{r}}\hat{\boldsymbol{r}}), \quad r > a \tag{7.102}$$

$$\boldsymbol{E}_0^-(\boldsymbol{r}) = \frac{3\mu^-}{\mu^+\eta^2 + 2\mu^-}\hat{\boldsymbol{a}}, \quad r < a \tag{7.103}$$

$$\boldsymbol{H}_0^+(\boldsymbol{r}) = Y^+\hat{\boldsymbol{b}} + Y^+\frac{\mu^+ - \mu^-}{2\mu^+ + \mu^-}\frac{a^3}{r^3}\hat{\boldsymbol{b}} \cdot (\widetilde{\boldsymbol{I}} - 3\hat{\boldsymbol{r}}\hat{\boldsymbol{r}}), \quad r > a \tag{7.104}$$

and

$$\boldsymbol{H}_0^-(\boldsymbol{r}) = Y^+\frac{3\mu^+}{2\mu^+ + \mu^-}\hat{\boldsymbol{b}}, \quad r < a. \tag{7.105}$$

Recalling that in the lossless case

$$\eta^2 = \frac{\epsilon^-\mu^-}{\epsilon^+\mu^+} \tag{7.106}$$

while the scattering amplitudes (3.217)–(3.221) are

$$
\boldsymbol{g}_e(\hat{\boldsymbol{r}}, \hat{\boldsymbol{k}}) = -ik^3a^3 \left[\frac{\mu^+\eta^2 - \mu^-}{\mu^+\eta^2 + 2\mu^-} \hat{\boldsymbol{r}} \times (\hat{\boldsymbol{r}} \times \hat{\boldsymbol{a}}) - \frac{\mu^+ - \mu^-}{2\mu^+ + \mu^-} \hat{\boldsymbol{r}} \times \hat{\boldsymbol{b}} \right] + O(k^4) \tag{7.107}
$$

and

$$
\boldsymbol{g}_m(\hat{\boldsymbol{r}}, \hat{\boldsymbol{k}}) = ik^3a^3Y^+ \left[\frac{\mu^+ - \mu^-}{2\mu^+ + \mu^-} \hat{\boldsymbol{r}} \times (\hat{\boldsymbol{r}} \times \hat{\boldsymbol{b}}) + \frac{\mu^+\eta^2 - \mu^-}{\mu^+\eta^2 + 2\mu^-} \hat{\boldsymbol{r}} \times \hat{\boldsymbol{a}} \right] + O(k^4) \tag{7.108}
$$

The differential scattering cross section (3.222) is

$$
\begin{aligned}
\sigma(\hat{\boldsymbol{r}}) = 4\pi k^4a^6 \Bigg[& \left(\frac{\mu^+\eta^2 - \mu^-}{\mu^+\eta^2 + 2\mu^-} \right)^2 (1 - (\hat{\boldsymbol{r}} \cdot \hat{\boldsymbol{a}})^2) \\
& + \left(\frac{\mu^+ - \mu^-}{2\mu^+ + \mu^-} \right)^2 (1 - (\hat{\boldsymbol{r}} \cdot \hat{\boldsymbol{b}})^2) - \frac{2(\mu^+\eta^2 - \mu^-)(\mu^+ - \mu^-)}{(\mu^+\eta^2 + 2\mu^-)(2\mu^+ + \mu^-)} \hat{\boldsymbol{r}} \cdot \hat{\boldsymbol{k}} \Bigg] \\
& + O(k^6)
\end{aligned} \tag{7.109}
$$

and the (total) scattering cross section (3.223) is

$$
\sigma_s = \frac{8\pi}{3} k^4a^6 \left[\left(\frac{k^+\eta^2 - \mu^-}{\mu^+\eta^2 + 2\mu^-} \right)^2 + \left(\frac{\mu^+ - \mu^-}{2\mu^+ + \mu^-} \right)^2 \right] + O(k^6) \tag{7.110}
$$

In the lossless case σ_a vanishes and

$$
\sigma_e = \sigma_s \tag{7.111}
$$

7.EM.4 *The lossy transmission problem*

The Rayleigh approximations in (3.117) and (3.118) are

$$
\boldsymbol{E}_0^+(\boldsymbol{r}) = \hat{\boldsymbol{a}} - \frac{a^3}{r^3} \hat{\boldsymbol{a}} \cdot (\widetilde{\boldsymbol{I}} - 3\hat{\boldsymbol{r}}\hat{\boldsymbol{r}}), \quad r > a \tag{7.112}
$$

$$
\boldsymbol{E}_0^-(\boldsymbol{r}) = \boldsymbol{0}, \quad r < a \tag{7.113}
$$

$$\boldsymbol{H}_0^+(\boldsymbol{r}) = Y^+\hat{\boldsymbol{b}} + Y^+\frac{\mu^+ - \mu^-}{2\mu^+ + \mu^-}\frac{a^3}{r^3}\hat{\boldsymbol{b}} \cdot (\widetilde{\boldsymbol{I}} - 3\hat{\boldsymbol{r}}\hat{\boldsymbol{r}}), \quad r > a \tag{7.114}$$

$$\boldsymbol{H}_0^-(\boldsymbol{r}) = Y^+\frac{3\mu^+}{2\mu^+ + \mu^-}\hat{\boldsymbol{b}}, \quad r < a \tag{7.115}$$

The scattering amplitudes (3.217)–(3.221) are

$$g_e(\hat{\boldsymbol{r}}, \hat{\boldsymbol{k}}) = -ik^3a^3\left[\hat{\boldsymbol{r}} \times (\hat{\boldsymbol{r}} \times \hat{\boldsymbol{a}}) - \frac{\mu^+ - \mu^-}{2\mu^+ + \mu^-}\hat{\boldsymbol{r}} \times \hat{\boldsymbol{b}}\right] + O(k^4) \tag{7.116}$$

and

$$g_m(\hat{\boldsymbol{r}}, \hat{\boldsymbol{k}}) = ik^3a^3Y^+\left[\hat{\boldsymbol{r}} \times \hat{\boldsymbol{a}} + \frac{\mu^+ - \mu^-}{2\mu^+ + \mu^-}\hat{\boldsymbol{r}} \times (\hat{\boldsymbol{r}} \times \hat{\boldsymbol{b}})\right] + O(k^4) \tag{7.117}$$

The differential scattering cross section (3.222) is

$$\sigma(\hat{\boldsymbol{r}}) = 4\pi k^4 a^6\left[1 - (\hat{\boldsymbol{r}} \cdot \hat{\boldsymbol{a}})^2 + \left(\frac{\mu^+ - \mu^-}{2\mu^+ + \mu^-}\right)^2(1 - (\hat{\boldsymbol{r}} \cdot \hat{\boldsymbol{r}})^2) - 2\frac{\mu^+ - \mu^-}{2\mu^+ + \mu^-}\hat{\boldsymbol{r}} \cdot \hat{\boldsymbol{k}}\right] + O(k^6) \tag{7.118}$$

and the (total) scattering cross section (3.223) is

$$\sigma_s = \frac{8\pi}{3}k^4a^6\left[1 + \left(\frac{\mu^+ - \mu^-}{2\mu^+ + \mu^-}\right)^2\right] + O(k^6) \tag{7.119}$$

The remainder in (7.119) is of order k^6 rather than k^5 because the far field coefficients A_n in (3.222) are real. The first terms in the low frequency expansions of the fields are not sufficient to evaluate the first term in the absorption cross section. However, the optical theorem (2.59) shows that

$$\sigma_a = O(k^2) \tag{7.120}$$

7.E Elasticity

7.E.1 *The rigid problem*

The first two terms in the low frequency expansion (3.228) of the elastic displacement are

$$\boldsymbol{u}_0^+(\boldsymbol{r}) = \left(1 - \frac{a}{r}\right)(\alpha_p\hat{\boldsymbol{k}} + \alpha_s\hat{\boldsymbol{b}}) + \frac{1 - \tau^2}{2(2 + \tau^2)}\left(\frac{a}{r} - \frac{a^3}{r^3}\right)(\widetilde{\boldsymbol{I}} - 3\hat{\boldsymbol{r}}\hat{\boldsymbol{r}}) \cdot (\alpha_p\hat{\boldsymbol{k}} + \alpha_s\hat{\boldsymbol{b}}), \quad r > a \tag{7.121}$$

$$u_1^+ = \left(1 - \frac{a^3}{r^3}\right)\left(a_p\hat{k} + \frac{a_s}{\tau}\hat{b}\right)(\hat{k}\cdot r)$$
$$- \left(1 - \frac{a}{r}\right)\frac{\tau^3 + 2}{\tau(\tau^2+2)}a(a_p\hat{k} + a_s\hat{b})$$
$$- \frac{(\tau^2-1)(\tau^3+2)}{2\tau(\tau^2+2)^2}a\left(\frac{a^3}{r^3} - \frac{a}{r}\right)(\tilde{I} - 3\hat{r}\hat{r})\cdot(a_p\hat{k} + a_s\hat{b})$$
$$+ \frac{3(\tau^2-1)}{2(2\tau^2+3)}\left\{\left(\frac{a^5}{r^5} - \frac{a^3}{r^3}\right)(\tilde{I} - 5\hat{r}\hat{r})\cdot\left(a_p\hat{k} + \frac{a_s}{\tau}\hat{b}\right)(\hat{k}\cdot r)\right.$$
$$\left. + \left(\frac{a^5}{r^5} - \frac{a^3}{r^3}\right)\hat{k}\left(a_p\hat{k} + \frac{a_s}{\tau}\hat{b}\right)\cdot r + \left(\frac{a^5}{r^5} - \frac{a^3}{r^3}\right)a_p r\right\} \tag{7.122}$$

The scattering amplitudes (3.277)–(3.285) are

$$g_r(\hat{r}, \hat{k}) = -\frac{3\tau^3}{\tau^2+2}\left[i\frac{k_p a}{\tau} + \frac{\tau^3+2}{\tau^2+2}\left(\frac{k_p a}{\tau}\right)^2\right](a_p\hat{k} + a_s\hat{b})\cdot\hat{r} + O(k^3) \tag{7.123}$$

$$g_\theta(\hat{r}, \hat{k}) = -\frac{3}{\tau^2+2}\left[i\frac{k_p a}{\tau} + \frac{\tau^3+2}{\tau^2+2}\left(\frac{k_p a}{\tau}\right)^2\right](a_p\hat{k} + a_s\hat{b})\cdot\hat{\theta} + O(k^3) \tag{7.124}$$

$$g_\varphi(\hat{r}, \hat{k}) = -\frac{3}{\tau^2+2}\left[i\frac{k_p a}{\tau} + \frac{\tau^3+2}{\tau^2+2}\left(\frac{k_p a}{\tau}\right)^2\right](a_p\hat{k} + a_s\hat{b})\cdot\hat{\varphi} + O(k^3) \tag{7.125}$$

In (7.121)–(7.125) we use the parameter

$$\tau = \tau^+ = \frac{k_p}{k_s} = \frac{c_s^+}{c_p^+} \tag{7.126}$$

for simplicity.

The differential scattering cross section, (3.286), may be obtained from (7.123)–(7.125) while the (total) scattering cross section is

$$\sigma_s = 12\pi a^2\frac{\tau(2+\tau^3)}{(2+\tau^2)^2}\frac{\alpha_p^2 + \alpha_s^2}{\alpha_p^2 + \tau\alpha_s^2} + O(k_p^2) \tag{7.127}$$

In the case of rigid scatterers the absorption cross section, σ_a, vanishes and

$$\sigma_e = \sigma_s \tag{7.128}$$

In particular, for P incidence, ($\alpha_s = 0$), we have

$$\sigma_s^{(p)} = 12\pi a^2\frac{\tau(2+\tau^3)}{(2+\tau^2)^2} + O(k_p^2) \tag{7.129}$$

and for S incidence, $(\alpha_p = 0)$, we have

$$\sigma_s^{(s)} = 12\pi a^2 \frac{(2+\tau^3)}{(2+\tau^2)^2} + O(k_p^2) \tag{7.130}$$

Observe that the Rayleigh approximation of the scattering cross section for scattering by a rigid sphere, (7.127), is proportional to the area of a large circle and the constant of proportionality depends on the ratio $\tau = c_s/c_p$ of the phase velocities and the amplitudes α_p and α_s of the incident wave. Furthermore, since $\tau < 1$ it follows that transverse excitation, $(\alpha_p = 0)$, results in stronger scattering than longitudinal excitation, $(\alpha_s = 0)$.

7.E.2 *The cavity problem*

For the cavity problem, the first term in the low frequency expansion is simply the first term in the expansion of the incident field and it is necessary to evaluate the next term to see scattering effects. Thus, the Rayleigh approximation is

$$\boldsymbol{u}_0^+(\boldsymbol{r}) = \alpha_p \hat{\boldsymbol{k}} + \alpha_s \hat{\boldsymbol{b}}, \quad r > a \tag{7.131}$$

while

$$\begin{aligned}
\boldsymbol{u}_1^+(\boldsymbol{r}) &= \left(\alpha_p \hat{\boldsymbol{k}} + \frac{\alpha_s}{\tau}\hat{\boldsymbol{b}}\right)\hat{\boldsymbol{k}}\cdot\boldsymbol{r} \\
&+ \frac{1}{9-4\tau^2}\left[5\tau^2\frac{a^3}{r^3} + 3(1-\tau^2)\frac{a^5}{r^5}\right]\left[2\alpha_p\hat{\boldsymbol{k}}\hat{\boldsymbol{k}} + \frac{\alpha_s}{\tau}(\hat{\boldsymbol{k}}\hat{\boldsymbol{b}} + \hat{\boldsymbol{b}}\hat{\boldsymbol{k}})\right]\cdot\boldsymbol{r} \\
&+ \frac{3\alpha_p}{4\tau^2(9-4\tau^2)}\left[(3-12\tau^2+4\tau^4)\frac{a^3}{r^3} + 4\tau^2(1-\tau^2)\frac{a^5}{r^5}\right]\boldsymbol{r} \\
&+ \frac{15(1-\tau^2)}{9-4\tau^2}\left(\frac{a^3}{r^3} - \frac{a^5}{r^5}\right)\left[\alpha_p\hat{\boldsymbol{k}}\cdot\hat{\boldsymbol{r}}\boldsymbol{r}\hat{\boldsymbol{r}}\cdot\hat{\boldsymbol{k}} + \frac{\alpha_s}{\tau}\hat{\boldsymbol{k}}\cdot\hat{\boldsymbol{r}}\boldsymbol{r}\hat{\boldsymbol{r}}\cdot\hat{\boldsymbol{b}}\right], \quad r > a
\end{aligned} \tag{7.132}$$

and the scattering amplitudes (3.298)–(3.306) are

$$\begin{aligned}
g_r(\hat{\boldsymbol{r}}, \hat{\boldsymbol{k}}) = ia^3k_p^3 &\left[\frac{9-16\tau^2-8\tau^4}{4\tau^2(9-4\tau^2)}\alpha_p - \frac{1}{3}(\alpha_p\hat{\boldsymbol{k}} + \alpha_s\hat{\boldsymbol{b}})\cdot\hat{\boldsymbol{r}}\right. \\
&\left. + \frac{10\tau^2}{9-4\tau^2}\left(\alpha_p\hat{\boldsymbol{k}}\cdot\hat{\boldsymbol{r}}\hat{\boldsymbol{r}}\cdot\hat{\boldsymbol{k}} + \frac{\alpha_s}{\tau}\hat{\boldsymbol{k}}\cdot\hat{\boldsymbol{r}}\hat{\boldsymbol{r}}\cdot\hat{\boldsymbol{b}}\right)\right] + O(k_p^4)
\end{aligned} \tag{7.133}$$

$$g_\theta(\hat{\boldsymbol{r}}, \hat{\boldsymbol{k}}) = ia^3k_s^3 \left[-\frac{1}{3}(\alpha_p \hat{\boldsymbol{k}} + \alpha_s \hat{\boldsymbol{b}}) \cdot \hat{\boldsymbol{\theta}} + \frac{5\tau}{9-4\tau^2}\alpha_p \hat{\boldsymbol{k}} \cdot (\hat{\boldsymbol{r}}\,\hat{\boldsymbol{\theta}} + \hat{\boldsymbol{\theta}}\hat{\boldsymbol{r}}) \cdot \hat{\boldsymbol{k}} + \frac{5}{9-4\tau^2}\alpha_s \hat{\boldsymbol{k}} \cdot (\hat{\boldsymbol{r}}\,\hat{\boldsymbol{\theta}} + \hat{\boldsymbol{\theta}}\hat{\boldsymbol{r}}) \cdot \hat{\boldsymbol{b}} \right] + O(k_p^4) \tag{7.134}$$

$$g_\varphi(\hat{\boldsymbol{r}}, \hat{\boldsymbol{k}}) = ia^3k_s^3 \left[-\frac{1}{3}(\alpha_p \hat{\boldsymbol{k}} + \alpha_s \hat{\boldsymbol{b}}) \cdot \hat{\boldsymbol{\varphi}} + \frac{5\tau}{9-4\tau^2}\alpha_p \hat{\boldsymbol{k}} \cdot (\hat{\boldsymbol{r}}\hat{\boldsymbol{\varphi}} + \hat{\boldsymbol{\varphi}}\hat{\boldsymbol{r}}) \cdot \hat{\boldsymbol{k}} + \frac{5}{9-4\tau^2}\alpha_s \hat{\boldsymbol{k}} \cdot (\hat{\boldsymbol{r}}\hat{\boldsymbol{\varphi}} + \hat{\boldsymbol{\varphi}}\hat{\boldsymbol{r}}) \cdot \hat{\boldsymbol{b}} \right] + O(k_p^4) \tag{7.135}$$

where τ is given by (7.126)

The differential scattering cross section may be computed using the expressions for the scattering amplitudes and the definition (3.308).

The (total) scattering cross section is

$$\sigma_s = \frac{4\pi k_p^4 a^6}{(9-4\tau^2)^2(\alpha_p^2 + \tau\alpha_s^2)} \left[\frac{(9-16\tau^2-8\tau^4)^2}{16\tau^4}\alpha_p^2 + \frac{5}{3}(9-16\tau^2-8\tau^4)\alpha_p^2 + \frac{20(2+3\tau^5)}{3\tau}\alpha_p^2 + \frac{10(3+2\tau^5)}{3\tau^3}\alpha_s^2 + \frac{(2+\tau^3)(9-4\tau^2)^2}{27\tau^3}(\alpha_p^2+\alpha_s^2) \right] + O(k_p^6) \tag{7.136}$$

In this case again σ_a vanishes. Hence

$$\sigma_e = \sigma_s \tag{7.137}$$

7.E.3 *The transmission problem*

In the case of an elastic scatterer, the first term in the low frequency expansion is again merely the first term in the expansion of the incident field

$$\boldsymbol{u}_0^+(\boldsymbol{r}) = \alpha_p \hat{\boldsymbol{k}} + \alpha_s \hat{\boldsymbol{b}}, \quad r > a \tag{7.138}$$

$$\boldsymbol{u}_0^-(\boldsymbol{r}) = \alpha_p \hat{\boldsymbol{k}} + \alpha_s \hat{\boldsymbol{b}}, \quad r < a \tag{7.139}$$

In order for any scattering effects to be seen we need the next terms which are somewhat complicated. To simplify the presentation we introduce the following notational definitions

$$N_1 = \frac{5\mu^+(\lambda^+ + 2\mu^+)}{15\mu^+(\lambda^+ + 2\mu^+) - 2(3\lambda^+ + 8\mu^+)(\mu^+ - \mu^-)} \tag{7.140}$$

$$N_2 = \frac{\lambda^+ + 2\mu^+}{3(\lambda^- + 2\mu^-) + 4(\mu^+ - \mu^-)} \tag{7.141}$$

$$N_3 = \frac{3(\lambda^+ + \mu^+)(\mu^+ - \mu^-)}{15\mu^+(\lambda^+ + 2\mu^+) - 2(3\lambda^+ + 8\mu^+)(\mu^+ - \mu^-)} \tag{7.142}$$

In terms of these quantities we have

$$\begin{aligned} \boldsymbol{u}_1^+(\boldsymbol{r}) = {} & \left[\alpha_p \hat{\boldsymbol{k}}\hat{\boldsymbol{k}} + \frac{\alpha_s}{\tau}\hat{\boldsymbol{b}}\hat{\boldsymbol{k}}\right] \cdot \boldsymbol{r} \\ & + \alpha_p \frac{a^3}{r^3}[(3N_1 - 1)\hat{\boldsymbol{k}}\hat{\boldsymbol{k}} + (N_2 - N_1)\widetilde{\boldsymbol{I}}] \cdot \boldsymbol{r} \\ & + \frac{\alpha_s}{2\tau}\frac{a^3}{r^3}(3N_1 - 1)(\hat{\boldsymbol{b}}\hat{\boldsymbol{k}} + \hat{\boldsymbol{k}}\hat{\boldsymbol{b}}) \cdot \boldsymbol{r} \\ & + \alpha_p N_3 \left(\frac{a^5}{r^5} - \frac{a^3}{r^3}\right)(2\hat{\boldsymbol{k}}\hat{\boldsymbol{k}} + \widetilde{\boldsymbol{I}} - 5\hat{\boldsymbol{k}} \cdot \hat{\boldsymbol{r}}\hat{\boldsymbol{r}} \cdot \hat{\boldsymbol{k}}\widetilde{\boldsymbol{I}}) \cdot \boldsymbol{r} \\ & + \frac{\alpha_s}{\tau} N_3 \left(\frac{a^5}{r^5} - \frac{a^3}{r^3}\right)(\hat{\boldsymbol{b}}\hat{\boldsymbol{k}} + \hat{\boldsymbol{k}}\hat{\boldsymbol{b}} - 5\hat{\boldsymbol{k}} \cdot \hat{\boldsymbol{r}}\hat{\boldsymbol{r}} \cdot \hat{\boldsymbol{b}}\widetilde{\boldsymbol{I}}) \cdot \boldsymbol{r}, \qquad r > a \end{aligned} \tag{7.143}$$

$$\begin{aligned} \boldsymbol{u}_1^-(\boldsymbol{r}) = {} & \alpha_p[3N_1\hat{\boldsymbol{k}}\hat{\boldsymbol{k}} + (N_2 - N_1)\widetilde{\boldsymbol{I}}] \cdot \boldsymbol{r} \\ & + \frac{\alpha_s}{2\tau}[(3N_1 + 1)\hat{\boldsymbol{b}}\hat{\boldsymbol{k}} + (3N_1 - 1)\hat{\boldsymbol{k}}\hat{\boldsymbol{b}}] \cdot \boldsymbol{r}, \qquad r < a \end{aligned} \tag{7.144}$$

and τ is given by (7.126).

These expressions simplify considerably in some special cases. For example, when $\mu^+ = \mu^-$ (i.e. the two media have the same sheer modulus) the expressions (7.140)–(7.142) become

$$N_1 = \frac{1}{3} \tag{7.145}$$

$$N_2 = \frac{1}{3}\frac{\lambda^+ + 2\mu^+}{\lambda^- + 2\mu^-} = \frac{1}{3}\frac{\rho^+}{\rho^-}\left(\frac{c_p^+}{c_p^-}\right)^2 \tag{7.146}$$

$$N_3 = 0 \tag{7.147}$$

and (7.143) and (7.144) reduce to

$$\begin{aligned} \boldsymbol{u}_1^+(\boldsymbol{r}) = {} & \left[\alpha_p \hat{\boldsymbol{k}}\hat{\boldsymbol{k}} + \frac{\alpha_s}{\tau}\hat{\boldsymbol{b}}\hat{\boldsymbol{k}}\right] \cdot \boldsymbol{r} \\ & + \frac{\alpha_p}{3}\frac{\lambda^+ - \lambda^-}{\lambda^- + 2\mu^-}\left(\frac{a}{r}\right)^3 \boldsymbol{r}, \qquad r > a \end{aligned} \tag{7.148}$$

$$\boldsymbol{u}_1^-(\boldsymbol{r}) = \left[\alpha_p \hat{\boldsymbol{k}}\hat{\boldsymbol{k}} + \frac{\alpha_s}{\tau}\hat{\boldsymbol{b}}\hat{\boldsymbol{k}}\right] \cdot \boldsymbol{r} + \frac{\alpha_p}{3}\frac{\lambda^+ - \lambda^-}{\lambda^- + 2\mu^-}\boldsymbol{r}, \quad r < a \tag{7.149}$$

When $\lambda^+ + \mu^+ = 0$, (7.140)–(7.142) become

$$N_1 = \frac{\mu^+}{\mu^+ + 2\mu^-} \tag{7.150}$$

$$N_2 = \frac{\mu^+}{3(\lambda^- + 2\mu^-) + 4(\mu^+ - \mu^-)} \tag{7.151}$$

$$N_3 = 0 \tag{7.152}$$

with resulting simplification in (7.143) and (7.144). No significant reduction occurs when $\lambda^+ = \lambda^-$ or when $\lambda^+ + 2\mu^+ = \lambda^- + 2\mu^-$.

The scattering amplitudes, (5.236) and (5.239), become

$$\begin{aligned} g_r(\hat{\boldsymbol{r}}, \hat{\boldsymbol{k}}) = \frac{ia^3k_p^3}{3}\Bigg[& \frac{3(\lambda^+ - \lambda^-)N_2 + 2(\mu^+ - \mu^-)(N_2 - N_1)}{\lambda^+ + 2\mu^+}\alpha_p \\ & + \left(\frac{\rho^-}{\rho^+} - 1\right)(\alpha_p \hat{\boldsymbol{k}} + \alpha_s \tau \hat{\boldsymbol{b}}) \cdot \hat{\boldsymbol{r}} \\ & + 6N_1 \frac{\mu^+ - \mu^-}{\lambda^+ + 2\mu^+}\left(\alpha_p \hat{\boldsymbol{k}} \cdot \hat{\boldsymbol{r}}\hat{\boldsymbol{r}} \cdot \hat{\boldsymbol{k}} + \frac{\alpha_s}{\tau}\hat{\boldsymbol{k}} \cdot \hat{\boldsymbol{r}}\hat{\boldsymbol{r}} \cdot \hat{\boldsymbol{b}}\right)\Bigg] + O(k_p^4) \end{aligned} \tag{7.153}$$

$$\begin{aligned} g_\theta(\hat{\boldsymbol{r}}, \hat{\boldsymbol{k}}) = \frac{ia^3k_s^3}{3}\Bigg\{ & \left(\frac{\rho^-}{\rho^+} - 1\right)\left(\alpha_p \hat{\boldsymbol{k}} + \frac{\alpha_s}{\tau}\hat{\boldsymbol{b}}\right) \cdot \hat{\boldsymbol{\theta}} \\ & + 3N_1\tau \frac{\mu^+ - \mu^-}{\mu^+}\left[\alpha_p \hat{\boldsymbol{k}} \cdot (\hat{\boldsymbol{r}}\hat{\boldsymbol{\theta}} + \hat{\boldsymbol{\theta}}\hat{\boldsymbol{r}}) \cdot \hat{\boldsymbol{k}} + \frac{\alpha_s}{\tau^+}\hat{\boldsymbol{k}} \cdot (\hat{\boldsymbol{r}}\hat{\boldsymbol{\theta}} + \hat{\boldsymbol{\theta}}\hat{\boldsymbol{r}}) \cdot \hat{\boldsymbol{b}}\right]\Bigg\} \\ & + O(k_p^4) \end{aligned} \tag{7.154}$$

and

$$\begin{aligned} g_\varphi(\hat{\boldsymbol{r}}, \hat{\boldsymbol{k}}) = \frac{ia^3k_s^3}{3}\Bigg\{ & \left(\frac{\rho^-}{\rho^+} - 1\right)\left(\alpha_p \hat{\boldsymbol{k}} + \frac{\alpha_s}{\tau}\hat{\boldsymbol{b}}\right) \cdot \hat{\boldsymbol{\varphi}} \\ & + 3N_1\tau \frac{\mu^+ - \mu^-}{\mu^+}\left[\alpha_p \hat{\boldsymbol{k}} \cdot (\hat{\boldsymbol{r}}\hat{\boldsymbol{\varphi}} + \hat{\boldsymbol{\varphi}}\hat{\boldsymbol{r}}) \cdot \hat{\boldsymbol{k}} + \frac{\alpha_s}{\tau}\hat{\boldsymbol{k}} \cdot (\hat{\boldsymbol{r}}\hat{\boldsymbol{\varphi}} + \hat{\boldsymbol{\varphi}}\hat{\boldsymbol{r}}) \cdot \hat{\boldsymbol{b}}\right]\Bigg\} \\ & + O(k_p^4) \end{aligned} \tag{7.155}$$

Observe that the radial amplitude g_r involves monopole, dipole and quadrupole terms, whereas the angular amplitudes g_θ and g_φ involve only dipole and quadrapole terms. The dipole terms, in all three amplitudes g_r, g_θ and g_φ, disappear if the mass densities in the two media are the same. The quadrapole terms disappear if the shear moduli do not differ in V^+ and V^-. Finally, the monopole term in g_r vanishes whenever both $\mu^+ = \mu^-$ and $\lambda^+ = \lambda^-$. Consequently, if $\rho^+ = \rho^-$ and $\mu^+ = \mu^-$, then the radial amplitude is of order k_p^3, while the angular amplitudes are of higher order.

Similar remarks can be made for the special cases of P and S incidence. The monopole term in g_r is solely due to P incidence, and it disappears for purely S incident wave when $\alpha_p = 0$.

The differential scattering cross section may be calculated from (5.244) and (5.237)–(5.239) and the expressions (7.153)–(7.155) for the scattering amplitudes.

The (total) scattering cross section is given by

$$
\begin{aligned}
\sigma_s = \frac{4\pi k_p^4 a^6}{27(\alpha_p^2 + \tau\alpha_s^2)} & \left\{ \left(\frac{\rho^-}{\rho^+} - 1\right)^2 \left(1 + \frac{2}{\tau^3}\right)\left(\alpha_p^2 + \frac{\alpha_s^2}{\tau^2}\right) \right. \\
& + 3\left[\frac{3(\lambda^+ - \lambda^-)N_2 + 2(\mu^+ - \mu^-)(N_2 - N_1)}{\lambda^+ + 2\mu^+}\right]^2 \alpha_p^2 \\
& + 12(\mu^+ - \mu^-)N_1 \frac{3(\lambda^+ - \lambda^-)N_2 + 2(\mu^+ - \mu^-)(N_2 - N_1)}{(\lambda^+ + 2\mu^+)^2}\alpha_p^2 \\
& + \frac{36}{5}N_1^2\left(\frac{\mu^+ - \mu^-}{\lambda^+ + 2\mu^+}\right)^2\left(3\alpha_p^2 + \frac{\alpha_s^2}{\tau^2}\right) \\
& \left. + \frac{18}{5}\frac{N_1^2}{\tau}\left(\frac{\mu^+ - \mu^-}{\mu^+}\right)^2\left(4\alpha_p^2 + 3\frac{\alpha_s^2}{\tau^2}\right)\right\} + O(k_p^6)
\end{aligned}
\tag{7.156}
$$

As before, there is no absorption as long as the Lamé constants are real (cf. (2.193)) hence σ_a vanishes and

$$\sigma_e = \sigma_s \tag{7.157}$$

In the special case of equal sheer moduli, $\mu^+ = \mu^-$, expression (7.156) becomes

$$
\begin{aligned}
\sigma_s = \frac{4\pi k_p^4 a^6}{27(\alpha_p^2 + \tau\alpha_s^2)} & \left[\left(\frac{\rho^-}{\rho^+} - 1\right)^2 \left(1 + \frac{2}{\tau^3}\right)\left(\alpha_p^2 + \frac{\alpha_s^2}{\tau^2}\right) \right. \\
& \left. + 3\left(\frac{\lambda^+ - \lambda^-}{\lambda^- + 2\mu^-}\right)^2 \alpha_p^2 \right] + O(k_p^6)
\end{aligned}
\tag{7.158}
$$

7.PS Point sources in acoustics

The following results are obtained in Dassios and Kamvyssas (1995, 1997) and in Dassios *et al* (1999). For a source located at the point

$$\boldsymbol{r}_0 = (0, 0, r_0) \tag{7.159}$$

we consider the incident field to be given by the **modified point source**

$$u^i_{\boldsymbol{r}_0}(\boldsymbol{r}) = \frac{r_0}{|\boldsymbol{r}-\boldsymbol{r}_0|} e^{ik(|\boldsymbol{r}-\boldsymbol{r}_0|-r_0)} \tag{7.160}$$

The **normalization** or **modification** is chosen so that the point source becomes a plane wave as the point source recedes to infinity, i.e.

$$\lim_{r_0\to\infty} u^i_{\boldsymbol{r}_0}(\boldsymbol{r}) = e^{-ik\hat{\boldsymbol{r}}_0\cdot\boldsymbol{r}} \tag{7.161}$$

which is a plane wave propagating in the direction $-\hat{\boldsymbol{r}}_0$.

7.PS.1 *The Dirichlet problem*

For modified point source incidence upon an acoustically soft sphere of radius a, the first few terms in the low frequency expansion (3.2) are

$$u_0^+(\boldsymbol{r}) = \frac{r_0}{|\boldsymbol{r}-\boldsymbol{r}_0|} - \frac{arr_0}{|a^2\boldsymbol{r}-r^2\boldsymbol{r}_0|}, \quad r > a, \ \boldsymbol{r} \neq \boldsymbol{r}_0 \tag{7.162}$$

$$u_1^+(\boldsymbol{r}) = (r_0 - a)\left(1 - \frac{a}{r}\right) - \frac{r_0^2}{|\boldsymbol{r}-\boldsymbol{r}_0|}$$

$$\frac{arr_0^2}{|a^2\boldsymbol{r}-r^2\boldsymbol{r}_0|}, \quad r > a, \quad \boldsymbol{r} \neq \boldsymbol{r}_0 \tag{7.163}$$

$$\begin{aligned} u_2^+(\boldsymbol{r}) = {} & \frac{r_0}{|\boldsymbol{r}-\boldsymbol{r}_0|}[|\boldsymbol{r}-\boldsymbol{r}_0| - r_0]^2 + 2a^2 - ar - \frac{4a^3}{3r} \\ & + ar\sum_{n=1}^{\infty}\left[\frac{1}{2n-1} - 2\frac{r_0^2}{r^2}\frac{n-1}{2n-1}\right. \\ & \left. - 2\frac{a^2}{r^2}\frac{2n+1}{(2n-1)(2n+3)}\right]\left(\frac{a^2}{rr_0}\right)^n P_n(\cos\theta), \quad r > a, \quad \boldsymbol{r} \neq \boldsymbol{r}_0 \end{aligned} \tag{7.164}$$

The scattering amplitude (3.38) is

$$\begin{aligned} g(\hat{\boldsymbol{r}}) = {} & -ika - (ka)^2\left[1 + \frac{a}{r_0}P_1(\cos\theta)\right] \\ & + ik^3a^3\left[\frac{2}{3} + P_1(\cos\theta) + \frac{a^2}{3r_0^2}P_2(\cos\theta)\right] + O(k^4) \end{aligned} \tag{7.165}$$

The differential scattering cross section (3.46) is

$$\begin{aligned} \sigma(\hat{\boldsymbol{r}}) = {} & 4\pi a^2 + 4\pi k^2a^4\left[\left(1 + \frac{a}{r_0}P_1(\cos\theta)\right)^2\right. \\ & \left. - \frac{4}{3} - 2P_1(\cos\theta) - \frac{2a^2}{3r_0^2}P_2(\cos\theta)\right] + O(k^4) \end{aligned} \tag{7.166}$$

while the (total) scattering cross section is

$$\sigma_s = 4\pi a^2\left[1 - \frac{(ka)^2}{3}\left(1 - \frac{a^2}{r_0^2}\right)\right] + O(k^4) \tag{7.167}$$

The absorption cross section vanishes

$$\sigma_a = 0 \tag{7.168}$$

Therefore, the extinction cross section is

$$\sigma_e = \sigma_s \tag{7.169}$$

7.PS.2 *The Neumann problem*

When the surface of the sphere is acoustically hard, the first few terms of (3.2) are

$$u_0^+(\boldsymbol{r}) = \frac{r_0}{|\boldsymbol{r} - \boldsymbol{r}_0|} + \sum_{n=1}^{\infty} \frac{n}{n+1}\left(\frac{a}{r}\right)^{n+1}\left(\frac{a}{r_0}\right)^{n} P_n(\cos\theta), \quad r > a, \quad \boldsymbol{r} \neq \boldsymbol{r}_0 \tag{7.170}$$

$$u_1^+(\boldsymbol{r}) = r_0(1 - \boldsymbol{u}_0^+(\boldsymbol{r})), \quad r > a, \quad \boldsymbol{r} \neq \boldsymbol{r}_0 \tag{7.171}$$

$$\begin{aligned} u_2^+(\boldsymbol{r}) = {} & \frac{r_0}{|\boldsymbol{r} - \boldsymbol{r}_0|}[|\boldsymbol{r} - \boldsymbol{r}_0| - r_0]^2 \\ & - a^2 \sum_{n=1}^{\infty} \frac{n}{(n+1)(2n-1)}\left(\frac{a}{r}\right)^{n-1}\left(\frac{a}{r_0}\right)^{n} P_n(\cos\theta) \\ & + a^2 \sum_{n=0}^{\infty}\left[\frac{2n(n-1)}{(2n-1)(n+1)} + \frac{2(2n+1)(n^2+n-1)}{(2n-1)(n+1)^2(2n+3)}\frac{a^2}{r_0^2}\right] \\ & \times \left(\frac{a}{r}\right)^{n+1}\left(\frac{a}{r_0}\right)^{n-2} P_n(\cos\theta), \quad r > a, \quad \boldsymbol{r} \neq \boldsymbol{r}_0 \end{aligned} \tag{7.172}$$

The scattering amplitude (3.54) is

$$\begin{aligned} g(\hat{\boldsymbol{r}}) = {} & \frac{a}{2r_0}(ka)^2 P_1(\cos\theta) - i(ka)^3\left[\frac{1}{3} + \frac{1}{2}P_1(\cos\theta) + \frac{2a^2}{9r_0^2}P_2(\cos\theta)\right] \\ & - \frac{a}{r_0}(ka)^4\left[\frac{3}{20}P_1(\cos\theta) + \frac{2}{9}P_2(\cos\theta) + \frac{a^2}{20r_0^2}P_3(\cos\theta)\right] + O(k^5) \end{aligned} \tag{7.173}$$

The differential scattering cross section is

$$\begin{aligned} \sigma(\hat{\boldsymbol{r}}) = {} & \pi\frac{k^2a^6}{r_0^2}(P_1(\cos\theta))^2 + 4\pi k^4a^6\left[\left(\frac{1}{3} + \frac{1}{2}P_1(\cos\theta) + \frac{2a^2}{9r_0^2}P_2(\cos\theta)\right)^2\right. \\ & \left. - \frac{a^2}{r_0^2}\left(\frac{3}{20}P_1(\cos\theta) + \frac{2}{9}P_2(\cos\theta) + \frac{a^2}{20r_0^2}P_3(\cos\theta)\right)P_1(\cos\theta)\right] + O(k^6) \end{aligned} \tag{7.174}$$

The (total) scattering cross section is

$$\sigma_s = \pi a^2 (ka)^2 \left[\frac{a^2}{3r_0^2} + (ka)^2 \left(\frac{7}{9} - \frac{a^2}{5r_0^2} + \frac{16a^4}{405r_0^4} \right) \right] + O(k^6) \tag{7.175}$$

In this case, again σ_a vanishes and

$$\sigma_e = \sigma_s \tag{7.176}$$

7.PS.3 *The Robin problem*

The first few terms in the low frequency expansion (3.2) are found to be

$$u_0^+(\boldsymbol{r}) = \frac{r_0}{|\boldsymbol{r} - \boldsymbol{r}_0|} + \sum_{n=1}^{\infty} \frac{n}{n+1} \left(\frac{a}{r_0} \right)^n \left(\frac{a}{r} \right)^{n+1} P_n(\cos\theta), \quad r > a, \quad \boldsymbol{r} \neq \boldsymbol{r}_0 \tag{7.177}$$

$$\begin{aligned} u_1^+(\boldsymbol{r}) &= r_0(1 - u_0^+(\boldsymbol{r})) \\ &\quad + a\nu \sum_{n=0}^{\infty} \frac{2n+1}{(n+1)^2} \left(\frac{a}{r_0} \right)^n \left(\frac{a}{r} \right)^{n+1} P_n(\cos\theta), \quad r > a, \quad \boldsymbol{r} \neq \boldsymbol{r}_0 \end{aligned} \tag{7.178}$$

$$\begin{aligned} u_2^+(\boldsymbol{r}) &= \frac{r_0}{|\boldsymbol{r} - \boldsymbol{r}_0|} [|\boldsymbol{r} - \boldsymbol{r}_0| - r_0]^2 + 2\nu a^2 \left(1 + \frac{r_0}{r} \right) \\ &\quad - a^2 \sum_{n=1}^{\infty} \frac{n}{(n+1)(2n-1)} \left(\frac{a}{r_0} \right)^n \left(\frac{a}{r} \right)^{n-1} P_n(\cos\theta) \\ &\quad + 2a^2 \sum_{n=0}^{\infty} \frac{2n+1}{(n+1)^2} \left[\frac{n(n^2-1)}{4n^2-1} \left(\frac{a}{r_0} \right)^{n-2} - \nu \left(\frac{a}{r_0} \right)^{n-1} \right. \\ &\quad \left. + \left(\frac{n^2+n-1}{(2n-1)(2n+3)} + \frac{\nu^2}{n+1} \right) \left(\frac{a}{r_0} \right)^n \right] \left(\frac{a}{r} \right)^{n+1} P_n(\cos\theta), \quad r > a, \quad \boldsymbol{r} \neq \boldsymbol{r}_0 \end{aligned} \tag{7.179}$$

The scattering amplitude (3.69) is

$$\begin{aligned} g(\hat{\boldsymbol{r}}) &= (ka)^2 \left[-\nu + \frac{a}{2r_0} P_1(\cos\theta) \right] \\ &\quad + i(ka)^3 \left[-\frac{1}{3} - \nu^2 + \left(\frac{3\nu}{4} \frac{a}{r_0} - \frac{1}{2} \right) P_1(\cos\theta) - \frac{2a^2}{9r_0^2} P_2(\cos\theta) \right] \\ &\quad + (ka)^4 \left[\nu + \nu^2 + \nu^3 + \left(\frac{3\nu}{4} - \frac{3}{20} \frac{a}{r_0} - \frac{3\nu^2}{8} \frac{a}{r_0} \right) P_1(\cos\theta) \right. \\ &\quad \left. + \left(\frac{5\nu}{27} \frac{a^2}{r_0^2} - \frac{2a}{9r_0} \right) P_2(\cos\theta) - \frac{a^3}{20r_0^3} P_3(\cos\theta) \right] + O(k^5) \end{aligned} \tag{7.180}$$

The differential scattering cross section (3.76) is

$$\sigma(\hat{\boldsymbol{r}}) = 4\pi k^2 a^4 \left(-\nu + \frac{a}{2r_0} P_1(\cos\theta) \right)^2 + 4\pi k^4 a^6 \Bigg\{ \left[-\frac{1}{3} - \nu^2 + \left(\frac{3\nu a}{4r_0} - \frac{1}{2} \right) P_1(\cos\theta) - \frac{2a^2}{9r_0^2} P_2(\cos\theta) \right]^2 + \left(-2\nu + \frac{a}{r_0} P_1(\cos\theta) \right) \left[\nu + \nu^2 + \nu^3 + \left(\frac{3\nu}{4} - \frac{3a}{20r_0} - \frac{3\nu^2 a}{8r_0} \right) P_1(\cos\theta) + \left(\frac{5\nu a^2}{27r_0^2} - \frac{2a}{9r_0} \right) P_2(\cos\theta) - \frac{a^3}{20r_0^3} P_3(\cos\theta) \right] \Bigg\} + O(k^6) \tag{7.181}$$

The (total) scattering cross section (3.77) is

$$\sigma_s = 4\pi a^2 (ka)^2 \left[\nu^2 + \frac{1}{12} \frac{a^2}{r_0^2} \right] + 4\pi a^2 (ka)^4 \left[\frac{7}{36} - \frac{4\nu^2}{3} - 2\nu^3 - \nu^4 + \frac{\nu^2}{16} \frac{a^2}{r_0^2} - \frac{a^2}{20r_0^2} + \frac{4a^4}{405r_0^4} \right] + O(k^6) \tag{7.182}$$

The absorption cross section (3.78) does not vanish for the Robin problem and is given by

$$\sigma_a = 4\pi a^2 \nu \sum_{n=0}^{\infty} \frac{2n+1}{(n+1)^2} \left(\frac{a}{r_0} \right)^{2n} + 4\pi a^2 \nu (ka)^2 \left[-(1+\nu)^2 + \sum_{n=1}^{\infty} \frac{2n+1}{(n+1)^2(2n-1)} \left(\frac{a}{r_0} \right)^{2n-2} - \sum_{n=1}^{\infty} \frac{2n+1}{(n+1)^3} \left(\frac{n-1}{2n-1} + \frac{\nu^2}{n+1} \right) \left(\frac{a}{r_0} \right)^{2n} \right] + O(k^4) \tag{7.183}$$

Finally, the extinction cross section is

$$\sigma_e = 4\pi a^2 \nu \sum_{n=0}^{\infty} \frac{2n+1}{(n+1)^2} \left(\frac{a}{r_0} \right)^{2n} + 4\pi a^2 (ka)^2 \left[-\nu - \nu^2 - \nu^3 + \frac{a^2}{12r_0^2} + \nu \sum_{n=1}^{\infty} \frac{2n+1}{(n+1)^2(2n-1)} \left(\frac{a}{r_0} \right)^{2n-2} - \nu \sum_{n=1}^{\infty} \frac{2n+1}{(n+1)^3} \left(\frac{n-1}{2n-1} + \frac{\nu^2}{n+1} \right) \left(\frac{a}{r_0} \right)^{2n} \right] + O(k^4) \tag{7.184}$$

7.PS.4 *The transmission problem*

Results for the lossy transmission problem will be given. Results for the lossless case are easily found by setting $\delta^- = 0$. The first few terms of (3.2) and (3.3) are

$$u_0^+(\boldsymbol{r}) = \frac{r_0}{|\boldsymbol{r}-\boldsymbol{r}_0|} + \left(1-\frac{\rho^+}{\rho^-}\right)\sum_{n=1}^{\infty}\frac{n\rho^-}{n\rho^+ + n\rho^- + \rho^-}\left(\frac{a}{r_0}\right)^n\left(\frac{a}{r}\right)^{n+1} P_n(\cos\theta), \quad r > a, \quad \boldsymbol{r} \neq \boldsymbol{r}_0 \tag{7.185}$$

$$u_0^-(\boldsymbol{r}) = \sum_{n=0}^{\infty}\frac{(2n+1)\rho^-}{n\rho^+ + n\rho^- + \rho^-}\left(\frac{r}{r_0}\right)^n P_n(\cos\theta), \quad r < a \tag{7.186}$$

$$u_1^+(\boldsymbol{r}) = r_0\left(1-\frac{r_0}{|\boldsymbol{r}-\boldsymbol{r}_0|}\right) - \left(1-\frac{\rho^+}{\rho^-}\right) r_0 \sum_{n=1}^{\infty}\frac{n\rho^-}{n\rho^+ + n\rho^- + \rho^-}\left(\frac{a}{r_0}\right)^n\left(\frac{a}{r}\right)^{n+1} P_n(\cos\theta) + c^+\delta^-\gamma^-\frac{\rho^+}{\rho^-}\sum_{n=1}^{\infty}\frac{n(2n+1)(\rho^-)^2}{(n\rho^+ + n\rho^- + \rho^-)^2}\left(\frac{a}{r_0}\right)^n\left(\frac{a}{r}\right)^{n+1} P_n(\cos\theta), \quad r > a, \quad \boldsymbol{r} \neq \boldsymbol{r}_0 \tag{7.187}$$

$$u_1^-(\boldsymbol{r}) = c^+\delta^-\gamma^-\frac{\rho^+}{\rho^-}\sum_{n=1}^{\infty}\frac{n(2n+1)(\rho^-)^2}{(n\rho^+ + n\rho^- + \rho^-)^2}\left(\frac{r}{r_0}\right)^n P_n(\cos\theta) - r_0\sum_{n=1}^{\infty}\frac{(2n+1)\rho^-}{n\rho^+ + n\rho^- + \rho^-}\left(\frac{r}{r_0}\right)^n P_n(\cos\theta), \quad r < a \tag{7.188}$$

$$u_2^+(\boldsymbol{r}) = \frac{r_0}{|\boldsymbol{r}-\boldsymbol{r}_0|}(|\boldsymbol{r}-\boldsymbol{r}_0| - r_0)^2 - \left(1-\frac{\rho^+}{\rho^-}\right)a^2\sum_{n=1}^{\infty}\frac{n}{2n-1}\frac{\rho^-}{n\rho^+ + n\rho^- + \rho^-}\left(\frac{a}{r_0}\right)^n\left(\frac{a}{r}\right)^{n-1} P_n(\cos\theta) + \sum_{n=0}^{\infty} c_n\frac{(2n+1)(\rho^-)^2}{(n\rho^+ + n\rho^- + \rho^-)^2}\left(\frac{a}{r_0}\right)^n\left(\frac{a}{r}\right)^{n+1} P_n(\cos\theta), \quad r > a, \quad \boldsymbol{r} \neq \boldsymbol{r}_0 \tag{7.189}$$

$$u_2^-(\boldsymbol{r}) = -2r_0^2 + \frac{\gamma^-\rho^-}{\gamma^+\rho^+}r_0^2 \sum_{n=0}^{\infty} \frac{2n+1}{2n+3} \frac{\rho^-}{n\rho^+ + n\rho^- + \rho^-} \left(\frac{r}{r_0}\right)^{n+2} P_n(\cos\theta)$$
$$+ \sum_{n=0}^{\infty} d_n \frac{(2n+1)(\rho^-)^2}{(n\rho^+ + n\rho^- + \rho^-)^2} \left(\frac{r}{r_0}\right)^n P_n(\cos\theta), \quad r < a \tag{7.190}$$

where

$$c_n = \frac{2n(n-1)}{(2n-1)(2n+1)} \frac{n\rho^+ + n\rho^- + \rho^-}{\rho^-} \left(1 - \frac{\rho^+}{\rho^-}\right) r_0^2$$
$$+ \frac{2n^2(\rho^+ c^+ \delta^- \gamma^-)^2}{(n\rho^+ + n\rho^- + \rho^-)\rho^-} - 2nc^+\delta^-\gamma^- \frac{\rho^+}{\rho^-} r_0$$
$$+ \frac{2a^2}{(2n-1)(2n+3)} \left[(2n+3)n\left(\frac{\rho^+}{\rho^-} + 1\right) - \left(\frac{n\rho^+ + n\rho^- + \rho^-}{\rho^-}\right)^2 \right.$$
$$\left. - (2n-1)\frac{\gamma^-}{\gamma^+} \right], \quad n \geq 0 \tag{7.191}$$

and

$$d_n = \left[\frac{n\rho^+ + n\rho^- + \rho^-}{(2n-1)\rho^-} - \frac{(n+2)\rho^+ + (n+1)\rho^-}{(2n+3)\rho^+} \frac{\gamma^-}{\gamma^+} \right] a^2$$
$$+ \frac{2(n-1)}{2n-1} \frac{n\rho^+ + n\rho^- + \rho^-}{\rho^-} r_0^2$$
$$+ \frac{2n^2(\rho^+ c^+ \delta^- \gamma^-)^2}{(n\rho^+ + n\rho^- + \rho^-)\rho^-} - 2nc^+\delta^-\gamma^- \frac{\rho^+}{\rho^-} r_0, \quad n \geq 0 \tag{7.192}$$

The scattering amplitude (3.33) is

$$g(\hat{\boldsymbol{r}}) = -k^2 a^2 \frac{\rho^+ - \rho^-}{\rho^+ + 2\rho^-} \frac{a}{r_0} P_1(\cos\theta)$$
$$+ ik^3a^3 \left[\frac{1}{3}\left(\frac{\gamma^-}{\gamma^+} - 1\right) + \frac{\rho^+ - \rho^-}{\rho^+ + 2\rho^-} P_1(\cos\theta) \right.$$
$$\left. + 3\frac{c^+\delta^-\gamma^-\rho^+\rho^-}{(\rho^+ + 2\rho^-)^2} \frac{1}{r_0} P_1(\cos\theta) + \frac{2}{3} \frac{\rho^+ - \rho^-}{2\rho^+ + 3\rho^-} \left(\frac{a}{r_0}\right)^2 P_2(\cos\theta) \right] + O(k^4) \tag{7.193}$$

The differential scattering cross section is

$$\sigma(\hat{\boldsymbol{r}}) = 4\pi \frac{a^6}{r_0^2} k^2 \left(\frac{\rho^+ - \rho^-}{\rho^+ + 2\rho^-}\right)^2 (P_1(\cos\theta))^2 + O(k^4) \tag{7.194}$$

and the (total) scattering cross section is

$$\sigma_s = \frac{4\pi a^6}{3r_0^2}\left(\frac{\rho^+ - \rho^-}{\rho^+ + 2\rho^-}\right)^2 k^2 + O(k^4) \tag{7.195}$$

The absorption cross section, which does not vanish for lossy scatterers, is

$$\sigma_a = 4\pi a c^+ \delta^- \gamma^- \frac{\rho^+}{\rho^-} \sum_{n=1}^{\infty} \frac{n(2n+1)(\rho^-)^2}{(n\rho^+ + n\rho^- + \rho^-)^2}\left(\frac{a}{r_0}\right)^{2n} (1 + e_n k^2) + O(k^4) \tag{7.196}$$

where

$$\begin{aligned} e_n = {} & \frac{r_0^2}{2n-1} - \frac{a^2}{2n-1}\frac{n\rho^+ + n\rho^- - \rho^-}{n\rho^+ + n\rho^- + \rho^-} \\ & - \left(\frac{nc^+\delta^-\gamma^-\rho^+}{n\rho^+ + n\rho^- + \rho^-}\right)^2 + \frac{2a^2}{2n+3}\frac{\gamma^-}{\gamma^+}\frac{\rho^-}{n\rho^+ + n\rho^- + \rho^-}, \quad n \geq 1 \end{aligned} \tag{7.197}$$

Finally, the extinction cross section is found from the definition

$$\sigma_e = \sigma_a + \sigma_s \tag{7.198}$$

where σ_a and σ_s are given by (7.196) and (7.195), respectively.

It is interesting to note the dramatic difference in the structure of the coefficients in the low frequency expansions when the incident field changes from a plane wave to a point source in acoustics. For plane wave incidence, each coefficient of the expansion in the near field is expressed in terms of a finite number of **multipoles**, whereas for point source incidence, each coefficient requires an infinite number. This difference is much less pronounced in the far field where each coefficient for point source incidence requires only one more multipole term than for plane wave incidence.

8

THE ELLIPSOID

8.0 Basics

In this chapter we present explicit results for low frequency scattering by a **triaxial ellipsoid** with **semi-axes** $a_1 > a_2 > a_3 > 0$. Results for rotationally symmetric ellipsoids are easily obtained by setting $a_2 = a_3$ (**prolate spheroid**) or $a_1 = a_2$ (**oblate spheroid**). Results for the sphere may be recovered by a limiting process from either a prolate or an oblate setting, but this is not always a trivial calculation due to the occurrence of indeterminate forms. No such difficulties occur in the reduction of ellipsoids to spheroids. The first low frequency result on scattering by an ellipsoid was obtained by Rayleigh (1897*a*) and it is based on dimensional arguments. The first systematic way to obtain low frequency approximations for scattering by an ellipsoid is due to Stevenson (1953*b*). From the middle 1960s onwards many results have been obtained for ellipsoids mainly by B. Sleeman, E. G. Lawrence, G. Dassios, K. Kiriaki, C. Athanasiadis, A. Charalambopoulos, S. K. Datta and others. All the results are given for plane wave incidence with propagation vector $\hat{\boldsymbol{k}}$. Unlike the case of the sphere, where the radial symmetry of the scatterer allowed us to restrict the incident direction to the x_3 axis without loss of generality, the asymmetry of the ellipsoid requires that $\hat{\boldsymbol{k}}$ not be so restricted in order to obtain the most general results.

Then the incident fields are as follows.

Acoustics:

$$u^i(\boldsymbol{r}) = e^{ik\hat{\boldsymbol{k}}\cdot\boldsymbol{r}} \tag{8.1}$$

Electromagnetics:

$$\boldsymbol{E}^i(\boldsymbol{r}) = \hat{\boldsymbol{a}}e^{ik\hat{\boldsymbol{k}}\cdot\boldsymbol{r}} \tag{8.2}$$

$$\boldsymbol{H}^i(\boldsymbol{r}) = Y^+\hat{\boldsymbol{b}}e^{ik\hat{\boldsymbol{k}}\cdot\boldsymbol{r}} \tag{8.3}$$

$$\hat{\boldsymbol{a}} \times \hat{\boldsymbol{b}} = \hat{\boldsymbol{k}} \tag{8.4}$$

Elasticity:

$$\boldsymbol{u}^i(\boldsymbol{r}) = \alpha_p\hat{\boldsymbol{k}}e^{ik_p\hat{\boldsymbol{k}}\cdot\boldsymbol{r}} + \alpha_s\hat{\boldsymbol{b}}e^{ik_s\hat{\boldsymbol{k}}\cdot\boldsymbol{r}} \tag{8.5}$$

$$\hat{\boldsymbol{k}} \cdot \hat{\boldsymbol{b}} = 0 \tag{8.6}$$

We will try to avoid any confusion between the components of electric polarization vector $\hat{\boldsymbol{a}}$ and the semiaxes a_i, by always using the notation $\hat{\boldsymbol{a}} \cdot \hat{\boldsymbol{x}}_i$ for the polarization vector components and reserving a_i for the semiaxes. Similarly, the rectangular components of $\hat{\boldsymbol{k}}$ will be donated by $\hat{\boldsymbol{k}} \cdot \hat{\boldsymbol{x}}_i$. The surface S of the ellipsoid is given in rectangular coordinates by the equation

$$\frac{x_1^2}{a_1^2} + \frac{x_2^2}{a_2^2} + \frac{x_3^2}{a_3^2} = 1, \qquad 0 < a_3 < a_2 < a_1 \tag{8.7}$$

The **semifocal distances** h_1, h_2 and h_3 are defined by

$$\left.\begin{aligned} h_1^2 &= a_2^2 - a_3^2 \\ h_2^2 &= a_1^2 - a_3^2 \\ h_3^2 &= a_1^2 - a_2^2 \end{aligned}\right\} \tag{8.8}$$

and in terms of these parameters the **ellipsoidal coordinates** (ρ, μ, ν) (Hobson 1955), in which the solutions of the basic potential problems are found, are defined by

$$x_1 = \frac{1}{h_2 h_3} \rho \mu \nu \tag{8.9}$$

$$x_2 = \frac{1}{h_1 h_3} \sqrt{\rho^2 - h_3^2} \sqrt{\mu^2 - h_3^2} \sqrt{h_3^2 - \nu^2} \tag{8.10}$$

$$x_3 = \frac{1}{h_1 h_2} \sqrt{\rho^2 - h_2^2} \sqrt{h_2^2 - \mu^2} \sqrt{h_2^2 - \nu^2} \tag{8.11}$$

where

$$0 \leq \nu^2 \leq h_3^2 \leq \mu^2 \leq h_2^2 \leq \rho^2 < \infty$$

It is standard practice to use the Greek letter μ for both an 'angular' ellipsoidal variable and the Lamé constant in elasticity but the context should remove any possible confusion. In terms of these ellipsoidal coordinates, equation (8.7) for the surface of the ellipsoid S is written as

$$\rho = a_1 \tag{8.12}$$

In general, the outward unit normal vector $\hat{\boldsymbol{\rho}}$ which plays the role of the 'radial' vector in ellipsoidal coordinates is given by

$$\hat{\boldsymbol{\rho}} = \left(\frac{\partial}{\partial \rho} \boldsymbol{r}\right) \left|\frac{\partial}{\partial \rho} \boldsymbol{r}\right|^{-1} = \frac{\rho \sqrt{\rho^2 - h_2^2} \sqrt{\rho^2 - h_3^2}}{\sqrt{\rho^2 - \mu^2} \sqrt{\rho^2 - \nu^2}} \sum_{n=1}^{3} \frac{\hat{\boldsymbol{x}}_n \hat{\boldsymbol{x}}_n}{\rho^2 - a_1^2 + a_n^2} \cdot \boldsymbol{r} \tag{8.13}$$

On the surface $\rho = a_1$, the unit normal, directed into V^+ is

$$\hat{\boldsymbol{n}} = \hat{\boldsymbol{\rho}} = \frac{a_1 a_2 a_3}{\sqrt{a_1^2 - \mu^2} \sqrt{a_1^2 - \nu^2}} \sum_{n=1}^{3} \frac{\hat{\boldsymbol{x}}_n \hat{\boldsymbol{x}}_n}{a_n^2} \cdot \boldsymbol{r} \bigg|_{\rho = a_1} \tag{8.14}$$

and the normal derivative is

$$\frac{\partial}{\partial n} = \frac{a_2 a_3}{\sqrt{a_1^2 - \mu^2}\sqrt{a_1^2 - \nu^2}} \frac{\partial}{\partial \rho} \tag{8.15}$$

The element of the surface area on $\rho = a_1$ is

$$ds = \frac{(\mu^2 - \nu^2)\sqrt{a_1^2 - \mu^2}\sqrt{a_1^2 - \nu^2}}{\sqrt{\mu^2 - h_3^2}\sqrt{h_2^2 - \mu^2}\sqrt{h_3^2 - \nu^2}\sqrt{h_2^2 - \nu^2}}\, d\mu\, d\nu \tag{8.16}$$

We introduce the following notation for the **elliptic integrals** in terms of which the results are expressed

$$I_0^1(\rho) = \frac{1}{2} \int_{\rho^2 - a_1^2}^{+\infty} \frac{dx}{\sqrt{x + a_1^2}\sqrt{x + a_2^2}\sqrt{x + a_3^2}} \tag{8.17}$$

$$I_0^1 = I_0^1(a_1) \tag{8.18}$$

$$I_1^n(\rho) = \frac{1}{2} \int_{\rho^2 - a_1^2}^{+\infty} \frac{dx}{(x + a_n^2)\sqrt{x + a_1^2}\sqrt{x + a_2^2}\sqrt{x + a_3^2}} \tag{8.19}$$

$$I_1^n = I_1^n(a_1), \quad n = 1, 2, 3 \tag{8.20}$$

$$I_2^1(\rho) = \frac{1}{2} \int_{\rho^2 - a_1^2}^{+\infty} \frac{dx}{(x + \Lambda)^2\sqrt{x + a_1^2}\sqrt{x + a_2^2}\sqrt{x + a_3^2}} \tag{8.21}$$

$$I_2^1 = I_2^1(a_1) \tag{8.22}$$

$$I_2^2(\rho) = \frac{1}{2} \int_{\rho^2 - a_1^2}^{+\infty} \frac{dx}{(x + \Lambda')^2\sqrt{x + a_1^2}\sqrt{x + a_2^2}\sqrt{x + a_3^2}} \tag{8.23}$$

$$I_2^2 = I_2^2(a_1) \tag{8.24}$$

$$I_2^{n+k}(\rho) = \frac{1}{2} \int_{\rho^2 - a_1^2}^{+\infty} \frac{dx}{(x + a_n^2)(x + a_k^2)\sqrt{x + a_1^2}\sqrt{x + a_2^2}\sqrt{x + a_3^2}} \tag{8.25}$$

$$I_2^{n+k} = I_2^{n+k}(a_1), \quad \text{for } n, k = 1, 2, 3 \text{ and } n \neq k \tag{8.26}$$

where the **separation constants** Λ, Λ' are

$$\left.\begin{array}{c}\Lambda\\ \Lambda'\end{array}\right\} = \frac{1}{3}\sum_{n=1}^{3} a_n^2 \pm \frac{1}{3}\sqrt{\sum_{n=1}^{3}\left(a_n^4 - \frac{a_1^2 a_2^2 a_3^2}{a_n^2}\right)} \tag{8.27}$$

The **Lamé functions** used in the sequel are the interior functions

$$E_0^1(x) = 1 \tag{8.28}$$

$$E_1^n(x) = (x^2 - a_1^2 + a_n^2)^{1/2}, \quad n = 1, 2, 3 \tag{8.29}$$

$$E_2^1(x) = x^2 - a_1^2 + \Lambda \tag{8.30}$$

$$E_2^2(x) = x^2 - a_1^2 + \Lambda' \tag{8.31}$$

$$E_2^{n+k}(x) = (x^2 - a_1^2 + a_n^2)^{1/2}(x^2 - a_1^2 + a_k^2)^{1/2} \tag{8.32}$$

where $(n,\ k) = (1,\ 2), (2,\ 3), (3,\ 1)$ and the exterior functions

$$F_n^m(x) = (2n+1)E_n^m(x)I_n^m(x) \tag{8.33}$$

The corresponding **Lamé products**, better known as **ellipsoidal harmonics**, are given in terms of Cartesian coordinates on S

$$\mathbb{E}_0^1(\rho, \mu, \nu) = 1 \tag{8.34}$$

$$\mathbb{E}_1^n(\rho, \mu, \nu) = \frac{h_1 h_2 h_3}{h_n} x_n, \quad n = 1, 2, 3 \tag{8.35}$$

$$\mathbb{E}_1^n(\rho, \mu, \nu) = (\Lambda - a_1^2)(\Lambda - a_2^2)(\Lambda - a_3^2)\left(\sum_{n=1}^{3}\frac{x_n^2}{\Lambda - a_n^2} + 1\right) \tag{8.36}$$

$$\mathbb{E}_2^2(\rho, \mu, \nu) = (\Lambda' - a_1^2)(\Lambda' - a_2^2)(\Lambda' - a_3^2)\left(\sum_{n=1}^{3}\frac{x_n^2}{\Lambda' - a_n^2} + 1\right) \tag{8.37}$$

$$\mathbb{E}_2^{n+k}(\rho, \mu, \nu) = h_1^2 h_2^2 h_3^2 \frac{x_n}{h_n}\frac{x_k}{h_k}, \quad n, k = 1, 2, 3 \text{ and } n \neq k \tag{8.38}$$

The volume of the ellipsoid is given by

$$|V^-| = \frac{4\pi}{3} a_1 a_2 a_3 \tag{8.39}$$

and its surface area can be expressed as (Dassios 1990*a*)

$$|S| = \frac{\pi}{a_1a_2a_3}\sum_{n=1}^{3}(a_1^2+a_2^2+a_3^2-a_n^2)\int_0^\infty \frac{dx}{(x+a_n^{-2})\sqrt{x+a_1^{-2}}\sqrt{x+a_2^{-2}}\sqrt{x+a_3^{-2}}} \tag{8.40}$$

Moreover, we record the following useful identities

$$a_1a_2a_3(I_1^1+I_1^2+I_1^3) = 1 \tag{8.41}$$

$$a_1^2I_1^1 + a_2^2I_1^2 + a_3^2I_1^3 = I_0^1 \tag{8.42}$$

$$I_2^1 = \frac{1}{2\Lambda a_1a_2a_3} - \frac{1}{2}\sum_{n=1}^{3}\frac{I_1^n}{\Lambda - a_n^2} \tag{8.43}$$

$$I_2^2 = \frac{1}{2\Lambda' a_1a_2a_3} - \frac{1}{2}\sum_{n=1}^{3}\frac{I_1^n}{\Lambda' - a_n^2} \tag{8.44}$$

$$h_1^2I_2^5 = I_1^3 - I_1^2 \tag{8.45}$$

$$h_2^2I_2^4 = I_1^3 - I_1^1 \tag{8.46}$$

$$h_3^2I_2^3 = I_1^2 - I_1^1 \tag{8.47}$$

which connect the nine elliptic integrals I_0^1 and I_1^n, for $n = 1, 2, 3$ and I_2^k for $k = 1, 2, 3, 4, 5$. In fact, if any two of these are known the other seven can be expressed in terms of these two. For the case of spheroids, where $a_2 = a_3$, the elliptic integral I_0^1 can be found to be

$$I_0^1 = \begin{cases} \dfrac{1}{\sqrt{a_1^2-a_3^2}}\cosh^{-1}\dfrac{a_1}{a_3}, & a_1 > a_2 = a_3 \\[2ex] \dfrac{1}{\sqrt{a_1^2-a_3^2}}\cos^{-1}\dfrac{a_3}{a_1}, & a_1 = a_2 > a_3 \end{cases} \tag{8.48}$$

where $a_1 > a_2 = a_3$ describes a prolate spheroid and $a_1 = a_2 > a_3$ describes an oblate spheroid. Then relations (2.41)–(2.47) can be used to express all the other elliptic integrals in terms of I_0^1. This reduction to spheroidal geometry is straightforward in contrast to the much more difficult reduction to the spherical case.

Before presenting explicit low frequency results for each of the boundary and transmission problems for the ellipsoid, we record the realizations of the various quantities introduced in Chapter 5 which may be used to obtain most of the leading order terms.

The conductor potential (5.8):

$$\phi^c(\boldsymbol{r}) = \frac{I_0^1(\rho)}{I_0^1}, \quad \rho > a_1 \tag{8.49}$$

Capacity (5.9):

$$C = \frac{1}{I_0^1} \tag{8.50}$$

Constant *c* (5.16):

$$\boldsymbol{c} = \boldsymbol{0} \tag{8.51}$$

The basic potential functions (5.13), (5.20), (5.26), (5.31) and (5.32):

$$\boldsymbol{\Phi}(\boldsymbol{r}) = \boldsymbol{r} \cdot \sum_{n=1}^{3} \frac{I_1^n(\rho)}{I_1^n} \hat{\boldsymbol{x}}_n \hat{\boldsymbol{x}}_n, \quad \rho > a_1 \tag{8.52}$$

$$\boldsymbol{\Psi}(\boldsymbol{r}) = \boldsymbol{r} \cdot \sum_{n=1}^{3} \frac{a_1 a_2 a_3 I_1^n(\rho)}{a_1 a_2 a_3 I_1^n - 1} \hat{\boldsymbol{x}}_n \hat{\boldsymbol{x}}_n, \quad \rho > a_1 \tag{8.53}$$

$$\left.\begin{aligned} \boldsymbol{v}^+(\boldsymbol{r}) &= \boldsymbol{r} \cdot \sum_{n=1}^{3} \frac{(1-\beta) a_1 a_2 a_3 I_1^n(\rho)}{(1-\beta) a_1 a_2 a_3 I_1^n - 1} \hat{\boldsymbol{x}}_n \hat{\boldsymbol{x}}_n, \quad \rho > a_1 \\ \boldsymbol{v}^-(\boldsymbol{r}) &= \boldsymbol{r} \cdot \sum_{n=1}^{3} \frac{1}{(1-\beta) a_1 a_2 a_3 I_1^n - 1} \hat{\boldsymbol{x}}_n \hat{\boldsymbol{x}}_n, \quad \rho < a_1 \end{aligned}\right\} \tag{8.54}$$

$$\left.\begin{aligned} \boldsymbol{\Phi}^+(\boldsymbol{r}) &= \boldsymbol{r} \cdot \sum_{n=1}^{3} \frac{\beta a_1 a_2 a_3 I_1^n(\rho)}{(\beta-1) a_1 a_2 a_3 I_1^n + 1} \hat{\boldsymbol{x}}_n \hat{\boldsymbol{x}}_n, \quad \rho > a_1 \\ \boldsymbol{\Phi}^-(\boldsymbol{r}) &= \boldsymbol{r} \cdot \sum_{n=1}^{3} \frac{a_1 a_2 a_3 I_1^n - 1}{(\beta-1) a_1 a_2 a_3 I_1^n + 1} \hat{\boldsymbol{x}}_n \hat{\boldsymbol{x}}_n, \quad \rho < a_1 \end{aligned}\right\} \tag{8.55}$$

$$\left.\begin{aligned} \boldsymbol{\Psi}^+(\boldsymbol{r}) &= \boldsymbol{r} \cdot \sum_{n=1}^{3} \frac{a_1 a_2 a_3 I_1^n(\rho)}{(1-\beta) a_1 a_2 a_3 I_1^n - 1} \hat{\boldsymbol{x}}_n \hat{\boldsymbol{x}}_n, \quad \rho > a_1 \\ \boldsymbol{\Psi}^-(\boldsymbol{r}) &= \boldsymbol{r} \cdot \sum_{n=1}^{3} \frac{a_1 a_2 a_3 I_1^n}{(1-\beta) a_1 a_2 a_3 I_1^n - 1} \hat{\boldsymbol{x}}_n \hat{\boldsymbol{x}}_n, \quad \rho < a_1 \end{aligned}\right\} \tag{8.56}$$

Observe that $\boldsymbol{\Phi}^+$ reduces to $\boldsymbol{\Phi}$ in the limit as $\beta \to \infty$ and $\boldsymbol{\Psi}^+$ reduces to $\boldsymbol{\Psi}$ as $\beta \to 0$.

The polarization tensor for an isolated conductor (5.17):

$$\widetilde{\boldsymbol{Q}} = \sum_{n=1}^{3} Q_n \hat{\boldsymbol{x}}_n \hat{\boldsymbol{x}}_n, \quad Q_n = \frac{4\pi}{3} \frac{1 - a_1 a_2 a_3 I_1^n}{I_1^n} \tag{8.57}$$

The electric polarizability tensor (5.18):

$$\widetilde{\boldsymbol{P}} = \sum_{n=1}^{3} P_n \hat{\boldsymbol{x}}_n \hat{\boldsymbol{x}}_n, \quad P_n = \frac{4\pi}{3 I_1^n} \tag{8.58}$$

The coefficients P_n in the electric polarizability tensor are not to be confused with Legendre polynomials.

The virtual mass tensor (5.21):

$$\widetilde{\boldsymbol{W}} = \sum_{n=1}^{3} W_n \hat{\boldsymbol{x}}_n \hat{\boldsymbol{x}}_n, \quad W_n = \frac{4\pi}{3} \frac{a_1^2 a_2^2 a_3^2 I_1^n}{1 - a_1 a_2 a_3 I_1^n} \tag{8.59}$$

The magnetic polarizability tensor (5.22):

$$\widetilde{\boldsymbol{M}} = \sum_{n=1}^{3} M_n \hat{\boldsymbol{x}}_n \hat{\boldsymbol{x}}_n, \quad M_n = \frac{4\pi}{3} \frac{a_1 a_2 a_3}{1 - a_1 a_2 a_3 I_1^n} \tag{8.60}$$

The general polarizability tensor (5.27):

$$\widetilde{\boldsymbol{X}}(\beta) = \sum_{n=1}^{3} X_n(\beta) \hat{\boldsymbol{x}}_n \hat{\boldsymbol{x}}_n, \quad X_n(\beta) = \frac{4\pi}{3} \frac{a_1 a_2 a_3 (1-\beta)}{(1-\beta) a_1 a_2 a_3 I_1^n - 1} \tag{8.61}$$

We will present the results in terms of the elliptic integrals I_0^1, I_1^n and I_2^n. However, all results may be converted to physical and geometric quantities given above through the connection formulae (8.41)–(8.47), the expressions for volume (8.39) and surface area (8.40) and the identities

$$I_0^1 = \frac{1}{C} \tag{8.62}$$

$$I_1^n = \frac{4\pi}{3} P_n = \frac{1}{a_1 a_2 a_3} - \frac{4\pi}{3 M_n}, \quad n = 1, 2, 3 \tag{8.63}$$

Note that as the ellipsoid degenerates to a sphere of radius a, in the tripole limit $a_1 \to a$, $a_2 \to a$, $a_3 \to a$, we obtain that

$$I_0^1 \to \frac{1}{a} \tag{8.64}$$

$$I_1^n \to \frac{1}{3a^3}, \quad n = 1, 2, 3 \tag{8.65}$$

and all the quantities above recover trivially the corresponding quantities for the sphere. Nevertheless, the reduction of a result from the ellipsoidal to the spherical geometry could be very difficult whenever intedeterminate forms are involved, since the above three limiting processes do not always commute.

8.A Acoustics

8.A.1 *The Dirichlet problem*

When a plane acoustic wave (8.1) is incident upon an acoustically soft ellipsoid the first few terms in the low frequency expansion (3.2) are found to be

$$u_0^+(\boldsymbol{r}) = 1 - \frac{I_0^1(\rho)}{I_0^1}, \quad \rho > a_1 \tag{8.66}$$

$$u_1^+(\boldsymbol{r}) = \frac{1}{I_0^1}\left(\frac{I_0^1(\rho)}{I_0^1} - 1\right) + \sum_{n=1}^{3}\left(1 - \frac{I_1^n(\rho)}{I_1^n}\right)\hat{\boldsymbol{k}}\cdot\hat{\boldsymbol{x}}_n\hat{\boldsymbol{x}}_n\cdot\boldsymbol{r}, \quad \rho > a_1 \tag{8.67}$$

The scattering amplitude (3.38) is

$$g(\hat{\boldsymbol{r}}, \hat{\boldsymbol{k}}) = -ik\frac{1}{I_0^1} - k^2\left(\frac{1}{I_0^1}\right)^2 + O(k^3) \tag{8.68}$$

To the order of approximation available the differential scattering cross section $\sigma(\hat{\boldsymbol{r}})$ is equal to the (total) scattering cross section σ_s and is given by

$$\sigma_s = 4\pi\left(\frac{1}{I_0^1}\right)^2 + O(k^2) \tag{8.69}$$

In this case, the absorption cross section σ_a vanishes and the extinction cross section σ_e is equal to σ_s. Observe that it is necessary to know the k^3 term in the expansion of g in order to obtain the complete k^2 term in the expansion of σ_s. The k^2 term of g will contribute to, but not completely determine, the k^2 term in σ_s. For higher order terms see Sleeman (1967*a,b*), Dassios (1980) and Athanasiadis (1994*a*).

8.A.2 *The Neumann problem*

When the surface of the ellipsoid is acoustically hard the first few terms of (3.2) are

$$u_0^+(\boldsymbol{r}) = 1, \quad \rho > a_1 \tag{8.70}$$

$$u_1^+(\boldsymbol{r}) = \sum_{n=1}^{3}\left(1 + \frac{a_1a_2a_3I_1^n(\rho)}{1 - a_1a_2a_3I_1^n}\right), \quad \rho > a_1 \tag{8.71}$$

The imaginary and real parts of the scattering amplitude (3.54) are

$$\operatorname{Im} g(\hat{\boldsymbol{r}}, \hat{\boldsymbol{k}}) = k^3\frac{a_1a_2a_3}{3}\left[\sum_{n=1}^{3}\frac{\hat{\boldsymbol{k}}\cdot\hat{\boldsymbol{x}}_n\hat{\boldsymbol{x}}_n\cdot\hat{\boldsymbol{r}}}{1 - a_1a_2a_3I_1^n} - 1\right] + O(k^5) \tag{8.72}$$

$$\operatorname{Re} g(\hat{\boldsymbol{r}}, \hat{\boldsymbol{k}}) = -k^6\frac{a_1^2a_2^2a_3^2}{27}\left[\sum_{n=1}^{3}\frac{\hat{\boldsymbol{k}}\cdot\hat{\boldsymbol{x}}_n\hat{\boldsymbol{x}}_n\cdot\hat{\boldsymbol{r}}}{(1 - a_1a_2a_3I_1^n)^2} + 3\right] + O(k^8) \tag{8.73}$$

The differential scattering cross section is

$$\sigma(\hat{\boldsymbol{r}}) = k^4\frac{4\pi a_1^2a_2^2a_3^2}{9}\left[\sum_{n=1}^{3}\frac{\hat{\boldsymbol{k}}\cdot\hat{\boldsymbol{x}}_n\hat{\boldsymbol{x}}_n\cdot\hat{\boldsymbol{r}}}{1 - a_1a_2a_3I_1^n} - 1\right]^2 + O(k^6) \tag{8.74}$$

and the (total) scattering cross section is

$$\sigma_s = k^4\frac{4\pi a_1^2a_2^2a_3^2}{27}\left[\sum_{n=1}^{3}\frac{(\hat{\boldsymbol{k}}\cdot\hat{\boldsymbol{x}}_n)^2}{(1 - a_1a_2a_3I_1^n)^2} + 3\right] + O(k^6) \tag{8.75}$$

Observe that only the Im g contributes to the cross sections to this order. The absorption cross section σ_a vanishes and the extinction cross section σ_e is equal to σ_s. For higher order terms see Dassios (1977), where the hard scatterer is obtained as a limiting case of the penetrable scatterer and Athanasiadis (1994*b*).

8.A.3 *The Robin problem*

The first term in the low frequency expansion (3.2) is

$$u_0^+(\boldsymbol{r}) = 1, \quad \rho > a \tag{8.76}$$

The scattering amplitude is

$$g(\boldsymbol{r}, \hat{\boldsymbol{k}}) = -k^2 \frac{\nu}{4\pi}|S| + O(k^3) \tag{8.77}$$

where $|S|$ is given in (2.40). The differential scattering cross section and the (total) scattering cross section are equal to this order and are given by

$$\sigma(\hat{\boldsymbol{r}}) = \sigma_s = k^2 \frac{\nu^2}{4\pi|S|^2} + O(k^4) \tag{8.78}$$

The absorption cross section does not vanish in this case and is

$$\sigma_a = \nu|S| + O(k^2) \tag{8.79}$$

Hence the extinction cross section is equal to the absorption cross section to this order, that is

$$\sigma_e = \nu|S| + O(k^2) \tag{8.80}$$

Higher order terms are not available in analytic form due to the scale factor in the normal derivative, see (8.15).

8.A.4 *The lossless transmission problem*

The first few terms in the expansions (3.2) and (3.3) are

$$u_0^+(\boldsymbol{r}) = 1, \quad \rho > a_1 \tag{8.81}$$

$$u_0^-(\boldsymbol{r}) = 1, \quad \rho < a_1 \tag{8.82}$$

$$u_1^+(\boldsymbol{r}) = \sum_{n=1}^{3} \left[1 - \frac{(\beta - 1)a_1 a_2 a_3 I_1^n(\rho)}{1 + (\beta - 1)a_1 a_2 a_3 I_1^n} \right] \hat{\boldsymbol{k}} \cdot \hat{\boldsymbol{x}}_n \hat{\boldsymbol{x}}_n \cdot \boldsymbol{r}, \quad \rho > a_1 \tag{8.83}$$

$$u_1^-(\boldsymbol{r}) = \sum_{n=1}^{3} \frac{\hat{\boldsymbol{k}} \cdot \hat{\boldsymbol{x}}_n \hat{\boldsymbol{x}}_n \cdot \boldsymbol{r}}{1 + (\beta - 1)a_1 a_2 a_3 I_1^n}, \quad \rho < a_1 \tag{8.84}$$

where in this lossless case

$$\beta = \frac{\rho^+}{\rho^-} \tag{8.85}$$

The imaginary and real parts of the scattering amplitude are

$$\operatorname{Im} g(\hat{\boldsymbol{r}}, \hat{\boldsymbol{k}}) = k^3 \frac{a_1 a_2 a_3}{3} \left[(\beta\eta^2 - 1) - (\beta - 1) \sum_{n=1}^{3} \frac{\hat{\boldsymbol{k}} \cdot \hat{\boldsymbol{x}}_n \hat{\boldsymbol{x}}_n \cdot \hat{\boldsymbol{r}}}{1 + (\beta - 1) a_1 a_2 a_3 I_1^n} \right] + O(k^5) \tag{8.86}$$

$$\operatorname{Re} g(\hat{\boldsymbol{r}}, \hat{\boldsymbol{k}}) = -k^6 \frac{a_1^2 a_2^2 a_3^2}{27} \left[3(\beta\eta^2 - 1)^2 + (\beta - 1)^2 \sum_{n=1}^{3} \frac{\hat{\boldsymbol{k}} \cdot \hat{\boldsymbol{x}}_n \hat{\boldsymbol{x}}_n \cdot \hat{\boldsymbol{r}}}{[1 + (\beta - 1) a_1 a_2 a_3 I_1^n]^2} \right] + O(k^8) \tag{8.87}$$

The differential scattering cross section is

$$\sigma(\hat{\boldsymbol{r}}) = k^4 \frac{4\pi a_1^2 a_2^2 a_3^2}{9} \left[(\beta\eta^2 - 1) - (\beta - 1) \sum_{n=1}^{3} \frac{\hat{\boldsymbol{k}} \cdot \hat{\boldsymbol{x}}_n \hat{\boldsymbol{x}}_n \cdot \hat{\boldsymbol{r}}}{1 + (\beta - 1) a_1 a_2 a_3 I_1^n} \right] + O(k^6) \tag{8.88}$$

and the (total) scattering cross section is

$$\sigma_s = k^4 \frac{4\pi a_1^2 a_2^2 a_3^2}{27} \left[3(\beta\eta^2 - 1)^2 + (\beta - 1)^2 \sum_{n=1}^{3} \frac{(\hat{\boldsymbol{k}} \cdot \hat{\boldsymbol{x}}_n)^2}{[1 + (\beta - 1) a_1 a_2 a_3 I_1^n]^2} \right] + O(k^6) \tag{8.89}$$

For lossless scatterers the absorption cross section vanishes and the extinction cross section σ_e is equal to σ_s. For higher order terms see Dassios (1977) and Athanasiadis (1994*a,b*).

8.A.5 *The lossy transmission problem*

In this case, the first few terms of the low frequency expansion (3.2) and (3.3) are

$$u_0^+(\boldsymbol{r}) = 1, \quad \rho > a_1 \tag{8.90}$$

$$u_0^-(\boldsymbol{r}) = 1, \quad \rho < a_1 \tag{8.91}$$

$$u_1^+(\boldsymbol{r}) = \sum_{n=1}^{3} \left[1 - \frac{(\rho^+ - \rho^-) a_1 a_2 a_3 I_1^n(\rho)}{\rho^- + (\rho^+ - \rho^-) a_1 a_2 a_3 I_1^n} \right] \hat{\boldsymbol{k}} \cdot \hat{\boldsymbol{x}}_n \hat{\boldsymbol{x}}_n \cdot \boldsymbol{r}, \quad \rho > a_1 \tag{8.92}$$

$$u_1^-(\boldsymbol{r}) = \sum_{n=1}^{3} \frac{\rho^-}{\rho^- + (\rho^+ - \rho^-) a_1 a_2 a_3 I_1^n} \hat{\boldsymbol{k}} \cdot \hat{\boldsymbol{x}}_n \hat{\boldsymbol{x}}_n \cdot \boldsymbol{r}, \quad \rho < a_1 \tag{8.93}$$

The scattering amplitude is

$$g(\hat{\boldsymbol{r}}, \hat{\boldsymbol{k}}) = ik^3 \frac{a_1 a_2 a_3}{3} \left[\left(\frac{\gamma^-}{\gamma^+} - 1 \right) \right.$$
$$\left. - \left(\frac{\rho^+}{\rho^-} - 1 \right) \sum_{n=1}^{3} \frac{\rho^-}{\rho^- + (\rho^+ - \rho^-) a_1 a_2 a_3 I_1^n} \hat{\boldsymbol{k}} \cdot \hat{\boldsymbol{x}}_n \hat{\boldsymbol{x}}_n \cdot \hat{\boldsymbol{r}} \right] + O(k^4) \quad (8.94)$$

The differential scattering cross section is

$$\sigma(\hat{\boldsymbol{r}}) = k^4 \frac{4\pi a_1^2 a_2^2 a_3^2}{9} \left[\left(\frac{\gamma^-}{\gamma^+} - 1 \right) \right.$$
$$\left. - \left(\frac{\rho^+}{\rho^-} - 1 \right) \sum_{n=1}^{3} \frac{\rho^-}{\rho^- + (\rho^+ - \rho^-) a_1 a_2 a_3 I_1^n} \hat{\boldsymbol{k}} \cdot \hat{\boldsymbol{x}} \hat{\boldsymbol{x}} \cdot \hat{\boldsymbol{r}} \right]^2 + O(k^6) \quad (8.95)$$

and the (total) scattering cross section is

$$\sigma_s = k^4 \frac{4\pi a_1^2 a_2^2 a_3^2}{27} \left[3 \left(\frac{\gamma^-}{\gamma^+} - 1 \right)^2 \right.$$
$$\left. + \left(\frac{\rho^+}{\rho^-} - 1 \right)^2 \sum_{n=1}^{3} \left(\frac{\rho^-}{\rho^- + (\rho^+ - \rho^-) a_1 a_2 a_3 I_1^n} \right)^2 (\hat{\boldsymbol{k}} \cdot \hat{\boldsymbol{x}}_n)^2 \right] + O(k^6) \quad (8.96)$$

In this lossy case the absorption cross section is nonvanishing and is given by

$$\sigma_a = k^2 \frac{4\pi a_1 a_2 a_3}{3} \frac{\rho^+}{\rho^-} c^+ \delta^- \gamma^- \sum_{n=1}^{3} \left[\frac{\rho^-}{\rho^- + (\rho^+ - \rho^-) a_1 a_2 a_3 I_1^n} \right]^2 (\hat{\boldsymbol{k}} \cdot \hat{\boldsymbol{x}}_n)^2 + O(k^4) \quad (8.97)$$

Since the k^4 term of σ_a is not explicitly known the complete k^4 term in the extinction cross section is also not known, although σ_s will make a contribution. Thus, the only result we can state with accuracy is

$$\sigma_e = k^2 \frac{4\pi a_1 a_2 a_3}{3} \frac{\rho^+}{\rho^-} c^+ \delta^- \gamma^- \sum_{n=1}^{3} \left[\frac{\rho^-}{\rho^- + (\rho^+ - \rho^-) a_1 a_2 a_3 I_1^n} \right]^2 (\hat{\boldsymbol{k}} \cdot \hat{\boldsymbol{x}}_n)^2 + O(k^4) \quad (8.98)$$

Higher order terms can be obtained using Twersky's technique (Dassios 1989*b*).

8.EM Electromagnetics

8.EM.1 *The perfect conductor problem*

The first term in the expansions (3.117) is

$$\boldsymbol{E}_0^+(\boldsymbol{r}) = \hat{\boldsymbol{a}} - \hat{\boldsymbol{a}} \cdot \sum_{n=1}^{3} \frac{1}{I_1^n} \Bigg[I_1^n(\rho)\hat{\boldsymbol{x}}_n\hat{\boldsymbol{x}}_n - \frac{x_n}{(\rho^2 - a_1^2 + a_n^2)\sqrt{\rho^2-\mu^2}\sqrt{\rho^2-\nu^2}}\hat{\boldsymbol{x}}_n\hat{\boldsymbol{\rho}}\Bigg], \quad \rho > a_1 \tag{8.99}$$

$$\boldsymbol{H}_0^+(\boldsymbol{r}) = Y^+\hat{\boldsymbol{b}} + Y^+\hat{\boldsymbol{b}} \cdot \sum_{n=1}^{3} \frac{a_1a_2a_3}{1 - a_1a_2a_3I_1^n} \Bigg[I_1^n(\rho)\hat{\boldsymbol{x}}_n\hat{\boldsymbol{x}}_n - \frac{x_n}{(\rho^2 - a_1^2 + a_n^2)\sqrt{\rho^2-\mu^2}\sqrt{\rho^2-\nu^2}}\hat{\boldsymbol{x}}_n\hat{\boldsymbol{\rho}}\Bigg], \quad \rho > a_1 \tag{8.100}$$

The scattering amplitude is found to be

$$g_e(\hat{\boldsymbol{r}}, \hat{\boldsymbol{k}}) = -ik^3\frac{a_1a_2a_3}{3}\sum_{n=1}^{3}\Bigg[\frac{1}{a_1a_2a_3I_1^n}\hat{\boldsymbol{r}} \times (\hat{\boldsymbol{r}} \times \hat{\boldsymbol{x}}_n)(\hat{\boldsymbol{x}}_n \cdot \hat{\boldsymbol{a}}) - \frac{1}{1 - a_1a_2a_3I_1^n}(\hat{\boldsymbol{r}} \times \hat{\boldsymbol{x}}_n)(\hat{\boldsymbol{x}}_n \cdot \hat{\boldsymbol{b}})\Bigg] + O(k^5) \tag{8.101}$$

$$g_m(\hat{\boldsymbol{r}}, \hat{\boldsymbol{k}}) = ik^3\frac{a_1a_2a_3}{3}Y^+\sum_{n=1}^{3}\Bigg[\frac{1}{a_1a_2a_3I_1^n}(\hat{\boldsymbol{r}} \times \hat{\boldsymbol{x}}_n)(\hat{\boldsymbol{x}}_n \cdot \hat{\boldsymbol{a}}) + \frac{1}{1 - a_1a_2a_3I_1^n}\hat{\boldsymbol{r}} \times (\hat{\boldsymbol{r}} \times \hat{\boldsymbol{x}}_n)(\hat{\boldsymbol{x}}_n \cdot \hat{\boldsymbol{b}})\Bigg] + O(k^5) \tag{8.102}$$

The differential scattering cross section is obtained from its definition (2.113) or from (5.159) while the (total) scattering cross section is

$$\sigma_s = k^4\frac{8\pi a_1^2a_2^2a_3^2}{27}\sum_{n=1}^{3}\Bigg[\left(\frac{1}{a_1a_2a_3I_1^n}\right)^2(\hat{\boldsymbol{x}}_n \cdot \hat{\boldsymbol{a}})^2 + \left(\frac{1}{1 - a_1a_2a_3I_1^n}\right)^2(\hat{\boldsymbol{x}}_n \cdot \hat{\boldsymbol{b}})^2\Bigg] = O(k^6) \tag{8.103}$$

For perfect conductors the absorption cross section σ_a vanishes, hence the extinction cross section σ_e is equal to σ_s. For higher order terms we refer to Stevenson (1953*b*) and Kiriaki and Athanasiadis (1989).

8.EM.2 *The impedance problem*

While the general low frequency formalism for the impedance problem has not been developed, it is possible to obtain the lowest order terms for the ellipsoidal scatterer. These are

$$\boldsymbol{E}_0^+(\boldsymbol{r}) = \hat{\boldsymbol{a}} - \hat{\boldsymbol{a}} \cdot \sum_{n=1}^{3} \frac{1}{I_1^n} \Bigg[I_1^n(\rho)\hat{\boldsymbol{x}}_n\hat{\boldsymbol{x}}_n - \frac{x_n}{(\rho^2 - a_1^2 + a_n^2)\sqrt{\rho^2-\mu^2}\sqrt{\rho^2-\nu^2}}\hat{\boldsymbol{x}}_n\hat{\boldsymbol{\rho}}\Bigg], \quad \rho > a_1 \tag{8.104}$$

$$\boldsymbol{H}_0^+(\boldsymbol{r}) = Y^+\hat{\boldsymbol{b}} - Y^+\hat{\boldsymbol{b}} \cdot \sum_{n=1}^{3} \frac{1}{I_1^n} \Bigg[I_1^n(\rho)\hat{\boldsymbol{x}}_n\hat{\boldsymbol{x}}_n - \frac{x_n}{(\rho^2 - a_1^2 + a_n^2)\sqrt{\rho^2-\mu^2}\sqrt{\rho^2-\nu^2}}\hat{\boldsymbol{x}}_n\hat{\boldsymbol{\rho}}\Bigg], \quad \rho > a_1 \tag{8.105}$$

The scattering amplitudes are

$$\boldsymbol{g}_e(\hat{\boldsymbol{r}}, \hat{\boldsymbol{k}}) = -\frac{ik^3}{3}\sum_{n=1}^{3}\frac{1}{I_1^n}[\hat{\boldsymbol{r}} \times (\hat{\boldsymbol{r}} \times \hat{\boldsymbol{x}}_n)(\hat{\boldsymbol{x}}_n \cdot \hat{\boldsymbol{a}}) + (\hat{\boldsymbol{r}} \times \hat{\boldsymbol{x}}_n)(\hat{\boldsymbol{x}}_n \cdot \hat{\boldsymbol{b}})] + O(k^4) \tag{8.106}$$

$$\boldsymbol{g}_m(\hat{\boldsymbol{r}}, \hat{\boldsymbol{k}}) = \frac{ik^3}{3}Y^+\sum_{n=1}^{3}\frac{1}{I_1^n}[(\hat{\boldsymbol{r}} \times \hat{\boldsymbol{x}}_n)(\hat{\boldsymbol{x}}_n \cdot \hat{\boldsymbol{a}}) - \hat{\boldsymbol{r}} \times (\hat{\boldsymbol{r}} \times \hat{\boldsymbol{x}}_n)(\hat{\boldsymbol{x}}_n \cdot \hat{\boldsymbol{b}})] + O(k^4) \tag{8.107}$$

The differential scattering cross section is given by (2.113) and the (total) scattering cross section is

$$\sigma_s = k^4\frac{8\pi}{27}\sum_{n=1}^{3}\left(\frac{1}{I_1^n}\right)^2[(\hat{\boldsymbol{x}}_n \cdot \hat{\boldsymbol{a}})^2 + (\hat{\boldsymbol{x}}_n \cdot \hat{\boldsymbol{b}})^2] + O(k^6) \tag{8.108}$$

Using (2.119) to compute the absorption cross section we find that we do not have enough terms in the low frequency expansion to calculate even the first term and all we can infer is that

$$\sigma_a = O(k^2) \tag{8.109}$$

in which case the extinction cross section is also known only to be $O(k^2)$.

8.EM.3 *The lossless transmission problem*

The first terms in the low frequency expansion of the total fields in both V^+ and V^- are

$$\boldsymbol{E}_0^+(\boldsymbol{r}) = \hat{\boldsymbol{a}} - \hat{\boldsymbol{a}} \cdot \sum_{n=1}^{3} \frac{(\mu^+\eta^2 - \mu^-)a_1a_2a_3}{(\mu^+\eta^2 - \mu^-)a_1a_2a_3I_1^n + \mu^-} \cdot \Bigg[I_1^n(\rho)\hat{\boldsymbol{x}}_n\hat{\boldsymbol{x}}_n - \frac{x_n}{(\rho^2 - a_1^2 + a_n^2)\sqrt{\rho^2-\mu^2}\sqrt{\rho^2-\nu^2}}\hat{\boldsymbol{x}}_n\hat{\boldsymbol{\rho}}\Bigg], \quad \rho > a_1 \tag{8.110}$$

$$\boldsymbol{E}_0^-(\boldsymbol{r}) = \hat{\boldsymbol{a}} \cdot \sum_{n=1}^{3} \frac{\mu^-}{(\mu^+\eta^2 - \mu^-)a_1a_2a_3I_1^n + \mu^-} \hat{\boldsymbol{x}}_n\hat{\boldsymbol{x}}_n, \quad \rho < a_1 \tag{8.111}$$

$$\boldsymbol{H}_0^+(\boldsymbol{r}) = Y^+\hat{\boldsymbol{b}} + Y^+\hat{\boldsymbol{b}} \cdot \sum_{n=1}^{3} \frac{(\mu^+ - \mu^-)a_1a_2a_3}{\mu^+ - (\mu^+ - \mu^-)a_1a_2a_3I_1^n} \left[I_1^n(\rho)\hat{\boldsymbol{x}}_n\hat{\boldsymbol{x}}_n \right.$$

$$\left. - \frac{x_n}{(\rho^2 - a_1^2 + a_n^2)\sqrt{\rho^2 - \mu^2}\sqrt{\rho^2 - \nu^2}} \hat{\boldsymbol{x}}_n\hat{\boldsymbol{\rho}} \right], \quad \rho > a_1 \tag{8.112}$$

$$\boldsymbol{H}_0^-(\boldsymbol{r}) = Y^+\hat{\boldsymbol{b}} \cdot \sum_{n=1}^{3} \frac{\mu^+}{\mu^+ - (\mu^+ - \mu^-)a_1a_2a_3I_1^n} \hat{\boldsymbol{x}}_n\hat{\boldsymbol{x}}_n, \quad \rho < a_1 \tag{8.113}$$

The scattering amplitudes are

$$\boldsymbol{g}_e(\hat{\boldsymbol{r}}, \hat{\boldsymbol{k}}) = -ik^3 \frac{a_1a_2a_3}{3} \sum_{n=1}^{3} \left[\frac{(\mu^+\eta^2 - \mu^-)}{(\mu^+\eta^2 - \mu^-)a_1a_2a_3I_1^n + \mu^-} \hat{\boldsymbol{r}} \times (\hat{\boldsymbol{r}} \times \hat{\boldsymbol{x}}_n)(\hat{\boldsymbol{x}}_n \cdot \hat{\boldsymbol{a}}) \right.$$

$$\left. + \frac{\mu^+ - \mu^-}{(\mu^+ - \mu^-)a_1a_2a_3I_1^n - \mu^+} (\hat{\boldsymbol{r}} \times \hat{\boldsymbol{x}}_n)(\hat{\boldsymbol{x}}_n \cdot \hat{\boldsymbol{b}}) \right] + O(k^4) \tag{8.114}$$

$$\boldsymbol{g}_m(\hat{\boldsymbol{r}}, \hat{\boldsymbol{k}}) = ik^3 \frac{a_1a_2a_3}{3} Y^+ \sum_{n=1}^{3} \left[\frac{(\mu^+\eta^2 - \mu^-)}{(\mu^+\eta^2 - \mu^-)a_1a_2a_3I_1^n + \mu^-} (\hat{\boldsymbol{r}} \times \hat{\boldsymbol{x}}_n)(\hat{\boldsymbol{x}}_n \cdot \hat{\boldsymbol{a}}) \right.$$

$$\left. - \frac{(\mu^+ - \mu^-)}{(\mu^+ - \mu^-)a_1a_2a_3I_1^n - \mu^+} \hat{\boldsymbol{r}} \times (\hat{\boldsymbol{r}} \times \hat{\boldsymbol{x}}_n)(\hat{\boldsymbol{x}}_n \cdot \hat{\boldsymbol{b}}) \right] + O(k^4) \tag{8.115}$$

The differential scattering cross section is given in terms of $\boldsymbol{g}_e(\hat{\boldsymbol{r}}, \hat{\boldsymbol{k}})$ in (2.113) and the (total) scattering cross section is

$$\sigma_s = k^4 \frac{8\pi a_1^2a_2^2a_3^2}{27} \sum_{n=1}^{3} \left[\left(\frac{(\mu^+\eta^2 - \mu^-)}{(\mu^+\eta^2 - \mu^-)a_1a_2a_3I_1^n + \mu^-} \right)^2 (\hat{\boldsymbol{x}}_n \cdot \hat{\boldsymbol{a}})^2 \right.$$

$$\left. + \left(\frac{(\mu^+ - \mu^-)}{(\mu^+ - \mu^-)a_1a_2a_3I_1^n - \mu^+} \right)^2 (\hat{\boldsymbol{x}}_n \cdot \hat{\boldsymbol{b}})^2 \right] + O(k^6) \tag{8.116}$$

The absorption cross section vanishes for a lossless scatterer and therefore

$$\sigma_e = \sigma_s \tag{8.117}$$

For higher order terms see Stevenson (1953*b*), Kiriaki and Athanasiadis (1989) and Athanasiadis (1991*b*).

8.EM.4 *The lossy transmission problem*

For lossy scatterers, where $\sigma^- > 0$, the leading low frequency approximations are

$$\boldsymbol{E}_0^+(\boldsymbol{r}) = \hat{\boldsymbol{a}} - \hat{\boldsymbol{a}} \cdot \sum_{n=1}^{n} \frac{1}{I_1^n} \left[I_1^n(\rho)\hat{\boldsymbol{x}}_n\hat{\boldsymbol{x}}_n - \frac{x_n}{(\rho^2 - a_1^2 + a_n^2)\sqrt{\rho^2-\mu^2}\sqrt{\rho^2-\nu^2}} \hat{\boldsymbol{x}}_n\hat{\boldsymbol{\rho}} \right], \quad \rho > a_1 \tag{8.118}$$

$$\boldsymbol{E}_0^-(\boldsymbol{r}) = \boldsymbol{0}, \quad \rho < a_1 \tag{8.119}$$

$$\boldsymbol{H}_0^+(\boldsymbol{r}) = Y^+\hat{\boldsymbol{b}} + Y^+\hat{\boldsymbol{b}} \cdot \sum_{n=1}^{3} \frac{(\mu^+ - \mu^-)a_1a_2a_3}{\mu^+ - (\mu^+ - \mu^-)a_1a_2a_3I_1^n} \left[I_1^n(\rho)\hat{\boldsymbol{x}}_n\hat{\boldsymbol{x}}_n - \frac{x_n}{(\rho^2 - a_1^2 + a_n^2)\sqrt{\rho^2-\mu^2}\sqrt{\rho^2-\nu^2}} \hat{\boldsymbol{x}}_n\hat{\boldsymbol{\rho}} \right], \quad \rho > a_1 \tag{8.120}$$

$$\boldsymbol{H}_0^-(\boldsymbol{r}) = Y^+\hat{\boldsymbol{b}} \cdot \sum_{n=1}^{3} \frac{\mu^+}{\mu^+ - (\mu^+ - \mu^-)a_1a_2a_3I_1^n} \hat{\boldsymbol{x}}_n\hat{\boldsymbol{x}}_n, \quad \rho < a_1 \tag{8.121}$$

The scattering amplitudes in this case are

$$\boldsymbol{g}_e(\hat{\boldsymbol{r}}, \hat{\boldsymbol{k}}) = -ik^3 \frac{a_1a_2a_3}{3} \sum_{n=1}^{3} \left[\frac{1}{a_1a_2a_3I_1^n} \hat{\boldsymbol{r}} \times (\hat{\boldsymbol{r}} \times \hat{\boldsymbol{x}}_n)(\hat{\boldsymbol{x}}_n \cdot \hat{\boldsymbol{a}}) + \frac{(\mu^+ - \mu^-)}{(\mu^+ - \mu^-)a_1a_2a_3I_1^n - \mu^+} (\hat{\boldsymbol{r}} \times \hat{\boldsymbol{x}}_n)(\hat{\boldsymbol{x}}_n \cdot \hat{\boldsymbol{b}}) \right] + O(k^4) \tag{8.122}$$

$$\boldsymbol{g}_m(\hat{\boldsymbol{r}}, \hat{\boldsymbol{k}}) = ik^3 \frac{a_1a_2a_3}{3} Y^+ \sum_{n=1}^{3} \left[\frac{1}{a_1a_2a_3I_1^n} (\hat{\boldsymbol{r}} \times \hat{\boldsymbol{x}}_n)(\hat{\boldsymbol{x}}_n \cdot \hat{\boldsymbol{a}}) - \frac{(\mu^+ - \mu^-)}{(\mu^+ - \mu^-)a_1a_2a_3I_1^n - \mu^+} \hat{\boldsymbol{r}} \times (\hat{\boldsymbol{r}} \times \hat{\boldsymbol{x}}_n)(\hat{\boldsymbol{x}}_n \cdot \hat{\boldsymbol{b}}) \right] + O(k^4) \tag{8.123}$$

Expression (8.122) can be used in (2.113) to calculate the differential scattering cross section, while the (total) scattering cross section is

$$\sigma_s = \frac{8\pi}{27} k^4 a_1^2 a_2^2 a_3^2 \sum_{n=1}^{3} \left[\left(\frac{1}{a_1a_2a_3I_1^n} \right)^2 (\hat{\boldsymbol{x}}_n \cdot \hat{\boldsymbol{a}})^2 + \left(\frac{(\mu^+ - \mu^-)}{(\mu^+ - \mu^-)a_1a_2a_3I_1^n - \mu^+} \right)^2 (\hat{\boldsymbol{x}}_n \cdot \hat{\boldsymbol{b}})^2 \right] + O(k^6) \tag{8.124}$$

There are not enough terms given in order to calculate the absorption cross section although we know from (5.197) that

$$\sigma_a = O(k^2) \tag{8.125}$$

8.E Elasticity

8.E.1 *The rigid problem*

The Rayleigh approximation for a rigid ellipsoid is given by

$$\boldsymbol{u}_0^+(\boldsymbol{r}) = (a_p\hat{\boldsymbol{k}} + a_s\hat{\boldsymbol{b}}) \cdot \sum_{n=1}^{3}\left(1 - \frac{L_0^n(\rho)}{L_0^n}\right)\hat{\boldsymbol{x}}_n\hat{\boldsymbol{x}}_n - \frac{(\tau^2-1)(\rho^2-a_1^2)}{\sqrt{\rho^2-\mu^2}\sqrt{\rho^2-\nu^2}}$$

$$\times\, \hat{\boldsymbol{\rho}}(a_p\hat{\boldsymbol{k}} + a_s\hat{\boldsymbol{b}}) \cdot \sum_{n=1}^{3} \frac{\hat{\boldsymbol{x}}_n\hat{\boldsymbol{x}}_n}{(\rho^2 - a_1^2 + a_n^2)L_0^n} \cdot \boldsymbol{r}, \quad \rho > a_1 \tag{8.126}$$

where

$$L_0^n(\rho) = (\tau^2-1)a_n^2 I_1^n(\rho) - (\tau^2+1)I_0^1(\rho), \quad n = 1,2,3 \tag{8.127}$$

$$L_0^n = L_0^n(a_1), \quad n = 1,2,3 \tag{8.128}$$

and as in the case of the sphere, for simplicity we set

$$\tau = \tau^+ = \frac{k_p}{k_s} = \frac{c_s^+}{c_p^+} \tag{8.129}$$

The first-order approximation is

$$\boldsymbol{u}_1^+(\boldsymbol{r}) = (a_p\hat{\boldsymbol{k}} + a_s\hat{\boldsymbol{b}}) \cdot \widetilde{\boldsymbol{A}}_1(\rho) + \left(a_p\hat{\boldsymbol{k}} + \frac{a_s}{\tau}\hat{\boldsymbol{b}}\right)\hat{\boldsymbol{k}} : \widetilde{\widetilde{\boldsymbol{B}}}_1(\rho) \cdot \boldsymbol{r}$$

$$+ \left(a_p\hat{\boldsymbol{k}} + \frac{a_s}{\tau}\hat{\boldsymbol{b}}\right)\hat{\boldsymbol{k}} : \widetilde{\boldsymbol{A}}_3(\rho) + \frac{\rho^2 - a_1^2}{\sqrt{\rho^2-\mu^2}\sqrt{\rho^2-\nu^2}}\hat{\boldsymbol{\rho}}\Big[(a_p\hat{\boldsymbol{k}} + a_s\hat{\boldsymbol{b}}) \cdot \widetilde{\boldsymbol{A}}_2(\rho) \cdot \boldsymbol{r}$$

$$+ \left(a_p\hat{\boldsymbol{k}} + \frac{a_s}{\tau}\hat{\boldsymbol{b}}\right)\hat{\boldsymbol{k}} : \widetilde{\widetilde{\boldsymbol{B}}}_2(\rho) : \boldsymbol{r}\boldsymbol{r}\Big], \quad \rho > a_1 \tag{8.130}$$

where the dyadic functions $\widetilde{\boldsymbol{A}}_1(\rho), \widetilde{\boldsymbol{A}}_2(\rho)$ and $\widetilde{\boldsymbol{A}}_3(\rho)$ are given by

$$\widetilde{\boldsymbol{A}}_1(\rho) = \frac{2(\tau^3+2)}{3\tau}\sum_{n=1}^{3}\left(1 - \frac{L_0^n(\rho)}{L_0^n}\right)\frac{\hat{\boldsymbol{x}}_n\hat{\boldsymbol{x}}_n}{L_0^n} \tag{8.131}$$

$$\widetilde{\boldsymbol{A}}_2(\rho) = -\frac{2(\tau^3+2)(\tau^2-1)}{3\tau}\sum_{n=1}^{3}\frac{\hat{\boldsymbol{x}}_n\hat{\boldsymbol{x}}_n}{(\rho^2 - a_1^2 + a_n^2)(L_0^n)^2} \tag{8.132}$$

$$\widetilde{\boldsymbol{A}}_3(\rho) = \frac{\tau^2-1}{6(\Lambda-\Lambda')\Delta}\sum_{n=1}^{3}\Big[(\tau^2-1)^2\Lambda\Lambda' I_2^1 I_2^2 \widetilde{\boldsymbol{I}}$$

$$+ 3\tau^4 \frac{I_1^1 I_1^2 I_1^3}{I_1^n}\hat{\boldsymbol{x}}_n\hat{\boldsymbol{x}}_n + \tau^2(\tau^2-1)\widetilde{\boldsymbol{Z}}_n\Big]$$

$$\times\left[\frac{(\Lambda-a_1^2)(\Lambda-a_2^2)(\Lambda-a_3^2)}{(\Lambda-a_n^2)(\rho^2-a_1^2+\Lambda)^2}\right.$$
$$\left.-\frac{(\Lambda'-a_1^2)(\Lambda'-a_2^2)(\Lambda'-a_3^2)}{(\Lambda'-a_n^2)(\rho^2-a_1^2+\Lambda')^2}\right] \tag{8.133}$$

with

$$\Delta=-\tau^2(\tau^2-1)^2\frac{\Lambda\Lambda' I_2^1 I_2^2}{a_1a_2a_3}-3\tau^6 I_1^1 I_1^2 I_1^3-\tau^4(\tau^2-1)I_1^1I_1^2I_1^3\sum_{n=1}^{3}\frac{M_{nn}}{I_1^n} \tag{8.134}$$

$$M_{nk}(\rho)=\frac{1}{\Lambda-\Lambda'}\left[\frac{\Lambda(\Lambda-a_1^2)(\Lambda-a_2^2)(\Lambda-a_3^2)}{(\Lambda-a_n^2)(\Lambda-a_k^2)}I_2^1(\rho)\right.$$
$$\left.-\frac{\Lambda'(\Lambda'-a_1^2)(\Lambda'-a_2^2)(\Lambda'-a_3^2)}{(\Lambda'-a_n^2)(\Lambda'-a_k^2)}I_2^2(\rho)\right],\quad n,k=1,2,3 \tag{8.135}$$

$$M_{nk}=M_{nk}(a_1),\quad n,\ k=1,2,3 \tag{8.136}$$

$$\widetilde{\boldsymbol{Z}}_1=(I_1^2M_{33}+I_1^3M_{22})\hat{\boldsymbol{x}}_1\hat{\boldsymbol{x}}_1-I_1^3M_{12}\hat{\boldsymbol{x}}_2\hat{\boldsymbol{x}}_2-I_1^2M_{31}\hat{\boldsymbol{x}}_3\hat{\boldsymbol{x}}_3 \tag{8.137}$$
$$\widetilde{\boldsymbol{Z}}_2=-I_1^3M_{12}\hat{\boldsymbol{x}}_1\hat{\boldsymbol{x}}_1+(I_1^1M_{33}+I_1^3M_{11})\hat{\boldsymbol{x}}_2\hat{\boldsymbol{x}}_2-I_1^1M_{32}\hat{\boldsymbol{x}}_3\hat{\boldsymbol{x}}_3 \tag{8.138}$$
$$\widetilde{\boldsymbol{Z}}_3=-I_1^2M_{31}\hat{\boldsymbol{x}}_1\hat{\boldsymbol{x}}_1-I_1^1M_{32}\hat{\boldsymbol{x}}_2\hat{\boldsymbol{x}}_2+(I_1^1M_{22}+I_1^2M_{11})\hat{\boldsymbol{x}}_3\hat{\boldsymbol{x}}_3 \tag{8.139}$$

and the tetradic functions $\widetilde{\widetilde{\boldsymbol{B}}}_1(\rho)$ and $\widetilde{\widetilde{\boldsymbol{B}}}_2(\rho)$, are given by

$$\widetilde{\widetilde{\boldsymbol{B}}}_1(\rho)=\widetilde{\boldsymbol{I}}\widetilde{\boldsymbol{I}}+\frac{\tau^2}{\Delta}\sum_{n=1}^{3}\left[(\tau^2-1)^2\Lambda\Lambda' I_2^1I_2^2\widetilde{\boldsymbol{I}}\right.$$
$$\left.+3\tau^4\frac{I_1^1I_1^2I_1^3}{I_1^n}\hat{\boldsymbol{x}}_n\hat{\boldsymbol{x}}_n+\tau^2(\tau^2-1)\widetilde{\boldsymbol{Z}}_n\right]I_i^n(\rho)\hat{\boldsymbol{x}}_n\hat{\boldsymbol{x}}_n$$
$$+\frac{\tau^2-1}{3\Delta}\sum_{n=1}^{3}\sum_{\substack{k=1\\k\neq n}}^{3}\left[(\tau^2-1)^2\Lambda\Lambda' I_2^1I_2^2\widetilde{\boldsymbol{I}}\right.$$
$$\left.+3\tau^4\frac{I_1^1I_1^2I_1^3}{I_1^n}\hat{\boldsymbol{x}}_n\hat{\boldsymbol{x}}_n+\tau^2(\tau^2-1)\widetilde{\boldsymbol{Z}}_n\right]M_{nk}(\rho)\hat{\boldsymbol{x}}_k\hat{\boldsymbol{x}}_k$$
$$-\sum_{n=1}^{3}\sum_{\substack{k=1\\k\neq n}}^{3}\frac{1}{\Delta_{nk}}\left[(\tau^2-1)\frac{a_k^2}{a_n^2-a_k^2}(I_1^n(\rho)I_1^k\right.$$
$$\left.-I_1^nI_1^k(\rho))\hat{\boldsymbol{x}}_k\hat{\boldsymbol{x}}_n\hat{\boldsymbol{x}}_k\hat{\boldsymbol{x}}_n+\Delta_{nk}(\rho)\hat{\boldsymbol{x}}_k\hat{\boldsymbol{x}}_n\hat{\boldsymbol{x}}_k\hat{\boldsymbol{x}}_n\right] \tag{8.140}$$

$$\widetilde{\widetilde{\boldsymbol{B}}}_2(\rho) = \frac{\tau^2 - 1}{6(\Lambda - \Lambda')\Delta}\sum_{n=1}^{3}\sum_{k=1}^{3}\left[\frac{(\Lambda - a_1^2)(\Lambda - a_2^2)(\Lambda - a_3^2)}{(\Lambda - a_n^2)(\Lambda - a_k^2)(\rho^2 - a_1^2 + \Lambda)^2}\right.$$
$$\left. - \frac{(\Lambda' - a_1^2)(\Lambda' - a_2^2)(\Lambda' - a_3^2)}{(\Lambda' - a_n^2)(\Lambda' - a_k^2)(\rho^2 - a_1^2 + \Lambda')^2}\right]\left[(\tau^2 - 1)^2 \Lambda\Lambda' I_2^1 I_2^2 \widetilde{\boldsymbol{I}}\right.$$
$$\left. + 3\tau^4 \frac{I_1^1 I_1^2 I_1^3}{I_1^n}\hat{\boldsymbol{x}}_n\hat{\boldsymbol{x}}_n + \tau^2(\tau^2 - 1)\widetilde{\boldsymbol{Z}}_n\right]\hat{\boldsymbol{x}}_n\hat{\boldsymbol{x}}_n$$
$$+ (\tau^2 - 1)\sum_{n=1}^{3}\sum_{\substack{k=1 \\ k\neq n}}^{3}\frac{I_1^k}{(\rho^2 - a_1^2 + a_n^2)(\rho^2 - a_1^2 + a_k^2)\Delta_{nk}}\hat{\boldsymbol{x}}_n\hat{\boldsymbol{x}}_k\hat{\boldsymbol{x}}_k\hat{\boldsymbol{x}}_n \quad (8.141)$$

with

$$\Delta_{nk}(\rho) = (\tau^2 + 1)I_1^k I_1^n(\rho) + (\tau^2 - 1)\frac{a_n^2 I_1^n I_1^n(\rho) - a_k^2 I_1^k I_1^k(\rho)}{a_n^2 - a_k^2} \quad (8.142)$$

$$\Delta_{nk} = \Delta_{nk}(a_1), \quad n, k = 1, 2, 3, \quad n \neq k \quad (8.143)$$

Note that in the expression (8.130) the direction of incidence, the direction of polarization and the observation point are explicit, while the dyadics $\widetilde{\boldsymbol{A}}_1, \widetilde{\boldsymbol{A}}_2$ and $\widetilde{\boldsymbol{A}}_3$ and the tetradics $\widetilde{\widetilde{\boldsymbol{B}}}_1$ and $\widetilde{\widetilde{\boldsymbol{B}}}_2$ involve only the ellipsoidal variable ρ, the parameter τ that controls the physics and the constants a_1, a_2 and a_3, which describe the geometry of the scatterer.

The scattering amplitudes (3.277)–(3.285) are

$$g_r(\hat{\boldsymbol{r}}; \hat{\boldsymbol{k}}) = \tau^3(a_p\hat{\boldsymbol{k}} + a_s\hat{\boldsymbol{b}})\hat{\boldsymbol{r}} : \widetilde{\boldsymbol{\Lambda}} + O(k_s^3) \quad (8.144)$$

$$g_\theta(\hat{\boldsymbol{r}}; \hat{\boldsymbol{k}}) = (a_p\hat{\boldsymbol{k}} + a_s\hat{\boldsymbol{b}})\hat{\boldsymbol{\theta}} : \widetilde{\boldsymbol{\Lambda}} + O(k_s^3) \quad (8.145)$$

$$g_\varphi(\hat{\boldsymbol{r}}; \hat{\boldsymbol{k}}) = (a_p\hat{\boldsymbol{k}} + a_s\hat{\boldsymbol{b}})\hat{\boldsymbol{\varphi}} : \widetilde{\boldsymbol{\Lambda}} + O(k_s^3) \quad (8.146)$$

where

$$\widetilde{\boldsymbol{\Lambda}} = \sum_{n=1}^{3}\left[\left(\frac{2ik_s}{L_0^n}\right) + \frac{\tau^3 + 2}{3}\left(\frac{2ik_s}{L_0^n}\right)^2\right]\hat{\boldsymbol{x}}_n\hat{\boldsymbol{x}}_n \quad (8.147)$$

The differential scattering cross section can be obtained directly from (3.286) and the (total) scattering cross section is

$$\sigma_s = \frac{16\pi\tau(\tau^3 + 2)}{3(a_p^2 + \tau a_s^2)}\sum_{n=1}^{3}\left[\frac{(a_p\hat{\boldsymbol{k}} + a_s\hat{\boldsymbol{b}}) \cdot \hat{\boldsymbol{x}}_n}{L_0^n}\right]^2 + O(k_s^2) \quad (8.148)$$

A rigid scatterer has a vanishing absorption cross section and therefore

$$\sigma_e = \sigma_s \quad (8.149)$$

No higher order terms have been obtained for the rigid ellipsoid.

8.E.2 *The cavity problem*

As shown in (5.206) the leading approximation for the cavity ellipsoid is just the contribution of the incident wave, which from (2.5) is

$$\boldsymbol{u}_0^+(\boldsymbol{r}) = a_p\hat{\boldsymbol{k}} + a_s\hat{\boldsymbol{b}} \tag{8.150}$$

Then the first approximation sees the ellipsoidal cavity in the following complicated way, where the directions of the incidence and polarization of the incident wave as well as the observation point are explicit

$$\begin{aligned}\boldsymbol{u}_1^+(\boldsymbol{r}) &= \left(a_p\hat{\boldsymbol{k}} + \frac{a_s}{\tau}\hat{\boldsymbol{b}}\right)\hat{\boldsymbol{k}}\cdot\boldsymbol{r} + \left(a_p\hat{\boldsymbol{k}} + \frac{a_s}{\tau}\hat{\boldsymbol{b}}\right)\hat{\boldsymbol{k}} : \widetilde{\widetilde{\boldsymbol{Q}}}_1(\rho)\cdot\boldsymbol{r}\\ &\quad + \left(a_p\hat{\boldsymbol{k}} + \frac{a_s}{\tau}\hat{\boldsymbol{b}}\right)\hat{\boldsymbol{k}} : (\widetilde{\boldsymbol{P}}(\rho) + \widetilde{\widetilde{\boldsymbol{Q}}}_2(\rho) : \boldsymbol{rr})\frac{\hat{\boldsymbol{\rho}}}{\sqrt{\rho^2-\mu^2}\sqrt{\rho^2-\nu^2}}\end{aligned} \tag{8.151}$$

The functions $\widetilde{\boldsymbol{P}}$, $\widetilde{\widetilde{\boldsymbol{Q}}}_1$ and $\widetilde{\widetilde{\boldsymbol{Q}}}_2$ depend only on the semiaxes a_1, a_2 and a_3, the parameters μ^+ and λ^+ and ρ. The dyadic function $\widetilde{\boldsymbol{P}}$ used here is not the electric polarizability tensor but rather is given by

$$\begin{aligned}\widetilde{\boldsymbol{P}}(\rho) &= \frac{(\tau^2-1)(\rho^2-a_1^2)}{2(\Lambda-\Lambda')}\sum_{n=1}^{3}\left[\frac{(\Lambda-a_1^2)(\Lambda-a_2^2)(\Lambda-a_3^2)}{(\Lambda-a_n^2)(\rho^2-a_1^2+\Lambda)^2}\right.\\ &\quad \left. - \frac{(\Lambda'-a_1^2)(\Lambda'-a_2^2)(\Lambda'-a_3^2)}{(\Lambda'-a_n^2)(\rho^2-a_1^2+\Lambda')^2}\right]\widetilde{\boldsymbol{D}}_n\end{aligned} \tag{8.152}$$

with

$$\widetilde{\boldsymbol{D}}_n = \frac{1}{\mu^+h_nD}\sum_{k=1}^{3}(-1)^{n+k+1}D_{kn}(\lambda^+\widetilde{\boldsymbol{I}} + 2\mu^+\hat{\boldsymbol{x}}_k\hat{\boldsymbol{x}}_k), \quad n = 1, 2, 3 \tag{8.153}$$

$$D = \det[\theta_{nk}] \tag{8.154}$$

D_{nk} is the minor determinant of $[\theta_{nk}]$ corresponding to the nk-entry $n, k = 1, 2, 3$.

$$\theta_{nk} = \frac{3\lambda^+I_1^k - 2(\lambda^+ + \mu^+)M_{nk}}{(\lambda^+ + 2\mu^+)h_k} + \delta_{nk}\frac{3}{h_k}\left(\frac{2\mu^+I_1^k}{\lambda^+ + 2\mu^+} - \frac{1}{a_1a_2a_3}\right), \quad n, k = 1, 2, 3 \tag{8.155}$$

and M_{nk} is given in (8.135) and (8.136). The tetradic functions $\widetilde{\widetilde{\boldsymbol{Q}}}_1(\rho)$ and $\widetilde{\widetilde{\boldsymbol{Q}}}_2(\rho)$ are given by

$$\widetilde{\widetilde{\boldsymbol{Q}}}_1(\rho) = 3\tau^2\sum_{n=1}^{3} I_1^n(\rho)\widetilde{\boldsymbol{D}}_n\hat{\boldsymbol{x}}_n\hat{\boldsymbol{x}}_n + (\tau^2-1)\sum_{n=1}^{3}\sum_{\substack{k=1\\k\neq n}}^{3} M_{nk}(\rho)\widetilde{\boldsymbol{D}}_k\hat{\boldsymbol{x}}_n\hat{\boldsymbol{x}}_n$$
$$+\sum_{n=1}^{3}\sum_{\substack{k=1\\k\neq n}}^{3}\frac{N_{nk}(\rho)}{1-N_{nk}-N_{kn}}(\hat{\boldsymbol{x}}_n\hat{\boldsymbol{x}}_k\hat{\boldsymbol{x}}_k\hat{\boldsymbol{x}}_n + \hat{\boldsymbol{x}}_k\hat{\boldsymbol{x}}_n\hat{\boldsymbol{x}}_k\hat{\boldsymbol{x}}_n) \tag{8.156}$$

$$\widetilde{\widetilde{\boldsymbol{Q}}}_2(\rho) = \frac{(\tau^2-1)(\rho^2-a_1^2)}{2(\Lambda-\Lambda')}\sum_{n=1}^{3}\sum_{k=1}^{3}\left[\frac{(\Lambda-a_1^2)(\Lambda-a_2^2)(\Lambda-a_3^2)}{(\Lambda-a_k^2)(\Lambda-a_n^2)(\rho^2-a_1^2+\Lambda)^2}\right.$$
$$\left.-\frac{(\Lambda'-a_1^2)(\Lambda'-a_2^2)(\Lambda'-a_3^2)}{(\Lambda'-a_k^2)(\Lambda'-a_n^2)(\rho^2-a_1^2+\Lambda')^2}\right]\widetilde{\boldsymbol{D}}_k\hat{\boldsymbol{x}}_n\hat{\boldsymbol{x}}_n$$
$$-\frac{(\tau^2-1)(\rho^2-a_1^2)a_1a_2a_3}{2}$$
$$\times\sum_{n=1}^{3}\sum_{\substack{k=1\\k\neq n}}^{3}\frac{\hat{\boldsymbol{x}}_n\hat{\boldsymbol{x}}_k\hat{\boldsymbol{x}}_k\hat{\boldsymbol{x}}_n + \hat{\boldsymbol{x}}_k\hat{\boldsymbol{x}}_n\hat{\boldsymbol{x}}_k\hat{\boldsymbol{x}}_n}{(1-N_{nk}-N_{kn})(\rho^2-a_1^2+a_n^2)(\rho^2-a_1^2+a_k^2)} \tag{8.157}$$

with

$$N_{nk}(\rho) = a_1a_2a_3[\tau^2 I_1^n(\rho) + (1-\tau^2)a_k^2 I_2^{n+k}(\rho)] \tag{8.158}$$
$$N_{nk} = N_{nk}(a_1), \quad n,\ k = 1, 2, 3 \tag{8.159}$$

The scattering amplitudes (5.220)–(5.223) are

$$g_r(\hat{\boldsymbol{r}};\hat{\boldsymbol{k}}) = \frac{1}{3}ik_p^3a_1a_2a_3\Bigg\{-(a_p\hat{\boldsymbol{k}}+a_s\hat{\boldsymbol{b}})\cdot\hat{\boldsymbol{r}}$$
$$+(1-2\tau^2)\left(a_p\hat{\boldsymbol{k}}+\frac{a_s}{\tau}\hat{\boldsymbol{b}}\right)\hat{\boldsymbol{k}} : (\widetilde{\boldsymbol{I}}+\widetilde{\widetilde{\boldsymbol{Q}}}_1(a_1):\widetilde{\boldsymbol{I}})$$
$$+2\tau^2\left(a_p\hat{\boldsymbol{k}}+\frac{a_s}{\tau}\hat{\boldsymbol{b}}\right) : (\hat{\boldsymbol{r}}\hat{\boldsymbol{r}}+\widetilde{\widetilde{\boldsymbol{Q}}}_1(a_1):\hat{\boldsymbol{r}}\hat{\boldsymbol{r}})\Bigg\} + O(k_s^4) \tag{8.160}$$

$$g_\theta(\hat{\boldsymbol{r}};\hat{\boldsymbol{k}}) = \frac{1}{3}ik_s^3a_1a_2a_3\Bigg\{\tau(\hat{\boldsymbol{r}}\times\hat{\boldsymbol{\theta}})\cdot\left[\left(a_p\hat{\boldsymbol{k}}+\frac{a_s}{\tau}\hat{\boldsymbol{b}}\right)\times\hat{\boldsymbol{k}}+\boldsymbol{g}_1(a_1)\right]$$
$$-(a_p\hat{\boldsymbol{k}}+a_s\hat{\boldsymbol{b}})\cdot\hat{\boldsymbol{\theta}}+2\tau\left(a_p\hat{\boldsymbol{k}}+\frac{a_s}{\tau}\hat{\boldsymbol{b}}\right)\hat{\boldsymbol{k}} : (\hat{\boldsymbol{\theta}}\hat{\boldsymbol{r}}+\widetilde{\widetilde{\boldsymbol{Q}}}_1(a_1):\hat{\boldsymbol{\theta}}\hat{\boldsymbol{r}})\Bigg\}$$
$$+O(k_s^4) \tag{8.161}$$

$$g_{\varphi}(\hat{\boldsymbol{r}};\hat{\boldsymbol{k}}) = \frac{1}{3} i k_s^3 a_1 a_2 a_3 \left\{ \tau(\hat{\boldsymbol{r}} \times \hat{\boldsymbol{\varphi}}) \cdot \left[\left(a_p \hat{\boldsymbol{k}} + \frac{a_s}{\tau} \hat{\boldsymbol{b}} \right) \times \hat{\boldsymbol{k}} + \boldsymbol{g}_1(a_1) \right] \right.$$
$$\left. - (a_p \hat{\boldsymbol{k}} + a_s \hat{\boldsymbol{b}}) \cdot \hat{\boldsymbol{\varphi}} + 2\tau \left(a_p \hat{\boldsymbol{k}} + \frac{a_s}{\tau} \hat{\boldsymbol{b}} \right) \hat{\boldsymbol{k}} : (\hat{\boldsymbol{\varphi}}\hat{\boldsymbol{r}} + \widetilde{\widetilde{\boldsymbol{Q}}}_1(a_1) : \hat{\boldsymbol{\varphi}}\hat{\boldsymbol{r}}) \right\}$$
$$+ O(k_s^4) \tag{8.162}$$

where

$$\boldsymbol{g}_1(a_1) = a_1 a_2 a_3 \left(a_p \hat{\boldsymbol{k}} + \frac{a_s}{\tau} \hat{\boldsymbol{b}} \right) \hat{\boldsymbol{k}} : \left[\frac{h_3^2 I_2^3}{1 - N_{12} - N_{21}} (\hat{\boldsymbol{x}}_1 \hat{\boldsymbol{x}}_2 + \hat{\boldsymbol{x}}_2 \hat{\boldsymbol{x}}_1)(\hat{\boldsymbol{x}}_1 \times \hat{\boldsymbol{x}}_2) \right.$$
$$+ \frac{h_2^2 I_2^4}{1 - N_{13} - N_{31}} (\hat{\boldsymbol{x}}_1 \hat{\boldsymbol{x}}_3 + \hat{\boldsymbol{x}}_3 \hat{\boldsymbol{x}}_1)(\hat{\boldsymbol{x}}_1 \times \hat{\boldsymbol{x}}_3)$$
$$\left. + \frac{h_1^2 I_2^5}{1 - N_{23} - N_{32}} (\hat{\boldsymbol{x}}_2 \hat{\boldsymbol{x}}_3 + \hat{\boldsymbol{x}}_3 \hat{\boldsymbol{x}}_2)(\hat{\boldsymbol{x}}_2 \times \hat{\boldsymbol{x}}_3) \right] \tag{8.163}$$

In terms of these scattering amplitudes the differential scattering cross section (5.224) can be calculated, while the (total) scattering cross section (5.225) is

$$\sigma_s = \frac{4\pi a_1^2 a_2^2 a_3^2 \tau^3 k_s^4}{135(a_p^2 + \tau a_s^2)} \left[\frac{5(\tau^3 + 2)}{\tau^2} (a_p^2 + a_s^2) \right.$$
$$+ 4(\tau^5 + 4) \left\| \left(a_p \hat{\boldsymbol{k}} + \frac{a_s}{\tau} \hat{\boldsymbol{b}} \right) \hat{\boldsymbol{k}} + \left(a_p \hat{\boldsymbol{k}} + \frac{a_s}{\tau} \hat{\boldsymbol{b}} \right) \hat{\boldsymbol{k}} : \widetilde{\widetilde{\boldsymbol{Q}}}_1(a_1) \right\|^2$$
$$+ (28\tau^5 - 40\tau^3 + 15\tau - 8) \left| \left(a_p \hat{\boldsymbol{k}} + \frac{a_s}{\tau} \hat{\boldsymbol{b}} \right) \hat{\boldsymbol{k}} : \left(\widetilde{\boldsymbol{I}} \right. \right.$$
$$\left. \left. + (\tau^2 - 1) \sum_{n=1}^{3} \sum_{k=1}^{3} M_{nk} \widetilde{\boldsymbol{D}}_k + 3\tau^2 \sum_{n=1}^{3} I_1^n \widetilde{\boldsymbol{D}}_n \right) \right|^2$$
$$\left. - 10 \left| \left(a_p \hat{\boldsymbol{k}} + \frac{a_s}{\tau} \hat{\boldsymbol{b}} \right) \times \hat{\boldsymbol{k}} + \boldsymbol{g}_1(\boldsymbol{a}_1) \right|^2 \right] + O(k_s^6) \tag{8.164}$$

where the norm of the dyadic is indicated by a double vertical line. The absorption cross section vanishes and therefore

$$\sigma_e = \sigma_s \tag{8.165}$$

No higher order terms have been obtained for the ellipsoidal cavity problem.

8.E.3 *The transmission problem*

As in the case of a cavity the zeroth order approximation for a penetrable ellipsoid is just the incident field

$$\boldsymbol{u}_0^+(\boldsymbol{r}) = a_p\hat{\boldsymbol{k}} + a_s\hat{\boldsymbol{b}}, \quad \rho > a_1 \tag{8.166}$$

$$\boldsymbol{u}_0^-(\boldsymbol{r}) = a_p\hat{\boldsymbol{k}} + a_s\hat{\boldsymbol{b}}, \quad \rho < a_1 \tag{8.167}$$

The first-order approximation written in a form that shows explicitly the directions of propagation and polarization of the incident wave is

$$\begin{aligned}\boldsymbol{u}_1^+(\boldsymbol{r}) = &\left(a_p\hat{\boldsymbol{k}} + \frac{a_s}{\tau}\hat{\boldsymbol{b}}\right)\hat{\boldsymbol{k}}\cdot\hat{\boldsymbol{r}} + \left(a_p\hat{\boldsymbol{k}} + \frac{a_s}{\tau}\hat{\boldsymbol{b}}\right)\hat{\boldsymbol{k}} : \widetilde{\widetilde{\boldsymbol{P}}}_1(\rho)\cdot\boldsymbol{r}\\ &+ \frac{\tau^2-1}{2}\left(a_p\hat{\boldsymbol{k}} + \frac{a_s}{\tau}\hat{\boldsymbol{b}}\right)\hat{\boldsymbol{k}}\\ &: [\widetilde{\boldsymbol{X}}(\rho) + \widetilde{\widetilde{\boldsymbol{P}}}_2(\rho) : \boldsymbol{rr}]\frac{\hat{\boldsymbol{\rho}}}{\sqrt{\rho^2-\mu^2}\sqrt{\rho^2-\nu^2}}, \quad \rho > a_1\end{aligned} \tag{8.168}$$

and

$$\boldsymbol{u}_1^-(\boldsymbol{r}) = \left(a_p\hat{\boldsymbol{k}} + \frac{a_s}{\tau}\hat{\boldsymbol{b}}\right)\hat{\boldsymbol{k}} : \widetilde{\widetilde{\boldsymbol{P}}}_3\cdot\boldsymbol{r}, \quad \rho < a_1 \tag{8.169}$$

Again $\widetilde{\boldsymbol{X}}$, $\widetilde{\widetilde{\boldsymbol{P}}}_1$ and $\widetilde{\widetilde{\boldsymbol{P}}}_2$ are functions of ρ and they depend on the semiaxes a_1, a_2 and a_3 and the parameters $\mu^\pm$ and $\lambda^\pm$. They assume the following forms

$$\begin{aligned}\widetilde{\boldsymbol{X}}(\rho) = \frac{1}{\Lambda-\Lambda'}\sum_{n=1}^{3}\Bigg[&1 - \frac{\Lambda(\Lambda-a_1^2)(\Lambda-a_2^2)(\Lambda-a_3^2)}{(\Lambda-a_n^2)(\rho^2-a_1^2+\Lambda)^2}\\ &+ \frac{\Lambda'(\Lambda'-a_1^2)(\Lambda'-a_2^2)(\Lambda'-a_3^2)}{(\Lambda'-a_n^2)(\rho^2-a_1^2+\Lambda')^2}\Bigg]\widetilde{\boldsymbol{A}}_n^n\end{aligned} \tag{8.170}$$

Note that here $\widetilde{\boldsymbol{X}}$ does not represent the general polarizability tensor

$$\begin{aligned}\widetilde{\widetilde{\boldsymbol{P}}}_1(\rho) = &\frac{3}{2}\sum_{n=1}^{3}\sum_{k=1}^{3} I_1^n(\rho)\widetilde{\boldsymbol{A}}_k^n[(\tau^2+1)\hat{\boldsymbol{x}}_k\hat{\boldsymbol{x}}_n + (\tau^2-1)\hat{\boldsymbol{x}}_n\hat{\boldsymbol{x}}_k]\\ &+ (\tau^2-1)\sum_{n=1}^{3}\sum_{k=1}^{3} M_{nk}(\rho)\widetilde{\boldsymbol{A}}_n^n\hat{\boldsymbol{x}}_k\hat{\boldsymbol{x}}_k\\ &- \frac{3}{2}(\tau^2-1)\sum_{n=1}^{3}\sum_{\substack{k=1\\k\neq n}}^{3} a_n^2 I_2^{n+k}(\rho)\widetilde{\boldsymbol{A}}_n^k(\hat{\boldsymbol{x}}_k\hat{\boldsymbol{x}}_n + \hat{\boldsymbol{x}}_n\hat{\boldsymbol{x}}_k)\end{aligned} \tag{8.171}$$

and

$$\widetilde{\widetilde{P}}_2(\rho) = -3\sum_{n=1}^{3}\sum_{k=1}^{3}\frac{1}{\rho^2 - a_1^2 + a_n^2}\widetilde{A}_k^n\hat{x}_k\hat{x}_n - \sum_{n=1}^{3}\sum_{k=1}^{3}\Phi_{nk}(\rho)\widetilde{A}_n^n\hat{x}_k\hat{x}_k$$
$$+3\sum_{n=1}^{3}\sum_{\substack{k=1\\k\neq n}}^{3}\frac{a_n^2}{(\rho^2 - a_1^2 + a_n^2)(\rho^2 - a_1^2 + a_k^2)}\widetilde{A}_n^k\hat{x}_n\hat{x}_k \tag{8.172}$$

On the other hand, $\widetilde{\widetilde{P}}_3$ is a constant tetradic that depends only on $a_1, a_2, a_3, \mu^{\pm}$ and $\lambda^{\pm}$

$$\widetilde{\widetilde{P}}_3 = \sum_{n=1}^{3}\left[3\tau^2 I_1^n\widetilde{A}_n^n + \hat{x}_n\hat{x}_n + (\tau^2 - 1)\sum_{k=1}^{3}M_{nk}\widetilde{A}_k^k\right]\hat{x}_n\hat{x}_n + \sum_{n=1}^{3}\sum_{\substack{k=1\\k\neq n}}^{3}\left[3I_1^k\widetilde{A}_n^k + \hat{x}_k\hat{x}_n\right.$$
$$\left.+\frac{3}{2}(\tau^2 - 1)(I_1^n - a_k^2 I_2^{n+k})(\widetilde{A}_n^k + \widetilde{A}_k^n)\right]\hat{x}_n\hat{x}_k \tag{8.173}$$

where the expressions for $M_{nk}(\rho)$ and M_{nk} are given in (8.135) and (8.136), respectively

$$\Phi_{nk}(\rho) = -\sqrt{\rho^2 - h_2^2}\sqrt{\rho^2 - h_3^2}\frac{d}{d\rho}M_{nk}(\rho)$$
$$= \frac{1}{\Lambda - \Lambda'}\left[\frac{\Lambda(\Lambda - a_1^2)(\Lambda - a_2^2)(\Lambda - a_3^2)}{(\Lambda - a_n^2)(\Lambda - a_k^2)(\rho^2 - a_1^2 + \Lambda)^2}\right.$$
$$\left.-\frac{\Lambda'(\Lambda' - a_1^2)(\Lambda' - a_2^2)(\Lambda' - a_3^2)}{(\Lambda' - a_n^2)(\Lambda' - a_k^2)(\rho^2 - a_1^2 + \Lambda')^2}\right], \quad n, k = 1, 2, 3 \tag{8.174}$$

and the constant dyadics $\widetilde{A}_n^k$ are given by

$$\widetilde{A}_k^k = \sum_{n=1}^{3}(-1)^{n+k+1}\frac{\Delta_{nk}}{\Delta}[(\lambda^+ - \lambda^-)\widetilde{I} + 2(\mu^+ - \mu^-)\hat{x}_n\hat{x}_n], \quad k = 1, 2, 3 \tag{8.175}$$

and

$$\widetilde{A}_n^k = \frac{\mu^+ - \mu^-}{3Z_n^k}\left[(2\tau^2 - 1)I_1^n - \frac{1}{a_1a_2a_3}\right](\hat{x}_n\hat{x}_k + \hat{x}_k\hat{x}_n) \tag{8.176}$$

for $n, k = 1, 2, 3$ and $n \neq k$. Also,

$$\Delta = \det[H_{nk}] \tag{8.177}$$

with

$$H_{nk} = 2(\tau^2 - 1)\left[\frac{\mu^+ a_1 a_2 a_3}{a_k^2 \Lambda \Lambda'} + (\mu^+ - \mu^-) M_{nk} + \frac{\mu^+ a_n^2}{a_1}\left(\frac{d}{d\rho} M_{nk}(\rho)\right)\Big|_{\rho=a_1}\right]$$
$$- 3\lambda^- \tau^2 I_1^k + 3\delta_{nk}\left[(\mu^+ - 2\mu^- \tau^2) I_1^n - \frac{\mu^+(2\tau^2 - 1)}{a_1 a_2 a_3}\right], \quad n,\ k = 1, 2, 3 \tag{8.178}$$

and Δ_{nk} is the nk-minor of the matrix $[H_{nk}]$. Finally,

$$Z_n^k = Z_k^n = \mu^+\left[(2\tau^2 - 1)^2 I_1^n I_1^k - \frac{1}{a_1^2 a_2^2 a_3^2}\right]$$
$$+ (\mu^+ - \mu^-)\left[(2\tau^2 - 1) I_1^n - \frac{1}{a_1 a_2 a_3}\right][(\tau^2 - 1) a_n^2 I_2^{n+k} - \tau^2 I_1^k]$$
$$+ (\mu^+ - \mu^-)\left[(2\tau^2 - 1) I_1^k - \frac{1}{a_1 a_2 a_3}\right][(\tau^2 - 1) a_k^2 I_2^{n+k} - \tau^2 I_1^n] \tag{8.179}$$

for every $n,\ k = 1, 2, 3$.

The scattering amplitudes (5.236)–(5.239) are

$$g_r(\hat{\boldsymbol{r}}; \hat{\boldsymbol{k}}) = i k_p^3 \frac{a_1 a_2 a_3}{3}\left\{\left(\frac{\rho^-}{\rho^+} - 1\right)(a_p \hat{\boldsymbol{k}} + a_s \hat{\boldsymbol{b}}) \cdot \hat{\boldsymbol{r}}\right.$$
$$+ \frac{1}{\lambda^+ + 2\mu^+}\left(a_p \hat{\boldsymbol{k}} + \frac{a_s}{\tau}\hat{\boldsymbol{b}}\right)\hat{\boldsymbol{k}} : \widetilde{\widetilde{\boldsymbol{P}}}_3 : [(\lambda^+ - \lambda^-)\widetilde{\boldsymbol{I}}$$
$$\left. + 2(\mu^+ - \mu^-)\hat{\boldsymbol{r}}\hat{\boldsymbol{r}}]\right\} + O(k_p^4) \tag{8.180}$$

$$g_\theta(\hat{\boldsymbol{r}}; \hat{\boldsymbol{k}}) = i k_s^3 \frac{a_1 a_2 a_3}{3}\left\{\left(\frac{\rho^-}{\rho^+} - 1\right)\left(a_p \hat{\boldsymbol{k}} + \frac{a_s}{\tau}\hat{\boldsymbol{b}}\right) \cdot \hat{\boldsymbol{\theta}}\right.$$
$$\left. + \tau\left(1 - \frac{\mu^-}{\mu^+}\right)\left(a_p \hat{\boldsymbol{k}} + \frac{a_s}{\tau}\hat{\boldsymbol{b}}\right)\hat{\boldsymbol{k}} : \widetilde{\widetilde{\boldsymbol{P}}}_3 : (\hat{\boldsymbol{r}}\hat{\boldsymbol{\theta}} + \hat{\boldsymbol{\theta}}\hat{\boldsymbol{r}})\right\} + O(k_p^4) \tag{8.181}$$

$$g_\varphi(\hat{\boldsymbol{r}}; \hat{\boldsymbol{k}}) = i k_s^3 \frac{a_1 a_2 a_3}{3}\left\{\left(\frac{\rho^-}{\rho^+} - 1\right)\left(a_p \hat{\boldsymbol{k}} + \frac{a_s}{\tau}\hat{\boldsymbol{b}}\right) \cdot \hat{\boldsymbol{\varphi}}\right.$$
$$\left. + \tau\left(1 - \frac{\mu^-}{\mu^+}\right)\left(a_p \hat{\boldsymbol{k}} + \frac{a_s}{\tau}\hat{\boldsymbol{b}}\right)\hat{\boldsymbol{k}} : \widetilde{\widetilde{\boldsymbol{P}}}_3 : (\hat{\boldsymbol{r}}\hat{\boldsymbol{\varphi}} + \hat{\boldsymbol{\varphi}}\hat{\boldsymbol{r}})\right\} + O(k_p^4) \tag{8.182}$$

Substitution of (8.180)–(8.182) into (5.244) gives the differential scattering cross section. The (total) scattering cross section (5.245) is

$$\sigma_s = \frac{4\pi\tau^3 k_s^4 a_1^2 a_2^2 a_3^2}{135(\mu^+)^2(a_p^2 + \tau a_s^2)}\left[\frac{5(\mu^+)^2(3\tau^3+2)}{3\tau^2}\left(1 - \frac{\rho^-}{\rho^+}\right)^2 (a_p^2 + a_s^2)\right.$$
$$+ 2(10\tau^5 + 3)(\mu^+ - \mu^-)^2 \widetilde{\boldsymbol{Q}} : (\widetilde{\boldsymbol{Q}} + \widetilde{\boldsymbol{Q}}^{\top}) + 15\tau^5(\lambda^+ - \lambda^-)^2(\operatorname{tr}\widetilde{\boldsymbol{Q}})^2$$
$$\left. + 4(\tau^5 - 1)(\mu^+ - \mu^-)^2(\operatorname{tr}\widetilde{\boldsymbol{Q}})^2 + 20\tau^2(\mu^+ - \mu^-)(\lambda^+ - \lambda^-)\operatorname{tr}\widetilde{\boldsymbol{Q}}\right] + O(k_s^6) \tag{8.183}$$

where here $\widetilde{\boldsymbol{Q}}$ is not the polarization tensor for an isolated conductor but is given by

$$\widetilde{\boldsymbol{Q}} = \left(a_p \hat{\boldsymbol{k}} + \frac{a_s}{\tau}\hat{\boldsymbol{b}}\right)\hat{\boldsymbol{k}} : \widetilde{\widetilde{\boldsymbol{P}}}_3 \tag{8.184}$$

and $\widetilde{\widetilde{\boldsymbol{P}}}_3$ is given in (8.173).

In this case, the absorption cross section vanishes and

$$\sigma_e = \sigma_s \tag{8.185}$$

No higher order terms have been obtained for the penetrable elastic ellipsoid.

BIBLIOGRAPHY

Abramowitz, M. and Stegun, I. A. (1965). *The handbook of mathematical functions*. Dover, New York.

Achenbach, J. D. (1973). *Wave propagation in elastic solids*. North-Holland, Amsterdam.

Achenbach, J. D., Sotiropoulos, D. A., and Zhu, H. (1987). Characterization of cracks from ultrasonic scattering data. *ASME Journal of Applied Mechanics*, **54**, 754–62.

Achenbach, J. D., Kitahara, M., Mikata, Y., and Sotiropoulos, D. A. (1988). Reflection and transmission of plane waves by a layer of compact inhomogeneities. *Pure and Applied Geophysics*, **128**, 101–19.

Achenbach, J. D., Sotiropoulos, D. A., and Zhang, C. (1989). Effects of near-tip inhomogeneities on scattering of ultrasonic waves by cracks. *Metallurgical Transactions*, **20A**, 4–17.

Ahluwalia, D. S., Kriegsmann, G. A., and Reiss, E. L. (1985). Scattering of low-frequency acoustic waves by baffled membranes and plates. *Journal of the Acoustical Society of America*, **78**, 682–7.

Ahner, J. F. (1975). The exterior Dirichlet problem for the Helmholtz equation. *Journal of Mathematical Analysis and Applications*, **52**, 415–29.

Ahner, J. F. (1978). The exterior Robin problem for the Helmholtz equation. *Journal of Mathematical Analysis and Applications*, **66**, 37–54.

Ahner, J. F. (1982). On constructing the solution to the exterior Dirichlet problem for the Helmholtz equation. *Journal of Mathematical Analysis and Applications*, **90**, 45–57.

Ahner, J. F. (1985). The exterior Dirichlet problem for the Laplace equation. *Mathematical Methods in the Applied Sciences*, **7**, 461–9.

Ahner, J. F. and Arenstorf, R. F. (1986). On the eigenvalues of the electrostatic integral operator. *Journal of Mathematical Analysis and Applications*, **117**, 187–97.

Ahner, J. F. and Hsiao, G. C. (1976). On the two-dimensional exterior boundary-value problems of elasticity. *SIAM Journal on Applied Mathematics*, **31**, 677–85.

Ahner, J. F. and Kleinman, R. E. (1973). The exterior Neuman problem for the Helmholtz equation. *Archives for Rational Mechanics and Analysis*, **52**, 26–43.

Ahner, J. F. and Weiner, H. W. (1991). On the exterior Laplace equation problem with Robin boundary condition. *Journal of Mathematical Analysis and Applications*, **157**, 127–46.

Alawneh, A. D. and Kanwal, R. P. (1977). Singularity methods in mathematical physics. *SIAM Review*, **19**, 437–71.

Alekseev, V. N. (1995). Low-frequency sound scattering from a sphere moving in ideal liquid. *Akusticheskii Zurnal*, **41**, 375–80.

Ammari, H. and Nédélec, J.-C. (1997). Propagation d'ondes électromagnétiques à basse fréquences. *Comptes Rendus Académie des Sciences, Paris Série I Mathématique*, **325**, 797–802.

Ammari, H., Laouadi, M., and Nédélec, J.-C. (1998). Low frequency behavior of solutions to electromagnetic scattering problems in a chiral media. *SIAM Journal on Applied Mathematics*, **58**, 1022–42.

Angell, T. S. and Kleinman, R. E. (1987). Polarizability tensors in low–frequency inverse scattering. *Radio Science*, **22**, 1120–26.

Apostolopoulos, T. and Dassios, G. (1992). A parallel algorithm for solving the inverse scattering problem. *Journal of Computational and Applied Mathematics*, **42**, 63–77.

Apostolopoulos, T., Kiriaki, K., and Polyzos, D. (1990). The inverse scattering problem for a rigid ellipsoid in linear elasticity. *Inverse Problems*, **6**, 1–9.

Ar, E. (1967). On the low-frequency acoustical scattering of a plane wave from a soft spindle at nose-on incidence. *Quarterly of Applied Mathematics*, **25**, 285–92.

Ar, E. and Kleinman R. E. (1966). The exterior Neumann problem for the three-dimensional Helmholtz equation. *Archives for Rational Mechanics and Analysis*, **23**, 218–36.

Arnaoudov, Y., Dassios, G., and Hadjinicolaou, M. (1999). The resistive coated sphere in the presence of a point generated wave field. *Mathematical Methods in the Applied Sciences*, **22**, 73–90.

Arnaoudov, Y., Dassios, G., and Kostopoulos, V. (1998). The soft and the hard coated sphere within a point source wave field. *Journal of the Acoustical Society of America*, **104**, 1929–42.

Asvestas, J. S. and Kleinman, R. E. (1969*a*). Low-frequency scattering by spheroids and disks 1. Dirichlet problem for a prolate spheroid. *Journal of the Institute of Mathematics and its Applications*, **6**, 42–56.

Asvestas, J. S. and Kleinman, R. E. (1969*b*). Low-frequency scattering by spheroids and disks 2. Neumann problem for a prolate spheroid. *Journal of the Institute of Mathematics and its Applications*, **6**, 57–75.

Asvestas, J. S. and Kleinman, R. E. (1970). Low-frequency scattering by spheroids and disks 3. Oblate spheroids and disks. *Journal of the Institute of Mathematics and its Applications*, **6**, 157–63.

Asvestas, J. S. and Kleinman, R. E. (1971). Low-frequency scattering by perfectly conducting obstacles. *Journal of Mathematical Physics*, **12**, 795–811.

Athanasiadis, C. (1989). Acoustic scattering theory for a multi-layered scatterer in low-frequency. *Mathematica Balkanica*, **3**, 337–46.

Athanasiadis, C. (1991*a*). Low-frequency electromagnetic scattering theory for a multi-layered scatterer. *Quarterly Journal of Mechanics and Applied Mathematics*, **44**, 55–67.

Athanasiadis, C. (1991*b*). The dielectric ellipsoidal in the presence of a low-frequency electromagnetic wave. *Journal of the Institute of Mathematics and Computer Sciences*, **4**, 225–40.

Athanasiadis, C. (1992). The low-frequency magnetic field of an electromagnetic scatterer with a perfectly conducting core. *Journal of the Institute of Mathematics and Computer Sciences*, **5**, 245–64.

Athanasiadis, C. (1994*a*). The multi-layered ellipsoid with a soft core in the presence of a low-frequency acoustic wave. *Quarterly Journal of Mechanics and Applied Mathematics*, **47**, 441–59.

Athanasiadis, C. (1994*b*). The hard-core multi-layered ellipsoid in a low-frequency acoustic field. *International Journal of Engineering Science*, **32**, 1351–59.

Athanasiadis, C. and Stratis, I. G. (1993). On an infinitely stratified scatterer in the presence of a low-frequency electromagnetic plane wave. *The Arabian Journal for Science and Engineering*, **18**, 14–47.

Athanasiadis, C. and Stratis, I. G. (1995). Low-frequency acoustic scattering by an infinitely stratified scatterer. *Rendiconti di Mathematica*, **15**, 133–52.

Athanasiadis, C. and Stratis, I. G. (1997). The conductive problem for Maxwell's equations at low frequencies. *Applied Mathematics Letters*, **10**, 101–5.

Atkinson, F. V. (1949). On Sommerfeld's radiation condition. *Philosophical Magazine*, Series 7, **40**, 645–51.

Babich, V. M. (1988). Scattering of long electromagnetic waves and the integral characteristics of scatterers. *Soviet Journal of Communication Technology and Electronics*, **33**, 12–17.

Baker, B. B. and Copson, E. T. (1939). *The mathematical theory of Huygens' principle*. Oxford University Press, Oxford.

Barton, G. (1989). *Elements of Green's functions and propagation. Potentials, diffusion and waves*. Oxford University Press, New York.

Beard, C. I., Kays, T. H., and Twersky, V. (1962). Mid-field forward scattering. *Journal of Applied Physics*, **33**, 2851–67.

Beckmann, P. and Spizzichino, A. (1963). *The scattering of electromagnetic waves from rough surfaces*. Pergamon Press, Oxford.

Bencheikh, L. (1995). Low frequency scattering of elastic waves by a cavity using a matched asymptotic expansion method. *Journal of the Australian Mathematical Society*, **B37**, 99–120.

Ben-Menahem, A. and Singh, S. J. (1981). *Seismic waves and sources*. Springer-Verlag, Berlin.

Berger, N. and Twersky, V. (1991). Coherent propagation of sound in correlated distributions of resonant spherical scatterers. *Journal of the Acoustical Society of America*, **89**, 604–16.

Berger, N., Lucas, R. J., and Twersky, V. (1991). Polydisperse scattering theory and comparison with data for red blood cells. *Journal of the Acoustical Society of America*, **89**, 1394–401.

Boersma, J. and Anthonissen, M. J. H. (1996). Calculations in mathematics on low-frequency diffraction by a circular disk. *Applied Computational Electromagnetics Society Journal*, **11**, 47–56.

Bohren, C. F. and Huffman, D. R. (1983). *Absorption and scattering of light by small particles*. Wiley, New York.

Boiko, A. I. (1972). Angular dependence of the field in the diffraction of a plane wave by a sphere of small radius. *Soviet Physics Acoustics*, **18**, 107–9.

Bojarski, N. N. (1982). Low-frequency inverse scattering. *IEEE Transactions on Antennas and Propagation*, **AP-30**, 775–8.

Bojarski, N. N. (1983). Bistatic low-frequency inverse scattering. *Journal of the Acoustical Society of America*, **73**, 733–5.

Bouwkamp, C. J. (1954). Diffraction theory. *Reports on Progress in Physics*, **17**, 35–100.

Bouwkamp, C. J. (1965). Note on diffraction by a circular aperture. *Acta Physica Polonica*, **27**, 37–9.

Bowman, J. J., Senior, T. B. A., and Uslenghi, P. L. E. (1969). *Electromagnetic and acoustic scattering by simple shapes*. North-Holland, Amsterdam.

Box, M. A. and McKellar, B. H. J. (1982). Long-wavelength limit of scattering from a lossy dielectric sphere. *Journal of the Optical Society of America*, **72**, 1090–1.

Brand, L. (1947). *Vector and tensor analysis*. Wiley, New York.

Bruggeman, J. C. (1987). The propagation of low-frequency sound in two-dimensional duct system with T joints and right angle bends: Theory and experiment. *Journal of the Acoustical Society*, **82**, 1045–51.

Burke, J. E. (1964). Low-frequency approximations for scattering by penetrable elliptic cylinders. *Journal of the Acoustical Society of America*, **36**, 2059–70.

Burke, J. E. (1966*a*). Low-frequency scattering by soft spheroids. *Journal of the Acoustical Society of America*, **39**, 826–31.

Burke, J. E. (1966*b*). Long-wavelength scattering by hard spheroids. *Journal of the Acoustical Society of America*, **40**, 325–30.

Burke, J. E. (1966*c*). Note on spheroidal wave functions. *Journal of Mathematical Physics*, **45**, 425–31.

Burke, J. E. (1968). Scattering by penetrable spheroids. *Journal of the Acoustical Society of America*, **43**, 871–75.

Burke, J. E. and Twersky, V. (1964). On scattering of waves by an elliptic cylinder and by a semielliptic protuberance on a ground plane. *Journal of the Optical Society of America*, **54**, 732–44.

Burke, J. E. and Twersky, V. (1966*a*). On scattering of waves by the infinite grating of elliptic cylinders. *IEEE Transactions on Antennas and Propagation*, **AP-14**, 465–80.

Burke, J. E. and Twersky, V. (1966*b*). On scattering and reflection by elliptically striated surfaces. *Journal of the Acoustical Society of America*, **40**, 883–95.

Butler, C. M. (1985). General solutions of the narrow strip (and slot) integral equations. *IEEE Transactions on Antennas and Propagation*, **AP-33**, 1085–90.

Butler, C. M. and Wilton, D. R. (1980). General analysis of narrow strips and slots. *IEEE Transactions on Antennas and Propagation*, **AP-28**, 42–8.

Cakoni, F. and Dassios, G. (1998). The coated thermoelastic body within a low frequency elastodynamic field. *International Journal of Engineering Science*, **36**, 1815–38.

Cakoni, F. and Dassios, G. The Atkinson–Wilcox theorem in thermoelasticity. *Quarterly of Applied Mathematics*. (In press).

Calderon, A. P. (1954). The multipole expansion of radiation fields. *Journal of Rational Mechanics and Analysis*, **3**, 523–37.

Casey, K. F. (1981). Low-frequency electromagnetic penetration of loaded apertures. *IEEE Transactions on Electromagnetic Compatibility*, **EMC-23**, 367–77.

Charalambopoulos, A. (1995*a*). The reconstruction of the surface of scatterers with continuous curvature via low-frequency moments. *IMA Journal of Applied Mathematics*, **54**, 171–201.

Charalambopoulos, A. (1995*b*). Inverse scattering for an acoustically soft scatterer in the low-frequency region. *International Journal of Engineering Science*, **33**, 599–609.

Charalambopoulos, A. and Dassios, G. (1992). Inverse scattering via low-frequency moments. *Journal of Mathematical Physics*, **32**, 4206–16.

Charalambopoulos, A. and Gintides, D. (1988). On the analyticity of the solution of the equation of elasticity in the exterior region of a cavity with respect to the wave number. *Zeitschrift für Angewandte Mathematik und Mechanik*, **78**, 1–8.

Charalambopoulos, A. and Kiriaki, K. (1992). Characterization of functions as radiation patterns in linear elasticity. *Mathematical Methods in the Applied Sciences*, **15**, 547–58.

Charalambopoulos, A. and Kiriaki, K. (1993). A method for solving the inverse elastic scattering problem via low-frequency moments. *Wave Motion*, **18**, 213–26.

Charalambopoulos, A., Dassios, G., and Hadjinicolaou, M. (1988). An analytic solution for low-frequency scattering by two soft spheres. *SIAM Journal on Applied Mathematics*, **58**, 370–86.

Chew, H. and Kerker, M. (1976). Abnormally low electromagnetic scattering cross section. *Journal of the Optical Society of America*, **66**, 445–49.

Chinnery, P. A., Humphrey, V. F., and Zhang, J. (1997). Low-frequency acoustic scattering by a cube: Experimental measurements and theoretical predictions. *Journal of the Acoustical Society of America*, **101**, 2571–82.

Ciarlet, P. G. (1988). *Mathematical elasticity. Vol. I: Three-dimensional elasticity*. North-Holland, Amsterdam.

Collin, R. E. (1981). Rayleigh scattering and power conservation. *IEEE Transactions on Antennas and Propagation*, **AP-29**, 795–8.

Collins, W. D. (1962). Some scalar diffraction problems for a spherical cap. *Archives for Rational Mechanics and Analysis*, **10**, 249–66.

Collins, W. D. (1965). On the solution of some axisymmetric boundary value problems by means of integral equations. VI. Further scalar diffraction problems. *Proceedings of the London Mathematical Society*, **15**, 167–92.

Colton, D. (1980). Remarks on the inverse scattering problem for low-frequency acoustic waves. *Journal of Differential Equations*, **37**, 374–81.

Colton, D. (1982). Stable methods for determining the surface impedance of an obstacle from low-frequency field data. *Applicable Analysis*, **14**, 61–70.

Colton, D. and Kleinman, R.E. (1980). The direct and inverse scattering problems for an arbitrary cylinder: Dirichlet boundary conditions. *Proceedings of the Royal Society of Edinburgh*, **86A**, 29–42.

Colton, D. and Kress, R. (1980). Iterative methods for solving the exterior Dirichlet problem for the Helmholtz equation with applications to the inverse scattering problem for low-frequency acoustic waves. *Journal of Mathematical Analysis and Applications*, **77**, 60–72.

Colton, D. and Kress, R. (1983). *Integral equation methods in scattering theory*. Wiley–Interscience, New York.

Colton, D. and Kress, R. (1998). *Inverse acoustic and electromagnetic scattering theory*, (2nd edn). Springer-Verlag, Berlin.

Courant, R. and Hilbert, D. (1962). *Methods of mathematical physics Vol. I, II*. Wiley, New York.

Darling, D. A. and Senior, T. B. A. (1965). Low-frequency expansions for scattering by separable and nonseparable bodies. *Journal of the Acoustical Society of America*, **37**, 228–34.

Dassios, G. (1976). Long-wavelength acoustical scattering for resistive bodies. In *International congress on applied mathematics* (ed. J. Mittas and G. Tsagas), pp. 632–48. Balkan Union of Mathematicians, Thessaloniki.

Dassios, G. (1977). Convergent low-frequency expansions for penetrable scatterers. *Journal of Mathematical Physics*, **18**, 126–37.

Dassios, G. (1980). Second order low-frequency scattering for the soft ellipsoid. *SIAM Journal on Applied Mathematics*, **38**, 373–81.

Dassios, G. (1981). Scattering of acoustic waves by a coated pressure-release ellipsoid. *Journal of the Acoustical Society of America*, **70**, 176–85.

Dassios, G. (1982). Low-frequency scattering theory for a penetrable body with an impenetrable core. *SIAM Journal on Applied Mathematics*, **42**, 272–80.

Dassios, G. (1987). The inverse scattering problem for the soft ellipsoid. *Journal of Mathematical Physics*, **28**, 2858–62.

Dassios, G. (1988*a*). The Atkinson–Wilcox expansion theorem for elastic waves. *Quarterly of Applied Mathematics*, **46**, 285–99.

Dassios, G. (1988*b*). On the harmonic radius and the capacity of an inverse ellipsoid. *Journal of Mathematical Physics*, **29**, 835–6.

Dassios, G. (1989*a*). Optimal geometrical and physical bounds for elastic Rayleigh scattering. In *Elastic wave propagation*, IUTAM–IUPAP Symposium in Galway (ed. M. F. McCarthy and M. A. Hayes), pp. 405–10. North-Holland, Amsterdam.

Dassios, G. (1989*b*). Low-frequency expansions for lossy scatterers. *International Journal of Engineering Science*, **27**, 723–6.

Dassios, G. (1990*a*). On a physical characterization of the surface of an ellipsoid. *International Journal of Engineering Science*, **28**, 1205–8.

Dassios, G. (1990*b*). Low-frequency moments in inverse scattering theory. *Journal of Mathematical Physics*, **31**, 1691–2.

Dassios, G. and Grillakis, M. (1983). Equipartition of energy in scattering theory. *SIAM Journal on Applied Mathematics*, **14**, 915–24.

Dassios, G. and Kamvyssas, G. (1995). Point source excitation in direct and inverse scattering. The soft and the hard small sphere. *IMA Journal of Applied Mathematics*, **55**, 67–84.

Dassios, G. and Kamvyssas, G. (1997). The impedance scattering problem for a point source field. The small resistive sphere. *Quarterly Journal of Mechanics and Applied Mathematics*, **50**, 321–32.

Dassios, G. and Kiriaki, K. (1984). The low-frequency theory of elastic wave scattering. *Quarterly of Applied Mathematics*, **42**, 225–48.

Dassios, G. and Kiriaki, K. (1986). The rigid ellipsoid in the presence of a low-frequency elastic wave. *Quarterly of Applied Mathematics*, **43**, 435–56.

Dassios, G. and Kiriaki, K. (1987). The ellipsoidal cavity in the presence of a low–frequency elastic wave. *Quarterly of Applied Mathematics*, **44**, 709–35.

Dassios, G. and Kiriaki, K. (1991). Size orientation and thickness identification of an ellipsoidal shell. In *Inverse problems and imaging* (ed. G. F. Roach), pp. 38–48. Pitman, London.

Dassios, G. and Kleinman, R. E. (1989*a*). On Kelvin inversion and low-frequency scattering. *SIAM Review*, **31**, 565–85.

Dassios, G. and Kleinman, R. E. (1989*b*). On the capacity and Rayleigh scattering for non convex bodies. *Quarterly Journal of Mechanics and Applied Mathematics*, **42**, 467–75.

Dassios, G. and Kleinman, R. E. (1999). Half space scattering problems at low frequencies. *IMA Journal of Applied Mathematics*, **62**, 61–79.

Dassios, G. and Kostopoulos, V. (1988). The scattering amplitudes and cross-sections in the theory of thermoelasticity. *SIAM Journal on Applied Mathematics*, **48**, 79–98; Errata: **49**, 1283–4 (1989).

Dassios, G. and Kostopoulos, V. (1990*a*). On Rayleigh expansions in thermoelastic scattering. *SIAM Journal on Applied Mathematics*, **50**, 1300–24.
Dassios, G. and Kostopoulos, V. (1990*b*). Thermoelastic Rayleigh scattering by a rigid ellipsoid. *Computational and Applied Mathematics*, **9**, 153–73.
Dassios, G. and Kostopoulos, V. (1994). Scattering of elastic waves by a small thermoelastic body. *International Journal of Engineering Science*, **32**, 1593–1603.
Dassios, G. and Lucas, R. J. (1994). An inverse problem in low-frequency scattering by an ellipsoidally embossed surface. *Wave Motion*, **20**, 33–9.
Dassios, G. and Lucas, R. J. (1996). Inverse scattering for the penetrable ellipsoid and ellipsoidal boss. *Journal of the Acoustical Society of America*, **99**, 1877–82.
Dassios, G. and Lucas, R. J. (1998). Electromagnetic imaging of ellipsoids and ellipsoidal bosses. *Quarterly Journal of Mechanics and Applied Mathematics*, **51**, 413–26.
Dassios, G. and Miloh, T. Rayleigh scattering for the Kelvin-inverted ellipsoid. *Quarterly of Applied Mathematics*. (In press).
Dassios, G. and Payne, L. (1988). Estimates for low-frequency elastic scattering by a rigid body. *Journal of Elasticity*, **20**, 161–80.
Dassios, G. and Payne, L. (1989). Energy bounds for Rayleigh scattering by an elastic cavity. *Journal of Mathematical Analysis and Applications*, **138**, 106–28.
Dassios, G. and Sleeman, B. (1991). A note on the reconstruction of ellipsoids from the X-ray transform. *IMA Journal of Mathematics Applied in Medicine and Biology*, **8**, 141–7.
Dassios, G., Kiriaki, K., and Polyzos, D. (1987). On the scattering amplitudes for elastic waves. *Zeitschrift für Angewandte Mathematik und Physik*, **38**, 856–73.
Dassios, G., Kiriaki, K., and Kostopoulos, V. (1991). Inverse thermoelastic Rayleigh scattering by a rigid ellipsoid. In *Inverse problems and imaging*, (ed. G. F. Roach), pp. 49–67. Pitman, London.
Dassios, G., Kiriaki, K., and Polyzos, D. (1995). Scattering theorems for complete dyadic fields. *International Journal of Engineering Science*, **33**, 269–77.
Dassios, G., Hadjinicolaou, M., and Kamvyssas, G. (1999). Direct and inverse scattering for point source fields. The penetrable small sphere. *Journal of Applied Mathematics and Mechanics* (ZAMM), **79**, 303–16.
Datta, S. K. (1970). The diffraction of a plane compressional elastic wave by a rigid circular disk. *Quarterly of Applied Mathematics*, **28**, 1–14.
Datta, S. K. (1977). Diffraction of plane elastic waves by ellipsoidal inclusions. *Journal of the Acoustical Society of America*, **61**, 1432–7.
Datta, S. K. and Sabina, F. J. (1986). Matched asymptotic expansions applied to diffraction of elastic waves. In *Low and high frequency asymptotics* (ed. V. K. Varadan and V. V. Varadan), pp. 71–264. North-Holland, Amsterdam.
Datta, S. K. and Shah, A. H. (1982). Scattering of SH waves by embedded cavities. *Wave Motion*, **4**, 265–83.
Datta, S. K., Shah, A. H., and Fortunko, C. M. (1982). Diffraction of medium and long wavelength horizontally polarized shear waves by edge cracks. *Journal of Applied Physics*, **53**, 2895–903.
De Hoop, A. T. (1958). On the plane-wave extinction cross-section of an obstacle. *Applied Scientific Research*, **7**, 463–9.
De Hoop, A. T. (1960). A reciprocity theorem for the electromagnetic field scattered by an obstacle. *Applied Scientific Research*, **8**, 135–40.
De Hoop, A. T. (1995). *Handbook of radiation and scattering of waves*. Academic Press, London.
De Meulenaere, F. and Van Bladel, J. (1977). Polarizability of some small apertures. *IEEE Transactions on Antennas and Propagation*, **AP-25**, 198–205.
De Smedt, R. (1981). Low frequency scattering through an aperture in a rigid screen EM—some numerical results. *Journal of Sound and Vibration*, **75**, 371–86.

De Smedt, R. and Van Bladel, J. (1980). Magnetic polarizability of some small apertures. *IEEE Transactions on Antennas and Propagation*, **AP-28**, 703–7.

Ducomet, B. (1992). Diffusion électromagnétique à basse fréquence par un réseau de cylinders diélectriques. *Annales de L'Institut Henri Poincaré—Physique Théorique*, **57**, 183–202.

Efimov, S. P. and Muratov, R. Z. (1978). Scattering theory interference theorems in vector problems of low-frequency diffraction (Russian). *Doklady Akademii Nauk SSSR*, **241**, 1315–18.

Eubanks, R. E. and Sternberg, E. (1956). On the completeness of the Boussinesq–Papkovich stress functions. *Journal of Rational Mechanics and Analysis*, **5**, 735–46.

Fabrikant, V. I. (1986). Sound penetration through an arbitrary shaped aperture in a rigid screen: Analytical determination of the quadratic terms in low-frequency expansion. *Journal of the Acoustical Society of America*, **80**, 1438–46.

Favorin, V. M. (1979). The behavior, for $k \to 0$, of Green's function of the third exterior boundary-value problem for the two-dimensional Helmholtz equation $\Delta u + k^2 u = f$. *Differentsialnye Uravneniya*, **15**, 709–16. (translation 497–502).

Feliziani, M. (1992). Numerical solutions of low-frequency scattering problems. *IEEE Transactions on Magnetics*, **28**, 1224–7.

Felsen, L. B. and Marcuvitz, N. (1973). *Radiation and scattering of waves*. Prentice-Hall, Englewood Cliffs.

Fichera, G. (1972). Existence theorems in elasticity. In *Handbuch der physik*, **Band VI a/2**. Springer-Verlag, Berlin.

Friedlander, F. G. (1958). *Sound pulses*. Cambridge University Press, Cambridge.

Gibbs, J. W. and Wilson, E. B. (1901). *Vector analysis*. Yale University Press, New Haven.

Gintides, D. and Kiriaki, K. (1992). Low-frequency acoustic scattering by a hard inverse prolate spheroid. *Quarterly Journal of Mechanics and Applied Mathematics*, **45**, 231–44.

Goel, G. C. and Jain, D. L. (1981). Scattering of plane waves by a penetrable elliptic cylinder. *Journal of the Acoustical Society of America*, **69**, 371–9.

Gradshteyn, I. S. and Ryzhik, I. M. (1965). *Table of integrals, series, and products*. Academic Press, New York.

Gray, G. A. and Kleinman, R. E. (1985). The integral equation method in electromagnetic scattering. *Journal of Mathematical Analysis and Applications*, **107**, 455–77.

Günter, N. M. (1967). *Potential theory and its applications to basic problems of mathematical physics*. Ungar, New York.

Gurtin, M. (1972). The linear theory of elasticity. In *Handbuch der physik*, **Band VI a/2**, Springer-Verlag, Berlin.

Hansen, T. B. (1993). Uniqueness theorems for static integral equations and calculations of constants for 2-D low-frequency scattering. *IEEE Transactions on Antennas and Propagation*, **41**, 1726–31.

Hansen, T. B. and Yaghjian, A. D. (1992). Low-frequency scattering from two-dimensional reflect conductors. *IEEE Transactions on Antennas and Propagation*, **40**, 1389–402.

Hariharan, S. I. and MacCamy, R. C. (1986). Low-frequency acoustic and electromagnetic scattering. *Applied Numerical Mathematics*, **2**, 29–35.

Harper, E. Y. (1969). Diffraction of plane acoustic waves and pulses as a singular perturbation problem. *Journal of Mathematical Physics*, **10**, 1795–803.

Herrick, D. F. and Senior, T. B. A. (1977*a*). Low-frequency scattering by rectangular dielectic particles. *Journal of Applied Physics*, **13**, 175–83.

Herrick, D. F. and Senior, T. B. A. (1977*b*). The dipole moments of a dielectric cube. *IEEE Transactions on Antennas and Propagation*, **AP-25**, 590–2.

Herzfeld, K. F. (1930). The scattering of sound-waves by small elastic spheres. *Philosophical Magazine*, **9**, 741–51.

Hill, R. N., Kleinman, R. E., and Pfaff, E. W. (1973). Convergent long-wavelength expansions method for two-dimensional scattering problems. *Canadian Journal of Physics*, **51**, 1541–64.

Hobson, E. W. (1955). *The theory of spherical and ellipsoidal harmonics*. Chelsea, New York.

Hönl, H., Maue, A. W., and Westpfahl, K. (1961). Theorie der Beugung. In *Handbuch der physik*, **Band XXV/1**. Springer-Verlag, Berlin.

Hsiao, G. C. (1991). Variational methods for boundary integral equations and inverse problems. In *Integral equations and inverse problems* (eds V. Petkov and R. Lazarov), pp. 96–106. Longman, New York.

Hsiao, G. C. and Kleinman, R. E. (1988). On a unified characterization of capacity. In *Potential theory* (ed. J. Kral, J. Lukes, I. Netuka, and J. Vesely), pp. 103–19. Plenum, New York.

Hsiao, G. C. and Wendland, W. L. (1987). On the low-frequency asymptotics of the exterior 2-D Dirichlet problem in dynamic elasticity. In *Inverse and ill-posed problems* (ed. H. Engl and C. Groetsch), pp. 461–82. Academic Press, New York.

Hsiao, G. C. and Wendland, W. L. (1989). Exterior boundary value problems in elastodynamics. In *Elastic wave propagation*, IUTAM–IUPAP Symposium in Galway (ed. M. F. McCarthy and M. A. Hayes), pp. 545–50. North-Holland, Amsterdam.

Hsiao, G. C. and Wendland, W. L. (1991). On the low frequency asymptotics for exterior elasticity problems. In *Integral equations and inverse problems* (ed. V. Petkov and R. Lazarov) pp. 115–27. Longman, New York.

Hudson, J. A. (1980). *The excitation and propagation of elastic waves*. Cambridge University Press, Cambridge.

Humphrey, A. T. (1995). Lord Rayleigh—the last of the great Victorian polymaths. *IMA Bulletin*, **31**, 113–20.

Hurd, R. A. (1961). A note on the diffraction of a scalar wave by a small circular aperture. *Canadian Journal of Physics*, **39**, 1065–70.

Hurd, R. A. (1979). Low-frequency scattering by a slit in an impedance plane. *Canadian Journal of Physics*, **75**, 1039–45.

Hurd, R. A. and Hayashi, Y. (1980). Low-frequency scattering by a slit in a conducting plane. *Radio Science*, **15**, 1171–8.

Jackson, J. D. (1962). *Classical electrodynamics*. Wiley, New York.

Jain, D. L. and Kanwal, R. P. (1970). Acoustic diffraction by a rigid annular disk. *Journal of Engineering Mathematics*, **4**, 219–28.

Jain, D. L. and Kanwal, R. P. (1971*a*). An integral equation method for solving mixed boundary value problems. *SIAM Journal on Applied Mathematics*, **20**, 642–58.

Jain, D. L. and Kanwal, R. P. (1971*b*). Electromagnetic diffraction by a thin conducting annular disk. *Journal of Mathematical Physics*, **12**, 723–36.

Jain, D. L. and Kanwal, R. P. (1972*a*). Acoustic diffraction of a plane wave by two coplanar parallel perfectly soft or rigid strips. *Canadian Journal of Physics*, **50**, 928–39.

Jain, D. L. and Kanwal, R. P. (1972*b*). Diffraction of elastic waves by two coplanar Griffith cracks in an infinite elastic medium. *International Journal of Solids and Structures*, **8**, 961–75.

Jain, D. L. and Kanwal, R. P. (1972*c*). Diffraction of elastic waves by two coplanar and parallel rigid strips. *International Journal of Engineering Science*, **10**, 925–37.

Jain, D. L. and Kanwal, R. P. (1975*a*). Scattering of acoustic electromagnetic and elastic SH waves by two-dimensional obstacles. *Annals of Physics*, **91**, 1–39.

Jain, D. L. and Kanwal, R. P. (1975*b*). An integral equation perturbation technique in applied mathematics—II. *Applicable Analysis*, **4**, 297–329.

Jain, D. L. and Kanwal, R. P. (1978). Scattering of P and S waves by spherical inclusions and cavities. *Journal of Sound and Vibration*, **57**, 171–202.

Jain, D. L. and Kanwal, R. P. (1979*a*). Scattering of elastic P and SV waves by a rigid elliptic cylindrical inclusion. *International Journal of Engineering Science*, **17**, 941–54.

Jain, D. L. and Kanwal, R. P. (1979*b*). Scattering of elastic waves by cylindrical flows and inclusions. *Journal of Applied Physics*, **50**, 4067–109.

Jain, D. L. and Kanwal, R. P. (1980). Scattering of elastic waves by an elastic sphere. *International Journal of Engineering Science*, **18**, 1117–27.

Jain, D. L. and Kanwal, R. P. (1985). Low-frequency scattering pattern of an arbitrary penetrable obstacle. *International Journal of Engineering Science*, **23**, 435–48.
Johnson, E. R. (1990). The low-frequency scattering of Kelvin waves by stepped topography. *Journal of Fluid Mechanics*, **215**, 23–44.
Johnson, E. R. (1993). Low-frequency scattering of Kelvin waves by continuous topography. *Journal of Fluid Mechanics*, **248**, 173–201.
Jones, D. S. (1953). Diffraction by a thick semi-infinite plate. *Proceedings of the Royal Society of London*, **217**, 153–75.
Jones, D. S. (1955). On the scattering cross section of an obstacle. *Philosophical Magazine*, **46**, 957–62.
Jones, D. S. (1965). Some remarks on diffraction by a disc. *Acta Physica Polonica*, **27**, 137–45.
Jones, D. S. (1976). Scattering of sound by a lifting boundary. *Quarterly Journal of Mechanics and Applied Mathematics*, **29**, 429–55.
Jones, D. S. (1979*a*). Acoustic radiation of long wavelength. *Proceedings of the Royal Society of Edinburgh*, **83A**, 245–54.
Jones, D. S. (1979*b*). Low-frequency electromagnetic radiation. *Journal of the Institute of Mathematics and its Applications*, **23**, 421–47.
Jones, D. S. (1979*c*). *Methods in electromagnetic wave propagation*. Oxford University Press, Oxford.
Jones, D. S. (1980). The scattering of long electromagnetic waves. *Quarterly Journal of Mechanics and Applied Mathematics*, **33**, 105–22.
Jones, D. S. (1981). Low-frequency scattering in elasticity. *Quarterly Journal of Mechanics and Applied Mathematics*, **34**, 431–51.
Jones, D. S. (1982). Crosspolarisation in scattering at low-frequencies. *Proceedings of the IEEE*, **129**, 327–30.
Jones, D. S. (1983*a*). Low-frequency scattering by a body in lubricated contact. *Quarterly Journal of Mechanics and Applied Mathematics*, **36**, 111–38.
Jones, D. S. (1983*b*). The second correction in low-frequency scattering by a body in lubricated contact. *Quarterly Journal of Mechanics and Applied Mathematics*, **36**, 223–36.
Jones, D. S. (1985). Scattering by inhomogeneous dielectric particles. *Quarterly Journal of Mechanics and Applied Mathematics*, **38**, 135–55.
Jones, D. S. (1986). *Acoustic and electromagnetic waves*. Oxford University Press, Oxford.
Jones, D. S. (1987). Note on low-frequency waves. *Radio Science*, **27**, 1219–24.
Jones, D. S. and Noble, B. (1961). The low-frequency scattering by a perfectly conducting strip. *Proceedings of the Cambridge Philosophical Society*, **57**, 364–66.
Junger, M. C. (1981). Rayleigh scattering by compressible, movable bodies of revolution. *Journal of the Acoustical Society of America*, **69**, 1568–72.
Kamvyssas, G. (1998). The spherical scatterer in the presence of a low frequency point generated wave field (in Greek). Unpublished D. Phil. thesis. University of Patras.
Kanwal, R. P. (1967). Theory of diffraction and matched asymptotic expansions. *Journal of Mathematical Physics*, **8**, 821–2.
Kanwal, R. P. (1970). An integral equation perturbation technique in applied mathematics. *Journal of Mathematics and Mechanics*, **19**, 625–56.
Karam, M. A., LeVine, D. M., Antar, Y. M. M., and Stogryn, A. (1995). Improvement of the Rayleigh approximation for scattering from small scatterer. *IEEE Transactions on Antennas and Propagation*, **43**, 681–8.
Katsevich, A. I. and Ramm, A. G. (1996). Approximate inverse geophysical scattering on a small body. *SIAM Journal on Applied Mathematics*, **56**, 192–218.
Keller, J. B., Kleinman, R. E., and Senior, T. B. A. (1972). Dipole moments in Rayleigh scattering. *Journal of the Institute of Mathematics and its Applications*, **9**, 14–22.
Kellogg, O. D. (1953). *Foundations of potential theory*. Dover, New York.
Kerker, M. (1975). Invisible bodies. *Journal of the Optical Society of America*, **64**, 376–9.

Kerr, F. H. (1992). Scattering of a plane elastic wave by spherical elastic inclusions. *International Journal of Engineering Science*, **30**, 169–86.
King, R. W. P. and Wu, T. T. (1959). *The scattering and diffraction of waves*. Harvard University Press, Cambridge.
Kiriaki, K. (1982). Low-frequency scattering theory for a penetrable body in an elastic medium. *Bulletin of the Greek Mathematical Society*, **23**, 33–53.
Kiriaki, K. (1989). Low-frequency expansions for a penetrable ellipsoidal scatterer in an elastic medium. *Journal of Engineering Mathematics*, **23**, 295–314.
Kiriaki, K. and Athanasiadis, C. (1986). Low-frequency electromagnetic scattering theory for a dielectric. *Bulletin of the Greek Mathematical Society*, **27**, 47–59.
Kiriaki, K. and Athanasiadis, C. (1988). Electromagnetic scattering theory for a dielectric with a perfect conductor core in low-frequencies. *Mathematica Balkanica*, **2**, 64–77.
Kiriaki, K. and Athanasiadis, C. (1989). Low-frequency scattering by an ellipsoidal dielectric with a confocal ellipsoidal perfect conductor core. *Mathematica Balkanica*, **3**, 370–89.
Kiriaki, K. and Kostopoulos, V. (1993). The ellipsoidal cavity in the presence of a low frequency thermoelastic wave. In *Mathematical and numerical aspects of wave propagation*, SIAM Conference (ed. R. E. Kleinman *et al.*), pp. 286–95. SIAM, Philadelphia.
Kiriaki, K. and Polyzos, D. (1988). The low-frequency scattering theory for a penetrable scatterer with an impenetrable core in an elastic medium. *International Journal of Engineering Science*, **26**, 1143–60.
Kiriaki, K., Polyzos, D., and Valavanidis, M. (1997). Low-frequency scattering of coated spherical obstacles. *Journal of Engineering Mathematics*, **31**, 379–95.
Kirsch, A. (1985). The Robin problem for the Helmholtz equation as a singular perturbation problem. *Numerical Functional Analysis and Optimization*, **8**, 1–20.
Kirsch, A. (1989). Surface gradients and continuity properties for some integral operators in classical scattering theory. *Mathematical Methods in the Applied Sciences*, **11**, 789–804.
Kittapa, R. and Kleinman, R. E. (1975). Acoustic scattering by penetrable homogeneous objects. *Journal of Mathematical Physics*, **16**, 421–32.
Kleinman, R. E. (1965*a*). The Dirichlet problem for the Helmholtz equation. *Archives for Rational Mechanics and Analysis*, **18**, 205–29.
Kleinman, R. E. (1965*b*). The Rayleigh region. *Proceedings of the IEEE*, **53**, 848–56.
Kleinman, R. E. (1966). Low frequency methods in classical scattering theory. The Technical University of Denmark, Lyngby.
Kleinman, R. E. (1967*a*). Far field scattering at low-frequencies. *Applied Scientific Research*, **18**, 1–8.
Kleinman, R. E. (1967*b*). Low-frequency solution of electromagnetic scattering problems. In *Electromagnetic wave theory* (ed. J. Brown), pp. 891–905. Pergamon Press, Oxford.
Kleinman, R. E. (1969). Scattering theory. In URSI Symposium on Electromagnetic Waves. *Alta Frequenza*, **38**, 324–6.
Kleinman, R. E. (1973). Dipole moments and near field potentials. *Applied Scientific Research*, **27**, 335–40.
Kleinman, R. E. (1976). Iterative solutions of boundary value problems. In *Function theoretic methods for partial differential equations. Lecture Notes in Mathematics*, **561**, 298–313. Springer-Verlag, Berlin.
Kleinman, R. E. (1978). Low-frequency electromagnetic scattering. In *Electromagnetic scattering* (ed. P. L. E. Uslenghi), pp. 1–28. Academic Press, New York.
Kleinman, R. E. (1980). Some applications of functional analysis in classical scattering. In *Mathematical methods and applications of scattering theory*, (ed. J. A. DeSanto, A. W. Saenz, and W. W. Zachary). *Lecture Notes in Physics*, **130**, 17–25. Springer-Verlag, Berlin.
Kleinman, R. E. and Senior, T. B. A. (1972). Rayleigh scattering cross sections. *Radio Science*, **7**, 937–42.

Kleinman, R. E. and Senior, T. B. A. (1975). Low-frequency scattering by space objects. *IEEE Transactions on Aerospace and Electronic Systems*, **AES 11**, 672–5.

Kleinman, R. E. and Senior, T. B. A. (1986). Rayleigh scattering. In *Low and high frequency asymptotics*, (ed. V. K. Varadan and V. V. Varadan), pp. 1–70. North-Holland, Amsterdam.

Kleinman, R. E. and Vainberg, B. (1993). Full low frequency asymptotic expansion for elliptic equations of second order. In *Mathematical and numerical aspects of wave propagation*. (ed. R. Kleinman *et al.*), pp. 296–301. SIAM, Philadelphia.

Kleinman, R. E. and Vainberg, B. (1994). Full low frequency asymptotic expansion for second order elliptic equations in two dimensions. *Mathematical Methods in the Applied Sciences*, **17**, 989–1004.

Kleinman, R. E. and Wendland, W. L. (1977). On Neumann's method for the exterior Neumann problem for the Helmholtz equation. *Journal of Mathematical Analysis and Applications*, **57**, 170–202.

Knopoff, L. (1959*a*). Scattering of compression waves by spherical obstacles. *Geophysics*, **24**, 30–9.

Knopoff, L. (1959*b*). Scattering of shear waves by spherical obstacles. *Geophysics*, **24**, 209–19.

Knops, R. J. and Payne, L. E. (1971). Uniqueness theorems in linear elasticity. *Springer Tracts in Philosophy, Vol. 19*. Springer-Verlag, Berlin.

Kraus, J. D. (1988). *Antennas* (2nd edn). McGraw-Hill, New York.

Kress, R. (1979). On the limiting behaviour of solutions to boundary integral equations associated with time harmonic wave equations for small frequencies. *Mathematical Methods in the Applied Sciences*, **1**, 89–100.

Kress, R. (1987). On the low wave number asymptotics for the two–dimensional exterior Dirichlet problem for the reduced wave equation. *Mathematical Methods in the Applied Sciences*, **9**, 335–41.

Kress, R. (1989). *Linear integral equations*. Springer-Verlag, Berlin.

Kriegsmann, G. A. and Reiss, E. L. (1983). Low-frequency scattering by local inhomogeneities. *SIAM Journal on Applied Mathematics*, **43**, 923–34.

Ksienski, D. A. and Senior, T. B. A. (1985). Scattering by small thin dielectric particles. *Applied Physics*, **38**, 225–31.

Ksienski, A. A., Lin, Y.-T., and White, L. J. (1975). Low-frequency approach to target identification. *Proceedings of the IEEE*, **63**, 1651–60.

Kuo, E. Y. T. (1990). Low-frequency acoustic wave-scattering phenomena under ice cover. *IEEE Journal of Oceanic Engineering*, **15**, 361–72.

Kupradze, V. D. (1963). Dynamic problems in elasticity. In *Progress in solid mechanics III*. North-Holland, Amsterdam.

Kupradze, V. D. (1979). *Three-dimensional problems of the mathematical theory of elasticity and thermoelasticity*. North-Holland, Amsterdam.

Lakhtakia, A. (1990). Polarizability dyadics of small bianisotropic spheres. *Journal de Physique*, **51**, 2235–42.

Lakhtakia, A. (1991). Rayleigh scattering by a bianisotropic ellipsoid in a bisotropic medium. *International Journal of Electronics*, **7**, 1057–62.

Lakhtakia, A., Varadan, V. K., and Varadan, V. V. (1991). Low-frequency scattering by an imperfectly conducting sphere immersed in a DC magnetic field. *International Journal of Infrared and Millimeter Waves*, **12**, 1253–64.

Lawrence, E. G. (1970). Diffraction of elastic waves by a rigid inclusion. *Quarterly Journal of Mechanics and Applied Mathematics*, **23**, 389–97.

Lax, M. and Feshbach, H. (1948). Absorption and scattering for impedance boundary conditions on spheres and circular cylinders. *Journal of the Acoustical Society of America*, **20**, 108–24.

Leite, R. C. C. and Tai, C. T. (1964). First-order theory for oblate and prolate anisotropic artificial dielectrics. *IEEE Transactions on Microwave Theory and Techniques*, **MTT 12**, 117–22.

Leontovich, M. A. (1948). *Investigations of radiowave propagation. Part II*, pp. 5–12, USSR Academy of Science, Moscow.

Lighthill, J. (1980). *Waves in fluids*. Cambridge University Press, Cambridge.

Lin, W.-G. (1980). Low-frequency scattering of dielectric cylinders. *IEEE Transactions on Microwave Theory and Techniques*, **MTT 28**, 1199–204.

Lindell, I. V., Sihvola, A. H., Muinonen, K. O., and Barber, P. W. (1991). Scattering by a small object close to an interface. I. Exact image theory formulation. *Journal of the Optical Society of America*, **8**, 472–6.

Lindell, I. V., Sten, J.-C.-E., and Kleinman, R. E. (1994). Low frequency image theory for the dielectric sphere. *Journal of Electromagnetic Waves and Applications*, **8**, 295–313.

Liouville, J. (1845). Extrait d'une lettre de M. William Thomson. *Journal de Mathématiques Pures et Appliquées*, **10**, 364–67.

Logan, N. A. (1965). Survey of some early studies of the scattering of plane waves by a sphere. *Proceedings of the IEEE*, **53**, 773–85.

Love, A. E. H. (1944). *A treatise on the mathematical theory of elasticity*. Dover, New York.

Love, J. D. (1974). Long-wavelength acoustic scattering by a torus of arbitrary aspect ratio. *Journal of the Institute of Mathematics and its Applications*, **13**, 321–44.

Lucas, R. J. and Twersky, V. (1984). Coherent response to a point source irradiating a rough plane. *Journal of the Acoustical Society of America*, **76**, 1847–63.

Lucas, R. J. and Twersky, V. (1985). Low-frequency reflection and scattering by ellipsoidally embossed surfaces. *Journal of the Acoustical Society of America*, **78**, 1838–50.

Lucas, R. J. and Twersky V. (1986). Inversion of data for near grazing propagation over rough surfaces. *Journal of the Acoustical Society of America*, **80**, 1459–72.

Lucas, R. J. and Twersky, V. (1987*a*). Inversion of data for near grazing propagation over graveled surfaces. *Journal of the Acoustical Society of America*, **81**, 619–23.

Lucas, R. J. and Twersky, V. (1987*b*). Inversion of ultrasonic scattering data for red blood cell suspensions under different flow conditions. *Journal of the Acoustical Society of America*, **82**, 794–9.

Maanders, E. J. and Mittra, R. (1977). *Modern Topics in Electromagnetics and Antennas.* Lectures delivered at the Technical University of Eindhoven, Peter Peregrinus Ltd., Exeter.

MacCamy, R. C. (1965). Low-frequency acoustic oscillations. *Journal of the Institute of Mathematics and its Applications*, **23**, 247–56.

MacCamy, R. C. (1997). Low frequency expansions for two-dimensional interface scattering problems. *SIAM Journal on Applied Mathematics*, **57**, 1687–701.

Mal, A. K., Ang, D. D., and Knopoff, L. (1968). Diffraction of elastic waves by a rigid circular disc. *Proceedings of the Cambridge Philosophical Society*, **64**, 237–47.

Marsden, J. E. and Hughes, T. J. R. (1983). *Mathematical foundations of elasticity*. Prentice-Hall, Englewood Cliffs.

Maze, G. and Ripoche, J. (1983). Visualization of acoustic scattering by elastic cylinders at low *ka*. *Journal of the Acoustical Society of America*, **73**, 41–3.

Mei, K. and Van Bladel, J. (1963). Low frequency scattering by rectangular cylinders. *IEEE Transactions on Antennas and Propagation*, **AP-11**, 52–6.

Meixner, J. (1949). Die kantenbedingung in der theorie der bengung elekromagnetischer wellen an vollkommen leitenden ebenen schirmen. *Annalen der Physik*, **6**, 1–9.

Meixner, J. (1954). The behavior of electromagnetic fields at edges. *IEEE Transactions on Antennas and Propagation*, **20**, 442–6.

Mie, G. (1908). Beiträge zur optik trüben medien, speziell kolloidaler mettallösungen. *Annalen der Physik*, **25**, 377–442.

Mikhlin, S. G. (1970). *Mathematical physics. An advanced course*. North-Holland, Amsterdam.

Millar, R. F. (1960*a*). A note on diffraction by an infinite slit. *Canadian Journal of Physics*, **38**, 38–47.

Millar, R. F. (1960*b*). The scattering of a plane wave by a row of small cylinders. *Canadian Journal of Physics*, **38**, 272–389.

Millar, R. F. (1965). On a transformation of use in multiple scattering problems. *Proceedings of the Cambridge Philosophical Society*, **61**, 777–9.

Millar, R. F. (1984). On the completeness of the Papkovich potentials. *Quarterly of Applied Mathematics*, **41**, 385–6.

Moon, P. and Spencer, D. E. (1961). *Field theory handbook*. Springer-Verlag, Berlin.

Morse, P. M. and Feshbach, H. (1953). *Methods of theoretical physics, Vol. I, II*. McGraw-Hill, New York.

Morse, P. M. and Ingard, K. U. (1961). Linear acoustic theory. In *Handbuch der physik*, **Band XI/1**. Springer-Verlag, Berlin.

Mountain, R. D. and Birnbaum, G. (1982). Inhomogeneity size and shape determination from scattering of low-frequency sound waves. *Journal of Applied Physics*, **53**, 3581–4.

Müller, C. (1948). Die grundzüge einer mathematischen theorie electromagnetischer schwingungen. *Archiv der Mathematik*, **1**, 296–302.

Müller, C. (1955). Radiation patterns and radiation fields. *Journal of Rational Mechanics and Analysis*, **4**, 235–46.

Müller, C. (1956). Electromagnetic radiation patterns and sources. *IRE Transactions on Antennas and Propagation*, **AP-4**, 224–32.

Müller, C. (1969). *Foundations of the mathematical theory of electromagnetic waves*. Springer-Verlag, Berlin.

Müller, C. and Niemeyer, H. (1961). Greensche tensoren und asympotische gesetze der elektromagnetischen hohlraum schwingungen. *Archives for Rational Mechanics and Analysis*, **7**, 305–48.

Newman, E. H. and Tekin, I. (1994). An overview of the application of the method of moments to large bodies in electromagnetics. In *Large-scale structures in acoustics and electromagnetics*, pp. 204–20. National Academy Press, Washington.

Noble, B. (1962). Integral equation perturbation methods in low-frequency diffraction. In *Electromagnetic waves* (ed. R. E. Langer), pp. 323–60. The University of Wisconsin Press, Madison.

Onishi, K. (1990). Numerical identification in inverse scattering problems. In *Inverse problems in engineering sciences*, (ed. M. Yamaguti *et al.*), pp. 79–85. Springer-Verlag, Tokyo.

Ormsby, J. F. A. (1974). Surface currents, scattering, and applications in the Rayleigh region. *IEEE Transactions on Antennas and Propagation*, **AP-22**, 726–30.

Palmer, D. R. (1996). Rayleigh scattering from nonspherical particles. *Journal of the Acoustical Society of America*, **99**, 1901–12.

Pao, Y.-H. and Mow, C.-C. (1973). *Diffraction of elastic waves and dynamic stress concentrations*. The Rand Corporation, New York.

Picard, R. (1984*a*). On the low-frequency asymptotics in electromagnetic theory. *Zeitschrift für die Reine Angewandte Mathematik*, **354**, 50–73.

Picard, R. (1984*b*). The limit $w \to 0$ in the theory of time-harmonic Maxwell equations. In *Classical scattering* (ed. G. F. Roach) pp. 153–60. Shiva, Cheshire.

Pólya, G. and Szegö, G. (1951). *Isoperimetric inequalities in mathematical physics*. Princeton University Press, Princeton.

Power, D. C. (1988). Approximate analytic continuation of the Rayleigh series. *IEEE Transactions on Antennas and Propagation*, **AP-36**, 1652–4.

Rahmat-Samii, Y. and Mittra, R. (1977). Electromagnetic coupling through small apertures in a conducting screen. *IEEE Transactions on Antennas and Propagation*, **AP-25**, 180–7.

Ramm, A. G. (1980). *Theory and applications of some new classes of integral equations*. Springer-Verlag, New York.

Ramm, A. G. (1982). *Iterative methods for calculating static fields and wave scattering by small bodies*. Springer-Verlag, New York.

Ramm, A. G. (1985). Wave scattering by small bodies. *Reports on Mathematical Physics*, **21**, 69–77.

Ramm, A. G. (1986*a*). *Scattering by obstacles*. Reidel, Dordrecht.

Ramm, A. G. (1986*b*). Behavior of the solutions to exterior boundary value problems at low-frequencies. *Journal of Mathematical Analysis and Applications*, **117**, 561–9.

Ramm, A. G. (1987). Characterization of the low-frequency scattering data in the inverse problem of geophysics. *Inverse Problems*, **3**, L33–L35.

Ramm, A. G., Weaver, O. L., Weck, N., and Witsch, K. J. (1990). Dissipative Maxwell's equations at low frequencies. *Mathematical Methods in the Applied Sciences*, **13**, 305–22.

Rayleigh, Lord (1870). On the theory of resonance. *Philosophical Transactions*, **161**, 77–118.

Rayleigh, Lord (1876). On the approximate solution of certain problems relating to the potential. *Proceedings of the London Mathematical Society*, **7**, 70–5.

Rayleigh, Lord (1881). On the electromagnetic theory of light. *Philosophical Magazine*, **12—Fifth series**, 81–101.

Rayleigh, Lord (1897*a*). On the incidence of aerial and electric waves upon small obstacles in the form of ellipsoids or elliptic cylinders and on the passage of electric waves through a circular aperture in a conducting screen. *Philosophical Magazine*, **XLIV**, 28–52.

Rayleigh, Lord (1897*b*). On the passage of waves through apertures in plane screens, and allied problems. *Philosophical Magazine*, **43**, 259–72.

Rayleigh, Lord (1902). Some general theorems concerning forced vibrations and resonance. *Philosophical Magazine*, **3**, 97–117.

Rayleigh, Lord (1904*a*). On the electrical vibrations associated with thin terminated conducting rods. *Philosophical Magazine*, **8**, 105–7.

Rayleigh, Lord (1904*b*). On the open organ-pipe problem in two dimensions. *Philosophical Magazine*, **8**, 481–7.

Rayleigh, Lord (1907*a*). On the passage of sound through narrow slits. *Philosophical Magazine*, **14**, 153–61.

Rayleigh, Lord (1907*b*). On the light dispersed from fine lines ruled upon reflecting surfaces or transmitted by very narrow slits. *Philosophical Magazine*, **14**, 350–9.

Rayleigh, Lord (1912*a*). On the self-induction of electric currents in a thin anchor-ring. *Proceedings of the Royal Society of London*, **86**, 562–71.

Rayleigh, Lord (1912*b*). Electrical vibrations on a thin anchor-ring. *Proceedings of the Royal Society of London*, **87**, 193–202.

Rayleigh, Lord (1913*a*). The correction to the length of terminated rods in electrical problems. *Philosophical Magazine*, **25**, 1–9.

Rayleigh, Lord (1913*b*). On the approximate solution of certain problems relating to the potential. II. *Philosophical Magazine*, **26**, 195–9.

Rayleigh, Lord (1913*c*). On the passage of waves through fine slits in thin opaque screens. *Proceedings of the Royal Society of London*, **89**, 194–219.

Rayleigh, Lord (1915). The theory of the Helmholtz resonator. *Proceedings of the Royal Society of London*, **92**, 265–75.

Rayleigh, Lord (1916*a*). On the electrical capacity of approximate spheres and cylinders. *Philosophical Magazine*, **31**, 177–86.

Rayleigh, Lord (1916*b*). On the energy acquired by small resonators from incident waves of like period. *Philosophical Magazine*, **32**, 188–90.

Rayleigh, Lord (1918). Note on the theory of the double resonator. *Philosophical Magazine*, **36**, 231–4.

Rayleigh, Lord (1945). *The theory of sound, Vol. I, II*. Dover, New York.

Richardson, J. M. (1984). Scattering of elastic waves from symmetric inhomogeneities at low frequencies. *Wave Motion*, **6**, 325–36.

Roy, A. (1987). Diffraction of elastic waves by an elliptic crack—II. *International Journal of Engineering Science*, **25**, 155–69.

Roy, A. and Sabina, F. S. (1983). Low-frequency acoustic diffraction by a soft disk. *Journal of the Acoustical Society of America*, **73**, 1494–8.

Roy, A., Carey, W., Nicholas, M., Schindall, J., and Crum, L. A. (1992). Low-frequency scattering from submerged bubble clouds. *Journal of the Acoustical Society of America*, **92**, 2993.

Ruck, G. T., Barrick, D. E., Stuart, W. D., and Krichbaum, C. K. (1970). *Radar cross section handbook*, Vol. 1, 2. Plenum Press, New York.

Sabina, F. J. (1987). General formulas for the low-frequency acoustic scattering by a soft body or disk. *Journal of the Acoustical Society of America*, **81**, 1677–82.

Sabina, F. J. and Babich, V. M. (1993). Near and far scattering field of a low frequency plane harmonic wave by a rough half-plane. In *Mathematical and numerical aspects of wave propagation* (ed. R. Kleinman *et al.*) pp. 426–35. SIAM, Philadelphia.

Samokhin, A. B. (1993*a*). Low-frequency scattering of electromagnetic waves by three-dimensional dielectric bodies. *Journal of Communications Technology and Electronics*, **38**, 1064–9.

Samokhin, A. B. (1993*b*). Low-frequency electromagnetic scattering by three-dimensional dielectric bodies. *Radiotekhnika i Elekronika*, **38**, 219–25.

Sarabandi, K. and Senior, T. B. A. (1990). Low-frequency scattering from cylindrical structures at oblique incidence. *IEEE Transactions on Geosciences and Remote Sensing*, **25**, 879–85.

Schiffer, M. and Szegö, G. (1949). Virtual mass and polarization. *Transactions of the American Mathematical Society*, **67**, 130–205.

Senior, T. B. A. (1960*a*). Scalar diffraction by a prolate spheroid at low-frequencies. *Canadian Journal of Physics*, **38**, 1632–41.

Senior, T. B. A. (1960*b*). Impedance boundary conditions for imperfectly conducting surfaces. *Applied Scientific Research*, **8**, 418–36.

Senior, T. B. A. (1973). Low-frequency scattering. *Journal of the Acoustical Society of America*, **53**, 742–7.

Senior, T. B. A. (1976). Low-frequency scattering by a dielectric body. *Radio Science*, **11**, 477–82.

Senior, T. B. A. (1980). Effect of particle shape on low-frequency absorption. *Applied Optics*, **19**, 2483–85.

Senior, T. B. A. (1982). Low-frequency scattering by a perfectly conducting body. *Radio Science*, **17**, 741–6.

Senior, T. B. A. (1983). Low-frequency scattering by a metallic plate. *Electromagnetics*, **3**, 131–44.

Senior, T. B. A. and Ahlgren, D. J. (1973). Rayleigh scattering. *IEEE Transactions on Antennas and Propagation*, **AP-21**, 134.

Senior, T. B. A. and Ksienski, D. A. (1984). Determination of a vector potential. *Radio Science*, **19**, 603–7.

Senior, T. B. A. and Naor, M. (1984). Low-frequency scattering by a resistive plate. *IEEE Transactions on Antennas and Propagation*, **AP-32**, 272–5.

Senior, T. B. A. and Volakis, J. L. (1989*a*). Derivation and application of a class of generalized boundary conditions. *IEEE Transactions on Antennas and Propagation*, **AP-37**, 1566–72.

Senior, T. B. A. and Volakis, J. L. (1989*b*). Scattering by gaps and cracks. *IEEE Transactions on Antennas and Propagation*, **AP-37**, 744–50.

Senior, T. B. A. and Volakis, J. L. (1995). *Approximate boundary conditions in electromagnetics*. IEEE Press, Stevenage.

Senior, T. B. A. and Weil, H. (1982). On the validity of modeling Rayleigh scatterers by spheroids. *Applied Physics*, **29**, 117–24.

Senior, T. B. A. and Willis, T. M. III. (1982). Rayleigh scattering by dielectric bodies. *IEEE Transactions on Antennas and Propagation*, **AP-30**, 1271.

Senior, T. B. A., Sarabandi, K., and Natzke, J. R. (1990). Scattering by a narrow gap. *IEEE Transactions on Antennas and Propagation*, **AP-38**, 1102–10.

Silver, S. (1949). *Microwave antenna theory and design. M. I. T. radiation laboratory series, Vol. 12*. McGraw-Hill, New York.

Skolnik, M. I. (1980). *Introduction to radar systems* (2nd edn). McGraw-Hill, New York.

Sleator, F. B. (1960). A variational solution to the problem of scalar scattering by a prolate spheroid. *Journal of Mathematical Physics*, **39**, 105–20.

Sleeman, B. D. (1967*a*). The scalar scattering of a plane wave by an ellipsoid. *Journal of the Institute of Mathematics and its Applications*, **3**, 4–15.

Sleeman, B. D. (1967*b*). The low-frequency scalar Dirichlet scattering by a general ellipsoid. *Journal of the Institute of Mathematics and its Applications*, **3**, 291–312. Corrigendum (1969), **5**.

Sleeman, B. D. (1967*c*). The low-frequency scalar diffraction by an elliptic disc. *Proceedings of the Cambridge Philosophical Society*, **63**, 1273–80. Corrigendum (1970), **68**, 171–2.

Sleeman, B. D. (1982). The inverse problem of acoustic scattering. *IMA Journal of Applied Mathematics*, **29**, 113–42.

Smyshlyaev, V. P. and Willis, J. R. (1994). Linear and nonlinear scattering of elastic waves by microcracks. *Journal of the Mechanics and Physics of Solids*, **42**, 585–610.

Sommerfeld, A. (1912). Die Greensche funktion der schwingungsgleichung. *Jahresbericht der Deutschen Mathematiker—Vereinigung*, **21**, 309–53.

Sotiropoulos, D. A. and Achenbach, J. D. (1988). Reflection of elastic waves by a distribution of coplanar cracks. *Journal of the Acoustical Society of America*, **84**, 752–61.

Sotiropoulos, D. A., Achenbach, J. D., and Zhu, H. (1987). An inverse scattering method to characterize inhomogeneities in elastic solids. *Journal of Applied Physics*, **62**, 2771–9.

Stakgold, I. (1998). *Green's functions and boundary value problems* (2nd edn). Wiley, New York.

Steinbrunn, K. (1969). Über das verhalten stationärer elekromagnetischer wellenfelder für kleine frequenzen. *Journal of Mathematical Analysis and Applications*, **27**, 127–63.

Sten, J. C. E. (1997). Multiline singularities applied to low-frequency scattering by a prolate spheroid. *The International Journal for Computation and Mathematics in Electrical and Electronic Engineering*, **16**, 72–107.

Stevenson, A. F. (1953*a*). Solution of electromagnetic scattering problems as power series in the ratio (dimension of scatterer)/wavelength. *Journal of Applied Physics*, **24**, 1134–42.

Stevenson, A. F. (1953*b*). Electromagnetic scattering by an ellipsoid in the third approximation. *Journal of Applied Physics*, **24**, 1143–51.

Stoker, J. J. (1956). On radiation conditions. *Communications on Pure and Applied Mathematics*, **9**, 577–95.

Stratton, J. A. (1941). *Electromagnetic theory*. McGraw-Hill, New York.

Stratton, J. A. and Chu, L. J. (1939). Diffraction theory of electromagnetic waves. *Physical Review*, **56**, 99–107.

Stuwe, H.-C. and Werner, P. (1989). Remarks on the low-frequency asymptotics for the reduced wave equation and the Schrödinger equation in low-dimensional spaces. *Asymptotic Analysis*, **2**, 179–202.

Tai, C. T. (1952). Quasi-static solutions for diffraction of a plane electromagnetic wave by a small oblate spheroid. *Transactions of the Institute of Radio Engineering*, **PGAP-1**, 13–16.

Tan, T. H. (1976). Theorem on the scattering and the absorption cross-section for scattering of plane, time-harmonic, elastic waves. *Journal of the Acoustical Society of America*, **59**, 1265–7.

Tan, T. H. (1977). Reciprocity relations for scattering of plane, elastic waves. *Journal of the Acoustical Society of America*, **61**, 928–31.

Thomas, D. P. (1963). Diffraction by a spherical cap. *Proceedings of the Cambridge Philosophical Society*, **59**, 197–209.

Thomson, W. (1847). Extrait de deux lettres addressées à M. Liouville. *Journal de Mathematiques Pures Appliquees*, **12**, 256–64.

Ting, L. and Keller, J. B. (1977). Radiation from the open end of a cylindrical or conical pipe and scattering from the end of a rod or slab. *Journal of the Acoustical Society of America*, **61**, 1438–44.

Tolstoy, I. (1984). Smoothed boundary conditions, coherent low–frequency scatter, and boundary modes. *Journal of the Acoustical Society of America*, **75**, 1–22.

Twersky, V. (1953). Reflection coefficients for certain rough surfaces. *Journal of Applied Physics*, **24**, 659–60.

Twersky, V. (1954). Certain transmission and reflection theorems. *Journal of Applied Physics*, **25**, 859–62.

Twersky, V. (1956). On the scattering of waves by an infinite grating. *IRE Transactions on Antennas and Propagation*, **AP-4**, 330–45.

Twersky, V. (1960). On multiple scattering of waves. In *USA report* XIII General Assembly of the International Scientific Radio Union, pp. 715–30, London.

Twersky, V. (1962*a*). On scattering of waves by the infinite grating of circular cylinders. *IRE Transactions on Antennas and Propagation*, **AP-10**, 737–68.

Twersky, V. (1962*b*). Multiple scattering by arbitrary configurations in three dimensions. *Journal of Mathematics Physics*, **3**, 83–91.

Twersky, V. (1964*a*). Rayleigh scattering. *Applied Optics*, **3**, 1150–62.

Twersky, V. (1964*b*). Acoustic bulk parameters of random volume distributions of small scatterers. *Journal of the Acoustical Society of America*, **36**, 1314–29.

Twersky, V. (1967). Multiple scattering of electromagnetic waves by arbitrary configurations. *Journal of Mathematical Physics*, **8**, 589–610.

Twersky, V. (1975). Low-frequency coupling in the planar rectangular lattice. *Journal of Mathematical Physics*, **16**, 658–66.

Twersky, V. (1978). Constraint on the compound depolarization of aligned ellipsoids. *Journal of Mathematical Physics*, **19**, 2576–8.

Twersky, V. (1983*a*). Wavelength-dependent refractive and absorptive terms for propagation in small-spaced correlated distributions. *Journal of the Optical Society of America*, **73**, 1562–7.

Twersky, V. (1983*b*). Multiple scattering by correlated monolayers. *Journal of the Acoustical Society of America*, **73**, 68–84.

Twersky, V. (1983*c*). Scattering and nonscattering obstacles. *SIAM Journal on Applied Mathematics*, **43**, 711–25.

Twersky, V. (1983*d*). Reflection and scattering of sound by correlated rough surfaces. *Journal of the Acoustical Society of America*, **73**, 85–94.

Twersky, V. (1985*a*). Wavelength-dependent bulk parameters for coherent sound in correlated distributions of small-spaced scatterers. *Journal of the Acoustical Society of America*, **77**, 29–37.

Twersky, V. (1985*b*). Wavelength-dependent electromagnetic parameters for coherent propagation in correlated distributions of small-spaced scatterers. *Journal of Mathematical Physics*, **26**, 2208–17.

Uslenghi, P. L. E. (1980). Low-frequency inverse scattering by far–field measurements. *Alta Frequenza*, **49**, 260–3.

Van Bladel, J. (1962). Good conductors in low-frequency fields. *IRE Transactions on Antennas and Propagation*, **AP-10**, 625–33.

Van Bladel, J. (1963). Low-frequency scattering by cylindrical bodies. *Applied Scientific Research*, **B10**, 195–202.

Van Bladel, J. (1967). Low-frequency scattering through an aperture in a rigid screen. *Journal of Sound and Vibration*, **6**, 386–95.

Van Bladel, J. (1968*a*). Low-frequency scattering by hard and soft bodies. *Journal of the Acoustical Society of America*, **44**, 1069–73.

Van Bladel, J. (1968*b*). Low-frequency scattering through an aperture in a soft screen. *Journal of Sound and Vibration*, **8**, 186–95.

Van Bladel, J. (1969). Coupling through small apertures, with an application to Helmholtz' resonator. *Journal of the Acoustical Society of America*, **45**, 604–13.

Van Bladel, J. (1970*a*). Coupling through a small aperture in a waveguide. *Journal of the Acoustical Society of America*, **47**, 202–10.

Van Bladel, J. (1970*b*). Small-hole coupling of resonant cavities and waveguides. *Proceedings of the IEEE*, **117**, 1098–1103.

Van Bladel, J. (1971). Small holes in a waveguide wall. *Proceedings of the IEEE*, **118**, 43–50.

Van Bladel, J. (1972). Small apertures in cavities at low-frequencies. *Archiv für Elektronik und Übertragungstechnik*, **26**, 481–6.

Van Bladel, J. (1973). Circuit parameters from Maxwell's equations. *Applied Scientific Research*, **28**, 381–97.

Van Bladel, J. (1976). Scattering of low-frequency E-waves by dielectric cylinders. *IEEE Transactions on Antennas and Propagation*, **AP-24**, 255–8.

Van Bladel, J. (1977*a*). Low-frequency asymptotic techniques. In *Modern topics in electromagnetics and antennas*, 1–56. Peter Pereginus Ltd., London.

Van Bladel, J. (1977*b*). The multipole expansion revisited. *Archiv für Elektronik und Uebertragungstechnik*, **31**, 407–11.

Van Bladel, J. (1979). Field penetration through small apertures: The first-order correction. *Radio Science*, **14**, 319–31.

Van Bladel, J. (1988). Hierarchy of terms in a multipole expansion. *Electronic Letters*, **24**, 492–3.

Van Bladel, J. (1995). *Singular electromagnetic fields and sources*. IEEE and Oxford University Press, New York.

Van Bladel, J. and Butler, C. M. (1981). Aperture problems. In *Theoretical methods for determining the interaction of electromagnetic waves with structures*, pp. 117–72. Sijthoff en Noordhoff, Alphen aan den Rijn.

Van de Hulst, H. C. (1949). On the attenuation of plane waves by obstacles of arbitrary size and form. *Physica*, **15**, 740–6.

Van de Hulst, H. C. (1980). *Multiple light scattering. tables, formulas, and applications*, Vol. 1. Academic Press, New York.

Van de Hulst, H. C. (1981). *Light scattering by small particles*. Dover, New York.

Varadan, V. V. and Varadan, V. K. (1979). Low-frequency expansions for acoustic wave scattering using Waterman's T-matrix method. *Journal of the Acoustical Society of America*, **66**, 586–9.

Varadan, V. K. and Varadan, V. V. (1986). *Low and high frequency asymptotics*. North-Holland, Amsterdam.

Varadan, V. V., Lakhtakia, A., and Varadan, V. K. (1991). *Field representations and introduction to scattering*. North-Holland, Amsterdam.

Varatharajulu, V. (1977). Reciprocity relations and forward amplitude theorems for elastic waves. *Journal of Mathematical Physics*, **18**, 537–43.

Vekua, I. N. (1967). *New methods for solving elliptic equations*. North-Holland, Amsterdam.

Von Wolfersdorf, L. and Rost, L. (1991). On mathematical modelling of absorption and dispersion of seismic waves for low frequencies. *Mathematical Methods in the Applied Sciences*, **14**, 563–72.

Wait, J. R. (1992). Low frequency electromagnetic scattering from an island. *IEEE Transactions on Antennas and Propagation*, **40**, 439–42.

Wang, W. X. (1988). The potential for a homogeneous spheroid in a spheroidal coordinate system: I. At an exterior point. *Journal of Physics*, **A21**, 4245–50.

Weatherburn, C. E. (1947). *Elementary vector analysis*. Bell, London.

Weatherburn, C. E. (1948). *Advanced vector analysis*. Bell, London.

Weck, N. and Witsch, K. J. (1990*a*). The low-frequency limit of the exterior Dirichlet problem for the reduced wave equation. *Applicable Analysis*, **38**, 33–43.

Weck, N. and Witsch, K. J. (1990*b*). Low-frequency asymptotics for dissipative Maxwell's equations in bounded domains. *Mathematical Methods in the Applied Sciences*, **13**, 81–93.

Weck, N. and Witsch, K. J. (1991). Exterior Dirichlet problem for the reduced wave equation: Asymptotic analysis of low frequencies. *Communications in Partial Differential Equations*, **16**, 173–95.

Weck, N. and Witsch, K. J. (1992*a*). Exact low-frequency analysis for a class of exterior boundary problems for the reduced wave equation in two dimensions. *Journal of Differential Equations*, **100**, 312–40.

Weck, N. and Witsch, K. J. (1992*b*). Complete low frequency analysis for the reduced wave equation with variable coefficients in three dimensions. *Communications in Partial Differential Equations*, **17**, 1619–63.

Weck, N. and Witsch, K. J. (1992*c*). Exact low frequency analysis for a class of exterior boundary value problems for the reduced wave equation in higher dimensions. *Asymptotic Analysis*, **6**, 161–72.

Weil, H., Senior, T. B. A., and Willis III, T. M. (1985). Internal and near fields of small particles irradiated in spectral absorption bands. *Journal of the Optical Society of America*, **2**, 989–96.

Werner, P. (1962). Randwert probleme der mathematischen akustik. *Archives for Rational Mechanics and Analysis*, **10**, 29–66.

Werner, P. (1963). On the exterior boundary value problem of perfect reflection for stationary electromagnetic wave fields. *Journal of Mathematical Analysis and Applications*, **7**, 348–96.

Werner, P. (1966*a*). On the behaviour of stationary electromagnetic wave fields for small frequencies. *Journal of Mathematical Analysis and Applications*, **15**, 447–96.

Werner, P. (1966*b*). On an integral equation in electromagnetic diffraction theory. *Journal of Mathematical Analysis and Applications*, **14**, 445–62.

Werner, P. (1975). Über das verhalten elektromagnetischer felder für kleine frequenzen in mehrfach zusammenhängenden gebieten. I. *Journal für die Reine Angewandte Mathematik*, **278/279**, 365–97.

Werner, P. (1976). Über das verhalten elektromagnetischer felder für kleine frequenzen in mehrfach zusammenhängenden gebieten. II. *Journal für die Reine Angewandte Mathematik*, **280**, 89–121.

Werner, P. (1983). Spectral properties of the Laplace operator with respect to electric and magnetic boundary conditions. *Journal of Mathematical Analysis and Applications*, **92**, 1–65.

Werner, P. (1985). Zur asymptotik der wellengleichung und der wärmeleitungsg-leichung in zwei-dimensionalen außenräumen. *Mathematical Methods in the Applied Sciences*, **7**, 170–201.

Werner, P. (1986). Low-frequency asymptotics for the reduced wave equation in two-dimensional exterior spaces. *Mathematical Methods in the Applied Sciences*, **8**, 134–56.

Werner, P. (1988). Aperiodic electromagnetic waves in exterior domains. *Journal of Mathematical Analysis and Applications*, **135**, 112–64.

Werner, P. (1990). Zero resonances in local perturbations of parallel-plane waveguides. *Mathematical Methods in the Applied Sciences*, **13**, 111–36.

Weston, V. H. (1963). Theory of absorbers in scattering. *IEEE Transactions on Antennas and Propagation*, **AP-11**, 578–84.

Wilcox, C. H. (1956*a*). A generalization of theorems of Rellich and Atkinson. *Proceedings of the American Mathematical Society*, **7**, 271–6.

Wilcox, C. H. (1956*b*). An expansion theorem for electromagnetic fields. *Communications on Pure and Applied Mathematics*, **9**, 115–34.

Wilcox, C. H. (1959). Spherical means and radiation conditions. *Archives for Rational Mechanics and Analysis*, **3**, 133–48.

Wilcox, C. H. (1975). *Scattering theory for the d'Alembert equation in exterior domains*. Springer-Verlag, Berlin.

Williams, W. E. (1963). Note on a problem in elastodynamics. *Journal of the London Mathematical Society*, **38**, 119–22.

Williams, W. E. (1971). Some results for low-frequency Dirichlet scattering by arbitrary obstacles and their application to the particular case of the ellipsoid. *Journal of the Institute of Mathematics and its Applications*, **7**, 111–18.

Wolfe, P. (1975). Low-frequency diffraction by a hard strip. *SIAM Journal on Applied Mathematics*, **29**, 273–87.

Wong, S. K. and Bush, W. B. (1992). Low-frequency acoustic scattering by slender bodies of arbitrary cross sections. *Journal of the Acoustical Society of America*, **92**, 487–91.

Yamakawa, N. (1956). Investigation of the disturbance produced by spherical obstacles on the elastic waves (I). *Quarterly Journal of Seismology*, **21**, 1–12.

Ye, Z. (1997). Low-frequency acoustic scattering by gas-filled probate spheroids in liquids. *Journal of the Acoustical Society of America*, **101**, 1945–52.

Zabreyko, P. P., Koshelev, A. I., Krasnosel'skii, M. A., Mikhlin, S. G., Rakovshchik, L. S., and Stet'senko, V. Y. (1975). *Integral Equations—A Reference Text*. Noordhoff International Publishing, Leyden.

INDEX